Plant Fibre

Plant Fibre

Chemistry, Processing and Advanced Engineering Products

Edited by
Lihui Chen
and
Mizi Fan

WHITTLES PUBLISHING

Published by
Whittles Publishing Ltd.,
Dunbeath,
Caithness, KW6 6EG,
Scotland, UK
www.whittlespublishing.com

ISBN 978-184995-228-6

Printed by CPI Group (UK) Ltd, Croydon CR0 4YY

Contents

Contributors

Hassan Ahmad
College of Engineering, Design and Physical Sciences, Brunel University London, UK

Professor Shilin Cao
College of Material Engineering, Fujian Agriculture and Forestry University, China

Professor Lihui Chen
College of Material Engineering, Fujian Agriculture and Forestry University, China

Dr Nairong Chen
College of Material Engineering, Fujian Agriculture and Forestry University, China

Dr Dasong Dai
College of Material Engineering, Fujian Agriculture and Forestry University, China

Professor Mizi Fan
College of Engineering, Design and Physical Sciences, Brunel University London, UK

Professor Fang Huang
College of Material Engineering, Fujian Agriculture and Forestry University, China

Professor Liulian Huang
College of Material Engineering, Fujian Agriculture and Forestry University, China

Professor Jinguo Lin
College of Material Engineering, Fujian Agriculture and Forestry University, China

Dr Shan Lin
College of Material Engineering, Fujian Agriculture and Forestry University, China

Dr Wendi Liu
College of Material Engineering, Fujian Agriculture and Forestry University, China

Dr Yudong Lu
Fujian Key Laboratory of Polymer Materials, Fujian Normal University, China

Dr Omar Abo Madyan
College of Engineering, Design and Physical Sciences, Brunel University London, UK

Professor Qingxian Miao
College of Material Engineering, Fujian Agriculture and Forestry University, China

Dr Jiuping Rao
College of Material Engineering, Fujian Agriculture and Forestry University, China

Dr Yonghui Zhou
College of Engineering, Design and Physical Sciences, Brunel University London, UK

Preface

The amount of material consumed by humanity has passed 100bn tonnes per annum and continues to grow year by year. The contradiction of the limited reserves of petrochemical resources, rapid growth of energy and material demand, and global environmental concern turns interest towards the utilization of renewable resources and towards lignocellulosics. Natural fibres (lignocellulosic biomass) is the most abundant and renewable source of carbon accessible to humanity. Lignocellulose is a complex composite structure itself, primarily composed of three biopolymers, which when separated into its constituents, may be seen as cheap raw material for innovating into high-value fuels, fine chemicals and biomaterials, and when acting in concert, serve as an abundant material with a variety of interesting properties. Humanity has for many millennia had a strong and ongoing relationship with natural fibres for uses in fuels and materials. However, the common goods that drive society have rapidly changed and as such simply returning to preindustrial materials is out of the question. The derivatisation of cellulose to provide a wide range of industrial products has attracted major interest and there are many developments and much potential being exploited for natural fibres. A new world is opening up for this valuable resource, although commercially viable ones are as yet limited, with development being less efficient and effective than it should have been. Many challenges remain.

The massive amount of effort devoted over past decades to better understand the lignocellulose matrix, its components, and connection between its molecular structure and its material properties provides a vast literature base that is essential for successful development and utilization of natural fibre and its engineering composites. To an academic scientist involved in these investigations, it is clear that these research investments have been a dazzling success. This book considers structure, processing, properties and applications with a comprehensive examination and discussion of the structure, chemistry and behaviour of plant fibre. This is followed by basic information and a thorough understanding of functional materials directly derived from plant fibres, mainly including nanocellulose, physically and chemically treated plant fibre, cellulose film, cellulose textile, cellulose detection materials and nanocellulose aerogel. The processing, properties and applications of functional plant fibre composites are then considered, including nanocellulose composite, plant fibre/natural resin composite, wood- plastic composites and long fibre composites.

This book presents a basic and clear understanding of the most advanced developments of a wide range of functional plant fibre materials. The collation will hopefully provide an invaluable update on the science and technology of plant fibre, and important basic information of value to those interested in the advanced development and utilization of functional natural fibre materials. The book will serve as a valuable reference or text book challenging academics, research scholars and engineers to think beyond standard practices when processing and utilizing natural fibres, and creating novel natural fibre engineering materials.

Mizi Fan

Chapter 1
Introduction: A perspective on plant fibre and its advanced products

Professor Lihui Chen, Professor Shilin Cao and Professor Mizi Fan

1.1 Introduction

Currently, major fuel and polymeric materials servicing human needs are, in every field, mainly derived from petrochemical resources. At the current consumption rate, the world's proven availability of petrochemical resources will soon be depleted: oil and natural-gas resources can be maintained for about 50 years; coal about 150 years. The contradiction inherent in the limited reserves of global petrochemical resources and the rapid growth in human energy and material demand becomes increasingly pronounced. The consumption of petrochemical resources causes a serious release of harmful gases, including carbon dioxide, sulphur dioxide, nitrogen oxides and so on. Carbon dioxide gas can cause global warming, and sulphur dioxide and nitrogen oxide gases are related to acid rain pollution, which leads to irreversible damage to the global environment. The replacement for, and the alternative technologies for reducing the consumption of, petrochemicals have been intensively studied, and the development and utilisation of plant fibre resources are considered to be one of the strategies with the greatest potential for resolving this crisis [1–2].

Plant fibre is a rich resource, producing 1.5×10^{12} tons of natural celluloses annually through photosynthesis [3]. There are a number of varieties: wood fibre, bast fibres, bamboo fibre, and many other non-food crop fibres [4–5]. Plant fibres have been used to manufacture paper, provide raw materials for chemical products, food packaging materials, medical materials, agricultural film materials. Plant fibres are renewable and carbon negative and having many other merits, especially with regard to the requirements of sustainable development [6].

The development and utilisation of plant fibre resources have mainly concentrated on: (1) direct use for producing new energy and chemical products to replace those that have traditionally relied upon oil, natural gas, coal and other chemicals; (2) the full use of plant fibre resources for new materials or products. The low cost of plant fibres, the prospect of using them fully and effectively, and their relevance to environmentally friendly 'green' processes and products, indicate their limitless potential.

1.2 Basic concept and classification of plant fibres

Fibres are substances composed of continuous or discontinuous filaments. In animals and plants, fibre plays an important role in maintaining tissue. Wide use of natural fibres includes woven lines, thread and hemp rope, paper, and woven felts. Natural fibres can also be used to make other materials and other hybrid composite materials. Natural fibre can be classified, according to its source, into plant fibre, animal fibre and mineral fibre. Plant fibre can be extracted from the seeds, fruits, stems and leaves of plants, and is known as natural cellulosic fibre since the main chemical component of plant fibre is cellulose.

Plant fibre, based on its structure and tissue sources, is generally divided into eight categories:

1. Bast fibres, also known as soft fibre, located in the epidermis of the stem. Commercial bast fibre plants are ramie, marijuana, kenaf, jute and linen.
2. Leaf fibre, also known as hard fibre, produced from monocotyledon leaves. Leaf fibrous plants are abaca, sisal (mainly originating from of Mexico, where it is now widely cultivated). Others include tequila of the genus *Cymbidium*, Mauritius Ma, New Zealand hemp, *Salsola* and the bromeliads.
3. Seed fibre, a number of plant seed epidermal cell growth into single cell fibres, produced from dicotyledonous plants, such as cotton and kapok. They are now the main source for the Western Hemisphere for the production of fine cotton wool and Cotton Island cotton.
4. Fruit fibre, from some plants, such as Ji Bei, coconut clothing, loofah and other fruits obtained in the vascular bundle. This fibre is now used for aircraft noise insulation and furniture, as well as damping lining for armoured vehicles.
5. Root fibre – processed plant roots – such as produced in the Mexican grass broom with chaos grass used as an export broom.
6. Spikelets, grasses, and sorghum spikes at all levels of the lateral axis, fineness, rich in fibre; can be used for brooms.
7. Wood fibre, commonly used in papermaking, and divided into softwood and hardwood fibre, the former such as spruce, fir, hemlock and larch, and the latter such as poplar and birch.
8. Other fibres, such as Gramineae, Mans and other plants of the stalk, for weaving and paper.

Bamboo fibre has recently been developed as a new type of textile fibre with heat, breathable and other fine features. It is the fifth largest plant fibre after cotton, hemp, silk and wool, for natural plant fibres.

1.3 Preparation of plant fibre and its physical and chemical properties

The preparation of hemp fibre mainly uses the chemical degumming process. Ramie is soaked with acid and then boiled with alkaline. After the boiled ramie has been beaten by the beating machine (mallet-beating), rinsed with bleach and neutralised with acid, it is washed, dehydrated, oiled and dried to get the finished fibre. Wood, bamboo and grass fibre are mainly prepared by chemical, physical or a combination of processes. Chemical methods use chemical agents in specific treatments of plant raw materials, so that most of the lignin is dissolved and the fibres are separated from each other into pulp. Physical and chemical methods mainly use mechanical friction acting on the surface of the plant material and thereby generating a tearing effect. In general, the intercellular layer of lignification softens before the material is ground, torn and separated into a single fibre. In some cases, some of the hemicellulose in the raw material of the plant fibre is removed by mild pretreatment (impregnation or cooking) before the raw material is pulverised. The lignin is less dissolved or substantially undissolved, but the intercellular layer can be softened.

Plant fibres, such as hemp, wood and bamboo fibre, have many merits. For example, ramie fibre normally has a length of 20–250 mm, although the longest can reach more than 600 mm, and a width of 20–80 μm. Ramie fibre is long, full of shiny, dyed bright, not easy to fade and mouldy, and has high strength, decay resistance characteristics. Bamboo fibre is 1–3 mm in length and 8–30 μm in width. Bamboo fibre has good permeability, instantaneous water absorption, strong wear resistance and good dyeing properties, in addition to natural antibacterial and anti-ultraviolet functions. Wood fibre is 1–5 mm in length and 15–70 μm in width. Wood fibre has a good aspect ratio, good strength and other characteristics, and is the main raw material for paper-based composites.

Plant fibre generally consists of cellulose, hemicellulose and lignin, together with minor constituents, such as inorganic salts, sugars, alkaloids, tannins, pigments, starch and pectin. Inorganic ions may include nitrogen, sulphur, phosphorus, calcium, magnesium, iron, potassium, sodium, copper, zinc, manganese and chlorine. Cellulose is a homogeneous linear polymer compound which is linked by a D-glucopyranosyl group by a 1,4-β-glycosidic linker and is capable of acid hydrolysis, alkaline hydrolysis, oxidative degradation, esterification, etherification, graft copolymerisation, crosslinking and other chemical reactions. Hemicellulose is a non-homogeneous polysaccharide consisting of two or more monosaccharides (mainly glucose, xylosyl, mannosyl, galactosyl, arabinosyl), and the molecule is often branched. Similar to cellulose, hemicellulose is also capable of acidic hydrolysis, alkaline hydrolysis, oxidation degradation, esterification, etherification, graft copolymerisation, crosslinking and other reactions. Lignin is an aromatic macromolecule compound with a three-dimensional structure composed of phenylpropane structural units (i.e. C6–C3 units) through ether bonds and carbon–carbon bonds, which can be subjected to acid hydrolysis, thermal decomposition, oxidative degradation, reduction degradation, condensation reaction and polymerisation reactions.

1.4 Plant fibre based products

Plant fibre has been widely used in papermaking, textiles, water treatment, and for medical and construction products, because of its wide availabilty, renewable and biodegradable characteristics. Various papers with different end uses can be produced by altering the combination of chemical and mechanical pulp fibres. Chemical pulp fibre is relatively soft and strong, and conducive to improving the strength and whiteness of paper, with a wide range of applications, such as for electrostatic copy paper, paper bags, high-grade thin wrapping paper, carbonless copy paper, banknote paper, cigarette paper, medical paper, rust paper, security paper and napkins (Fig. 1.1). Mechanical pulp fibre is relatively hard, the bonding strength is poor, the whiteness

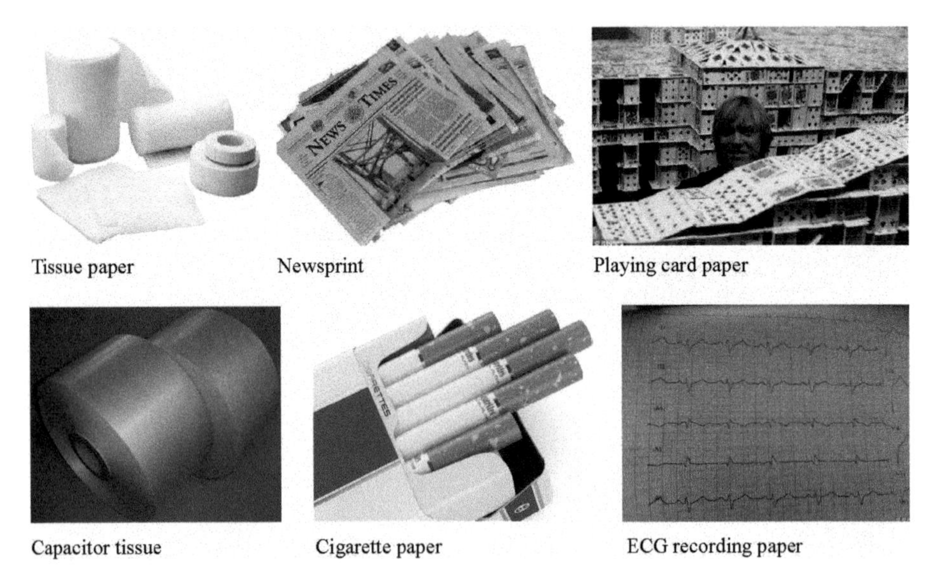

Figure 1.1 Examples of natural fibre based paper products.

is low, but it has the advantage of improving the loose thickness, stiffness, opacity and air permeability of the paper, which can be used alone for the preparation of newsprint. Mechanical fibres in a certain proportion of mixed preparation can produce coated paper, wallpaper base paper, playing cards, insulating paper, paper diapers, tissue paper and photographic base paper.

In the textile industry, plant fibres, such as viscose fibre, Lyocell fibre and their fibre fabrics, have been used for underwear, bedding, towels, bathrobes and bathroom supplies, non-woven fabrics, sanitary materials, daily decorations, and many other products. Bamboo viscose fibres feel smooth, are soft and unique, and have better hygroscopicity, breathablity, sagging along with natural antibacterial, deodorant and anti-ultraviolet properties, when compared with ordinary viscose fibres. Lyocell fibre produced by using N-methylmorpholine-N-oxide (NMMO) as a solvent has the comfort of cotton, the drapeness of viscose fibre, the tensile strength of polyester fibre, the feel of silk, and has gloss and other excellent characteristics. It is a new type of regenerated cellulose fibre. Lyocell fibres can also be blended with other chemical fibres and natural fibres to develop fabrics with better or specific properties.

In the medical industry, attempts have also been made to produce a variety of cellulose-based membranes with plant fibres, through immersion precipitation phase transformation, melt extrusion, thermal-induced phase separation and other methods, including cellulose microfiltration and ultrafiltration membranes [7], nanofiltration membranes [8, 9], and reverse osmosis membranes. Cellulose membranes made from plant fibres have a natural wettability because of the hydrogen bond between the membrane interface and the water molecules. The adsorption of proteins is extremely low, leading to superior resistance to contamination, low flux attenuation and rendering them easy to clean, compared to hydrophobic polymer membranes.

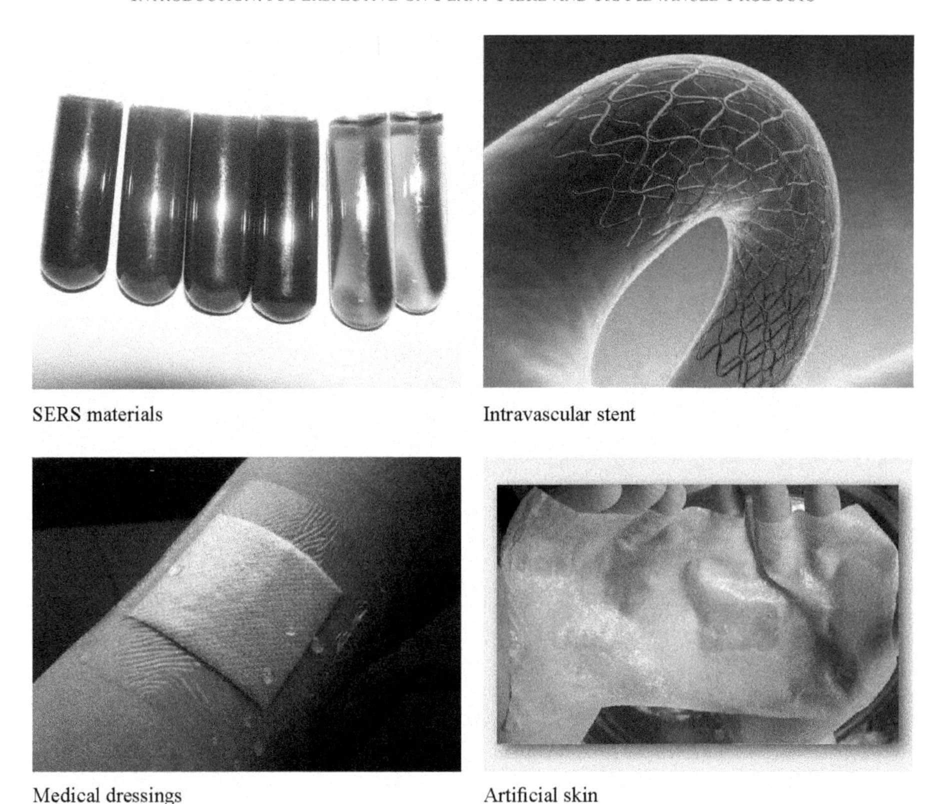

Figure 1.2 Examples of natural fibre based medical products.

Because of their good biocompatibility, plant fibres are widely used in the preparation of various medical products (Fig. 1.2), such as artificial skin, dialysis membranes, bioreactor and drug-release systems, tissue engineering, construction of scaffold materials for cartilage repair, dental implants, artificial blood vessels, nerve surgical dressings, and dural repair substances. In addition, plant fibre can also be used as a template for the synthesis of antimicrobial materials, metal nanomaterials and surface-enhanced Raman spectroscopy for medical testing of special materials.

In the field of construction, plant fibre can be used to manufacture a variety of composite materials, including fiberboard, bamboo heavy materials, wood–plastic composites and many laminate materials. The composites have been widely used in furniture manufacturing, building decoration, automobile and motorcycle manufacturing and in household appliances (Fig. 1.3).

One of the advanced natural fibre composites is plant fibre reinforced composites, in most cases called wood–plastic composite (WPC). WPC is produced by using the plant fibre as the main raw material, enhancing it through the interface fusion of the fibre–polymer, e.g. polyethylene, polypropylene, PVC and other thermoplastic matrices, supplemented by appropriate processing aids. WPC has the advantages of low cost,

Furniture

WPC decking

Composite garden material

Composite windows

Figure 1.3 Examples of natural fibre composites used in construction.

excellent mechanical properties, little water absorption, anti-ageing, good dimensional stability and biodegradability. It may replace all logs, plastic, steel, aluminum and other similar products in construction. WPC is not only able to use plant resources efficiently, having both the high strength and high elasticity of wood fibre materials, but it can also use recycled plastic to produce a new generation of wood materials superior to wood.

1.5 New developments in plant fibre composites

In recent years, many developments in plant fibre have taken place, especially in the nano, electronic and intelligent fields, such as nanomaterials, gel materials, optoelectronic materials and biomimetic intelligent materials. Plant fibres consist of a variety of cell morphologies, pore structures and chemical compositions, and are a class of structured polymer-based natural composite materials, ranging in scale from the centimetres of wood fiber and micron-grade wood cells down to nanoscale microfibres and cellulose molecules. The structured, complex and orderly multiscale graded structure lays a solid foundation for Bacon high-performance materials and the preparation of special multiscale structure new materials. In addition to their exquisite multiscale hierarchical structure, plant fibres also naturally form fine-graded porous structures with irregular round, square, elliptical and polygonal pores with sizes ranging from microns to nanometres. The shape, size and distribution characteristics of the tube wood cells provide a wide selection of raw materials for the design and preparation of various high value-added porous materials.

Nanocellulose from plant fibres can be prepared by physical methods, e.g. high-pressure homogenisation, mechanical grinding, ultrasonic and chemical methods, e.g. acid and alkali hydrolysis, and biological methods, e.g. cellulase hydrolysis [10, 11] According to the morphological characteristics, nanofibres can be divided into nano-crystalline cellulose (NCC) and nanofibrillated cellulose (NFC). Nanocellulose fibres have high purity, aspect ratio, Young's modulus, crystallinity, strength and low coefficient of thermal expansion, as well as light weight, biodegradable and renewable properties. Nanocellulose fibres have broad application prospects in a wide range of industrial sectors [12]. The current research focus is on the development of cellulose nanostructure orientation design and construction theory, and methods to regulate the nanostructures of cellulose at the molecular level, in order to achieve the design and trimming of cellulose nanostructures, and construct and assemble nanocellulose functional materials. The technologies are orientated towards more green and efficient approaches to research and development into high value-added celluloses [13].

Cellulose gel is another new development in plant fibres. Cellulose gel is a three-dimensional network structure constructed of cellulose or cellulose derivatives or with other polymers through crosslinking architectures. According to the type of cross-linking, gel size or gel medium, the cellulose gel can be divided into chemical gel or physical gel, microgel or nanogel and hydrogel. The construction of cellulose gel can be achieved by covalent or noncovalent bonds, i.e., intermolecular interactions. Cellulose gel is bio-compatible and biodegradable and can be widely used in the biomedical field [14, 15].

Dissolving plant fibres with cellulose solvents, such as ionic liquids, N-methyl-morpholine-N-oxide (NMMO) or sodium hydroxide/urea aqueous solutions, has recently led to the development of a flexible transparent nanocellulose film. The incorporation of inorganic oxide (e.g. indium tin oxide, zinc oxide) conductive materials with nanocellulose gives rise to flexible transparent conductive film as the base material for optoelectronic devices. Some of these optoelectronic materials prepared with plant fibres have photoluminescence or fluorescent properties and can be used in organic light-emitting diodes, organic thin-film transistors, security and packaging. Some of these materials also have piezoelectric and ion-transport effects for sensors, micro-electromechanical systems, artificial muscles, and microwave remote control drives. Films embedded with multi-walled carbon nanotubes can be used to prepare flexible lithium batteries, supercapacitors and paper-based energy storage devices. Conductive cellulose films can also be used to produce the electrode material for flexible solar cells, flexible displays and other relevant applications [16, 17].

In short, plant fibre is one of the most abundant resources in nature. It has good non-toxic, biodegradable and biocompatible properties and has a unique natural structure that is porous, flexible and has high mechanical strength. The continuing advent of new solvents, new technologies, and new ideas will greatly promote the development of plant fibre functional materials. It is believed that the application of advanced plant fibre materials will become more widely available, and its research will become more vigorous.

References

[1] Huber G W, Iborra S & Corma A. Synthesis of transportation fuels from biomass: Chemistry, catalysts, and engineering. *Chemical Reviews*, 2006 (106): 4044–4098.

[2] Valenzuela M B, Jones C W & Agrawal PK. Batch aqueous-phase reforming of woody biomass. *Energy Fuels*, 2006 (20): 1744–1752.

[3] Daiyong Y E, Huang H, Fu H & Chen H. Advances in cellulose chemistry. *Journal of Chemical Industry and Engineering*, 2006, 57 (8): 1782–1791.

[4] Fan M. Characterization and performance of elementary hemp fibres: Factors influencing tensile strength. *Bioresources*, 2010, 5 (4): 2307–2322.

[5] Fan M & Fu F. Introduction: A perspective – natural fibre composites in construction. In M Fan and F. Fu (eds) *Advanced High Strength Natural Fibre Composites in Construction* (pp. 1–20). Cambridge, UK: Woodhead Publishing Series (Elsevier), 2017.

[6] Wang S & Xia Y. Review on utilization of plant fiber. *Science and Technology of Food Industry*, 2006 (7): 202–205.

[7] Konovalova V, Guzikevich K, Burban A, Kujawski W, Jarzynka K & Kujawa J. Enhanced starch hydrolysis using α-amylase immobilized on cellulose ultrafiltration affinity membrane. *Carbohydrate Polymers*, 2016 (152): 710–717.

[8] Weng R, Chen L, Xiao H, Huang F, Lin S, Cao S & Huang L. Preparation and characterization of cellulose nanofiltration membrane through hydrolysis followed by carboxymethylation. *Fibers and Polymers*, 2017 (18): 1235–1242.

[9] Weng R, Chen L, Lin S, Zhang H, Wu H, Liu K, Cao S & Huang L. Preparation and characterization of antibacterial cellulose/chitosan nanofiltration membranes. *Polymers*, 2017 (9): 116–129.

[10] Jonoobi M, Oladi R, Davoudpour R, Oksman K, Dufresne A, Hamzeh Y & Davoodi R. Different preparation methods and properties of nanostructured cellulose from various natural resources and residues: A review. *Cellulose*, 2015 (22): 935–969.

[11] Mondal S. Preparation, properties and applications of nanocellulosic materials. *Carbohydrate Polymers*, 2017 (163): 301–316.

[12] Grishkewich N, Mohammed N, Tang J & Tam KC. Recent advances in the application of cellulose nanocrystals, *Colloid & Interface Science*, 2017 (29): 32–45.

[13] Kim J H, Shim B S, Kim H S, Lee Y J, Min S K, Jang D, Abas Z & Kim J. Review of nanocellulose for sustainable future materials. *International Journal of Precision Engineering and Manufacturing-Green Technology*, 2015, 2 (2): 197–213.

[14] Sangeetha N M & Maitra U. Supramolecular gels: Functions and uses. *Chemical Society Review*, 2005 (34): 821–836.

[15] Kang H, Liu R & Huang Y. Cellulose-based gels. *Macromolecular Chemistry & Physics*, 2016, 217 (12): 1322–1334.

[16] Zhu H, Luo W, Ciesielski P N, Fang Z, Zhu J Y, Henriksson G, Himmel M E & Hu L. Correction to wood-derived materials for green electronics, biological devices and energy applications, *Chemical Reviews*, 2016, 116 (16): 9305–9374.

[17] Jung Y H, Chang T H, Zhang H, Yao C, Zheng Q, Yang V W, Mi H, Kim M, Cho S J, Park D W, Jiang H, Lee J, Qiu Y, Zhou W, Cai Z, Gong S & Ma Z. High-performance green flexible electronics based on biodegradable cellulose nanofibril paper. *Nature Communications*, 2015 (6): 7170.

Chapter 2
The structure of plant fibres

Professor Jinguo Lin

This chapter overviews the resources and structure of natural fibres. These include fibres from wood (both hardwood and softwood), bamboo, bast fibre crops (jute, ramie, hemp and flax), bagasse and straw (rice and wheat). Cell morphology, characteristics and potential utilisation are also discussed.

2.1 Introduction

Plant fibres are usually extracted from wood, bamboo, straw and other non-food crops. Wood and bamboo fibres are produced from perennial trees (softwood and hardwood) and bamboo (scattered and clustered bamboo) respectively. Crop straw fibres are mainly from by-products of crops. The morphology characteristics of wood, bamboo and straw fibres share considerable similarities, which are described in this chapter.

2.2 Wood fibres

2.2.1 Resources for wood fibres

Global forest land, assessed by the United Nations Food and Agriculture Organization (FAO), is slightly over 4 billion hm [1], accounting for 31% of the total land area, which includes 264 million hm[2] planted forest (Table 2.1). The per capita forest area is 0.6 hm^2. Tropical forests and temperate forests each account for one-half the total forest area; developing and developed countries each account for one-half as well. The countries that own the most abundant global forest resources (more than half the global forest resource) are the Russian Federation, Brazil, Canada, the United States of America and China.

The global forest stock is about 386.4 billion m^3, of which Europe (including Russia) and South America each account for one-third. The countries ranked at the forefront of forest stock are Russia, Brazil, Australia, New Zealand, Papua New Guinea, Canada and the USA. Europe (including Russia) produces and exports a huge amount of timber. In recent years, Russia has become the country that produces and exports the largest quantity of logs in the world. Sweden, Austria, Germany and Finland are also important European countries for the production and export of timber. In North America, the United States and Canada are major world exporters of timber and timber products.

Table 2.1 Global forest and planted forest land (2016).

Regional separation	Forest area (thousand hm²)	Plantation area (thousand hm²)
Asia Pacific region	740 383	119 884
Africa	674 419	15 409
Europe	1 005 001	69 318
Latin America and Caribbean	890 782	14 952
Near East	122 327	15 082
North America	678 958	37 529
Global	4 032 905	264 084

2.2.2 Softwood fibre

2.2.2.1 Cell morphology of softwood fibre

Softwood fibre has fewer cell types with a relatively simple structure compared to that of hardwood fibre, mainly including axial tracheids, wood rays, axial parenchyma cells and epithelial parenchyma cells (Fig. 2.1). Axial tracheids are the main cells, which constitute 89–98% of the stem volume of softwood fibre (Figs 2.2 and 2.3).

Axial tracheids are thick-walled cells arranged along the trunk axis, closed at both ends, hollow inside, thin and long with pits in the cell wall. They are capable of transporting water and other nutrition. They provide mechanical support and dominate the wood's properties. Axial tracheids are radially stacked on the cross-section and the adjacent grains are slightly interleaved. Earlywood tracheids have blunt-broad,

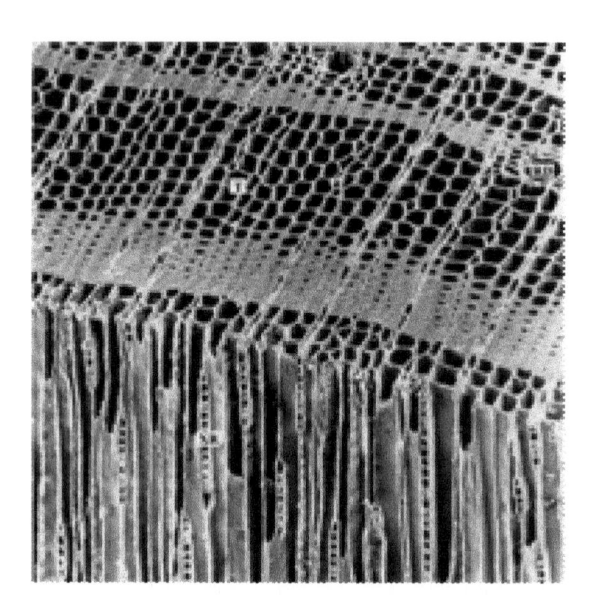

Figure 2.1 Stereoscopic image of the microscopic structure of softwood.

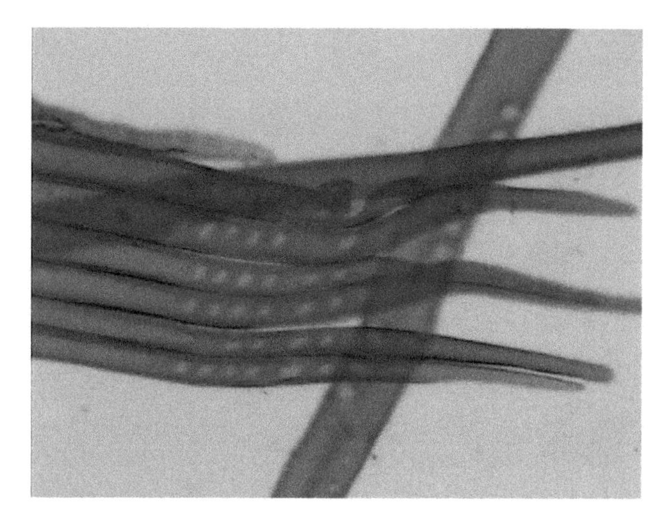

Figure 2.2 The end morphology of axial tracheids in Masson pine.

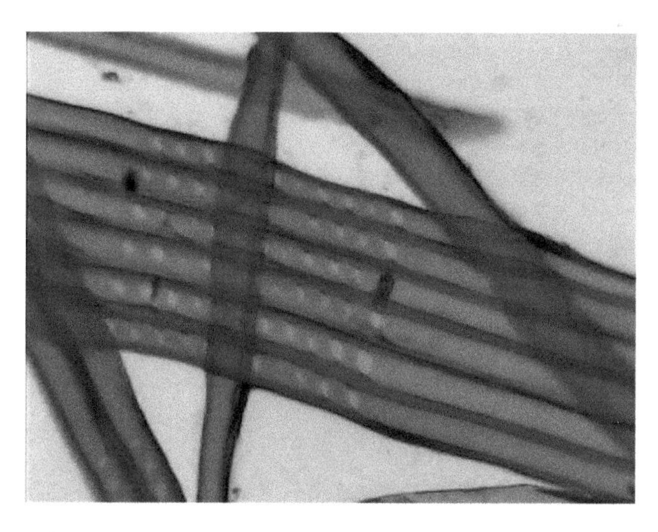

Figure 2.3 The middle morphology of axial tracheids in Masson pine.

large cavity and thin cell wall, and are quadrilateral or polygonal in cross-section; Latewood tracheids have tips at two ends and a thick cell wall, and are flat in cross-section. The average length of tracheids is 3–5 mm and the width is 15–80 μm. The length to width ratio is (75–20):1. The length of latewood tracheids is longer than that of earlywood tracheids. The change in thickness of tracheids from earlywood progressing to latewood varies, i.e. it could be slow such as with Chinese fir, but it can also be rapid, such as with Masson pine. The diameter in the tangential direction of axial tracheids determines the roughness of the fibre structure, i.e. the grain structure is less than 30 μm, the medium structure is 30–40 μm, the coarse structure is 45 μm or more.

2.2.2.2 Morphological characteristics of softwood fibre

Chinese fir (*Cunninghamia lanceolata* (Lamb.) Hook.) is a typical fast-growing softwood species in the south of China and hence taken as an example for discussion. It is found that a leading factor influencing plantation wood fibre morphology is the age of the trees, followed by site conditions and stand density. The influence of age on the morphological indices of Chinese fir fibre has been found to be very significant: except for the ratio of length to width, all fibre morphology indices increased with growing age [2].

Comparing the wood fibre morphology of Chinese fir plantations with *Pinus massoniana* [3–4], the average fibre length of Chinese fir plantation wood at age 17–23 years was shorter than that of *Pinus massoniana*: the former ranges from 3.00 mm to 3.19 mm whilst the latter is 3.61 mm. The width of Chinese fir fibres is also narrower than that of *Pinus massoniana* fibres, the former ranging from 44.38 μm to 45.34 μm with the latter being approximately 50 μm. The ratio of length to width was similar: the former ranges from 67.62 to 70.42, the latter is around 72. Moreover, the cell wall of Chinese fir fibres was thinner than their counterparts in *Pinus massoniana* fibres, with the former being 4 μm–5.28 μm and the latter 6.25 μm. The ratio of cell wall to cavity for Chinese fir is far less than that of Masson's pine, being 0.345–0.359 for the former and 0.64 for the latter. According to the trend of variation in the morphology index value of Chinese fir fibre, the ratio of length to width increased intensively from age 7 years to age 17 years (Table 2.2), but this increase slows down after the age of 17 years. This variation indicates that the stand has entered the process of maturity and wood fibre quality is more stable after the age 17 years, which is, in fact, the milestone for possible harvesting of the tree for use.

Table 2.2 Size of wood fibre under various conditions.

Factors	Level	Length (m)	Width (μm)	Ratio of length to width	Lumen diameter (μm)	Wall thickness (μm)	Ratio of wall to cavity
Tree age (a)	7	2.36	34.43	68.55	22.23	2.79	0.251
	12	2.72	41.43	65.65	26.07	3.99	0.306
	17	3.00	44.38	67.62	28.94	4.00	0.345
	23	3.19	45.34	70.42	29.37	5.28	0.359
Stand density (trees/hm^2)	1500–1800	2.81	41.25	68.22	26.64	4.28	0.317
	1815–2100	2.81	40.98	68.55	26.64	4.28	0.317
	2115–2400	2.82	41.42	68.08	26.66	4.24	0.314
	2415–2700	2.83	41.94	67.40	26.67	4.25	0.314
Site index (m)	16	2.75	40.21	68.40	26.59	4.19	0.311
	14	2.79	41.26	67.72	26.66	4.48	0.313
	12	2.83	41.70	67.97	26.65	4.30	0.318
	10	2.89	42.41	68.15	26.70	4.32	0.319

2.2.3 Hardwood fibre

2.2.3.1 Cell morphology of hardwood fibre

Most hardwood fibres have vessels, except a few species such as *Tetracentron sinense* Oliv. and *Trochodendron aralioides* Sieb. Hardwood contains several molecule types, mainly including vessel, wood fibre, axial parenchyma, wood rays, tracheids and gum canals. Its structure is complex with an irregular arrangement. Hardwood constitutes about 20% vessel members, about 50% wood fibre and 17% wood rays. Axial parenchyma is about 13%, and the shape, size and thickness of various cell walls vary considerably. According to the types of pit on the cell wall of wood fibre, there may be fibre tracheid with bordered pits and libriform fibre with pits.

Fibre tracheid is a cell with a small cavity, a thick wall and tips at two ends, and it is very similar to the latewood tracheid of softwoods. The cell wall has bordered pits like convex lenses and a crack-shaped pore. Tough libriform fibre is spindle-shaped, with slightly a sharpened tip, a thicker cell wall, narrower cavity and pits. According to the International Council on Wood Anatomy (IAWA), wood fibre length is divided into seven grades:

- very short <500 μm
- short 500–700 μm
- slightly shorter 700–900 μm
- medium 900–1600 μm
- slightly longer 1600–2200 μm
- long 2200–3000 μm
- extremely long >3000 μm

The fibre length of Chinese hardwood is generally 500–2000 μm, the average length is 1000 μm, placing it in the medium length grade, the diameter is 10–50 μm, and the wall thickness is 1–11 μm. For hardwood species with clear annual rings, the length of the latewood fibres is usually much longer than that of earlywood fibres, while there is no significant difference among tree species where the annual rings are not distinct. In the cross-section of the trunk, the radial variation of the average length of the wood fibre could be summarised as follows: the fibre around the pith is the shortest, it gradually grows in the juvenile wood, the rapid growth slows down in mature wood, and then it becomes relatively stable.

2.2.3.2 Morphological characteristics of Acacia trees wood fibre

Acacia genus Mimosaceae (Mimosaceae) *Acacia* (*Acacia*) has about 1200 species throughout the world. Arbor species account for about 10–20%. Major characteristics are its rapid growth and strong adaptability. China began introducing Acacia trees from the end of 1970s, and currently more than 20 species have been imported successfully, with a total planted area of more than 500,000 hm^2. Four Acacia species (*Acacia mangium, Acacia acuriculaeformis, Acacia crassicarpa* and *Acacia aulacocarpa*) have been introduced into China more successfully, and exhibit good growth,

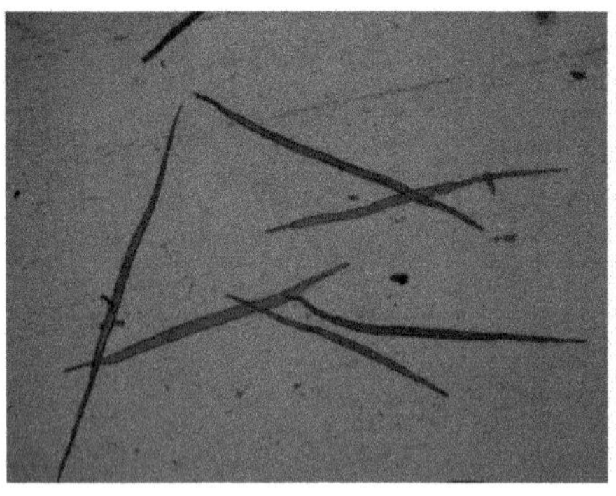
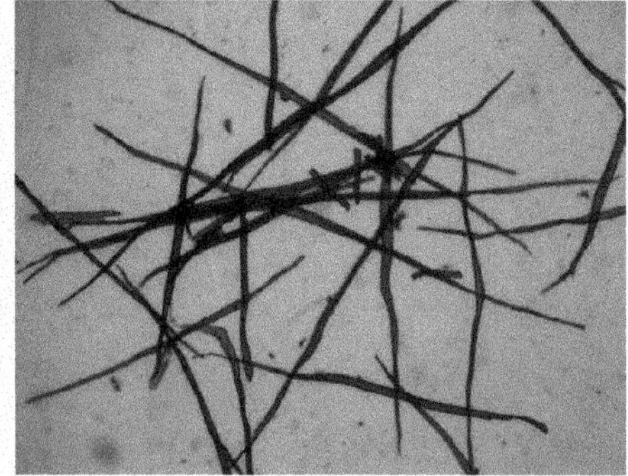

Figure 2.4 Morphology of wood fibre for Acacia trees.

Table 2.3 Fibre configuration of different species of Acacia wood.

Species	Ave. length (m)	Ave. width (μm)	Ratio: length to width	Lumen diameter (μm)	Cell wall thickness (μm)	Ratio: wall to cavity
Acacia mangium	0.887	20.02	48.38	12.72	7.35	0.57
Acacia crassicarpa	0.935	16.61	58.86	9.59	6.35	0.65
Acacia acuriculaeformis	0.902	16.93	55.17	9.28	7.56	0.84
Acacia mangium	0.977	16.42	60.36	10.59	6.35	0.74

strong performance adaptability, salt tolerance and growth capacity. These four species have been introduced in Fujian, Guangdong, Hainan, Guangxi, Yunnan and other southern provinces. Wood fibre morphology of 13-year-old *Acacia mangium, Acacia acuriculaeformis, Acacia crassicarpa* and *Acacia aulacocarpa* has been determined and presented in Table 2.3. It is apparent that, in general, wood fibres of these four Acacia species are relatively short and their average fibre length obeys the following sequence: *Acacia aulacocarpa* > *Acacia crassicarpa* > *Acacia acuriculaeformis* > *Acacia mangium*. Average fibre width obeys the following sequence: *Acacia mangium* > *Acacia acuriculaeformis* > *Acacia crassicarpa* > *Acacia aulacocarpa*. The ratio of length to width was *Acacia aulacocarpa* > *Acacia crassicarpa* > *Acacia acuriculaeformis* > *Acacia mangium* [5].

Poplar wood is a typical and important fibrous material. The fibre length of poplar wood is 635–1261 μm, the fibre width is 17.96–22.33 μm, cell diameter is 14.34–17.83 μm, double wall thickness is 3.55–5.35 μm, the ratio of wall to lumen is 0.22–0.34 μm, and the ratio of length to width is 37.81–60.89 μm [6]. It could be concluded that fibre length, fibre width and ratio of length to width of the aforementioned four Acacia wood fibres are close to those of poplar wood, that the cell diameter is smaller than that of poplar wood, whilst the double wall thickness and the ratio of wall to lumen are greater than those of poplar wood.

2.3 Bamboo fibres

2.3.1 Resources of bamboo fibres

There are about 78 genera and more than 1400 species of bamboo around the world [7]. Asia–Pacific Bamboo District, American Bamboo District and African Bamboo District are three major bamboo production regions in the world, of which the Asia–Pacific bamboo area has the most abundant bamboo species. In the Asia-Pacific region China is known as the 'Kingdom of bamboo', possessing considerable bamboo resources. There are about 48 genera and more than 500 species of bamboo in China, mainly distributed south of Yangtze River, with about 7.2 million hm^2 of total bamboo land, of which the plantation area has reached 5 million hm^2. The annual cutting of bamboo is 12 million tonnes, ranking it first in the world. Bamboo grows fast, generally taking 3 to 4 years to become lumber, which makes its more prominent as a fibre resource.

2.3.2 Cell morphology of bamboo fibre

Bamboo fibre is a special cell in bamboo structure, staying in stem in the form of bundle sheath or isolated vascular bundle with a content of about 40% of the whole bamboo stem. It mainly plays the role of mechanical support. The bamboo cell is slender and hard with a thin cell wall and smooth inner and outer walls. A small number of pits or bordered pits are sparsely and evenly distributed on radial and tangential walls of the bamboo fibre. The aperture of bamboo fibre is small, generally less than 1 μm. Bamboo fibre has evolved from the tracheids, derived from the daughter cells of parenchyma cells formed by longitudinal division, which is different from wood fibre and other monocotyledon plant fibres. A single bamboo fibre is slender, spindle-shaped, hollow inside with no natural twist (Fig. 2.5). The cell wall morphology of bamboo fibre mainly is of two different types: (1) a thick wall with small diameter, distributed in the fibre sheath inside; and (2) a thin wall with larger diameter, distributed in the fibrous sheath outside. Fibres with thick cell walls and small cavities are very hard, easy to cut off, not easily fibrillated, and they are not conducive to pulp and paper, but this feature is conducive to the production of structural materials.

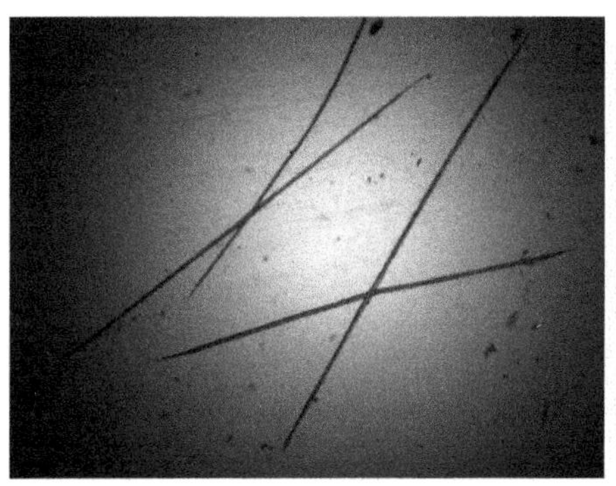
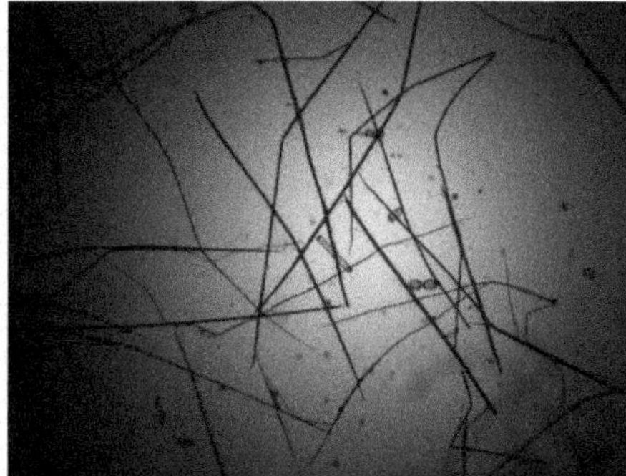

Figure 2.5 Cell morphology of bamboo fibre.

The length of bamboo fibre is usually 1.5–2.0 mm, the length to width ratio is > 100, the holocellulose content is more than 70%, the lignin content is about 25% (between that for softwood and hardwood), making it an ideal raw material for fibre materials.

2.3.3 Morphological characteristics of bamboo fibre

Phyllostachys heterocycla cv. *pubescens* is one of the most important economic bamboo species. The stalk is generally 8–15 m high but can attain 25 m, the diameter is generally 8–10 cm, but can be as large as 18 cm, whilst the length of the internode at the upper and lower positions is shorter than that at middle position of the stalk. *Phyllostachys heterocycla* cv. *pubescens* bamboo is tough, straight grained, hard and smooth, and its fibre content is around 30–35%. The effect of age and the vertical position of culm on the morphological characteristics of bamboo fibre is shown in Table 2.4 [8]. The average length of bamboo fibre at different ages is between 1659 μm and 2331 μm, the average width is between 12.00 μm and 15.07 μm, the average length to width ratio is between 132.77 and 196.96, and the average wall thickness is 3.25–5.33 μm. The average diameter of the cavity is between 1.17 μm and 2.08 μm, and the average ratio of wall to cavity is between 3.59 and 7.86.

Table 2.4 Morphological characteristics of *Dendrocalamus latiflorus* bamboo fibre vs. age and position.

		Morphology of fibre					
Bamboo culm height (m)	Age (year)	Length (μm)	Width (μm)	Ratio of length to width (μm)	Wall thickness (μm)	Cavity diameter (μm)	Ratio of wall to cavity
0.5	1	2036.12	13.52	136.38	3.25	1.46	4.80
	2	1989.39	14.93	139.62	4.28	1.72	6.97
	4	2238.83	14.79	156.93	4.58	1.34	7.05
	6	2254.70	15.07	154.24	4.52	1.57	6.36
2.5	1	1939.82	12.23	163.32	3.88	1.66	3.59
	2	2115.10	12.39	173.42	4.83	1.75	6.42
	4	2330.88	12.00	196.96	5.33	1.78	6.43
	6	2029.80	12.30	168.13	4.74	2.08	5.08
4.5	1	1996.05	12.79	155.74	3.57	1.96	4.12
	2	2018.39	13.24	157.15	4.25	1.36	6.62
	4	2263.13	13.57	170.90	4.64	1.66	7.67
	6	2298.18	13.96	166.67	4.14	1.63	6.48
6.5	1	1658.58	12.60	132.77	3.65	2.03	3.64
	2	1866.85	12.77	151.81	4.39	1.19	7.86
	4	1902.67	12.56	145.84	4.77	1.56	6.25
	6	1793.65	12.19	158.96	4.28	1.17	7.79

2.4 Morphological characteristics of other important bamboo fibres

2.4.1 *Dendrocalamus* latiflorus Munro

The effect of age and the height of the culm on the morphological characteristics of *Dendrocalamus latiflorus* bamboo fibre can be seen in Table 2.5 [8]. The average length of *Dendrocalamus latiflorus* bamboo fibre at different ages is between 1684 μm and 2263 μm, the average width is between 13.25 μm and 16.25 μm, the average ratio of length to width is between 113 and 170, the average wall thickness is between 3.21 μm and 4.62 μm, the average cavity diameter is between 1.40 μm and 2.84 μm, and the average ratio of wall to cavity is between 2.92 and 6.44.

2.4.2 *Dendrocalamopsis oldhami* (Munro) Keng f.

The effect of age and the height of the culm on the morphological characteristics of *Dendrocalamopsis oldhami* (Munro) Keng f. bamboo fibre is illustrated in Table 2.6 [8]. The average length of *Dendrocalamopsis oldhami* (Munro) Keng f. green bamboo fibre at different ages is between 1552 μm and 2097 μm, the average width is between

Table 2.5 Morphological characteristics of *Dendrocalamus latiflorus* bamboo fibre vs. age and position.

Bamboo culm height (m)	Age (year)	Morphology of fibre					
		Length (μm)	Width (μm)	Ratio of length to width	Wall thickness (μm)	Cavity diameter (μm)	Ratio of wall to cavity
0.5	1	1962.47	13.57	148.59	3.55	2.34	3.81
	2	2009.56	14.56	140.01	3.96	1.61	5.15
	4	2088.65	14.61	149.88	4.02	1.51	6.07
	6	1684.43	13.43	130.65	3.93	1.94	4.37
2.5	1	2072.52	15.88	139.10	3.82	2.54	3.13
	2	2049.94	15.17	143.63	4.61	2.63	3.97
	4	2161.51	15.82	143.28	4.62	1.72	5.93
	6	1860.84	16.25	118.31	4.25	2.17	4.42
4.5	1	2087.43	14.53	143.34	3.50	2.66	2.92
	2	2263.48	13.25	169.88	4.26	2.79	3.35
	4	2129.66	13.43	150.02	4.36	1.40	6.44
	6	1943.98	13.55	139.38	4.03	1.88	4.66
6.5	1	1974.54	15.45	131.80	3.21	1.88	3.46
	2	2118.01	13.75	156.37	4.43	2.84	3.51
	4	1955.85	14.71	139.02	4.39	1.81	5.52
	6	1722.93	15.07	120.86	3.64	2.48	3.34

Table 2.6 Morphological characteristics of *Dendrocalamopsis oldhami* bamboo fibre vs. age and position.

Bamboo culm height (m)	Age (year)	Morphology of fibre					
		Length (μm)	Width (μm)	Ratio of length to width	Wall thickness (μm)	Cavity diameter (μm)	Ratio of wall to cavity
0.5	1	1724.56	12.02	161.07	3.09	3.49	1.94
	2	1656.41	15.22	111.42	4.17	2.60	3.52
	4	1852.00	17.64	110.17	4.41	2.59	3.75
	6	1958.25	15.79	130.22	4.97	1.99	5.40
2.5	1	1768.72	11.87	173.80	2.59	3.00	2.06
	2	2043.98	14.77	132.23	4.07	2.34	4.89
	4	2097.13	16.80	132.01	4.27	2.43	4.92
	6	2088.42	15.15	145.04	4.70	1.55	6.33
4.5	1	1731.14	11.63	162.91	2.80	3.32	1.83
	2	1973.71	13.93	130.44	3.82	2.61	3.14
	4	2020.68	16.15	131.42	4.10	2.50	3.46
	6	2002.56	14.75	143.08	4.79	1.64	5.08
6.5	1	1765.55	11.94	164.14	2.72	3.26	1.87
	2	1552.17	14.01	113.60	4.51	2.59	4.23
	4	1584.28	16.27	101.47	4.84	1.68	4.85
	6	1599.62	14.69	119.47	5.17	1.52	6.99

11.63 μm and 17.64 μm, the average ratio of length to width is between 101 and 174, the average wall thickness is between 2.59 μm and 5.17 μm, the average cavity diameter is between 1.46 μm and 3.49 μm and the average ratio of wall to cavity is between 1.83 and 7.13.

2.4.3 *Neosinocalamus affinis* (Rendle) Keng f.

The effect of age and the height of the culm on the morphological characteristics of *Neosinocalamus affinis* bamboo fibre is showed in Table 2.7 [8]. The average length of 3- or 4-year-old *Neosinocalamus affinis* bamboo fibre is between 1687 μm and 2252 μm, the average width is between 11.97 μm and 15.94 μm, the average ratio of length to width is between 119.45 and 195.30, the average wall thickness is between 4.28 μm and 5.51 μm, the average cavity diameter is between 1.51 μm and 1.94 μm, and the average ratio of wall to cavity is between 4.70 and 7.09.

2.4.4 *Bambusa chungii* McClure

The effect of age and the height of the culm on the morphological characteristics of *Bambusa chungii* bamboo fibre can be seen in Table 2.8 [8]. The average length of

Table 2.7 Morphological characteristics of *Neosinocalamus affinis* bamboo fibre vs. age and position.

Bamboo culm height (m)	Age (year)	Morphology of fibre: Length (μm)	Width (μm)	Ratio of length to width	Wall thickness (μm)	Cavity diameter (μm)	Ratio of wall to cavity
0.5	3	1686.67	14.73	119.45	3.97	1.62	5.24
	4	1928.98	15.94	125.52	3.99	1.38	5.98
2.5	3	2084.38	13.17	161.28	4.28	1.62	5.66
	4	1833.00	14.13	132.83	5.08	1.76	6.23
4.5	3	2252.00	11.97	195.30	3.69	1.80	4.43
	4	2006.36	12.52	165.53	3.70	1.78	5.11
6.5	3	1976.62	12.92	155.67	4.13	1.64	5.24
	4	1943.50	13.74	150.23	5.05	1.63	6.64

Table 2.8 Morphological characteristics of *Bambusa chungii* bamboo fibre vs. age and position.

Bamboo culm height (m)	Age (year)	Morphology of fibre: Length (μm)	Width (μm)	Ratio of length to width	Wall thickness (μm)	Cavity diameter (μm)	Ratio of wall to cavity
0.5	3	1892.73	12.69	149.70	4.35	1.38	6.59
	4	1562.09	11.89	131.35	3.95	1.36	5.94
2.5	3	2128.13	12.24	174.68	3.88	1.45	5.70
	4	2044.37	11.83	161.50	3.86	1.38	5.76
4.5	3	1866.99	11.87	157.91	4.01	1.54	5.50
	4	1612.43	11.61	140.49	4.17	1.40	6.10
6.5	3	1912.00	13.49	150.93	4.19	1.38	6.37
	4	1900.42	12.32	158.65	3.88	1.23	6.63

3- or 4-year-old *Bambusa chungii* bamboo fibre is between 1562 μm and 2128 μm, the average width is between 11.61 μm and 13.49 μm, the average ratio of length to width is between 131.35 and 174.68, the average wall thickness is between 3.86 μm and 4.35 μm, the average cavity diameter is between 1.23 μm and 1.54 μm, and the average ratio of wall to cavity is between 5.50 and 6.63.

Table 2.9 Morphological characteristics of *Phyllostachys heterocyclada* bamboo fibre vs. age and position.

Bamboo culm height (m)	Age (year)	Morphology of fibre					
		Length (μm)	Width (μm)	Ratio of length to width	Wall thickness (μm)	Cavity diameter (μm)	Ratio of wall to cavity
0.5	3	1850.62	14.71	124.90	3.90	1.76	4.89
	4	2178.24	15.48	150.18	3.45	1.87	4.13
	5	1797.96	13.73	135.97	3.80	1.61	5.02
2.5	3	1817.30	14.10	136.29	4.35	1.58	6.15
	4	1787.61	13.43	130.32	4.30	1.59	5.64
	5	2073.08	14.45	150.71	4.34	1.47	5.69
4.5	3	1854.78	14.29	140.47	3.30	1.86	4.44
	4	1882.49	15.36	134.08	3.19	1.60	4.37
	5	1625.55	13.67	124.79	3.52	1.54	5.49
6.5	3	1661.37	14.78	117.50	3.97	1.48	5.53
	4	1691.31	16.52	107.66	3.30	1.28	5.46
	5	1708.66	12.61	142.15	3.83	1.10	6.74

2.4.5 Phyllostachys heterocyclada Oliver

The effect of age and vertical position of culm on the morphological characteristics of *Phyllostachys heterocyclada* bamboo fibre is illustrated in Table 2.9 [8]. The average length of 3-, 4- and 5-year-old *Phyllostachys heterocyclada* bamboo fibre is between 1626 μm and 2178 μm, the average width is between 12.61 μm and 16.52 μm, the average ratio of length to width is between 108 and 151, the average wall thickness is between 3.19 μm and 4.35 μm, the average cavity diameter is between 1.10 μm and 1.87 μm, and the average ratio of wall to cavity is between 4.13 and 6.14.

2.5 Bast fibres

2.5.1 Resources of bast fibres

Among natural fibres, bast fibre not only has very high strength and modulus but also the characteristics of being hard, friction resistant and corrosion resistant. In 2007, the global harvest area of bast fibre crops was 3.2 million hm^2 and output was 6.7 million tonnes. China has rich resources of hemp: the harvest area was about 613,000 hm^2 and output was about 1.11 million tonnes, which accounted for 19.3% and 16.5% of global harvested area and total output respectively (see Table 2.10).

Table 2.10 Major bast crop resources.

Bast	Whole world		China	
	Harvested area (*10⁴ hm²)	Total output (*10⁴ tonnes)	Harvested area (*10⁴ hm²)	Total output (*10⁴ tonnes)
Ramie	40.8	28	39.6	27.7
Flax	68.6	277.6	15.7	69.4
Hemp	5	3.7	1.1	4.1
Jute/kenaf	165	325	3.1	8.2
Sisal	38	33	1.8	1.6
Total	317.4	670.3	61.3	111.0

Source: Data for the world hemp resources were extracted from FAO. Data for domestic hemp resources were extracted from Chinese agriculture almanac 2006.

2.5.2 Cell morphology of bast fibre

Bast fibres are obtained from a variety of bast fibre plants, including phloem fibre and leaf fibre. Phloem fibre is produced on a vast scale worldwide. This type of fibre is soft and suitable for textile processing, commercially known as 'soft fibre'. In particular, ramie is of fine quality and has a long single fibre length and can be used in single fibre spinning. The single fibre length of other bast fibres is very short, and they are generally used in processed fibre spinning. The cell wall of bast fibre from Jute is lignified. The bast fibre of Jute is mostly short and has been used widely for spinning cordage and packing sacks. With flax the cell wall of the bast fibre is not lignified and the fibre thickness is similar to cotton fibre; it is widely used in textile materials. The bast fibre cell is long and spindle-shaped and the cell wall is thickened and highly lignified; the top part of the cell is bonded together to form a tough tissue. The cell length and width of different types of bast fibre is from 2 mm to 180 mm and from 5 μm to 50 μm respectively [9].

2.5.3 Morphological characteristics of flax fibre

The length of flax fibre is between 8.0 mm and 40.0 mm, and the average length is about 18 mm. The width of flax fibre is between 8.8 μm and 24.0 μm, with an average width of about 16 μm. The ratio of length to width is more than 1000. The outer surface of flax fibre is smooth; it is acuminate both ends. The cell cavity is very small, but the wall is thick. The cavity wall has obvious stria and a few pits. The structure of a single fibre is inconsistent in different parts of the flax stem. The single fibre cross-section of the root is oblate or round, and the cell wall is thin and multilevel. The pith is large and hollow. From the root of the 1/6 part to the middle of the stem, a single fibre is mostly triangular in cross-section with a thick cell wall. The single fibre cells of the stem tip are small, sometimes unmyelinated. However, the main body of the fibre is polygonal in section. The longitudinal surface morphology of flax fibre has the characteristics of uniform thickness, exhibiting a clear and regular bamboo-like cross [9].

2.5.4 The characteristics of ramie fibre

The length of ramie fibre is between 120 mm and 180 mm, the width is between 20 μm and 50 μm, and the ratio of length to width is about 2000. The shape of ramie fibre is irregular i.e. it can be stripes or transverse stripes and the shape of both ends is round or has a lance-like shape. The cross-sectional morphology of ramie fibre can be oval, flat, polygonal or other shapes. The cavity is oval or irregularly shaped. The cell wall is uniform, with radial stripes appearing zonal in the cross-section of the juvenile fibrous cell. The longitudinal surface morphology of ramie fibre may be summarised as follows: uneven thickness, cylindrical or flat ribbon, and unobvious turn. The fibre surface is smooth or with obvious vertical stripes, usually has nodules on both sides and the cracks intersected or inclined. The lignification of the fibre is very low (almost no lignin): therefore, the fibre is rich in toughness and elasticity and is not easy to break [10].

2.5.5 Morphological characteristics of jute fibre

Jute fibre cells adhere to each other to form a bundle, and each bundle is bonded together with 20 to 30 fibrous cells. The length of a fibre bundle is 2–3 m, and the single fibre length is 2–3 mm with a width of 15–25 μm. Its aspect ratio is about 100. There are no knots on the smooth fibre surface. There are about 10 polygonal holes in the cross-section of each jute fibre, mostly pentagonal or hexagonal. Cavities exist in the middle of the cross-section, and appear round or oval, wide or narrow. The longitudinal surface of jute fibre has a clear vertical structure but the cross-section is not obvious. The cell wall is neat and cylindrical and has no natural curl in it. The lignification of jute fibre is relatively high with greater lignin content compared to that of ramie or flax. Moreover, the fibre is extremely long, which makes it very tough [11].

2.6 Bagasse fibre

2.6.1 Resources of bagasse fibre

Bagasse is a rich agricultural and forestry waste. It is a fibrous residue obtained from sugarcane after crushing the sugar. It is the largest agricultural waste in the world [12]. The world's annual output of bagasse exceeds 100 million tonnes. China produces more than 6 million tonnes of sugarcane annually.

2.6.2 Morphological characteristics of bagasse fibre

The fibre morphology varies among various sugarcane varieties. Even fibres from the same species may exhibit some differences in morphology due to differences in geography, climate and growth stages. Generally, the length of bagasse fibre is 1000–2000 μm, the width is 14–28 μm, and the length to width ratio is 60–80. The ratio of wall to lumen is <1. Its characteristics are medium length, large width and small ratio of wall to lumen.

2.7 Crop straw fibres

2.7.1 Resources of crop straw

Straw usually refers to the remaining part of corn, wheat, rice and other crops after harvesting the seed, and is a renewable biological resource with many uses. The amount of straw produced from agriculture is significant: for example, 1 kg of rice can produce 1.5 kg of straw; 1 kg of wheat can produce 1.5 kg of straw; 1 kg of corn can produce 4 kg straw. According to United Nations Environment Programme statistics, the world's growing crops produce as much as 1.7 billion tonnes of straw each year, indicating extensive resources of straw. China's annual output of crop straw ranks first in the world at about 700 million tonnes, of which rice straw accounts for 27.5% and wheat straw 17.2%. Both are important fibre raw materials.

2.7.2 Morphology of crop straw fibre

2.7.2.1 Wheat straw fibre morphology

The length of wheat straw fibre is close to that for hardwood fibres (Table 2.11) but the width is smaller. The average fibre length is 1500 μm, the average width is 14 μm, and the average thickness is 3 μm. The internode fibre is the longest fibre and has the largest length/width ratio, while the knot fibre is the shortest with the smallest length/width ratio [13].

2.7.2.2 Rice straw fibre morphology

Rice straw is a type of herbaceous plant, the fibre from which is short and thin (Table 2.12). The average fibre length is <1000 μm, the average width is here only

Table 2.11 Fibre morphology of different parts of wheat straw.

	Length (μm)			Width (μm)		
Position	Average	Max.	Min.	Average	Max.	Min.
All	1320	2940	610	12.9	24.5	7.4
Internode	1520	2630	660	14.0	27.9	8.3
Spike-stalk	1200	2390	390	11.5	24.5	7.4
Sheath	1260	3310	440	14.7	34.3	8.8
Leaf	860	1470	240	12.1	19.6	6.4
Knot	470	1290	180	17.8	43.1	8.3

Table 2.12 Fibre morphology of rice straw.

	Length (mm)				Width (μm)						
Position	Average	Max.	Min.	Commonly	Average	Max.	Min.	Commonly	L/D ratio	Wall thickness (μm)	Cavity diameter (μm)
Stems	1.02	2.12	0.40	0.52–1.41	9.1	18.8	8.3	6.0–13.0	112	1.6	8.7

about 8 μm. The fibre wall has both obvious and unobvious pits and cross knots. The length and width of straw are measured by overhead projector, and the fibre wall thickness and cavity diameter of straw can be determined by light microscopy [13].

2.8 Conclusion and outlook

Plant fibre is a natural resource that can be regenerated in nature. The morphological characteristics of wood fibre, bamboo fibre and crop straw fibre share considerable similarities whilst also exhibiting some differences. The structural characteristics of plant fibre directly affect the quality of plant fibre products. Future scientific and technological work should focus on an understanding of the structural characteristics of plant fibre and their effect on plant fibre products.

References

[1] Xu Y. *Wood Science* [M]. Beijing: China Forestry Publishing House, 2006: 50–60.

[2] Nie S, Lin Jing & Lin S. A study on the variation fiber morphology of Chinese fir wood from plantation [J], *Journal of Northwest Forestry University*, 1998, 13 (4): 9–13.

[3] Wu Y. *Lignocellulosic Chemistry* [M]. Beijing: China Light Industry Press, 1991: 32–34.

[4] Nanjing Forestry University. *Handbook of Chemical Industry of Forest Products* [M]. Beijing: China Forestry Publishing House, 1984: 1–17.

[5] Chen X, Lin Su & Lin J. Study on wood pulping performances of four introduced Acacia species in southern Fujian [J]. *Journal of Forest and Environment*, 2006, 26 (2): 144–147.

[6] Zhou L, Liu S, Tian G & Liu Y. Radial variation of anatomical properties of Poplar clone107 [J]. *Journal of Northeast Forestry University*, 2010, 10: 1–4.

[7] Xu J, Zhao R & Fei B. Research on properties and utilization of bamboo in China [J]. *Wood Processing Machinery*, 2007 (3): 39–42.

[8] Liu Z. Study on fiber properties of chief bamboo species in Fujian province [D]. Fujian Agriculture and Forestry University, 2011.

[9] Zhao X, Wang Y, Huang H & Huang J. Research on identification of flax, ramie and hemp fiber [J]. *China Fiber Inspection*, 2010, 15: 65–67.

[10] Yang L. Comparison of property of bamboo and ramie fiber [J]. *Textile Science & Technology*, 2005, 8: 59–62.

[11] Chen J & Yu W. Study situation of research on the morphology and structure of jute fibers [J]. *Plant Fibers and Products*, 2005, 5: 254–258, 239.

[12] Trejo-Hernandez M R, Ortiz A & Okoh A I, Biodegradation of heavy crude oil Maya using spent compost and sugarcane bagasse wastes [J]. *Chemosphere*, 2007, 68 (5): 848–855.

[13] Liu Y & Zhao G. *Wood Science* [M]. Beijing: China Forestry Publishing House, 2012.

Chapter 3
The chemistry of plant fibres

Professor Shilin Cao and Professor Fang Huang

This chapter overviews the chemical composition of plant fibres from wood and graminaceous plants, including cellulose, hemicellulose, lignin, extractives and ash. The chemical structure and properties of cellulose, hemicellulose and lignin are comprehensively discussed to provide a solid platform for the better utilisation of plant fibres.

3.1 Introduction

Plant material is derived from the reaction between CO_2 in the air, water and sunlight, via photosynthesis, to produce carbohydrates that form the building blocks of biomass plant material. Typically, photosynthesis converts less than 1% of the available sunlight to the stored chemical energy [1]. The solar energy driving photosynthesis is stored in the chemical bonds of the structural components of plant material. If plant material is processed efficiently, either chemically or biologically, by extracting the energy stored in the chemical bonds and the subsequent 'energy' product combined with oxygen, the carbon is oxidised to produce CO_2 and water. The process is cyclical, as the CO_2 is then available to produce new biomass plant.

The value of a particular type of biomass plant depends on the chemical and physical properties of the large molecules from which it is made. For millennia humanity has exploited the energy stored in these chemical bonds, by burning biomass plants as fuel and by eating plants for the nutritional content of their sugar and starch. More recently, the fossilised biomass plant has been exploited as coal and oil. However, since it takes millions of years to convert biomass plants into fossil fuels, these are not renewable within a timescale mankind can use. Burning fossil fuels uses 'old' biomass plants and converts them into 'new' CO_2, which contributes to the 'greenhouse' effect and depletes a non-renewable resource. Burning new biomass plants contributes no new CO_2 to the atmosphere, because replanting the harvested biomass plants ensures that CO_2 is absorbed and returned for a cycle of new growth.

Numerous crops have been proposed or are being tested for commercial energy farming. Potential energy crops include woody crops and grasses/herbaceous plants (all perennial crops), starch and sugar crops and oilseeds.

In general, the characteristics of the ideal energy crop are [2, 3]:

- high yield (maximum production of dry matter per hectare),
- low energy input to produce,
- low cost,
- composition with the least contaminants,
- low nutrient requirements.

Desired characteristics will also depend on local climate and soil conditions. Water consumption can be a major constraint in many areas of the world and makes the drought resistance of the crop an important factor.

3.2 Composition of lignocellulosic biomass plants

Understanding the lignocellulosic biomass plant, particularly its chemical composition, is a prerequisite for developing effective pretreatment technologies to deconstruct its rigid structure, designing enzymes to liberate sugars, particularly cellulase to release glucose, from recalcitrant cellulose, as well as engineering microorganisms to convert sugars into ethanol and other bio-based chemicals. The main components of the lignocellulosic materials are cellulose, hemicellulose, lignin and a remaining smaller part (extractives and ash). The composition of lignocellulose depends greatly on its source. There is a significant variation in the lignin and (hemi) cellulose content of lignocellulose depending on whether it is derived from hardwood, softwood or grasses. Table 3.1 summarises the composition of lignocellulose encountered in the most common sources of biomass plants.

3.3 Physical and chemical characteristics of lignocellulosic materials

Lignocellulosic biomass material has a complex internal structure. The major components of lignocellulosic material are cellulose, hemicellulose and lignin, with intricate structures. To obtain a clear picture of the material, an analysis of the structure of each main component is made in this section, concluding with a description of the structure of lignocellulose itself. The physical properties of each component and how these components contribute to the behaviour of the complex structure are also addressed.

3.3.1 Cellulose

Cellulose is the β-1,4-polyacetal of cellobiose (4-O-β-D-glucopyranosyl-D-glucose). Cellulose is more commonly considered a polymer of glucose, because cellobiose consists of two molecules of glucose. The chemical formula of cellulose

Table 3.1 Typical lignocellulosic biomass composition (% dry basis) [4, 5, 6, 7].

	Cellulose	Hemicellulose	Lignin
Pine	43.3	20.5	28.3
Spruce	45.0	22.9	27.9
Douglas fir	45.0	19.2	30.0
Poplar	44.7	18.5	26.4
Eucalyptus	49.5	13.1	27.7
Corn stover	36.8	30.6	23.1
Miscanthus	39.3	24.8	22.7
Wheat straw	42.1	23.8	20.5
Switchgrass	42.0	25.2	18.1

Table 3.2 DP of native wood and non-woody cellulose after nitration using viscosmetric method [13].

Biomass species	DP
Trembling aspen	5000
Beech	4050
Red maple	4450
Eastern white cedar	4250
Eastern hemlock	3900
Jack pine	5000
Tamarack	4350
White spruce	4000
Balsam fir	4400
White birch	5500
Eucalyptus regnans	1510
Pinus radiata	3063
Sugarcane bagasse	925
Wheat straw	1045

is $(C_6H_{10}O_5)_n$. Many properties of cellulose depend on its degree of polymerisation (DP), i.e. the number of glucose units that make up one polymer molecule. The DP of cellulose varies from 5000 in native wood to 1000 in bleached wood pulp [8], and 500–1000 in herbaceous cellulose, as shown in Table 3.2. Each D-anhydroglucopyranose unit possesses hydroxyl groups at C_2, C_3 and C_6 positions, capable of undergoing the typical reactions known for primary and secondary alcohols. The molecular structure imparts cellulose with its characteristic properties: hydrophilicity, chirality, degradability, and broad chemical variability initiated by the high donor reactivity of hydroxyl groups.

The nature of the bonding between the glucose molecules (β-1,4 glycosidic) allows the polymer to be arranged in linear chains. Cellulose has a strong tendency to form intra- and inter-molecular hydrogen bonds by hydroxyl groups between the molecules of cellulose. The hydrogen bonds in turn result in the formation of an ultrastructure that is comprised of several parallel chains hydrogen-bonded to each other [9]. The coalescence of several polymer chains leads to the formation of microfibrils, which in turn are united to form fibers.

The hydrogen bonds in the linear cellulose chains promote aggregation into a crystalline structure and give cellulose a multitude of partially crystalline fibre structures and morphologies [10]. The average degree of crystallinity of native cellulose ranges from 50% to 70% [11, 12]. The ultrastructure of native cellulose (cellulose I) has been discovered to possess unexpected complexity in the form of two crystal phases: I_α and I_β [13]. Electron diffraction and nuclear magnetic resonance (NMR) studies have shown that cellulose I_α is an allomorph with triclinic unit cells, whereas cellulose I_β is an allomorph with two-chain monoclinic units [14]. The relative amounts of I_{α} and I_{β} have been found to vary between samples from different origins. The I_{α}-rich specimens have been found in the cell wall of some algae and in bacterial cellulose, whereas I_{β}-rich specimens have been found in cotton, wood, and ramie fibres [15, 16]. Native cellulose also contains para-crystalline and amorphous portion. Para-crystalline cellulose is loosely described as chain segments having more order and less mobility than amorphous chains segments but as being less ordered and more mobile than the crystalline domain [17, 18]. The presence of crystalline cellulose, with regions of less order, together with the size of the elementary fibrils, work together to produce interesting combinations of contrary properties such as stiffness and rigidity on the one hand and flexibility on the other hand [19].

There are different techniques to measure cellulose crystallinity. The frequently used method to determine the crystallinity of cellulose is based on X-ray diffraction (XRD). This method was used for 70–85% of the studies [20]. The crystallinity index (CrI) is usually employed to describe the crystalline degree of biomass and pulps [21, 22].

In addition to the XRD method, the Cross Polarisation/ Magic Angle Spinning ^{13}C Nuclear Magnetic Resonance (CP/MAS ^{13}C NMR) [23, 24] and Fourier-transform Infrared Spectroscopy (FTIR) [25] have been employed to determine CrI values. In the CP/MAS ^{13}C NMR method, the CrI can be determined from the area of crystalline and amorphous C4 signals using the following equation, based on the solid NMR analysis [26]:

Results for the cellulose crystallinity and structure of cellulose for hybrid poplar, Loblolly pine and switchgrass are given in Table 3.3. Cellulose from poplar is 63% crystalline with cellulose I_{β} as the predominant crystalline form. The less ordered amorphous region of poplar comprises 18% solvent inaccessible fibril surfaces. The intermediate para-crystalline form of cellulose also accounts for a significant proportion of poplar cellulose structure (Table 3.3). Crystallinity of poplar cellulose is comparable to that for Loblolly pine but is about 20% higher than for switchgrass (Table 3.3). The amorphous region of switchgrass cellulose is mostly in the form of inaccessible surfaces, however, which may hinder enzymatic hydrolysis.

In the FTIR methods, the ratio of amorphous to crystalline cellulose associates with the ratio of intensities of the bands at 900 cm^{-1} and 1098 cm^{-1} [30], and the crystalline cellulose polymorphs (I_{α}/I_{β}) ratio, can be measured by comparing the intensity of the bands at 750 cm^{-1} and 710 cm^{-1} [31, 32, 33].

In the cellulose crystallinity measurement, although different methods use different techniques, the basic principles are the same: CrI measures the relative fraction of crystalline cellulose in the combination of crystalline and amorphous fractions.

Table 3.3 Cellulose crystallinity and structure determined from CP/MAS ^{13}C NMR.

	Crystallinity	I_{α}	$I_{\alpha+\beta}$	Para-crystalline	I_{β}	Accessible fibril surface	Inaccessible fibril surface
Poplar [27]	63	5.0	14.2	31.1	19.8	10.2	18.3
Loblolly pine [28]	63	0.1	30.1	24.8	6.9	33.1	15.6
Switchgrass [29]	44	2.3	8.8	27.3	4.5	5.7	51.3

It has been shown that different measurement techniques give different CrI values; however, the order of crystallinity is relatively constant within each measurement [25]. Furthermore, it should be noted that the presence of lignin and hemicelluloses in the lignocellulosic biomass may cause interference with the intensities of FTIR bands or introduce overlaps in the peaks during the XRD and NMR measurements [34]. Thus, it is advisable to remove lignin and hemicelluloses in the biomass prior to the cellulose crystallinity measurement [35, 36]. Moreover, it has also been reported that drying the sample before analysis always causes a change in cellulose crystallinity [37, 38] Therefore, the cellulose sample for CrI analysis must be prepared carefully.

Crystalline cellulose has a very limited accessibility to water and chemicals. Chemical attack can therefore be expected to occur primarily on amorphous cellulose and crystalline surface. Cellulose is a relatively hygroscopic material absorbing 8–14% water under normal atmospheric conditions (20°C, 60% relative humidity) [39]. Nevertheless, it is insoluble in water, where it swells. Cellulose is also insoluble in dilute acid solutions at low temperature. The solubility of the polymer is strongly related to the degree of hydrolysis achieved. As a result, factors that affect the hydrolysis rate of cellulose also affect its solubility. Cellulose is soluble in concentrated acids but severe degradation of the polymer by hydrolysis occurs during the process. Extensive swelling of cellulose takes place in alkaline solutions, along with dissolution of the low molecular weight fractions of the polymer (DP < 200) [40].

3.3.1.1 Chemical properties of cellulose

3.3.1.1.1 TEMPO-mediated oxidation

(2,2,6,6-Tetramethylpiperidine-1-oxyl)-mediated (or TEMPO-mediated) oxidation of cellulose has been used to convert the hydroxymethyl groups present on their surface to their carboxylic form. This oxidation reaction, being highly discriminative of primary hydroxyl groups, is also 'green' and simple to implement. It involves the application of a stable nitroxyl radical, 2,2,6,6-tetramethylpiperidine-1-oxyl (TEMPO), in the presence of NaBr and NaOCl. The use of this technique has been the subject of a number of reports since it was first introduced by De Nooy *et al.* [41], who showed that only the hydroxymethyl groups of polysaccharides were oxidised, while the secondary hydroxyls remained unaffected. In fact, TEMPO-mediated oxidation of cellulose involves a topologically confined reaction sequence, and as a consequence of the 2-fold screw axis of the cellulose chain, only half the accessible hydroxymethyl groups are available to react, the other half being buried within the crystalline particles.

3.3.1.1.2 Cationisation

Positive charges can also be easily introduced on the surface of cellulose: for example, weak or strong ammonium containing groups, such as epoxypropyltrimethylammonium chloride (EPTMAC), can be grafted onto the cellulose surfaces [42]. Such surface cationisation proceeds via a nucleophilic addition of the alkali-activated cellulose hydroxyl groups to the epoxy moiety of EPTMAC and leads to stable aqueous suspensions of cellulose with unexpected thixotropic gelling properties. Shear birefringence was observed, but no liquid crystalline chiral nematic phase separation was detected for these cationic celluloses, most likely owing to the high viscosity of the suspension.

3.3.1.1.3 Esterification, silylation and other surface chemical modifications

Homogeneous and heterogeneous acetylation of cellulose extracted from *Valonia* and tunicate has been studied by Sassi *et al.* by using acetic anhydride in acetic acid [43]. Their ultrastructural study, carried out by TEM imaging and X-ray diffraction, showed that the reaction proceeded via a reduction of the diameters of the crystals, while only a limited reduction in cellulose length was observed. It has been suggested that the reaction involved a non-swelling mechanism that affected only the cellulose chains localised at the crystal surface. In the case of homogeneous acetylation, the partially acetylated molecules immediately partitioned into the acetylating medium as soon as they were sufficiently soluble, while in heterogeneous conditions, the cellulose acetate remained insoluble and surrounded the crystalline core of unreacted cellulose chains. The simultaneous occurrence of cellulose hydrolysis and acetylation of hydroxyl groups has been also reported. Fischer esterification of hydroxyl groups along with the hydrolysis of amorphous cellulose chains has been introduced as a viable one-pot reaction methodology that allows isolation of acetylated cellulose in a single-step process [44, 45].

Cellulose whiskers resulting from the acid hydrolysis of tunicate have been partially silylated by a series of alkyldimethylchlorosilanes, with the carbon backbone of the alkyl moieties ranging from a short carbon length of isopropyl to longer lengths represented by n-butyl, n-octyl, and ndodecyl [46]. Finally, coupling cellulose with N-octadecyl isocyanate, via a bulk reaction in toluene, has also been reported to enhance their dispersion in an organic medium and compatibility with polycaprolactone, which significantly improved the stiffness and ductility of the resultant nanocomposites [47].

3.3.1.1.4 Polymer grafting

Polymer grafting on the surface of cellulose has been carried out using two main strategies, namely, 'grafting-onto' and 'grafting-from' [48]. The 'grafting onto' approach involves attachment onto hydroxyl groups at the cellulose surface of presynthesised polymer chains by using a coupling agent. In the 'grafting from' approach, the polymer chains are formed by in situ surface-initiated polymerisation from immobilised initiators on the substrate.

The 'grafting onto' approach was used by Ljungberg *et al.* [49] to graft maleated polypropylene (PPgMA) onto the surface of tunicate-extracted cellulose. The resulting grafted nanocrystals showed very good compatibility and high adhesion when dispersed in atactic polypropylene. Araki *et al.* [50] and Vignon *et al.* [51] studied the grafting of amine-terminated polymers onto the surface of TEMPO-mediated oxidised cellulose by using a peptide coupling process catalysed by carbodiimide derivatives in water. The same approach has been implemented by Mangalam et al. [52] who grafted DNA oligomers on the surface of cellulose. The grafting of polycaprolactone having different molecular weights on the surface of cellulose has been achieved by using isocyanate-mediated coupling [53]. These authors reported that the grafted PCL chains were able to crystallise at the surface of cellulose when a high grafting density was reached. Similar efforts were made by Cao *et al.* [54], who reported the isocyanate-catalysed grafting of presynthesised water-borne polyurethane polymers via a one-pot process. Such crystallisation provoked cocrystallisations of the free chains of the respective polymer matrices during nanocellulose-based nanocomposite processing. Furthermore, this cocrystallisation phenomenon induced the formation of a co-continuous phase between the matrix and filler, which significantly enhanced the interfacial adhesion and consequently contributed to the improvement of mechanical strength of the resulted nanocomposites.

The 'grafting from' approach applied to nanocellulose was first reported by Habibi *et al.* [55], who grafted polycaprolactone onto the surface of nanocellulose via ring-opening polymerisation (ROP) using stannous octoate ($Sn(Oct)_2$) as a grafting and polymerisation agent. Likewise, Chen *et al.* [56] and Lin *et al.* [57] conducted similar grafting reactions under microwave irradiation to enhance the grafting efficiency. In situ polymerisation of furfuryl alcohol from the surface of cellulose whiskers was studied by Pranger *et al.* [58]. In this case, the polymerisation was catalysed by sulphonic acid residues from the nanocellulose surface. At elevated temperatures, the sulphonic acid groups were de-esterified and consequently released into the medium to catalyse the in situ polymerisation. Yi *et al.* [59] and Morandi *et al.* [60] propagated polystyrene brushes via atom transfer radical polymerisation (ATRP) on the surface of nanocellulose with ethyl 2-bromoisobutyrate as the initiator agent. Similarly, other vinyl monomers, mainly acrylic monomers such as N-isopropylacrylamide, were also polymerised from the surface of nanocellulose to produce thermoresponsive substrates [61]. Grafting of polyaniline from nanocellulose was achieved by in situ polymerisation of aniline onto nanocellulose in hydrochloric acid aqueous solution, via an oxidative polymerisation using ammonium peroxydisulfate as the initiator [62].

3.3.2 Hemicellulose

The term hemicellulose is a collective term. It is used to represent a family of polysaccharides that are found in the plant cell wall and have different composition and structure depending on their source and the extraction method. Unlike cellulose, hemicellulose is composed of combinations of pentose (xylose (Xyl), arabinose (Ara)), hexoses (mannose (Man), galactose (Gal) and glucose (Glu)); and it is frequently acetylated and has side-chain groups such as uronic acid and its 4-O-methyl ester. The chemical nature of hemicellulose varies among different species. In general, the main hemicelluloses of softwood are galactoglucomannans and arabinoglucuronoxylan, while in hardwood it is glucuronoxylan. Table 3.4 summarises the main structural features of hemicelluloses appearing in both softwood and hardwood.

Important aspects of the structure and composition of hemicellulose are the lack of crystalline structure, mainly due to the highly branched structure, and the presence of acetyl groups on the polymer chain. Hemicellulose extracted from plants possesses a high degree of polydispersity, polydiversity and polymolecularity (a broad range of size, shape and mass characteristics). However, the degree of polymerisation does not usually exceed 300 units whereas the minimum limit can be around 50 monomers, which figures are much lower than for cellulose.

Table 3.5 lists the monosaccharide of several lignocellulosic biomasses. It can be seen that with the exception of glucose, softwood contains more mannose than hardwood, while the latter has more xylose.

Table 3.4 Major hemicellulose components in softwood and hardwood [63, 64].

Wood	Hemicellulose type	Amount (% on wood)	Composition: Units	Molar ratio	Linkage	DP
SW	galactoglucomannans	1–15	β-D-Man*p*	4	1 → 4	100
			β-D-Glc*p*	1	1 → 4	
			β-D-Gal*p*	0.1	1 → 6	
			Acetyl	1		
	arabinoglucuronoxylan	7–10	β-D-Xylp	10	1 → 4	100
			4-O-Me-α-D-Glc*p*A	2	1 → 2	
			B-L-Araf	1.3	1 → 3	
HW	glucuronoxylan	15–30	β-D-Xyl*p*	10	1 → 4	200
			4-O-Me-α-D-Glc*p*A	1	1 → 2	
			Acetyl	7		
	glucomannan	2–5	β-D-Man*p*	1–2	1 → 4	200
			β-D-Glc*p*	1	1 → 4	

Table 3.5 Content of monosaccharides for several lignocellulosic biomasses.

	Composition (%, dry basis)				
	Arabinose	Galactose	Glucose	Mannose	Xylose
Loblolly pine [65]	1.4	2.1	48.0	7.3	9.7
Poplar [66]	0.4	0.7	45.5	2.7	17.0
Switchgrass [67]	2.1	0.6	43.7	trace	18.5
Miscanthus [68]	1.8	0.4	39.5	trace	19.0
Corn stover [69]	3.4	1.8	38.9	0.4	23.0
Wheat straw [70]	6.7	2.0	44.5	23.8	0.2
Rice straw [76]	4.2	1.4	52.5	12.0	0.1
Sugarcane bagasse [71]	3.4	2.0	n.a	28.6	0.2
Cotton stalk [72]	1.3	1.1	31.1	8.3	trace

n.a.: data not available

In addition, most sugar components in the hemicellulose can take part in the formation of lignin-carbohydrate complexes (LCC) by covalent linkages between lignin and carbohydrates [73, 74]. The most frequently suggested LCC linkages in native wood are benzyl ester, benzyl ether, and glycosidic linkages [75]. The benzyl ester linkage is alkali-labile and may therefore be hydrolysed during the alkaline pretreatment. The latter two linkages are alkali-stable and would survive from the hydrolysis during alkaline pretreatment.

3.3.2.1 Chemical properties of hemicelluloses and their derivatives

3.3.2.1.1 Xylan derivatives

Due to the lack of a commercial supply, as well as their low molecular weight and poor solubility, xylans have found little industrial utility, and interest in their modification has been rather low in comparison to commercially available polysaccharides such as cellulose or starch. With the aim of improving the functional properties of xylans and/or imparting new functionalities to them, various chemical modifications have been investigated during the past decade. Most of them have been presented in recent reviews [76].

Partial etherification of the beech wood MGX with p-carboxybenzyl bromide in aqueous alkali yielded fully water-soluble xylan ethers with DS up to 0.25 without significant depolymerisation; the Mw determined by sedimentation velocity was 27 000 g/mol. [77] By combining endo-β-xylanase digestion and various 1D- and 2D-NMR techniques, the distribution of the substituents was suggested to be blockwise rather than uniform. The derivatives exhibited remarkable emulsifying and protein foamstabilising activities. Through the reaction of the beech wood xylan and its sulphoethyl derivative with 1-bromododecane in DMSO at moderate reaction temperatures, further amphiphilic derivatives with remarkable tensioactive properties were prepared indicating their potential as biosurfactants [78].

Cationic groups are usually introduced into xylan, similarly to cellulose or starch, through the formation of ether bonds. In previous studies [79], investigations were directed towards xylan-rich waste materials, such as hardwood sawdust, corncobs, and sugar cane bagasse, by reacting with 3-chloro-2-hydroxypropyltrimethylammonium chloride (CHTMAC). Subsequent extraction steps with water and dilute alkali lead to fractionation into trimethyl-ammonium-2-hydroxypropyl (TMAHP) cellulose, TMAHP-hemicellulose, and lignin. More recently, the cationisation of xylans, isolated from beech wood, corncobs, rye bran and the viscose spent liquor was investigated [403]. The results indicated that the DS depended on the molar ratios of CHTMAC/xylan and NaOH/CHMTAC as well as on the xylan type used. The functionalisation pattern of the cationised xylans with DS 0.25–0.98 was characterised by ^{13}C NMR spectroscopy after hydrolytic chain degradation [80].

The TMAHP-MGX isolated from cationised aspen sawdust was reported to be applicable as a better additive; it significantly increased the tear strength of bleached spruce organosolv pulp [82]. The TMAHP derivatives prepared from isolated xylans were shown to improve papermaking properties and act as flocculants for pulp fibres at very low additions (~0.25%), very probably due to irreversible adsorption onto cellulose fibres [81]. The derivatives exhibit antimicrobial activity against some Gram-negative and Gram-positive bacteria, depending on the DS and xylan type [82].

Carboxylic acid esters of xylan are prepared under typical conditions used for polysaccharide esterification, i.e. activated carboxylic acid derivatives are allowed to react with the polymer both heterogeneously and homogeneously. Heterogeneous esterification of oak wood sawdust and wheat bran hemicelluloses with excess octanoyl chloride (without solvent) was described [83]. The separated liquid fraction contained, as well as lignin, degraded esterified hemicelluloses. Homogeneous acylation of xylan-rich hemicelluloses has mostly been carried out in DMF in combination with LiCl [84].

Under moderate reaction conditions, acylated derivatives with DS ranging from 0.18 to 1.71 were prepared from MGX of poplar wood [85] and wheat straw AGX [86]. The DMF/LiCl solvent was also used for acetylation of hemicellulose fractions isolated from wheat straw and poplar also in the presence of dimethylaminopyridine, DMAP, as a catalyst [87]. SEC measurements revealed that polymer degradation was low at reaction temperatures below 80 °C. The conversion of wheat straw and bagasse hemicelluloses with succinic anhydride in aqueous alkaline solutions yielded carboxyl groups containing derivatives with DS < 0.26 [88]. Application as thickening agents and metal-ion binders has been proposed for these derivatives. Various catalysts have been used to accelerate acetylation, succinylation and oleoylation of wheat straw and bagasse hemicelluloses [89].

By the treatment of oat spelt xylan with phenyl or tolyl isocyanate in pyridine the fully functionalised corresponding carbamates were prepared [90]. Xylan 3,5-dimethylphenylcarbamate showed higher recognition ability for chiral drugs compared to that of the same cellulose derivative [91].

In a most recent paper [92], the preparation of corn fibre arabinoxylan esters by reaction of the polymer with C_2– C_4 anhydrides using methanesulphonic acid as a catalyst is described. The water-insoluble derivatives with high molecular weight showed glass-transition temperatures from 61 °C to 138 °C, depending on the DS and substituent type. The products were thermally stable up to 200 °C. Above this temperature their stability rapidly decreased.

Xylan sulphates, known also as pentosan polysulphates (PPS), are permanently studied with regard to their biological activities [93]. Usually, sulphuric acid, sulphur trioxide, or chlorsulphonic acid are employed as sulphating agents alone or in combination with alcohols, amines or chlorinated hydrocarbons as reaction media [94].

Thermoplastic xylan derivatives have been prepared by in-line modification with propylene oxide of the xylan present in the alkaline extract of barley husks [95]. Following peracetylation of the hydroxypropylated xylan in formamide solution yielded the water-insoluble acetoxypropyl xylan. The thermal properties of the derivative qualify this material as a potential biodegradable and thermoplastic additive to melt-processed plastics. Xylan from oat spelts was oxidised to 2,3-dicarboxylic derivatives in a two-step procedure using $HIO_4/NaClO_2$ as oxidants [96].

The neutral X_m-type homoxylan from the seaweed *Palmaria decipiens* was oxidised with bromine solution [97]. Oxidation occurs preferentially at C-2, as shown by means of gel-permeation chromatography (GPC) after reductive cleavage, reduction, and peracetylation. Periodate oxidation of xylan-introduced dialdehyde functions, which gave ligands for the coordination of Cu(II) after reaction with p-chloroaniline [98]. Hydrophobic films were prepared from corn bran xylan in a two-step process [99]. The dialdehyde of the polymer prepared by periodate oxidation under controlled reaction conditions was subjected to reductive amination using laurylamine in aqueous medium. The DS of the dodecylamine-grafted xylan films ranged between 0.5 and 1.1, depending on the reaction conditions. The plastic behaviour at ambient temperature correlated with the glass-transition temperature around –30 °C. The product indicates potential application as a bioplastic [105].

3.3.2.1.2 Modification of Mannan-type hemicelluloses

The widely commercially exploited guar GaM has been the subject of some studies dealing with chemical or enzymic modifications aimed at extending the application range of this polysaccharide. Specific oxidation on the C-6 position of the Galp side-chain units was performed by β-galactosidase [100]. Grafting of polyacrylamide onto guar gum [101] and *Ipomoea* gum [102] in aqueous medium initiated by the potassium persulphate/ascorbic acid redox system was performed in the presence of atmospheric oxygen and Ag^{+} ions. After grafting, a tremendous increase in the viscosity of both gum solutions was achieved, and the grafted gums were found to be thermally more stable.

The hydroxypropyl derivative of guar GaM (HPG) was prepared with propylene oxide in the presence of an alkaline catalyst. HPG was subsequently etherified as such with docosylglycidyl ether in isopropanol and the presence of an alkaline catalyst [103]. The peculiar features of the long-chain hydrophobic derivatives were ascribed to a balance between inter- and intra-molecular interactions, which is mainly governed by the local stress field.

Due to the low solubility of the *Cassia* tora gum, composed mainly of GaM [104], the polymer was modified by graft copolymerisation of acrylamide onto the gum [105], cyanoethylation [106], and carboxymethylation [107]. The modified GaMs from Cassia tora seeds have been applied as a beater additive in paper-making [108]. The hydroxypropyl derivative of guar gum has already found application as a water-blocking agent in the formulation of explosive cartridges, as a processing aid in the mining and mineral industry, water-based paints, etc. [109].

From the GaM of the seed endosperm of *Adenanthera pavonina*, films were prepared by casting solutions of GaM/collagen blends and subsequent cross-linking with glutaraldehyde [110]. The thermal, dielectric, and piezoelectric properties of the films have been studied to develop new materials for electronic devices used for various biological applications. IR spectroscopy of the films indicated that no interaction or binding occurred between both polymers as the typical spectral bands showed no shift. The cross-linking reaction was found to reduce the swelling of the films as a function of the GaM concentration, which also influenced the dielectric properties.

The GGM-rich hemicelluloses, isolated from water-impregnated spruce chips by heat-fractionation [111], have been used as pre-polymers after modification with methacrylic functions [112]. Radical polymerisation of the modified hemicelullose with 2-hydroxyethyl methacrylate in water yielded elastic, soft, transparent, and easily swollen hydrogels.

3.3.2.1.3 Xyloglucan derivatives

From tamarind seed xyloglucan, carboxymethyl derivatives with different levels of DS were prepared in isopropanol medium [113]. The swelling power, solubility and tolerance to organic solvents of the derivatives increased with increasing DS. The interaction properties of the unmodified xyloglucan with calcium chloride and sodium tetraborate were found to be reversed upon carboxymethylation.

A range of derivatives of tamarind seed polysaccharide has been prepared and characterised (selected solution properties were examined) [114]. Following oxidation of the terminal Galp residues with galactose-oxidase, subsequent oxidation and reductive amination have been used to prepare a range of carboxylated and alkylaminated derivatives, respectively. Sulphated derivatives have been prepared by reaction with a sulphur trioxide-pyridine complex in DMF. Based on the dependence of [η] on ionic strength, carboxylated and sulphated derivatives were found to have characteristically stiffened backbones, as found previously for the native polysaccharide [115]. Binding of divalent cations to carboxylated derivatives is shown to be relatively weak, although polymer precipitation was noted in the presence of Pb^{2+}. Alkylaminated polysaccharides show only a modest decrease in surface and interfacial tension compared with the native polysaccharide, although significant foam formation and stabilisation was found for a nonylaminated sample. Following enzymic depolymerisation, this material showed a marked decrease in surface and interfacial tension suggesting that interfacial activity in alkylaminated tamarind polysaccharide is only apparent under disruptive solution conditions. The results of ^{1}H-NMR line-width and T1 measurements before and after depolymerisation indicate that this is due to solution viscosity rather than specific interaction effects.

3.3.2.1.4 Bioconversion of hemicellulose

Total biodegradation of xylan requires endo-β-1,4-xylanase, β-xylosidase, and several accessory enzymes, such as α-L-arabinofuranosidase, α-glucuronidase, acetylxylan esterase, ferulic acid esterase, and *p*-coumaric acid esterase, which are necessary for hydrolysing various substituted xylans. The endo-xylanase attacks the main chains of xylans, and β-xylosidase hydrolyses xylooligosaccharides to xylose. The α-arabinofuranosidase and a-glucuronidase remove the arabinose and 4-O-methyl glucuronic acid substituents, respectively, from the xylan backbone. The esterases hydrolyse the ester linkages between xylose units of the xylan and acetic acid (acetylxylan esterase) or between arabinose side-chain residues and phenolic acids, such as ferulic acid (ferulic acid esterase) and *p*-coumaric acid (*p*-coumaric acid esterase).

Many microorganisms, such as *Penicillium capsulatum* and *Talaromyces emersonii*, possess complete xylan-degrading enzyme systems [116]. Bachmann and McCarthy [117] reported significant synergistic interaction among endo-xylanase, β-xylosidase, α-arabinofuranosidase, and acetylxylan esterase of the thermophilic actinomycete *Thermomonospora fusca*. Synergistic action between depolymerising and side-group cleaving enzymes has been verified using acetylated xylan as a substrate [118]. Many xylanases do not cleave glycosidic bonds between xylose units that are substituted. The side chains must be cleaved before the xylan backbone can be completely hydrolysed [119]. On the other hand, several accessory enzymes only remove side chains from xylooligosaccharides. These enzymes require a partial hydrolysis of xylan before the side chains can be cleaved [120]. Although the structure of xylan is more complex than cellulose and requires several different enzymes with different specificities for complete hydrolysis, the polysaccharide does not form tightly packed crystalline structures like cellulose and is, thus, more accessible to enzymatic hydrolysis [121].

3.3.3 Lignin

Of the three major biopolymers that constitute wood, lignin is distinctly different from the other macromolecular polymers [122]. Lignin is an amorphous, cross-linked, three-dimensional polyphenolic polymer that is synthesised by enzymatic dehydrogenative polymerisation of 4-hydroxyphenyl propanoid units [123, 124]. The biosynthesis of lignin derives from the polymerisation of three types of phenylpropane units as monolignols: coniferyl, sinapyl, and *p*-coumaryl alcohol [125, 126].

The polymerisation process is initiated by an enzyme-catalysed oxidation of the monolignol phenolic hydroxyl groups to yield free radicals. A monolignol free radical can then couple with another monolignol free radical to generate a dilignol. Subsequent nucleophilic attack by water, alcohols, or phenolic hydroxyl groups on the benzyl carbon of the quinone methide intermediate restores the aromaticity of the benzene ring. The generated dilignols then undergo further polymerisation to form protolignin.

The relative abundance of these units depends on the contribution of a particular monomer to the polymerisation, and biosynthesis has been hotly debated in the literature and details of this complex pathway [27, 128, 129]. Hardwood and herbaceous lignins are composed mainly of syringyl (S) and guaiacyl (G) units with minor amounts of p-hydroxyphenyl (H), whereas softwood lignin is composed mainly of guaiacyl units and trace amounts of H [130, 131, 132, 133, 134, 135].

Among the inter-unit linkages in lignin structures, β-O-4 (β aryl ether) linkages are the most frequently occurring inter-unit linkage and are also the ones most easily cleaved by chemical processes such as pulping and biomass pretreatments. The other linkages β-5, β-β, 5-5, 4-O-5 and β-1 are all more resistant to chemical degradation [135]. Hardwood lignins with a higher proportion of S units have fewer β-5, 5-5 and 4-O linkages than softwood lignin with more G units [133].

Although the exact structure of protolignin is unknown, improvements in methods for identifying lignin-degradation products together with advances in spectroscopic methods have enabled scientists to elucidate the predominant structural features of lignin. Table 3.6 showed the typical abundance of common linkages and functional groups found in softwood lignin [71, 133].

Table 3.6 Proportions of different types of linkage connecting the phenylpropane units in softwood lignin.

Linkage type	Dimer structure	~Percentage
β-O-4	Phenylpropane β-aryl ether	50
β-5	phenylcoumaran	9–12
5-5	Biphenyl	15–25
5-5/α-O-4	Dibenzodioxicin	10–15
4-O-5	Diaryl ether	4
β-1	1,2-Diaryl propane	7
β-β	β-β-linked structures	2

Wet chemistry techniques, such as thioacidolysis and nitrobenzene oxidation, coupled with gas chromatography, have traditionally been used to study lignin structure. While these methods can be very precise for specific functional groups and structure moieties, each technique can only provide limited information and does not give a general picture of the entire lignin structure. The thioacidolysis procedure cleaves β-O-4 linkages in lignin, giving rise to monomers and dimers which are then used to calculate the S and G content. Similar formation can be obtained using nitrobenzene oxidation but it can lead to overestimating S/G ratios [134]. In general, the S/G ratio of lignin is a good indicator of its overall composition and response to pulping and biomass pretreatment. S/G ratios of lignin from different biomass species are summarised in Table 3.7. It can be seen that the S/G ratio of the hardwood species (i.e. poplar) is higher than the herbaceous biomass, such as switchgrass and *Miscanthus*. This is expected given the higher H contents in grass. In softwood species (i.e. Loblolly pine), the lignin is made up of 99% G units and a trace of H units.

The advantage of spectroscopic methods over degradation techniques is their ability to analyse the whole lignin structure and directly detect lignin moieties. The development of quantitative ^{13}C NMR for lignin analysis [141] was an important advance in lignin chemistry. Multidimensional NMR spectroscopy has also been successfully utilised to elucidate details of lignin structure [142]. While two-dimensional (2D) NMR can help provide unambiguous structural assignments, performing quantitative 2D NMR experiments requires special precautions and is typically only semi-quantitative at best [143]. A combination of quantitative ^{13}C and 2D HMQC NMR has been shown to provide comprehensive structural information on lignin from a variety of sources [149, 144]. For example, 80% of the side chains of eucalyptus lignin were estimated at the structural level using these methods. Other NMR techniques, such as phosphitylation of lignin hydroxyl groups, followed by ^{31}P NMR, ^{1}H and HSQC NMR, can also yield valuable structural information. The most frequently used procedure to isolate lignin for NMR analysis is ball milling wood to a fine meal, followed by lignin extraction with aqueous dioxane. This milled wood lignin (MWL) is regarded as being fairly similar to natural lignin in wood.

Table 3.7 The G/S/H lignin ratio of common biomass feedstock.

	G lignin	S lignin	H lignin	S/G ratio
Switchgrass [135]	51	41	8	0.80
Poplar [136]	32	68	trace	2.12
Miscanthus [137]	52	44	4	0.85
Corn stover [138]	51	3.6	46	0.07
Wheat straw [139]	49	46	5	0.94
Rice straw [145]	45	40	15	0.89
Sugarcane bagasse [140]	30	37	33	1.23
Loblolly pine [71]	99	–	1	–

The property of polydispersity, just as with hemicellulose, characterises lignin as well. The degree of polymerisation (DP) for softwood lignin is approximately 60–100 and the molecular weight is in excess of 10 000.

Lignin in wood behaves as an insoluble three-dimensional network. It plays an important role in the cell's endurance and development, as it affects the transport of water, nutrients and metabolites in the plant cell. It acts as a binder between cells, creating a composite material that has a remarkable resistance to impact, compression and bending.

Lignin is much less hydrophilic than either cellulose or hemicelluloses and it has a general effect of inhibiting water adsorption and fiber swelling. Solvents that have been identified to significantly dissolve lignin include low molecular alcohols, dioxane, acetone, pyridine, and dimethyl sulfoxide. Furthermore, it has been observed that, at elevated temperatures, thermal softening of lignin takes place, which allows depolymerisation reactions of an acidic or alkaline nature to accelerate.

3.3.3.1 Chemical decomposition and derivatives of lignin

3.3.3.1.1 Chemical decomposition

Chemical degradation of lignin is a field that has been well researched. Although a number of methods have been proposed, several methods are generally known. The following can be considered as major chemical degradation methods of lignin: acidolysis (hydrolysis), nitrobenzene oxidation, permanganate oxidation, hydrogenolysis, ozonisation, and thermolysis.

Acidolysis is closely related to hydrolysis, which is used for bio-ethanol production. The decomposition mechanisms and hydrolysis products of lignocellulosics under various conditions in hot-compressed water were discussed in order to establish preferable conditions among hydrolysis behaviour in hot-compressed water, acid, alkaline, and enzymatic hydrolysis [145].

It is known that nitrobenzene oxidation of lignins was originally introduced by Freudenberg [146] in order to investigate lignin structures. It was reported that nitrobenzene oxidation of softwood lignin gave vanillin as a major product and that hardwood lignins gave syringaldehyde as oxidation products [147]. Higuchi reported that grasses and their lignins gave 4-hydroxybenzaldehyde, vanillin and syringaldehyde as oxidation products by nitrobenzene [148].

Permanganate oxidation was introduced for the analysis of lignin structure by Freudenberg *et al.* [149]. Aromatic carboxylic acids such as 4-hydroxyphenolic acid, vanillic acid, and syringic acid were found by permanganate oxidation of lignins [150].

Sakakibara reported the hydrogenolysis of lignins in detail [151]. He mentioned that hydrogenation and hydrogenolysis are different processes and defined them as follows. Hydrogenation is the addition of hydrogen (H_2) to a multiple bond. Hydrogenolysis is a process in which cleavage of a carbon–carbon or carbon–hetero atom single bond is accomplished by reaction with hydrogen. Examples of monomeric aromatic products in the hydrogenolysis of lignins are as follows: 4-*n*-propylguaiacol, dihydro-coniferyl alcohol, 4-*n*-propylsyringol, dihydrosinapyl alcohol, 4-methylguaiacol, 4-methylsyringol, 4-ethylguaiacol and 4-ethylsyringol.

Ozone is used for the cleavage of carbon–carbon double and triple bonds. Accordingly, ozone is effective in the degradation of lignin and can be used for the bleaching of pulps. Various lignin model compounds were used to study the mechanism of lignin degradation. Examples of model compounds are mostly monomeric and dimeric lignin model compounds [152]. For example, ozonisation of lignin model compounds was carried out using vanillyl alcohol and veratoryl alcohols [153]. As the degradation product of the above model compounds, δ-lactone of the monomethyl ester of β-hydroxymethyl muconic acid was obtained. Accordingly, it was considered that demethylation of veratoryl alcohol occurred at the 4-position of the aromatic ring followed by aromatic ring opening between the carbon atoms of the 3- and 4-positions of the aromatic ring. In the case of the ozonisation of the same model compounds under basic conditions, δ-lactone of β-hydroxymethyl muconic acid was obtained as the degradation product. Accordingly, it was considered that demethylation occurred at the 3-position of the aromatic ring. It was found that oxidation of the side chain was much less than aromatic ring opening.

3.3.3.1.2 Lignin derivatives

In this section, novel polymers synthesised using lignin degradation products are described. Polystyrene derivatives having 4-hydroxyphenyl, guaiacyl, and syringyl groups [154]. Polystyrene derivatives are amorphous and glass transition was observed by DSC [155]. Due to the presence of the hydroxyl group, glass transition is observed at a higher temperature than for polystyrene. The T_g of poly(4-hydroxystyrene) ($M_n = 2.2 \times 10^5$, $M_w/M_n = 3.2$) was 182 °C, poly(3-methoxy-4-hydroxystyrene) ($M_n = 1.1 \times 10^5$, $M_w/M_n = 4.1$) was 142 °C and poly(3, 5-methoxy, 4-hydroxystyrene) ($M_n = 9.9 \times 10^4$, $M_w/M_n = 3.7$) was 108 °C [156]. It is clear that inter-molecular hydrogen bonding restricts molecular motion and the methoxyl group enhances the molecular mobility. Hydrogen bonds formed in polyhydroxystyrene derivatives are broken by water molecules and the T_g decreases at ca., 100 °C when each hydroxyl group is restrained by one water molecule [157].

Polyethers having 4-hydroxyphenyl or guaiacyl groups, poly(oxy-1,4 phenylene-carbonylmethylene and poly(oxy-2, methoxy-1,4-phenylene-carbonylmethylene) (R = OCH_3) were synthesised [158]. Polyacylhydrazone derivatives having guaiacyl groups in the repeating unit were synthesised using vanillin and dibromoalkane as starting materials. Polyesters having guaiacyl groups and spiro-dioxane rings in the main chain were prepared using vanillin and pentaerythritol as starting materials. Polybenzalazine has guaiacyl groups in each repeating unit. The obtained polymer was not soluble in dioxane, DMSO, DMF, m-cresol or chloroform. Melting temperature was observed at 205 °C in a DSC heating curve (10 °C min^{-1}). T_d was ca. 300 °C. Polybenzalazine is a stable crystalline polymer [159].

3.3.4 Ash content and inorganic element profiles

The inorganic elements present in biomass collectively constitute its ash content, act as a waste stream during its conversion to biofuels and are the source of biochar and

Table 3.8 Ash content and inorganic elements (wt% dry weight) of common biomass.

Biomass species	Ash (% dry wt.)	Inorganic elements (% dry wt.)				
		P	K	Na	Ca	Mg
Hybrid poplar [163]	1.80	0.06	0.21	0.01	0.56	0.04
Willow [169]	2.29	0.49	1.83	0.15	6.76	0.48
Oak [164]	n.a.	n.a.	0.09	0.01	0.08	0.02
Switchgrass [165]	4.30	0.05	0.07	0.02	0.62	0.05
Loblolly pine [71, 166]	0.30	0.01	0.04	n.a.	0.08	n.a.

n.a.: data not available.

slagging during thermomechanical conversion. Knowledge of the ash content and composition is essential regardless of the conversion pathway or end product. In addition, several studies have highlighted that soil productivity requirements may necessitate the return of this valuable inorganic resource to the soils [160, 161]. Also, some inorganic elements, such as P, K, Ca and Mg, act as macronutrients and a knowledge of their contents in the biomass can provide information on nutrient depletion of the soil, which can be used to maintain soil fertility in subsequent rotations [162]. A compilation of available data on ash content and selected inorganic element distributions of some softwood, hardwood and herbaceous biomass is given in Table 3.8. The data presented in Table 3.8 shows that while there is significant variation in ash content, ranging from 0.3% to 4.3%, the distribution of inorganic elements shows very little variation among the different species. The data shows that the ash content of hybrid poplar is slightly higher than softwood biomass (i.e. Loblolly pine), but substantially (four times) lower than other biofuel feedstocks such as switchgrass.

3.3.5 Extractive content

Non-structural material is often removed from biomass prior to chemical analysis due to its potential interference with analytical techniques. This includes solvent-soluble, non-volatile compounds such as fatty acids, resins, sterols, terpenes, waxes, etc., and usually comprises a minor proportion of the biomass. For large-scale lignocellulosic biorefinery operations, however, extractives can be a potential source of value-added co-products. The compounds present in the extractives fraction are a function of the solvent, which is usually ethanol, acetone, dichloromethane, or a mixture of ethanol/benzene or ethanol/toluene. The organic solvent extractive contents of some biomass species are presented in Table 3.9. It can be seen that switchgrass contains much higher extractive contents than the woody biomass since it includes a large amount of low-molecular weight carbohydrates besides the fatty acids, sterols and alkanols, etc. In addition, the extractives of softwood species (i.e. Scots pine) contain many more resin acids than the hardwood species (i.e. aspen).

Table 3.9 Extractive contents of selected lignocellulosic biomass (wt% dry weight).

	Switchgrass [167]	Wheat straw [168]	Aspen [169, 170]	Scots pine [171]
Extraction method	95% ethanol	Toluene/ethanol (2:1, v/v)	Acetone	Acetone
Total extractive yield	13.00	2.38	3.80	4.58
Carboxylic acids	1.56	0.76	0.24	0.21
Sugars and their derivatives	7.8	–	–	–
Alkanol	1.95	–	–	–
Glycerol	0.13	–	–	1.71
Alkane	0.26	–	–	–
Sterol	0.78	0.10	0.22	0.05
Wax	–	0.03	–	–
Phenolics	–	–	0.07	–
Flavonoids	–	–	0.07	–
Steryl triterpene esters	–	0.33	0.46	–
Lipids	–	–	0.08	–
Resin acids	–	0.56	–	0.33

3.3.6 Elemental composition

The major elemental composition of biomass on a gravimetric basis, which is commonly referred to as ultimate analysis, is very important in performing mass balances in biomass conversion processes. These results can also be used to calculate empirical molecular formulae. Elemental compositions for some common biomasses compiled from the literature are given in Table 3.10. The sulphur content of poplar wood is lower compared to wheat straw and switchgrass (Table 3.10), which is an advantageous feature regarding the strict environmental regulations limiting the sulphur content of

Table 3.10 Elemental contents of some biomass species.

	Ultimate analysis (% dry wt.)					
Biomass species	C	H	O	N	S	Si
Hybrid poplar [172]	50.20	6.06	40.40	0.60	0.02	–
Eastern cottonwood [173]	50.29	6.45	–	–	–	–
P. deltoides [72]	49.65	5.85	41.88	0.08	0.05	–
Corn stover [181]	43.65	5.56	43.31	0.61	0.01	–
Switchgrass [181]	47.75	5.75	42.37	0.74	0.08	–
Wheat straw [181]	43.20	5.00	39.40	0.61	0.11	–
Ponderosa pine [181]	49.25	5.99	44.36	0.06	0.03	–
Rice straw [174]	41.00	4.00	36.00	–	–	12.00

Table 3.11 Higher heating value (HHV) for common biomass feedstocks [175].

Biomass species	Heating value (dry) (MJ/kg)
Hybrid poplar	19.38
Black cottonwood	15.00
Ponderosa pine	20.02
Douglas fir	20.37
Corn stover	17.65
Wheat straw	17.51
Switchgrass	18.64

transportation fuels. As expected, there is not much variation in the elemental composition of different biomass species. It should be noted that the rice straw contains a higher amount of silicon compared with other biomass species.

3.4 Heating values

Heating value is the net enthalpy released upon reacting a material with oxygen under isothermal conditions. If water vapour formed during the reaction condenses at the end of the process, the latent enthalpy of condensation contributes to what is termed the higher heating value (HHV). These measurements are typically performed in the bomb calorimeter. Table 3.11 shows the HHVs of some common biomass species. It may be found that the softwoods (Ponderosa pine and Douglas fir) yield slightly higher HHVs than the hardwoods (hybrid poplar and cottonwood), herbaceous (switchgrass) biomass and agricultural residues (corn stover, wheat straw), as shown in Table 3.11.

3.5 Conclusion and future trends

Recently, with the increase in global pollution and a scarcity of petroleum sources, the demand for high-quality sustainable and degradable materials has gradually increased. As a sustainable resource, plant fibre is cheap, abundant, degradable, biologically compatible and has good physical strength. It can be derivatised to have photic, electrical, magnetic and biological properties. It could have adsorption, separation and catalyst functions in the new material area. The application of new plant fibres will meet the global economic, energy and new materials development.

Cellulose, hemicellulose and lignin can be isolated from plant fibres, which could be fabricated into cellulose-base, hemicellulose-base and lignin-base functionalised materials. The three relatively active hydroxyl groups in the β-D-glucopyranose unit of the cellulose molecule can be replaced by new functional groups. These derivatives can be used in food, coatings and the pharmaceutical area. It can be used in gene carrier, gels, isomers, photic materials and antioxidants. It can also be applied in amino acid detection, protein and drug carriers and as a catalyst in water treatment. Some radical polymerisation techniques, such as atom transfer radical polymerisation

(ATRP), reversible addition fragmentation chain transfer polymerisation (RAFT), single electron transfer living radical polymerisation (SET-LRP), could be used in cellulose free-radical polymerisation in different areas. These derivatives can be used in the biological pharmaceutical industry and biotechnologies, as well as for membrane fabrication. The porous cellulose membrane can be used in filters, photonics, moulding and cell cultivation. It can also be used as an absorbent for heavy metals, anti-bacterial, cellulose gels (hydrogels, aerogels, cellulose/silicane gels and cellulose/polyaniline).

The polysaccharides in hemicellulose can be modified through oxidation, hydrolysation, esterisation and etherisation. These derivatives can be used in food additives, copper additives, stabilisers and sweeteners, and emulsifiers. In the pulp and paper industry, hemicellulose can be used to enhance the physical strength, reduce the refining energy, and improve anti-bacterial ability. In chemical engineering, hemicellulose can be used for heat-resistant materials, anti-conductors, and piezoelectric materials. It can also be used to fabricate hydrogels. In the pharmaceutical industry, the konjac glucomannan can be used as a drug releaser, for cell treatment and emulation. Xylan can be used to reduce inflammation, enhance cell mitosis and work as anticancer. It can also be used to adjust the bacterial metabolism in the intestinal tract.

Lignin is a cross-linked racemic macromolecule. It has a three-dimensional structure with phenolic and aliphatic groups. In sulphite pulping, lignin is removed from wood pulp as sulphonates. These lignosulphonates have several uses. They can be used as dispersants in high-performance cement applications, water treatment formulations and textile dyes. They can also be used as additives in specialty oil field applications and agricultural chemicals. Lignin can also be used as a raw material for several chemicals, such as vanillin, DMSO, ethanol, xylitol sugar, and humic acid. In addition, lignin can be applied as an environmentally sustainable dust suppression agent for roads. Furthermore, lignin can be used in strengtheners, adhesives, fertilisers and food additives.

In summary, the naturally high molecular weight polymers in plant fibres can be further investigated and applied as high-value added materials in our daily life.

References

[1] P. McKendry, *Bioresource Technology* (2002), **83**, 37.

[2] H. J. Huang & S. Ramaswamy, *Biomass Bioenergy* (2009), **33**, 234.

[3] R. Kumar, S. Singh & O. V. Singh, *J. Industrial Microbiology and Biotechnology* (2008), **35**, 377.

[4] M. Galbe & G. Zacchi, Advances in Biochemical Engineering/Biotechnology, *Biofuels* (2007), **108**, 41.

[5] C. N. Hamelinck, G. V. Hooijdonk, & A. P. C. Faaij, *Biomass and Bioenergy* (2005), **28**, 384.

[6] S. J. Kim, M. Y. Kim, S. J. M.S. Jeong & I. M. Jang, *Industrial Crops and Products* (2012), **38**, 46.

[7] T. Ingram, K. Wormeyer, J. C. Lima, V. Bockemuhl, G. Antranikian, G. Brunner & I. Smirnova, *Bioresource Technology* (2011), **102**, 5221.

[8] B. B. Hallac, & A. J. Ragauskas, Biofuels, *Bioproducts & Biorefining* (2011), **5 (2)**, 215.

[9] J. Faulon, G. A. Carlson & P. G. Hatcher, *Organic Geochemistry* (1994), **21**, 1169.

[10] D. Klemm, B. Heublein, H. P. Fink & A. Bohn, *Angewandte Chemie* (2005), **44**, 3358.

[11] A. Thygesen, J. Oddershede, H. Lilholt, A. B. Thomsen & K. Stahl, *Cellulose*, **12 (6)**, 563 (2005).
[12] R. H. Newman, *Holzforschung* (2004), **58 (1)**, 91.
[13] R. H. Atalla & D. L. VanderHart, *Science* (1984) **223 (4633)**, 283.
[14] J. Sugiyama, R. Vuong & H. Chanzy, *Macromolecules* (1991), **24**, 4168.
[15] F. Horii, A. Hirai, & R. Kitamaru, *Macromolecules* (1987), **20 (6)**, 1440.
[16] J. Sugiyama, J. Persson, & H. Chanzy, *Macromolecules* (1991), **24 (6)**, 2461.
[17] P. T. Larsson, E. L. Hult, K. Wickholm, E. Pettersson & T. Iversen, *Solid State Nucl. Magn. Reson.* (1999), **15 (1)**, 31.
[18] C. H. Stephens, P. M. Whitmore, H. R. Morris & M. E. Bier, *Biomacromolecules* (2008), **9**, 1093.
[19] L. Wagberg & G. O. Annergren, Physicochemical characterization of papermaking fibers. In C. F. Baker (ed.), *The Fundamentals of Papermaking Materials: Transactions of the 11th Fundamental Re-search Symposium in Held at Cambridge*, Pira International, Surrey, UK, 1997, p. 1.
[20] S. Park, J. O. Baker, M. E. Himmel, P. A. Parilla & D. K. Johnson, *Biotechnol. Biofuels* (2010), **3**, 10.
[21] L. Segal, J. J. Creely, A. E. Martin Jr & C. M. Conrad, *Text. Res. J.* (1959), **29**, 786.
[22] X. Zhao, L. Wang & D. Liu, J. *Chem. Technol. Biot.* (2008), **83**, 950.
[23] R. Newman, *Solid State Nucl. Magn. Reson.* (1999), **15 (1)**, 21.
[24] S. Park, D. Johnson, C. Ishizawa, P. Parilla & M. Davis, *Cellulose* (2009), **16 (4)**, 641.
[25] S. Oh, D. Yoo, Y. Shin & G. Seo, *Carbohydr. Res.* (2005), **340 (3)**, 417.
[26] P. T. Larsson, U. Westermark & T. Iversen, *Carbohydr. Res.* (1995), **278**, 339.
[27] M. Foston, C. A. Hubbell, M. Davis & A. J. Ragauskas, *Bioenerg. Res.* (2009), **2**, 193.
[28] P. Sannigrahi, A. J. Ragauskas & S. J. Miller, *Bioenerg. Res.* (2008), **1**, 205.
[29] R. Samuel, Y. Pu, M. Foston & A. J. Ragauskas, *Cellulose* (2010), **1**, 85.
[30] L. Laureano-Perez, F. Teymouri, H. Alizadeh & B. E. Dale, *Appl. Biochem. Biotechnol.* (2005), **121–124**, 1081.
[31] C. Boisset, H. Chanzy, B. Henrissat, R. Lamed, Y. Shoham & E. A. Bayer, *Biochem. J.* (1999), **340**, 829.
[32] M. Wada & T. Okano, *Cellulose* (2001), **83**, 183.
[33] C. Tokoh, K. J. Takabe & M. Fujita, *Cellulose* (2002), **91**, 65.
[34] X. Zhao, L. Zhang & D. Liu, *Biofuels, Bioproducts & Biorefinin*, (2012), **6 (4)**, 465.
[35] Z. Yu, H. Jameel, H. M. Chang & S. Park, *Bioresource. Technol.* (2011), **102**, 9083.
[36] X. Zhao, E. van der Heide, T. Zhang & D. Liu, *BioResources* (2010), **5**, 1565.
[37] L. T. Fan, Y. H. Lee & D. H. Beardmore, *Biotechnol. Bioeng.*, (1981), **23**, 419.
[38] L. T. Fan, Y. H. Lee & D. H. Beardmore, *Biotechnol. Bioeng.* (1980), **22**, 177.
[39] P. F. H. Harmsen, W. J. J. Huijgen, L. M. Bermúdez López & R. R. C. Bakker, *Literature Review of Physical and Chemical Pretreatment Processes for Lignocellulosic Biomass,* Energy Research Center of the Netherlands, Petten, The Netherlands, 2010.
[40] H. Krassig & J. Schurz, in *Ullmann's Encyclopedia of Industrial Chemistry, 6th edition,* Wiley-VCH, Weinheim, Germany, 2002.
[41] A. E. J. de Nooy, A.C. Besemer & H. van Bekkum, *Recueil des Travaux Chimiques des Pays-Bas* (1994), **113**, 165.
[42] M. Hasani, E. D. Cranston, G. Westmana & D. Gray, *Soft Matter* (2008), **4**, 2238.
[43] J.-F Sassi, & H. Chanzy, *Cellulose* (1995), **2**, 111.
[44] B. Braun & J. R. Dorgan, *Biomacromolecules* (2009), **10**, 334.
[45] M. J. Sobkowicz, B. Braun & J. R. Dorgan, *Green Chem.* (2009), **11**,680.
[46] C. Gousse, H. Chanzy, G. Excoffier, L. Soubeyrand & E. Fleury, *Polymer* (2002), **43**, 2645.

[47] G. Siqueira, J. Bras & A. Dufresne, *Biomacromolecules* (2009), **10**, 425.
[48] B. Zhao & W. J. Brittain, *Prog. Polym. Sci.* (2000), **25**, 677.
[49] N. Ljungberg, C. Bonini, F. Bortolussi, C. Boisson, L. Heux & J. Y. Cavaille, *Biomacromolecules* (2005), **6**, 2732.
[50] J. Araki, M. Wada & S. Kuga, *Langmuir* (2001), **17**, 21.
[51] M. Vignon, S. Montanari & Y. Habibi, (Centre National de la Recherche Scientifique CNRS, Fr.) FR 2003/5195, (2004).
[52] A. P. Mangalam, J. Simonsen & A. S. Benight, *Biomacromolecules* (2009), **10**, 497.
[53] Y. Habibi & A. Dufresne, *Biomacromolecules* (2008), **9**, 1974.
[54] X. Cao, Y. Habibi & L.A. Lucia, *J. Mater. Chem.* (2009), **19**, 7137.
[55] Y. Habibi, A.-L. Goffin, N. Schiltz, E. Duquesne, P. Dubois & A. J. Dufresne, *Mater. Chem.* (2008), **18**, 5002.
[56] G. Chen, A. Dufresne, J. Huang & P. R. Chang, *Macromol. Mater.* Eng. (2009), **294**, 59.
[57] N. Lin, G. Chen, J. Huang, A. Dufresne & P. R. Chang, *J. Appl. Polym. Sci.* (2009), **113**, 3417.
[58] L. Pranger & R. Tannenbaum, *Macromolecules* (2008), **41**, 8682.
[59] J. Yi, Q. Xu, X. Zhang & H. Zhang, *Polymer* (2008), **49**, 4406.
[60] G. Morandi, L. Heath & W. Thielemans, *Langmuir* (2009), **25**, 8280.
[61] J. Zoppe, Y. Habibi & O. J. Rojas, Abstr. Pap., ACS Natl. Meet. (2008), **235**, CELL-057.
[62] C. H. Haigler. In C. H. Haigler & P. Weimer (eds), *Biosynthesis and Biodegradation of Cellulose* (pp.??–??). Marcel Dekker: New York, (1991).
[63] S. Willför, A. Sundberg, A. Pranovich, & B. Holmbom, *Wood Sci. Technol.* (2005), **39**, 601.
[64] S. Willför, A. Sundberg, J. Hemming, & B. Holmbom, *Wood Sci. Technol.* (2005), **39**, 245.
[65] F. Huang, P.M. Singh, & A. J. Ragauskas, *J. Agr. Food Chem.* (2011), **59**, 12910.
[66] USDOE-Office of Energy Efficiency and Renewable Energy. Biomass Feedstock and Composition Database (2006). Available at http://www1.eere.energy.gov/biomass/printable_versions/feedstock_databases.html (accessed September 1, 2012).
[67] Z. Hu, R. Sykes, M. F. Davis, E. C. Brummer & A. J. Ragauskas *Bioresource Technol.* (2010), **101**, 3253.
[68] T. de Vrije, G. G. de Haas, G. B. Tan, E. R. P. Keijsers & P.A.M. Claassen, *Intl. J. Hydrogen Energy* (2002), **27**, 1381.
[69] D. W. Templeton, A. D. Sluiter, T. K. Hayward, B. R. Hames & S. R. Thomas, *Cellulose* (2009), **16**, 621.
[70] J. X. Sun, F. C. Mao, X. F. Sun & and R. C. Sun, *J. Wood Chem. Technol.* (2004) **24 (3)**, 239.
[71] X. F. Sun, R. C., Sun, J. Tomkinson & M. S. Baird, *Carbohydrate Poly.* (2003), **53**, 483.
[72] R. A. Silverstein, Y. Chen, R. R. Sharma-Shivappa, M. D. Boyette & J. Osborne, *Bioresource Technol.* (2007), **98**, 3000.
[73] A. Barakat, H. Winter, C. Rondeau-Mouro, B. Saake, B. Chabbert & B. *Cathala, Planta* (2007), **226**, 267.
[74] M. Bunzel, J. Ralph, F. Lu, R. D. Hatfield & H. Steinhart, *J Agric. Food Chem.* (2004), **52**, 6496.
[75] M. Lawoko, G. Henriksson & G. Gellerstedt, *Holzforschung* (2006), **60**, 156.
[76] A. Ebringerová, & Z. Hromádková (1999) In: S. E. Harding (ed.) *Biotechnology and Genetic Engineering Reviews*. Intercept, (1999), vol. 16, p. 325.
[77] A. Ebringerová, J. Alföldi, Z. Hromádková, G. M. Pavlov & S. E. Harding, *Carbohydr. Polym.* (2000), **42**, 123.
[78] A. Ebringerová, I. Sroková, P. Talába, M. Kaˇcuráková & Z. Hromádková *J. Appl. Polym. Sci.* (1998), **67**, 1523.

[79] Th. Heinze, A. Koschella & A. Ebringerová. In: P. Gatenholm & M. Tenkanen (eds), *Hemicelluloses: Science and Technology*. ACS Symposium Series 864. American Chemical Society, Washington DC (2004), p. 312.
[80] A. Ebringerová, & Z. Hromádková, *Angew Makromol Chem* (1996), **24**, 97.
[81] M. Antal, A. Ebringerová, Z. Hromádková, I. Pikulík & M. Laleg *MickoMM* (1997) Papier 51:223
[82] A. Ebringerová, A. Belicová & L. Ebringer *J. Microbiol. Biotechnol.* (1994), **10**, 640.
[83] S. Thiebaud & M. E. Borredon, *Bioresource Technol.* (1998), **63**,139.
[84] J. M. Fang, R. C. Sun, P. Fowler, j. Tomkinson & C. A. S. Hill, *J. Appl. Polym. Sci.* (1999), **74**, 2301.
[85] R. C. Sun, J. M. Fang, J. Tomkinson & C. A. S. Hill, *J. Wood Chem. Technol.* (1999), **19**, 287.
[86] R. C. Sun , J. M. Fang & J. Tomkinson, *Polym. Degrad. Stabil.* (2000), **67**, 345.
[87] R. C. Sun , J. M. Fang, J. Tomkinson & G. L. Jones, *Ind. Crop Prod.* (1999), **10**, 209.
[88] R. C. Sun, X. F. Sun & X. J. Bing, *Appl. Polym. Sci.* (2002), **83**, 757.
[89] X. F. Sun, R. C. Sun, J. Tomkinson & M. S. Baird, *Carbohydr. Polym.* (2003), **53**, 483.
[90] M. Vincendon, *Macromol Chem* (1993), **194**, 321.
[91] Y. Okamoto, J. Noguchi & E. Yashima, *React. Funct. Polym.* (1998), **37**, 183.
[92] C. M. Buchanan, N. L. Buchanan, J. S. Debenham, P. Gatenholm, M. Jacobsson, M. C. Shelton, T. L. Watterson & M. D. Wood, *Carbohydr. Polym.* (2003), **52**, 345.
[93] S. J. Elliot, L. J. Striker, W. G. Stetler-Stevenson, T. A. Jaco & G. E. Striker, *J. Am. Soc. Nephrol.* (1999), **10**, 62.
[94] D. Klemm, B. Philipp, T. Heinze, U. Heinze & W. Wagenknecht, *Comprehensive Cellulose Chemistry.* Wiley-VCH, Weinheim (1998).
[95] W. G. Glasser, R.K Jain & M. A. Sjöstedt, *Biotechnol. Adv.* (1996), **14**, 605.
[96] S. Matsumura, M. Nishioka, S. Yoshikawa & S. Yoshikawa, *Macromol. Rapid Comm.* (1991), **12**, 89.
[97] J. R. Jerez, B. Matsuhiro & C. C. Urzua, *Cirbohydr. Polym.* (1993), **32**, 155.
[98] N. P. Barroso, J. Costamagna, B. Matsuhiro & M. Villagran, *Bol. Soc. Chil. Quim.* (1997), **42**, 301.
[99] E. Fredon, R. Granet, R. Zerrouki, P. Krausz, L. Saulnier, J.-F. Thibault, J. Rosier & C. Petit, *Carbohydr. Polym.* (2002), **49**, 1.
[100] M. R. Sierakowski, M. Milas, J. Desbrieres & M. Rinaudo, *Carbohydr. Polym.* (2000), **42**, 51.
[101] U. D. N. Bajpai, J. Alka & R. Sandeep (1990) *J. Appl. Polym. Sci.* (1990), 39, 2187.
[102] V. Singh, V. Srivastava, M. Pandey, R. Esthi & R. Sanghi, *Carbohydr. Polym.* (2003), **51**, 357.
[103] E. Lapasin, L. de Lorenzi, S. Pridl & G. Torriano (1995) *Carbohydr. Polym.* (1995), **28**, 195.
[104] P. L. Soni & R. Pa, *Trends Carbohydr. Chem.* (1996), **2**, 33.
[105] B. R. Sharma, V. Kumar & P. L. Soni, *J. Appl. Polym. Sci.* (2002), **86**, 3250.
[106] B. R. Sharma, V. Kumar & P. L. Soni, *Starch/Stärke* (2003), **55**, 38.
[107] B. R. Sharma, V. Kumar, P. L. Soni & P. J. Sharma, *J. Appl. Polym. Sci.* (2003), **89**, 3216.
[108] P.L. Soni, S. V. Singh & S. Naithani S, *Paper Intern.* (2000), **5**, 14.
[109] E. Lapasin, L. de Lorenzi, S. Pridl & G. Torriano, *Carbohydr Polym* (1995), **28**, 195.
[110] S. D. Figueiro, J. C. Goes, R. A. Moreira, A. S. B. Sombra, *Carbohydr. Polym.* (2004), **56**, 313.
[111] J. Lundqvist, A. Teleman, L. Junel, G. Zacchi, O. Dahlman, F. Tjerneld & H. Stålbrand, *Carbohydr. Polym.* (2002), **48**, 29.

[112] M. S. Lindblad, A. C. Albertsson & E. Ranucci. In P. Gatenholm & M. Tenkanen (eds), *Science and Technology*, ACS Symposium Series 864. American Chemical Society, Washington DC (2004), p. 347.
[113] H. Prabhanjan, *Starch/Stärke* (1989), **41**, 409.
[114] P. Lang, G. Masci, M. Dentini, V. Crescenzi, D. Cooke, M. J. Gidley, C. Fanutti & J. S. G. Reid, *Carbohydr. Polym.* (1992), **17**, 185.
[115] M. J. Gidley, P. J. Lillford, D. W. Rowlands, P. Lang, M. Dentini, V. Creszenzi, M. Edwards, C. Fanutti & J. S. G. Reid, *Carbohydr. Res.* (1991), **214**, 299.
[116] E. X. F. Filho, M. G. Touhy, J. Pulls, M. P. Coughlan, The xylan-degrading enzyme systems of Penicillium capsulatum and Talaromyces emersonii. *Biochem. Soc. Trans.* (1991), **19**, 25S.
[117] S. L. Bachmann & A. J. McCarthy, Purification and cooperative activity of enzymes constituting the xylan-degrading system of Thermomonospora fusca. *Appl Environ. Microbiol.* (1991), **57**, 2121–2130.
[118] K. Poutanen & J. Puls, The xylanolytic enzyme system of Trichoderma reesei. In G. Lewis & M. Paice (eds), *Biogenesis and Biodegradation of Plant Cell Wall Polymers*. American Chemical Society, Washington, DC (1989), pp. 630–640.
[119] S. F. Lee, C. W. Forsberg, Purification and characterization of an α-L-arabinofuranosidase from Clostridium acetobutylicum ATCC 824. *Can. J. Microbiol.* (1987), **33**, 1011–1016.
[120] K. Poutanen, M. Tenkanen, H. Korte & J. Puls, Accessory enzymes involved in the hydrolysis of xylans. In G. F. Leatham & M. E. Himmel (eds), *Enzymes in Biomass Conversion.* American Chemical Society, Washington, DC (1991), pp. 426–436.
[121] H. J. Gilbert & G. P. Hazlewood, Bacterial cellulases and xylanases. *J. Gen. Microbiol.* (1993), **139**, 187–194.
[122] G. Brunow, K. Lundquist, and G. Gellerstedt, Lignin. In E. Sjostrom & R. Alen (eds), *Analytical Methods in Wood Chemistry. Pulping and Papermaking,* Springer, Berlin (1999), p. 77.
[123] L. B. Davin & N. G. Lewis, *Curr. Opin. Biotechnol.* (2005), **16**, 407.
[124] C. Halpin, *Biotechnol. Genet. Eng. Rev.* (2004), **21**, 229.
[125] W. Boerjan, J. Ralph & M. Baucher, *Annu. Rev. Plant Biol.* (2003), **54**, 519.
[126] Y. Pu, N. Jiang & A. J. Ragauskas, *J. Wood Chem. Technol.* (2007), **27**, 23.
[127] J. Ralph, K. Lundquist, G. Brunow, F. Lu, H. Kim & P.F. Schatz, *Phytochem., Rev.* (2004), **3**, 29.
[128] N. G. Lewis & E. Yamamoto, *Annu. Rev. Plant Physiol. Plant Mol. Biol.*(1990), **41**, 455.
[129] W. Boerjan, J. Ralph & M. Baucher, *Annu. Rev. Plant Biol.* (2003), **54**, 519.
[130] Y. Pu, D. Zhang, P. M. Singh & A. J. Ragauskas, *Biofuels Bioprod. Bioref.* (2008), **2**, 58.
[131] A. J. Ragauskas, M. Nagy, D. H. Kim, C. A. Eckert, J. P. Hallett & C. L. Liotta, *Indust. Biotechnol.* (2006), **2**, 55.
[132] E. A. Capanema, M. Y. Balakshin & J. F. Kadla, *J. Agric. Food Chem.* (2005), **53**, 9639.
[133] F. S. Chakar & A. J. Ragauskas, *Ind. Crops Products* (2004), **20**, 131.
[134] J. J. Stewart, J. F. Kadla & S. D. Mansfield, *Holzforschung* (2006), **60**, 111.
[135] R. Samuel, Y. Pu, B. Raman & A. J. Ragauskas, *Appl. Biochem. Biotechnol.* (2010), **162**, 62.
[136] A. R., Robinson & S. D. Mansfield, *Plant J.* (2009), **58**, 706.
[137] R. El Hage, N. Brosse, L. Chrusciel, C. Sanchez, P. Sannigrahi & A. J., Ragauskas, *Poly. Degrada. Stab.*(2009), **94**, 1632.
[138] S. C. Fox & A. G. McDonald, *Bioresources* (2010), **5**, 990.
[139] A. Buranov & G. Mazza, *Indust. Crops and Products* (2008), **28**, 237.
[140] E. C. Ramires, J. D. Megiatto, C. Gardrat, A. Castellan & E. Frollini, *Biotechnol. Bioeng.* (2010), **107**, 612.

[141] D. Robert & D. Gagnaire, In *Quantitative Analysis of Lignins by 13C NMR. Proceedings of the International Symposium on Wood and Pulping Chemistry,* (Stockholm, Sweden, 1981), p. 86.
[142] E. Ammalahti, G. Brunow, A. Bardet, D. Robert & I. Kilpelainen, *J. Agric. Food Chem.* (1998), **46**, 5113.
[143] E. A. Capanema, M.Y . Balakshin & J. F. Kadla, *J. Agric. Food Chem.* (2004), **52**, 1850.
[144] E. A. Capanema, M. Y. Balakshin & J. F. Kadla, *J. Agric. Food Chem.* (2005), **53**, 9639.
[145] Y. Yu, X. Lou & H. Wu, Some recent advances in hydrolysis of biomass in hotcompressed water and its comparisons with other hydrolysis methods. *Energy Fuels* (2008), **22**, 46–60.
[146] K. Freudenberg, Uber lignin. *Angew. Chem.* (1939), **52**, 362–363.
[147] R. H. J. Creighton, R. D. Gibbs & H. Hibbert, Studies on lignin and related compounds. LXXV. Alkaline nitrobenzene oxidation of plant materials and application totaxonomic classification. *J. Am. Chem. Soc.* (1944), **66**, 32–37.
[148] T. Higuchi, Y. Ito, M. Shimada & I. Kawamura, Chemical properties of milled wood lignin of grasses. *Phytochemisty* (1967), **6**, 1551–1556.
[149] K. Freudenberg, A. Janson, E. Knopf & A. Haag, Zur Kenntnis des Lignins. *Chem. Ber.* (1936), **69**, 1415–1425.
[150] S. Larsson & G. E. Miksche, Gas chromatographic analysis of lignin oxidation products. The diphenyl ether linkage in lignin. *Acta Chem. Scand.* (1967), **21**, 1970–1971.
[151] A. Sakakibara, Hydrogenolysis. In S. Y. Lin & C. W. Dence (eds), *Methods in Lignin Chemistry,* Springer Series in Wood Science, Springer, Berlin (1992), pp. 350–368.
[152] K. V. Sarkanen, A. Islam & C. D. Anderson, Ozonation. In S. Y. Lin & C. W. Dence (eds), *Methods in Lignin Chemistry,* Springer Series in Wood Science, Springer, Berlin, (1992), pp. 387–406.
[153] H. Hatakeyama, T. Tonooka, J. Nakano & N. Migita, Ozonization of lignin model compounds. *Kogyo Kagaku Zasshi.* (1967), **70**, 2348–2352.
[154] H. Hatakeyama, E. Hayashi & T. Haraguchi, Biodegradation of poly(3-methoxy-4-hydroxy styrene). *Polym. J.* (1977), **18**, 759–763.
[155] K. Nakamura, T. Hatakeyama & H. Hatakeyama, Effect of substituent groups on hydrogen bonding of polyhydroxystyrene derivatives. *Polym. J.* (1983), **15**, 361–366.
[156] K. Nakamura, T. Hatakeyama & H. Hatakeyama, DSC Studies on hydrogen bonding ofpoly(4-hydroxy-3, 5-dimethoxystyrene) and related derivatives. *Polym. J.* (1986), **18**, 219–225.
[157] K. Nakamura, T. Hatakeyama & H. Hatakeyama, 2 Relationship between hydrogen bonding and bound water in polyhydroxystyrene derivatives. *Polymer* (1983), **24**, 871–876.
[158] S. Hirose, K. Nakamura & T. Hatakeyama, Molecular design of linear aromatic polymers derived from phenols related to lignin. In C. Schuerch (ed.), *Cellulose and Wood: Chemistry and Tecnology.* Wiley, NY (1989), pp. 1133–1144.
[159] S. Hirose, K. Nakamura & T. Hatakeyama, Isothermal crystallization of polybenzalazine derivatives synthesized from vanillin. *Sen-i Gakkaishi* (1986), **42**, T49–T53.
[160] R. Lal, *Science* (2004), **304**, 1623.
[161] L. K. Paine, T. L. Peterson, D. J. Undersander, K. C. Rineer, G. A. Bartlelt & S. A. Temple, *Biomass Bioenerg.* (1996), **10**, 231.
[162] T. W. Bowersox, *Wood Science* (1979), **11**, 257.
[163] P. J. Tharakan, T. A. Volk, L. P. Abrahamson & E. H. White, *Biomass Bioenerg.* (2003), **25**, 571.
[164] L. Allison, A. J. Ragauskas & J. S. Hsieh, *Tappi J.* (2000), **83 (8)**, 1.
[165] R. Fahmi, A. V. Bridgewater, L. I. Darvell, J. M. Jones, N. Yates & S. Thain, *Fuel* (2006), **86**, 1560.

[166] D. H. van Lear, J. B. Waide & M. J. Tueke, *Forest Sci.* (1984), **30 (2)**, 395.
[167] J. Yan, Z. Hu, Y. Pu, E. C. Brummer & A. J., Ragauskas, *Biomass Bioenergy* (2010), **34**, 48.
[168] R. C. Sun & J. Tomkinson, *J. Wood. Sci.* (2003), 49, 47.
[169] M. P. Fernandez, P. A., Watson & C. Breuil, *J. Chromato. A.* (2001), **922**, 225.
[170] A. D. Yanchunk, I. Spilda & M. M. Micko, *Wood Sci. Technol.* (1988), **22**, 67.
[171] J. Dorado, F. W. Classsen, T. A. van Beek, G. Lenon, J. B. P. A. Wijnberg & R. Sierra-Alvarez, *J. Biotechnol.*(2000), **80**, 231.
[172] T. R. Miles, R. W. Bryers, B. M. Jenkins & L. L. Oden, in Alkali deposits found in biomass power plants. A preliminary investigation of their extent and nature: National Renewable Energy Laboratory Report NREL/TP-433-8142, (Oakridge, TN, 1995).
[173] P. Chow & G. L. Rolfe, *Wood. Fiber Sci.* (1980), 21, 30.
[174] M. D. Summers, in *Fundamental Properties of Rice Straw in Comparison with Softwoods,* (University of California Davis, Davis, CA, 2000).
[175] R. C. Brown, *Biorenewable Resources: Engineering New Products from Agriculture,* Iowa State Press, Ames, IA, 2003.

Chapter 4

Plant fibre: Its behaviour

Professor Qingxian Miao, Dr Dasong Dai and Dr Jiuping Rao

This chapter discusses both single natural fibre and fibre bundles. It begins with the microstructure, surface chemistry, inherent defects and mechanical properties of single natural fibre. The chapter then provides an overview of both the chemical and mechanical pulping production of natural fibre bundles, for which various processing technologies and resultant characteristics and properties are discussed. This provides a fundamental database for the utilisation of natural fibres for various applications, which are presented in the succeeding chapters.

4.1 Introduction

All plants can be processed to single plant fibres and fibre bundles by different manufacturing methods. Single fibres and fibre bundles have various properties and applications depending on different sources and different processing methods. The properties of single plant fibres, including their mechanical properties, microfibril angle, defects, and surface properties are reviewed in this chapter. Both the microfibril angle and defects have a significant influence on the mechanical properties of single fibres. The surface properties of single fibres affect the bonding strength of fibre bundles.

In this chapter, the processing methods and properties of fibre bundles are also discussed. The fibre bundle can be obtained through chemical and mechanical processes. The chemical methods mainly include kraft and sulphite processes, whilst the mechanical process consists mainly of chemithermomechanical, thermo mechanical and alkaline peroxide mechanical processes. The resulting corresponding fibre bundles may also be known as chemical pulp and mechanical pulp. The pulps have different properties (e.g. brightness and strength) due to the different processes. Compared with chemical pulp, mechanical pulp has a higher yield, shorter fibre, and lower brightness and strength. Thus, the application of fibre bundles varies because of their different properties. In many cases, fibre bundles need to be bleached and purified to a high brightness. The bleaching methods commonly applied include the elemental chloride free process and the total chloride free process, which meet the strict laws and regulations of environment protection.

4.2 Single plant fibre

4.2.1 Mechanical properties of plant fibre

Plant fibres are, in general, suitable for reinforcing inorganic polymers, synthetic polymers and natural polymers due to their relative high strength, stiffness and low density [1] (see Table 4.1). The characteristic values for flax can reach a level close to those for

Table 4.1 Mechanical properties of non-wood lignocellulosic fibres.

Type of fibre	Fibre	Density (g cm^{-3})	Elongation (%)	Tensile strength (MPa)	Young's modulus (GPa)	References
Stem fibres	Bamboo	0.6–0.91	1.4	193–600	20.6–46.0	[5–7]
	Flax	1.5	1.2–3.2	345–2000	15–80	[8–10]
	Hemp	1.48	1.6	550–900	26–80	[3, 11, 12]
	Jute	1.3	1.16–1.5	393–800	13–55	[9, 13]
	Kenaf	1.45	1.6	157–930	22.1–60	[14–18]
	Ramie	1.5	1.2–3.8	400–938	61.4–128	[19]
Leaf fibres	Banana	0.72–0.88	2.0–3.34	161.8–789.3	7.6–9.4	[20]
	Pineapple	1.07	2.2	126.6	4.4	[21]
	Sisal	1.5	3.0–7.0	468–700	9.4–22	[19]
Fruit fibres	Coir	1.2	17–47	175	4.0–6.0	[2, 22]
	Oil palm	0.7–1.55	4–18	50–400	0.57–9.0	[23, 24]
Wood fibres	Softwood Kraft (spruce)	1.5	–	1000	18–40	[25]
	Hardwood Kraft (birch)	1.2	–	–	37.9	[26]
Synthetic fibres	E-glass	2.5	2.5	2000–3500	70	[2]
	S-glass	2.5	2.8	4570	86	[2]
	Aramide	1.4	3.3–3.7	3000–3150	63.0–67.0	[2]

E-glass fibres [2]. However, all natural fibres have a remarkably wider range of characteristic values than synthetic fibres (Table 4.1), which is one of their drawbacks and can be explained by the difference in their fibre structure. It is apparent from a comparison with wood fibres that non-wood fibres show similar mechanical properties. The fibre properties and structure are influenced by a number of conditions and vary with area of growth, climate, and the age of plant [3, 4]. Furthermore, the technical digestion of the fibre is another important factor that determines the structure and characteristics of the fibres.

4.2.2 Microfibril angle of plant fibre

The cell walls of natural fibres are formed of crystalline microfibrils based on cellulose, which are connected to a complete layer of lignin and hemicellulose. Numerous researchers have [27–32] tried to illustrate the correlation between the physical structure and the mechanical properties of natural fibres (some of the structural parameters of natural fibres are summarised in Table 4.2). In 1980, McLaughlin and Tait [33] firstly carried out a detailed statistical analysis of physical structure/property relations. They found a positive correlation between the mechanical properties (tensile strength

Table 4.2 Structural parameters of natural fibres.

Fibres	Microfibril angle (°)	Crystallinity index (%)	References
Bamboo	2–10	66.3	[6, 35]
Flax	5–10	42.9	[18, 35, 36]
Hemp	2.2–6	55–66.3	[18, 37–39]
Jute	8–8.1	50–58.9	[6, 18, 40, 41]
Kenaf	–	54.6	[42]
Ramie	7.5	69.4	[18, 35]
Banana	11	45–68	[6, 43, 44]
Pineapple	8–14	44–60	[44, 45]
Sisal	10–22	57	[18, 46]
Coir	39–49	27–33	[6, 44]
Oil palm	–	40	[38]

and Young's modulus) and the physical structure (cellulose content and microfibril angle (MFA)) [34].

The correlation between tensile strength and physical structure (i.e. cellulose content, MFA, fibre length and fibre diameter) were developed by Mukherjee and Satyanarayana [47] as shown in Eq. 4.1:

$$Y = \mathrm{K}X_c\cos\theta\left(\frac{L}{D}\right)^a + C \qquad \text{(Eq. 4.1)}$$

where, Y is the tensile strength, K and C are the constants for equation, X_c is the cellulose content, θ is the MFA, L is the length of fibers and D is the diameter of fibre. Three different equations (Equations 4.2, 4.3, 4.4) were developed to calculate fibre stiffness with respect to the fibre axis:

(1) Without deformation (MFA < 45°)

$$E_{|} = [X_{1C}E_{\|,C} + (1 - X_{1C})E_{NC}]\cos^2 X_2 = E^{*}\cos^2 X_2 \qquad \text{(Eq. 4.2)}$$

where $E_{\|C}$ is the Young's modulus of the fibre in the parallel direction, E_{NC} is the Young's modulus of the non-crystalline parts, X_{1C} is the content of the crystalline part in the fibre and X_2 is microfibril angle.

(2) Deformation (MFA > 45°)

$$E_{|} = K_{NC}{*}(1 - 2\cot^2 X_2)^2/(1 - X_{1C}) = K(1 - 2\cot^2 X_2)^2 \qquad \text{(Eq. 4.3)}$$

where, K_{NC} is the bulk modulus of the non-crystalline parts, X_{1C} is the content of the crystalline part in the fibre, and X_2 is microfibril angle.

(3) General formulation

$$E1 = \frac{E \times \cos^2 X_2 \times [K \times (1 - \cot^2 X_2)^2]}{E \times \cos^2 X_2 + K(1 - 2\cot^2 X_2)^2} \qquad \text{(Eq. 4.4)}$$

where, X_2 is microfibril angle and K is a constant for the equation.

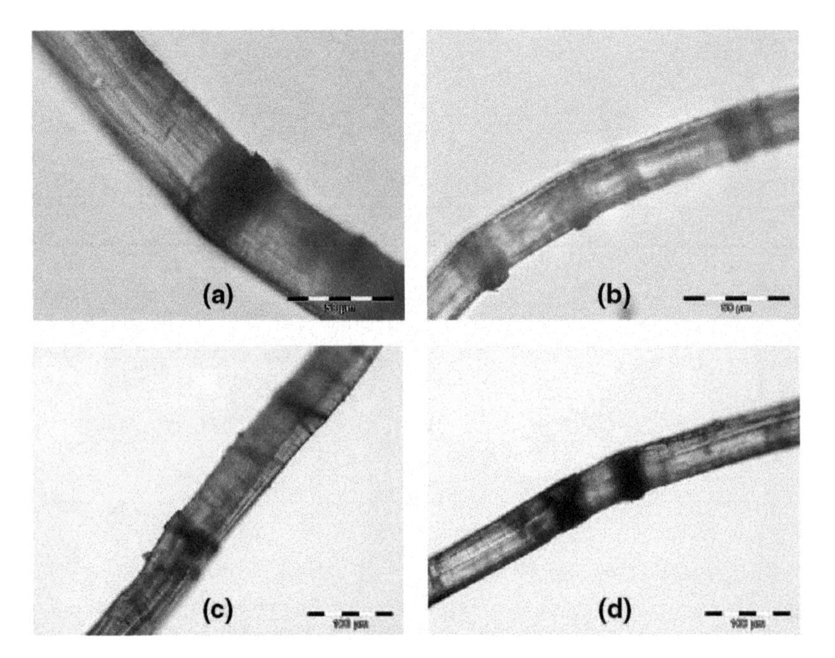

Figure 4.1 Defects of plant fibres: a = kink band (× 500 magnification); b = node (× 500 magnification); c = dislocation (× 200 magnification); d = slip plane (× 200 magnification).

4.2.3 Defects of plant fibre

As the weakest points in plant fibre, defects play a key role in the mechanical performance of fibres [48]. Defects in natural fibres are also called nodes, slip planes, micro-compressions, misaligned zones, amongst other names (see Fig. 4.1) [49]. By employing the finite element (FE) method, Nilsson and Gustafsson [48] investigated the effect of defects on the tensile performance of natural fibres and found that the elastic stiffness of fibre was greatly affected by deformations, and the S-shaped stress vs strain curve can be explained by non-linear geometric effects and the yielding of hemicellulose in the deformed areas. The FE analysis is in good agreement with previous work by Baley [50] about the initiative of cracks in the flax fibre, which in general appears around the deformation of fibres during a tensile loading. The tensile strength of the fibres is also influenced by the fineness of the fibre [51]. A decrease in fibre fineness gives rise to higher fibre strength. The tensile strength of natural fibres also depends on the test length of the specimens, which is of vital importance in terms of its reinforcing efficiency.

4.2.4 Surface properties of plant fibre

The splintering fracture (see Fig. 4.2 [52]) indicates that hemp fibres have high microfibril angle and low aspect ratio, while banana fibres have high cellulose content, smaller microfibril angle and high aspect ratio, exhibiting a cross-fracture mode with no pull-out microfibrils. The surface morphologies of fibres are shown in Fig. 4.2,

Figure 4.2 Fractural surface of hemp fibres.

Table 4.3 Surface properties of natural fibres.

Fibres	Surface area ($m^2 g^{-1}$)	γ^d dispersive surface energy ($m J^{m-2}$)	$(\zeta_0 - \zeta_\infty)/\zeta_0$ (mV)	$\zeta_{plateau}$ (mV)	References
Flax	0.31 ~ 0.79	23.85	0.88 ~ 0.95	–1.1 ~ –0.21	[53–56]
Hemp	0.75	31.6	0.91	–0.1	[54, 56, 57]
Sisal	1.63	32.9–33.3 48.35	0.76 ~ 0.88	–1.7 ~ –0.4	[54–56, 58]
Coir	0.48	45.05	0.22	–4.6~–3.8	[55, 58, 59]

which also indicate considerable hemicelluloses, lignin, waxes, oils and surface impurities on the fibre surfaces. The surface property is influenced by fibre morphology, extractive chemicals and the processing conditions, which in fact is the key property of natural fibres (some of the surface properties of non-wood lignocellulosic fibres are summarised in Table 4.3), as it can affect the interfacial adhesion with the resins on the surface of fibres and the mechanical properties of natural fibre-based composite. Due to the high polar feature of the surface, the fibres are also less compatible with non-polar resins. The combination of the inherently polar and hydrophilic natural fibres and the non-polar resins, therefore results in difficulties in compounding these materials, leading to inefficiencies of stress transfer. The use of different kinds of physical (i.e. corona discharge) and chemical surface treatment methods (i.e. coupling agents such as silanes) leads to changes in the surface structure of the fibres as well as changes in the surface properties.

4.3 Fibre bundles: chemical pulp fibres

4.3.1 The process of chemical pulp fibres

Chemical pulping employs chemicals to treat the fibrous plant materials under special conditions, aiming to remove lignin and separate the fibre bundle into pulp fibres. In this process, the lignin in the middle lamella, which contributes to the bonding of fibres, should be considerably dissolved out. Based on the removal of lignin, the fibre cell can be split or is ready to split. Chemical pulping is characterised by significant removal of lignin, pronounced retention of cellulose and partial removal of hemicelluloses (based on the pulp quality). It is well known that chemical pulping includes alkali pulping and sulphite pulping. The commercial alkali pulp occupies more than 95% of the total amount of the chemical pulp [60–62]. The main technological procedure consists of stock preparation, chemical cooking, washing, screening/cleaning and bleaching.

4.3.1.1 Kraft pulp fibres

Based on different cooking chemicals, alkali pulping can be divided into kraft pulping, soda pulping and sodium polysulphide pulping. Of these, kraft and soda pulping are the common pulping methods, especially kraft pulping which is the dominant pulping method. Kraft pulping is suitable for various raw materials, including not only most of the fibrous plant materials but also wood, waste wood, wood residue, saw powder and high resin content wood [61].

The kraft pulping method mainly contains batch and continuous cooking processes. Continuous cooking presents advantages, such as high production automation, balanced energy consumption, weak atmospheric pollution, high yield, and stable and uniform pulp quality. NaOH and Na_2S are the primary cooking chemicals for kraft pulping, and the chemicals' composition and dosage, liquor ratio, maximum cooking temperature and the time for increasing the temperature, are the basic technological parameters. Table 4.4 summarises the cooking conditions for the kraft pulping of various raw materials.

In recent years, new developments in chemical pulping have focused on high efficiency, high delignification selectivity, and low energy consumption and contamination [63]. In order to enhance the delignification selectivity in the cooking process, and thus increase the removal of lignin and lower the degradation of carbohydrates, researchers have paid more attention to extended delignification approaches. The practical application of extended delignification mainly involves transforming cooking from batch pulping and modified kraft cooking to continuous pulping. Another pulping approach is to introduce cooking additives, including anthraquinone and sodium polysulphide, which oxidise the thermal groups in cellulose and hemicelluloses chains (from aldehyde group to saccharic acid), thus stopping the peeling reaction and improving the cooking yield and pulp strength [64].

4.3.1.2 Sulphite pulp

As per distinct chemicals and pH, sulphite pulping can be divided into acid bisulphite pulping, hydrosulphite pulping, subacid hydrosulphite pulping, neutral hydrosulphite pulping and alkaline hydrosulphite pulping. Of these, acid hydrosulphite pulping is the most extensively used [61, 62].

Sulphite pulping presents mild cooking conditions and the target raw materials include wood, straw, bamboo and bagasse (low resin content), especially for acid hydrosulphite pulping. Sulphite pulp is generally characterised by low pulp strength. However, sulphite pulping carries the advantage that the cooking liquor can be recovered and utilised for the production of various high-value chemicals.

Table 4.4 Cooking conditions for kraft pulping for different raw materials.

Raw material	Alkali charge (%)	Sulphidity (%)	Liquor ratio	Max. temperature (°C)	Increasing temperature (minutes)	Heat preservation (minutes)	Yield (%)	Lignin content
Red pine	17.5 (Na_2O)	25–30	1:3	165	170	35	51 (Unscreened stock)	55–65 (Kappa number)
Spruce [5]	20 (Na_2O)	25	1:4	170	150	120	43 (Unscreened stock)	21.25 (Kappa number)
Bamboo	15.5 (NaOH)	16	1:2.1	160	120	120	–	20.2 (Kappa number)
Euca-Lyptus [6]	20 (NaOH)	35	1:4	150	60	90	48.6 (Screened stock)	17.2 (Kappa number)
Wheat straw [7]	13 (Na_2O)	11	1:6	160	115	0	52.12	10.77 ($KMnO_4$ number)
Bagasse [8]	10–13 (Na_2O)	15	1:5	164	210	30	–	8–15 (Kappa number)

Factors affecting sulphite pulping include chemical composition and dosage, pH and salt type, temperature and pressure, liquor ratio, the time for increasing the temperature and the retaining time at the highest temperature. Tables 4.5, 4.6 and 4.7 summarise the typical sulphite pulping conditions for wood and non-wood raw materials. The cooking additive for sulphite pulping is anthraquinone, which improves the delignification rate, decreases the cooking time, saves the carbohydrates, and increases the pulp yield and strength.

4.3.1.3 Chemical pulp bleaching

Bleaching is expected to improve the pulp brightness, thus meeting the paper quality requirements. The bleaching for chemical pulp is carried out by employing oxidants to degrade and remove the residual lignin from the brownstock, which reduces the number of chromophoric groups. By contrast, the bleaching for mechanical pulp is achieved using hydrogen peroxide and hyposulphite to react with lignin, whereby the structure of the chromophoric groups is changed [61, 62, 65, 66]. In comparison to mechanical pulp bleaching, the bleaching of chemical pulp presents a more complex technological processes.

Table 4.5 Cooking conditions for poplar acid calcium hydrosulphite pulping.

Raw material	Poplar
Chip filling (minutes)	50–60
Increasing to 100 °C (minutes)	80
Preserving between 100 °C and 112 °C (minutes)	90
Increasing from 112 °C to the highest temperature (minutes)	150
Heat preservation (minutes)	240
Max. temperature (°C)	138–140
Steam release (minutes)	50–60
Pressure (MPa)	0.245
Emptying pulp (minutes)	30
Total time (hours)	10–11

Table 4.6 Cooking conditions for softwood alkaline sodium hydrosulfite pulping.

Raw material	Cooking type	NaOH dosage (Na_2O based) %	Na_2SO_3 dosage (Na_2O based) %	Na_2CO_3 dosage (Na_2O based) %	Maximum temperature °C	Time (mins)	Kappa no.	Yield %
Softwood	Continuous cooking	11	11	4	172	120	72	50

Table 4.7 Cooking conditions for non-wood neutral hydrosulphite pulping.

Raw material	Wheat straw	Bamboo	Bagasse
$(NH_4)_2SO_3$ dosage (%)	12	20	20
NaOH dosage (%)	–	2(Na_2O)	3
$(NH_4)_2CO_3$ dosage (%)	1.5	–	–
pH	–	10–12	12
Liquor ratio	2.3	2	3.5
Pressure (MPa)	0.686	0.686	0.686
Idle time (minutes)	10	10	30
Heating time (minutes)	90	96	110
Heat preservation time (minutes)	130	270	120
Blow time (minutes)	30	45	–
Kappa number	14–16	14	9–12
Yield (%)	45–50 (screened pulp)	38 (unscreened pulp)	–

Table 4.8 Strength and optical properties of bleached softwood kraft pulp.

Bleaching method	TAPPI viscosity (mPa·s)	PFI Rev.	Tensile index ($N\cdot m\cdot g^{-1}$)	Tearing index ($mN\cdot m^2\cdot g^{-1}$)	Scattering coefficient ($m^2\cdot kg^{-1}$)
Traditional method	11.7	6500	89	7.7	16.6
TCF	11.7	6200	92	8.2	15.9

Owing to strengthened awareness of environmental protection concerns, element chlorine bleaching (chlorine and hypochlorite) is gradually being eliminated because of the hazardous substance, organic chloride, that is generated. The dominant chemical pulp bleaching processes are the element chlorine free approach (ECF, chlorine dioxide) and the total chlorine free approach (TCF, oxygen, ozone and hydrogen peroxide). In addition, biotechnology, such as xylanase-aided bleaching, has been widely applied in the bleaching process of chemical pulp.

In general, the bleaching process for chemical pulp contains several treatment procedures, in which different bleaching agents are sufficiently functioned. In the practical process, the origin of the raw materials, the property of unbleached pulp, the application of bleached pulp, and the cost of the chemicals should be considered. Advanced bleaching technological processes can produce superior pulp quality, including high target brightness and strength along with low energy and water consumption, as well as low pollution. It is well known that the common ECF bleaching contains OD(EO)DED, DEDED, OD(EO)D, OZED, amongst others. And the common TCF bleaching includes OZQP, OXZP, OZQP and OQPP. Table 4.8 summarises the physical and optical properties of bleached softwood kraft pulp.

4.3.2 Properties of chemical pulp fibre

4.3.2.1 Kraft pulp fibre

The age of trees has a significant effect on pulp properties, as shown in Table 4.9. The 17-year-old fir has the greatest potential for kraft pulping according to the strength properties of the resultant pulps under optimised cooking conditions [67].

The cooking process also has an effect on the productivity and properties of chemical pulp fibres. As shown in Table 4.10, the extended modified continuous cooking (EMCC) process has higher delignification selectivity than the conventional kraft

Table 4.9 Strength properties of Chinese fir from plantation at different tree ages.

Tree-age (years)	Beating degree (ºSR)	Breaking length (KM)	Burst index ($KPa\cdot m^2\cdot g^{-1}$)	Tearing index ($mN\cdot m^2\cdot g^{-1}$)
12	45.2	8.732	7.54	6.83
17	45.1	9.512	7.87	7.42
23	44.2	8.556	7.19	8.61

Table 4.10 Comparison between extended delignification and conventional kraft cooking of bamboo.

Sample	Kappa number	Viscosity ($mL \cdot g^{-1}$)	Yield (%)	Beating degree (°SR)	Tensile index ($N \cdot m^2 \cdot g^{-1}$)	Burst index ($KPa \cdot m^2 \cdot g^{-1}$)	Tearing index ($mN \cdot m^2 \cdot g^{-1}$)
EMCC	12.5	1031	44.8	22	56	3.5	17
CKC	12.5	920	40.1	22	50	2.4	15.8

cooking (CKC) process. EMCC bamboo pulp has higher yield and viscosity, as well as better strength properties at the same Kappa number and beating degree compared with CKC pulp [68].

Nowadays, element chlorine free bleaching (ECF) and total chlorine free bleaching (TCF) are the dominant bleaching processes for chemical pulp due to strict environment protection, and ECF is more popular than TCF. It should be noted that different ECF bleaching processes can lead to different bleached pulp properties even for the same brownstock, as shown in Table 4.11 [69].

The kraft pulping process results in low pulp yield, and the resulted kraft pulp has low bleachability but high pulp strength and wide application.

Softwood (pine and spruce) brown kraft pulp can be used for manufacturing sack paper, cable paper, condenser paper and packaging paper, and the bleached pulp can be employed to produce writing paper, offset paper, culture paper. Having short and thin fibres, hardwood kraft pulp can be used as a raw material to prepare printing paper with superior surface properties and opacity. Compared to softwood and hardwood pulp, straw pulp shows the lowest strength and can be used for fabricating general cultural paper.

Kraft pulping results in a decrease in cellulose viscosity and degree of polymerisation, and an increase in carboxyl groups. The fibre properties and chemical composition of raw materials significantly affects the physical strength of kraft pulp. The fibre length of the raw material is responsible for the pulp strength, e.g. longer fibre results in higher tearing strength but lower bursting strength. Generally, kraft pulp has high single fibre strength and a high fibrillation level, because the refining procedure leads to the increased fibre bonding and paper strength.

Table 4.11 Effect of bleaching processes on bamboo pulp properties.

Bleaching process	Brightness (%ISO)	Viscosity ($mL \cdot g^{-1}$)	Tearing index ($mN \cdot m^2 \cdot g^{-1}$)	Burst index ($KPa \cdot m^2 \cdot g^{-1}$)	Breaking length (m)
$OXD_1E_{OP}D_2$	86.4	803.6	10.3	4.17	6900
$OD_1E_OD_2$	84.7	901.8	11.3	4.46	6710
$OD_1E_{OP}D_2$	85.7	823.6	13.7	5.76	7420

Note: Where, the O is oxygen delignification, X is the hemicellulase treatment, D is the chloride dioxide bleaching, E is the alkaline treatment and P is the hydrogen peroxide bleaching.

4.3.2.2 Sulphite pulp

Sulphite pulp presents high pulp yield and brightness of brownstock, and its pulping process requires lower energy than kraft pulp to obtain the highest pulp strength. However, hydrosulphite pulp has lower strength than kraft pulp at the same beating degree. The strength of hardwood hydrosulphite pulp is lower than that of softwood hydrosulphite pulp, which is the result of its shorter fibre length.

The brownstock for sulphite pulping has high transparency and oil resistance, thus being a superior material for manufacturing translucent glass paper, oil-proof paper, flat envelopes, relief paper and newsprint furnish. The bleached sulphite pulp can be used to produce writing paper, double-sided offset paper, dictionary paper and cigarette paper.

4.4 Fibre bundles: mechanical pulp fibres

4.4.1 The mechanical pulp fibre process

Mechanical pulp refers to the pulp produced by mechanical processes. According to the level of mechanical treatment, mechanical pulping can be divided into three categories: mechanical pulping, chemi-mechanical pulping and semi-chemical pulping. Mechanical pulping relies on mechanical refining to treat fibrous materials. Negligible amounts of lignin are removed in the process, which results in high pulp yield (more than 95%). Chemi-mechanical pulping includes mechanical refining following chemical pretreatment, where large amounts of hemicelluloses are separated without the removal of lignin, which softens the middle layer lamella, thus being supportive for subsequent mechanical treatment. Semi-chemical pulping is similar to chemi-mechanical pulping with regards to the chemical pretreatment and the following mechanical treatment, but treatment is milder for the semi-chemical pulping than for chemi-mechanical pulping, thus resulting in a higher pulp yield for the former (65–84%) [61]. Generally, mechanical pulping includes thermal mechanical pulping (TMP), bleached chemical thermal mechanical pulping (BCTMP) and alkaline peroxide (APMP) pulping. CTMP and APMP are considered as high yield pulps. With regard to the raw materials used for mechanical pulping, wood is the dominant raw material followed by bamboo and straw.

4.4.1.1 Thermomechanical pulp (TMP)

The TMP process primarily includes chip washing, pre-heating, beating and refining. In the washing stage, the temperature of the washing water is 30–50 °C and the washing time is 1–2 minutes. The wood chips are then treated in screw chip feeder in order to remove the moisture and air, and transferred into the pre-heater under a pressure of 147–196 kPa and a temperature of 115–135 °C for 2–5 minutes [61]. Subsequently, defibrination of heating-based wood chips is carried out in the first disc with pressure and in the second disc without pressure. Finally, the wood chips are refined into pulp fibres. There are several influencing factors during TMP production, including the wood species, refining consistency, heating temperature and time, and disc gap. In some cases, the TMP must be bleached to a certain brightness for some

Table 4.12 Effect of bleaching methods on the properties of Masson TMP.

Bleaching method	Brightness (%ISO)
H_2O_2	72.6
FAS	58.2
H_2O_2-FAS	76.3
Control pulp	42.9

high-quality papers. With regard to the bleaching of mechanical pulp, hydrogen peroxide, sodium dithinoite, and formamidinesulfinic acid (FAS) are commonly used, of which hydrogen peroxide is the most popular. Table 4.12 lists the effect of different bleaching methods on the masson TMP properties [70].

The manufacturing system of TMP for fibreboard mainly includes a screw feeder, regurgitate controller, vertical pre-heating drum and disc. This is similar to the system for producing paper-based TMP, while the former just has a thermofiner and two disc refiner zones (pre-refining and principal refining zone). In the pre-refining zone, the taper in the disc is beneficial for the wood refining before the principal refining zone, while the surface of the disc in the principal refining zone is smooth with a gap of 0.1–0.2 mm between discs. Simultaneously, the cooking of chips is performed at 143–158 °C (at a saturated steam pressure of 0.4–0.6 MPa) for 3–5 minutes, and the heated chips are conveyed to the grinding zone by a screw conveyer. Due to friction, extrusion and curlating, the fibres are defibrillated into single fibres and fibre bundles and, finally, extruded.

4.4.1.2 Chemithermomechanical pulp (CTMP)

CTMP refers to the combination of chemical pretreatment with the TMP process. The application of chemical pretreatment is to soften the fibres, improve the pulp strength and decrease wood fragments. The main procedures of CTMP include: (1) heating the wood chips from the screening and washing stage; (2) impregnating the heated wood chips in cold alkaline sodium sulphite with steam heating to 80–90 °C for 30–60 minutes; (3) further heating the chips to 120–130 °C under pressure for several minutes; (4) refining the chips in two stages; and (5) thickening and bleaching the resultant pulp after latency, screening and separation [63]. Alkaline hydrogen peroxide is used at the bleaching stage, turning the CTMP into BCTMP.

The species of wood, and the soaking and refining processes, are the basic factors that affect CTMP production. Fresh wood chip is a preferred raw material for manufacturing CTMP, as it increases the length and width of the processed fibres and reduces the amount of fibre bundle. Pre-steaming treatment can significantly remove the air and enhance the temperature of wood chips, resulting in an improved penetration and diffusion of the chemical liquor into the chips at the impregnation stage. A screw extruder is used before the impregnating treatment in order to achieve uniform soakage. In practice, the compression ratio should be > 4:1. Pre-impregnation is performed in the sodium sulphite and the addition of sodium hydroxide is necessary for producing hardwood CTMP, where NaOH improves the fibre swelling and fibrillation, thus reducing the refining energy consumption. The brightness of the pulp is improved due to the increasing sulphonation. Simultaneously, the refining process leads to similar effect on the pulp properties of TMP and CTMP. However, CTMP produces higher brightness and bleachability than TMP.

4.4.1.3 Alkaline peroxide mechanical pulp (APMP)

APMP developed from BCTMP. The difference between the two is that the bleaching of BCTMP is carried out in a bleaching tower whilst APMP involves a combined pulping and bleaching process. The chemicals used for APMP are alkali and hydrogen peroxide without sodium sulphite [60]. The preconditioning refiner chemical alkaline peroxide mechanical pulp (P-RC APMP) process can be considered as modified APMP. For APMP, NaOH and H_2O_2 are applied to impregnate the wood chips at the preconditioning stage (P). The chemical treatment works in the first disc or between the first and second discs (RC) for bleaching the wood fibres. With traditional APMP, the whole bleaching process is finished before the refining treatment, whilst for the P-RC APMP it is finished in the refining disc and/or bleaching tower [60].

The typical technological process for APMP is as follows:

Wood chips → washing → steaming → 1#screw squeezer → 1#soaking tube → 1#reaction bin → 2# screw squeezer → 2# soaking tube → 2# reaction bin → 1#high consistency disc → thinning chest → 1#high consistency disc → latency chest → screening and cleaning → thickener → pulp.

In the above process, the screw squeezer treatments are the critical processes, where the compression ratio should be > 4:1 to realise the benefits of improving chemical sorption, saving chemical dosage, decreasing reaction time and increasing pulp brightness.

4.4.2 Properties of mechanical pulp fibre

4.4.2.1 TMP

Compared with traditional mechanical pulp, TMP results in a higher strength of fibres and a lower fibre bundle content due to the preheating treatment. The fraction of long fibres is possible, which reduces the fibre fragment content. Paper made from TMP is characterised by high bulk and coarseness and superior optical properties.

The popular application of TMP is for the manufacture of newsprint. Low freeness is desirable for improving the smoothness and surface strength. In general, about 50% of TMP output is for newsprint, and 20% is for magazine and coating paper, while only about 15% is for paperboard. Based on the high drainability and low fragment content, tissue paper is another promising application of TMP, and 50–60% of facial tissue is produced from high quality TMP. Table 4.13 compares the properties of spruce and poplar TMP.

Table 4.13 Comparison of properties of spruce and poplar TMP.

Raw material	Freeness (mL)	Energy ($kW·h·t^{-1}$)	Bursting index ($kPa·m^2·g^{-1}$)	Breaking length (km)	Tearing index ($mN·m^2·g^{-1}$)	Brightness (%)
Spruce	100	1620	2.36	4.2	8.6	53
Poplar	–	2280	1.2	2.5	3.7	53

The quality of thermomechanical pulp is significantly affected by the species of wood, wood moisture, cooking temperature and disc gap. Within a certain range, an increase in the cooking temperature largely reduces the bonding of fibres, thus resulting in lower mechanical damage from fibre defibrillation and saving fibre inherent morphology, which benefits the production of high-strength fibre composites. It should be noted that a temperature exceeding a certain value can considerably damage the fibre structure, which leads to a decrease in the physical strength of fibres. The molten lignin due to the high temperature can also cover the fibre surface, which results in a vitrification coating of lignin on the fibre surface, with resultant limited fibre fibrillation and low strength of fibre composites. As shown in Table 4.14, the strength of fibre composites increased when the temperature increased from 145 °C to 165 °C, while for a temperature exceeding 175°C strength was impaired. The fibre quality of TMP for fibreboard is characterised by fibre separation, screening distribution and loose coefficient. Fibre separation is an indirect factor that indicates the fibre quality, e.g. higher fibre separation results in higher fibre surface and lower drainability. Commonly, fibre separation is measured with a freeness tester (DS, ml) and a beating degree tester (°SR). Fibre screening distribution investigates the mass percentage of fibre passing through/retained on the specific screen, which is preferable to fibre separation as an indication of fibre quality. Table 4.15 shows the relationship between fibre separation and fibreboard quality.

4.4.2.2 BCTMP

Compared to TMP fibre, CTMP fibre has a longer average length and a lower content of fragments. In addition, CTMP is rougher and stiffer, with greater bleachability and better optical properties.

Table 4.14 Effect of cooking temperature on fibre yield, fibre drainability and fibreboard property.

Cooking temperature (°C)	Fibre yield (%)	Fibre drainability (s)	Fibreboard property	
			Density ($g·cm^{-3}$)	Static bending intensity (MPa)
145	96.2	14.3	0.72	19.3
155	94.1	18.5	0.71	24.5
165	92.7	19.1	0.72	27.6
175	90.3	19.4	0.71	25.2

Table 4.15 Effect of different raw fibres on the property of medium density fibreboard.

Tree species	Fiber drainability (s)	Fibre yield (%)	Screening value (28–200 mesh) (%)	Property of medium density fibreboard: Density ($g \cdot cm^{-3}$)	Static bending intensity (MPa)	Internal bonding strength (MPa)
Masson pine	11.3	88.8	58.7	0.72	21.4	0.3
Spruce	17.6	89.1	59.1	0.72	27.2	0.45
Aspen	18.7	90.1	60.2	0.73	29.3	0.89
Palm [71]	–	85–92	–	0.74	31.6	–
Moso bamboo	10.6	84.5	–	0.76	26.8	0.85

Table 4.16 Properties of Swedish aspen CTMP for newsprint.

	Unscreen	Screen		Unscreen	Screen
Freeness (mL)	98	89	Tensile index ($N \cdot m^2 \cdot g^{-1}$)	29.8	35
Fibre bundle content (%)	0.37	0.23	Dried elongation (%)	1.6	1.6
Wet tensile index ($N \cdot m^{-1}$)	58.1	72.2	Tearing index ($mN \cdot m^2 \cdot g^{-1}$)	3.0	3.9
Wet elongation (%)	6.7	4.8	Brightness (%)	66.6	63.5
Density ($kg \cdot m^{-3}$)		451	Optical coefficient ($m^2 \cdot kg^{-1}$)	56.0	55.0

The properties of CTMP can be widely modified based on the requirements of different products, which may include cultural paper, thinner paper, household paper and paperboard. Newsprint can be produced with 100% CTMP. Table 4.16 shows the properties of poplar CTMP under the following production conditions: 1.3% Na_2SO_3, 1.7% NaOH, 110 °C, 3 minutes, energy consumption 1800–1900 $kW \cdot h\ t^{-1}$.

There are distinct differences for CTMP from different trees. Table 4.17 shows the properties of four types of hardwood CTMP [72]. The production conditions are as follows: chemical pretreatment of 10% Na_2SO_3, 2% NaOH, 120°C, 120 minutes, liquor ratio 1:3; bleaching treatment of 5% pulp consistency, 1.5% H_2O_2, 0.1% EDTA, 3% Na_2SO_3, 0.5% NaOH, 60°C, 90 minutes.

The sufficient fibrillation derived from the refining treatment resulted in large ribbon fibre of CTMP, which can be used for the manufacture of paper with high strength and opacity and superior surface smoothness, even partly replacing the chemical pulp for printing paper. Additionally, high-strength CTMP can be used to produce low-density coating paper.

4.4.2.3 APMP

APMP consumes lower chemicals and energy than BCTMP and possesses higher bursting and tensile strength at the same freeness as BCTMP. Furthermore, compared to BCTMP, the same dosage of H_2O_2 results in greater brightness, opacity and a higher

Table 4.17 Properties of hardwood BCTMP.

Hardwood	*Sapium discolor* muell.	*Michelia macclurei*	*Alniphyllum fortunei*	*Castanopsis carlesii*
Yield (%)	68.34	73.46	83.56	80.50
Beating degree (°SR)	60.5	60	61.5	56.0
Basis weight ($g \cdot m^{-2}$)	59.1	82.0	65.5	81.5
Tensile index ($N \cdot m^2 \cdot g^{-1}$)	32.5	20.3	26.6	25.9
Tearing index ($mN \cdot m^2 \cdot g^{-1}$)	4.81	2.20	4.25	3.24
Burst index ($kPa \cdot m^2 \cdot g^{-1}$)	1.70	0.83	1.21	1.23
Brightness (%ISO) (Control pulp)	31.57	58.43	65.43	53.30
Brightness (%ISO) (Bleached pulp)	50.57	59.13	75.60	64.07

scattering coefficient. The cost of producing APMP is lower as well. For softwood, APMP and BCTMP have similar chemical and energy consumption, pulp yield and strength.

Other non-wood raw materials can also be applied to produce APMP, such as bamboo, cotton stalk and rice stalk. There are abundant bamboo resources in southern China. The chemical composites and fibre morphology of bamboo lie between hardwood and softwood. Bamboo has great potential for APMP production. Under optimal production conditions, the properties of modified green bamboo APMP are as follows: yield 74.24%, brightness 62.4%, break length 3.15 km, tearing index 5.83 $mN \cdot m^2 \cdot g^{-1}$ [73].

4.5 Conclusion and outlook

Plant single fibres and fibre bundles can be obtained through different processes. Compared with traditional processes, chemical processes provide high efficiency and high selectivity of delignification, lower energy consumption, lower environmental contamination, such as rapid displacement heating (RDH), displacement digester system (DDS), modified continuous cooking (MCC) and extended modified continuous cooking (EMCC); the mechanical process would develop with higher yield, higher brightness and strength, lower energy consumption and environmental contamination – for example, biomechanical pulp, preconditioning refiner chemical APMP. The properties of fibres can be correspondingly further improved with superior processes.

The application of plant fibres would be broadened with improved fibre properties. The fibre can be used for the production of many traditional materials, such as packaging materials, printing materials, building materials, industrial materials and other technical materials. The fibres could be further applied in the production of high-grade precision and advanced industrial products – for example, transparent fibre, photoelectric materials, ultra-light materials used in aircraft industry, and high-strength materials used by the military and in the drilling industry.

References

[1] Bledzki, A. K. and Gassan, J. (1996). Effect of coupling agents on the moisture absorption of natural fibre-reinforced plastics. *Angew. Makromol. Chem.*, **236**, 129-38.

[2] Bledzki, A. K. and Gassan, J. (1999). Composites reinforced with cellulose based fibres. *Prog. Polym. Sci.*, **24**, 221–74.

[3] Keller, A., Leupin, M., Mediavilla, V. and Wintermantel, E. (2001). Influence of the growth stage of industrial hemp on chemical and physical properties of the fibres. *Industrial Crops and Products*, **13**, 35–48.

[4] Mediavilla, V., Leupin, M. and Keller, A. (2001). Influence of the growth stage of industrial hemp on the yield formation in relation to certain fibre quality traits. *Industrial Crops and Products*, **13**, 49–56.

[5] Lakkad, S. C. and Patel, J. M. (1981). Mechanical properties of bamboo, a natural composite. *Fibre Science and Technology*, **14**, 319–22.

[6] Jain, S., Kumar, R. and Jindal, U. C. (1992). Mechanical behaviour of bamboo and bamboo composite. *J Mater Sci*, **27**, 4598–604.

[7] Ratna Prasad, A. V. and Mohana Rao, K. (2011). Mechanical properties of natural fibre reinforced polyester composites: Jowar, sisal and bamboo. *Mater. Des.*, **32**, 4658–63.

[8] Sridhar, M. K., Basavarajappa, G., Kasturi, S. G. and Balasubramanian, N. (1982). Evaluation of jute as a reinforcement in composites. *Indian J Fibre Text Res*, **7**, 87–92.

[9] Mohanty, A. K., Misra, M. and Hinrichsen, G. (2000). Biofibres, biodegradable polymers and biocomposites: An overview. *Macromolecular Materials and Engineering*, **276-277**, 1–24.

[10] Charlet, K., Jernot, J.-P., Breard, J. and Gomina, M. (2010). Scattering of morphological and mechanical properties of flax fibres. *Industrial Crops and Products*, **32**, 220–24.

[11] Wambua, P., Ivens, J. and Verpoest, I. (2003). Natural fibres: Can they replace glass in fibre reinforced plastics? *Composites Science and Technology*, **63**, 1259–64.

[12] Korte, S. (2006). Processing-Property Relationships of Hemp Fibre. Master of Engineering thesis, University of Canterbury, New Zealand.

[13] Hughes, J. M. (2000). On the mechanical properties of bast fibre reinforced thermosetting polymer matrix composites

[14] Liu, W., Drzal, L. T., Mohanty, A. K. and Misra, M. (2007). Influence of processing methods and fiber length on physical properties of kenaf fiber reinforced soy based biocomposites. *Composites Part B: Engineering*, **38**, 352–59.

[15] Ochi, S. (2008). Mechanical properties of kenaf fibers and kenaf/PLA composites. *Mechanics of Materials*, **40**, 446–52.

[16] Du, Y. (2010). An applied investigation of kenaf-based fiber/polymer composites as potential lightweight materials for automotive components. Ph.D. thesis, Mississippi State University.

[17] Elsaid, A., Dawood, M., Seracino, R. and Bobko, C. (2011). Mechanical properties of kenaf fiber reinforced concrete. *Construction and Building Materials*, **25**, 1991–2001.

[18] Mohanty, A. K., Misra, M. and Drzal, L. T. (2005). Plant fibers as reinforcement for green composites, In: A. Bismarck, S. Mishra and T. Lampke (eds), *Natural Fibers, Biopolymers, and Biocomposites*, CRC Press (Taylor & Francis).

[19] Mohanty, A. K., Misra, M. and Drzal, L. T. (2005). *Natural Fibers, Biopolymers, and Biocomposites*, CRC Press (Taylor & Francis).

[20] Merlini, C., Soldi, V. and Barra, G. M. O. (2011). Influence of fiber surface treatment and length on physico-chemical properties of short random banana fiber-reinforced castor oil polyurethane composites. *Polymer Testing*, **30**, 833–40.

[21] Arib, R. M. N., Sapuan, S. M., Ahmad, M., Paridah, M. T. and Zaman, H. (2006). Mechanical properties of pineapple leaf fibre reinforced polypropylene composites. *Materials & Design*, **27**, 391–96.

[22] Satyanarayana, K. G., Sukumaran, K., Mukherjee, P. S., Pavithran, C. and Pillai, S. G. K. (1990). Natural fibre-polymer composites. *Cem. Concr. Compos.*, **12**, 117–36.

[23] Hill, C. A. S. and Khalil, H. (2000). The effect of environmental exposure upon the mechanical properties of coir or oil palm fiber reinforced composites. *J. Appl. Polym. Sci.*, **77**, 1322–30.

[24] Shinoj, S., Visvanathan, R., Panigrahi, S. and Kochubabu, M. (2011). Oil palm fiber (OPF) and its composites: A review. *Industrial Crops and Products*, **33**, 7–22.

[25] Neagu, R., Gamstedt, E. and Lindström, M. (2006). Characterization methods for elastic properties of wood fibers from mats for composite materials. *Wood and Fiber Science*, **38**, 95–111.

[26] Neagu, R. C., Gamstedt, E. K. and Berthold, F. (2006). Stiffness contribution of various wood fibers to composite materials. *J. Compos. Mater.*, **40**, 663–99.

[27] Hearle, J. W. S. and Stevenson, P. J. (1963). Nonwoven fabric studies, Part III: The anisotropy of nonwoven fabrics. *Textile Research Journal*, **33**, 877–88.

[28] Rebenfeld, L. (1965). Morphological foundations of fiber properties. *Journal of Polymer Science Part C: Polymer Symposia*, **9**, 91–112.

[29] Cowdrey, D. R. and Preston, R. D. (1966). Elasticity and microfibrillar angle in the wood of Sitka spruce. *Proceedings of the Royal Society of London. Series B. Biological Sciences*, **166**, 245–72.

[30] Page, D. H., El-Hosseiny, F., Winkler, K. and Bain, R. (1972). The mechanical properties of single wood-pulp fibres. Part I: A new approach. *Pulp and Paper Magazine of Canada*, **73**, 72–77.

[31] Gordon, J. E. and Jeronimidis, G. (1974). Work of fracture of natural cellulose. *Nature*, **252**, 116.

[32] Treloar, L. R. G. (1977). Physics of textiles. *Physics Today*, **30**, 23–30.

[33] McLaughlin, E. C. and Tait, R. A. (1980). Fracture mechanism of plant fibres. *Journal of Materials Science*, **15**, 89–95.

[34] Thygesen, A. (2006). Properties of hemp fibre polymer composites: An optimisation of fibre properties using novel defibration methods and fibre characterisation. PhD thesis, Risø National Laboratory, Roskilde, Denmark.

[35] He, J., Tang, Y. and Wang, S. (2007). Differences in morphological characteristics of bamboo fibres and other natural cellulose fibres: Studies on X-ray diffraction, solid state ^{13}C-CP/MAS NMR, and second derivative FTIR spectroscopy data. *Iranian Polymer Journal*, **16**, 807–18.

[36] Bogoeva-Gaceva, G., Avella, M., Malinconico, M., Buzarovska, A., Grozdanov, A., Gentile, G. and Errico, M. E. (2007). Natural fiber eco-composites. *Polym. Compos.*, **28**, 98–107.

[37] Le Troedec, M., Sedan, D., Peyratout, C., Bonnet, J. P., Smith, A., Guinebretiere, R., Gloaguen, V. and Krausz, P. (2008). Influence of various chemical treatments on the composition and structure of hemp fibres. *Composites Part A: Applied Science and Manufacturing*, **39**, 514–22.

[38] Ouajai, S. and Shanks, R. A. (2005). Composition, structure and thermal degradation of hemp cellulose after chemical treatments. *Polym. Degrad. Stab.*, **89**, 327–35.

[39] Li, Y. and Pickering, K. L. (2008). Hemp fibre reinforced composites using chelator and enzyme treatments. *Composites Science and Technology*, **68**, 3293–98.

[40] Razera, I. A. T. and Frollini, E. (2004). Composites based on jute fibers and phenolic matrices: Properties of fibers and composites. *J. Appl. Polym. Sci.*, **91**, 1077–85.

[41] Saha, P., Manna, S., Chowdhury, S. R., Sen, R., Roy, D. and Adhikari, B. (2010). Enhancement of tensile strength of lignocellulosic jute fibers by alkali-steam treatment. *Bioresource Technology*, **101**, 3182–87.
[42] Öztürk, İ., Irmak, S., Hesenov, A. and Erbatur, O. (2010). Hydrolysis of kenaf (*Hibiscus cannabinus L.*) stems by catalytical thermal treatment in subcritical water. *Biomass Bioenergy*, **34**, 1578–85.
[43] Kumar, R., Choudhary, V., Mishra, S. and Varma, I. (2008). Banana fiber-reinforced biodegradable soy protein composites. *Frontiers of Chemistry in China*, **3**, 243–50.
[44] Venkateshwaran, N. and Elayaperumal, A. (2010). Banana fiber reinforced polymer composites: A review. *J. Reinf. Plast. Compos.*, **29**, 2387–96.
[45] Chand, N., Tiwary, R. K. and Rohatgi, P. K. (1988). Bibliography Resource structure properties of natural cellulosic fibres: An annotated bibliography. *Journal of Materials Science*, **23**, 381–87.
[46] de Paiva, J. M. F. and Frollini, E. (2006). Unmodified and modified surface sisal fibers as reinforcement of phenolic and lignophenolic matrices composites: Thermal analyses of fibers and composites. *Macromolecular Materials and Engineering*, **291**, 405–17.
[47] Mukherjee, P. S. and Satyanarayana, K. G. (1986). An empirical evaluation of structure-property relationships in natural fibres and their fracture behaviour. *J Mater Sci*, **21**, 4162–68.
[48] Nilsson, T. and Gustafsson, P. J. (2007). Influence of dislocations and plasticity on the tensile behaviour of flax and hemp fibres. *Composites Part A: Applied Science and Manufacturing*, **38**, 1722–28.
[49] Dai, D. and Fan, M. (2010). Characteristic and performance of elementary hemp fibre. *Materials Sciences and Applications*, **1**, 336–42.
[50] Baley, C. (2004). Influence of kink bands on the tensile strength of flax fibers. *Journal of Materials Science*, **39**, 331–34.
[51] Satyanarayana, K. G., Sukumaran, K., Mukherjee, P. S. and Pillai, S. G. K. (1986). Materials science of some lignocellulosic fibers. *Metallography*, **19**, 389–400.
[52] Dai, D. and Fan, M. (2011). Investigation of the dislocation of natural fibres by Fourier-transform infrared spectroscopy. *Vibrational Spectroscopy*, **55**, 300–6.
[53] Stamboulis, A., Baillie, C. A. and Peijs, T. (2001). Effects of environmental conditions on mechanical and physical properties of flax fibers. *Composites Part A: Applied Science and Manufacturing*, **32**, 1105–15.
[54] Bismarck, A., Aranberri-Askargorta, I., Springer, J., Lampke, T., Wielage, B., Stamboulis, A., Shenderovich, I. and Limbach, H. H. (2002). Surface characterization of flax, hemp and cellulose fibers; Surface properties and the water uptake behavior. *Polym. Compos.*, **23**, 872–94.
[55] Arbelaiz, A., Cantero, G., Fernandez, B., Mondragon, I., Ganan, P. and Kenny, J. M. (2005). Flax fiber surface modifications: Effects on fiber physico mechanical and flax/polypropylene interface properties. *Polym. Compos.*, **26**, 324–32.
[56] Baltazar-y-Jimenez, A. and Bismarck, A. (2007). Wetting behaviour, moisture up-take and electrokinetic properties of lignocellulosic fibres. *Cellulose*, **14**, 115–27.
[57] Park, J.-M., Quang, S. T., Hwang, B.-S. and DeVries, K. L. (2006). Interfacial evaluation of modified jute and hemp fibers/polypropylene (PP)-maleic anhydride polypropylene copolymers (PP-MAPP) composites using micromechanical technique and nondestructive acoustic emission. *Composites Science and Technology*, **66**, 2686–99.
[58] Cordeiro, N., Gouveia, C., Moraes, A. G. O. and Amico, S. C. (2011). Natural fibers characterization by inverse gas chromatography. *Carbohydrate Polymers*, **84**, 110–17.
[59] Bismarck, A., Mohanty, A. K., Aranberri-Askargorta, I., Czapla, S., Misra, M., Hinrichsen, G. and Springer, J. (2001). Surface characterization of natural fibers; surface properties and the water up-take behavior of modified sisal and coir fibers. *Green Chemistry*, **3**, 100–7.

[60] Zhang, H., Li, Z., Ni, Y. and Heiningen, V. (2012). *Fundamentals of Pulp Production*, Chemical Industry Press,

[61] Zhang, H. (2014). *Pulping Principles and Engineering*, China Light Industry Press,

[62] Smook, G. (1996). *Handbook for Pulp & Paper Technologists,* Tappi Press, Peachtree Corners, GA.

[63] Industry, C. T. A. o. P. (2011). *2011 China Papermaking Statistical Yearbook*,

[64] Casey, J. P. (1980). *Pulp and Paper: Chemistry and Chemical Technology*, Wiley-Interscience Publication,Hoboken, NJ.

[65] Gullichsen, J. and Fogelholm, C.-J. (2000). *Chemical Pulping*, Gummerus Printing, Helsinki, Finland.

[66] Dence, C. W. and Reeve, D. W. (1996). *Pulp Bleaching: Principles and Practice*, Tappi Press.

[67] Shaofan, N., Jinguo, L., Liping, B., Lihui, C. and Sizu, L. (1998). Evaluation of pulping properties of Chinese fir wood from plantation at different ages. *Journal of Fujian College of Forestry*, **18**, 87–91.

[68] Shilin, C., Zhan Huaiyu, Lihui, C. and Youhe, H. (2006). Comparison between extended delignification cooking and conventional kraft cooking of bamboo. *Chemistry and Industry of Forest Products*, **26**, 65–68.

[69] Liulian, H., Lihui, C., Jianchun, Z. and Junwen, P. (2006). ECF bleaching of bamboo kraft pulp. *Shanghai Papermaking*, **37**, 10–16.

[70] Huang, L. and Lihui, C. (2010). Study on bleaching properties of Pinus masson thermomechanical pulp. *Journal of Fujian College of Foretry*, **30**, 212–16.

[71] Qinzhi, Z., Daiquan, P., Ruoxuan, X., Jiuping, R. and Chunchu, W. (1998). Study on using oil palm (*Elaeis guineensis*) fiber to develop medium density fiberboard. *Journal of Fujian College of Forestry*, **18**, 170–73.

[72] Liuliang, H. (2000). Study on the CTMP properties of four kinds of broad-leaved trees. *Journal of Fujian College of Forestry*, **20**, 97–100.

[73] Yongwen, Z. (2005). Study on *Dendrocalamopsis oldhami* improved alkaline peroxide mechanical ulping. Masters dissertation, FuJian Agriculture And Forestry University.

Chapter 5
Nanocellulose: The next world-changing super material

Dr Dasong Dai

This chapter provides a systematic outline of nanocellulose that demonstrates its superior merits throughout the supply chain. The chapter starts with its abundant resources and the various processing technologies that have been or are being developed, including chemical, mechanical and hybrid systems. The chapter then presents a comprehensive characterisation and resulting properties of morphology, geometry, mechanical behaviour. There follows a discussion of the numerous novel applications of nanocellulose, including its use without modification for nanopaper, aerogel template and nanocomposites, and, with modification, for stabiliser and hydrophobic nanocomposites. Finally, the chapter offers an outlook on the further development of nanocellulose-based materials for its many industrial applications.

5.1 Introduction

Cellulose, extracted from plant fibre, is defined as a macromolecule, a nonbranched chain of variable length of 1-4-linked β-D-anhydroglucopyranose units. This natural polymer accompanies the hominisation of humans and has served humans for more than 10,000 years. The development of the textile industry in 19th century and the coming of the industrial revolution provided motivation for the use of cellulose in more industrial sectors. In addition this also gave rise to the formation of new discipline of cellulose. According to statistics from Web of Science, between 1900 and 2016, some 11,829 papers about cellulose were published, 75% of which were published in four main research areas: chemistry, materials science, polymer science and engineering. This wealth of research has steadily revealed more details of the nature and properties of cellulose, leading to it being used in more and more areas to meet the needs of humanity.

In the ten years covering 2007–2016, one branch of cellulose research, namely nanocellulose (NC), which is mainly extracted from plant fibre, is attracting more and more attention due to the material's outstanding properties. As shown in Fig. 5.1, research into NC within the field of cellulose research grew from 2.78% to 13.14% in the period 2007–16. This promising nanomaterial is stronger than steel and stiffer than Kevlar with lightweight. Most importantly, the potential market for the material is very large – estimated at more than 340 million tonnes consumption in the future. The history of NC research has not been a linear development. Its Boltzmann's development (Fig. 5.2) goes through four stages over the last

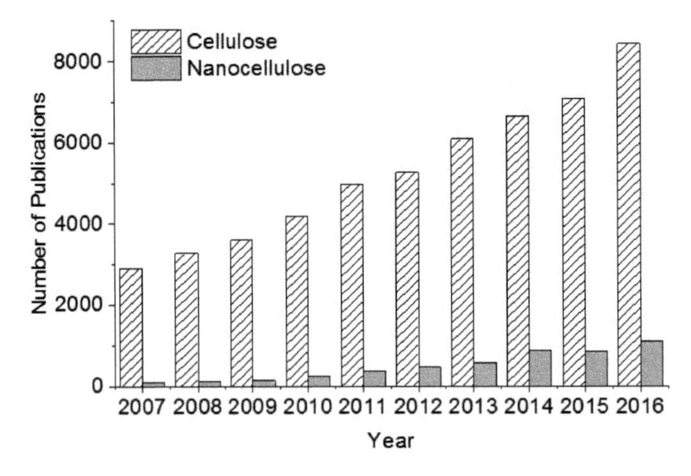

Figure 5.1 Comparison of cellulose and nanocellulose research, 2007–16.

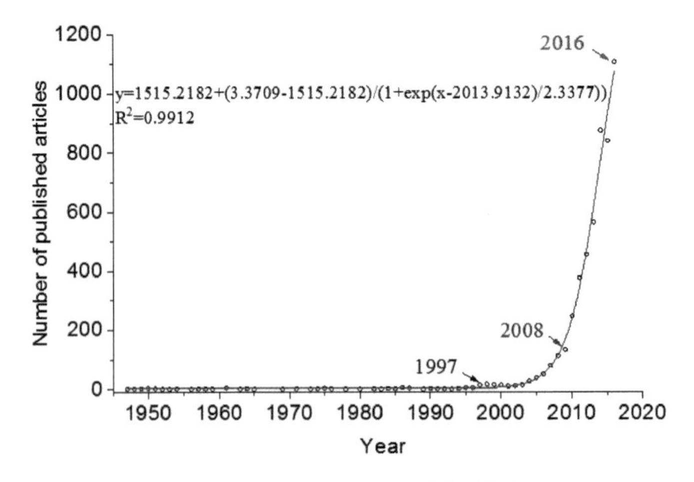

Figure 5.2 Annual publications about nanocellulose, 1947–2016.

70 years. In the first 50 years, namely the first stage, the number of articles published about NC annually was fewer than 10; in the second stage (from 1997 to 2007), the number of articles published annually grew from 11 to 81, showing rapid growth; in the third stage (from 2008 to 2015), the number of articles published annually grew sharply from 114 to 840; and in 2016, the number of articles published annually reached a figure of 1,107, a milestone indicating how NC research has become a major discipline within in cellulose science.

In 1947, Nickerson and Habrle [1] first employed hydrochloric and sulphuric acid hydrolysis to produce NC (cellulose sols) from cellulose materials. Two years later, inspired by Nickerson, Rånby [2] developed acid hydrolysis with pH control. He and his co-workers [3–6] then determined the size and crystal structure of NC using SEM and XRD respectively, and found that after acid hydrolysis the size of cellulose could be reduced to the nanoscale without changing the crystalline structure of cellulose. In 1953, Mukherjee and Woods [7] used milder acid hydrolysis to fabricate NC from various cellulose fibres (e.g. ramie, cotton and jute), and found that the nanoscale of cellulose could be produced in milder conditions. These investigations have led to noteworthy advances in our knowledge of the texture of cellulose at the nanoscale. However, research on NC then lost its way between 1953 and 1982 due to the lack of awareness about its novel properties. In 1982 and 1983, researchers [8–10] from Japan reported upon their systematic investigations into NC. A new form of NC was fabricated by using a high-pressure homogeniser and they found that this novel material had high water retention value [10], high chemical accessibility [9]. In addition, they also discussed the application of NC to food, paint, medical industrials and so on. These encouraging results set NC research back on course and researchers began to pay attention to this new form of cellulose. In 1987, Boldizar and colleagues [11] first employed NC as a filler to reinforce thermoplastics: the promising results showed that this nanomaterial had significant potential as a filler of thermoplastics. Unfortunately, their results did not stimulate a response from other researchers.

The remarkable work of Favier and his co-workers [12, 13] in 1995 marks a turning point in NC research; the excellent improvement in the mechanical properties of nanomaterials demonstrated the huge potential of NC in human life and took this research to a new stage. In the new century, greater funding was, e.g. Nanoforest, Nanocell and Horizon, provided from governments, which led to an explosive growth period in NC research and opened many more windows for the application of NC.

5.2 Source materials of NC from plant fibre

NC can be obtained from various sources that contain cellulose, e.g. plant fibres, bacterial cellulose and animal cellulose (Fig. 5.3). According to our statistics, of the reported publications more than 90% of them employ plant fibres as raw materials to fabricate NC due to their wide variety and distribution worldwide. The plant fibres mainly come from wood fibres, stem fibres (e.g. bamboo, flax, kenaf fibres), straw fibres, leaf fibres (e.g. banana, sisal fibres), seed fibres (e.g. cotton fibres), fruit fibres (e.g. coir fibres). Many different terminologies have been used in the last 70 years in the processing of NC, and more and more sources are used, Researchers have found that the original source is one of the main parameters that affect the morphology [14] and other properties [15] of NC.

5.2.1 Terminology of NC

NC is a general term that represents the nanoscale cellulosic material from cellulose-based sources [16]. Many terms have been used in the last 70 years, often inconsistently. These may include: (1) cellulose microfibril(s) (2) micro(-)fibrillar cellulose (3) micro(-)fibrillated cellulose (4) cellulose nano(-)fibril(s) (5) nano(-) fibrillar cellulose (6) nano(-)fibrillated cellulose (7) cellulose nano(-)crystal(s) (8) nanocrystalline cellulose (9) cellulose whisker (10) cellulose nano(-)whisker (11) nano(-)cellulose(s) (12) cellulose nano(-)fibre(s) (13) cellulose nano(-)fiber(s), and (14) cellulose nanoparticle(s).

Figure 5.3 Sources of nanocellulose: (a) plant cellulose; (b) animal cellulose; and (c) bacterial cellulose.

These terminologies make it very difficult for researchers to get a clear image of the research on cellulose nanomaterials. In the last 30 years, the various terminologies of NC have appeared at different periods, some becoming more and more popular, as shown in Fig. 5.4. The more popular terms are 'cellulose nanocrystal' and 'nanocellulose'. Some terms have passed their peak in popularity (e.g. cellulose microfibril). However, in the very early stages (before 1984), no exact term was used to describe this special nanomaterial (see Fig. 5.4). The special properties of NC have been subsumed within the field of cellulose. Nevertheless, scientists had made the wise assumption that cellulose and cellulose derivatives are generally dissolved as micelles. In the 1940s, scientists set out to prove this assumption. Nickerson and his co-workers [1, 17–22], Conrad and Scroggie [23], Battista and Coppick [24], Hermans and Weidinger [25] all employed acid hydrolysis to degrade cellulose with the aim of revealing the fine structure of cellulose, but they could only find glucose after acid hydrolysis. No exciting findings were reported until the end of 1940s. By using acid

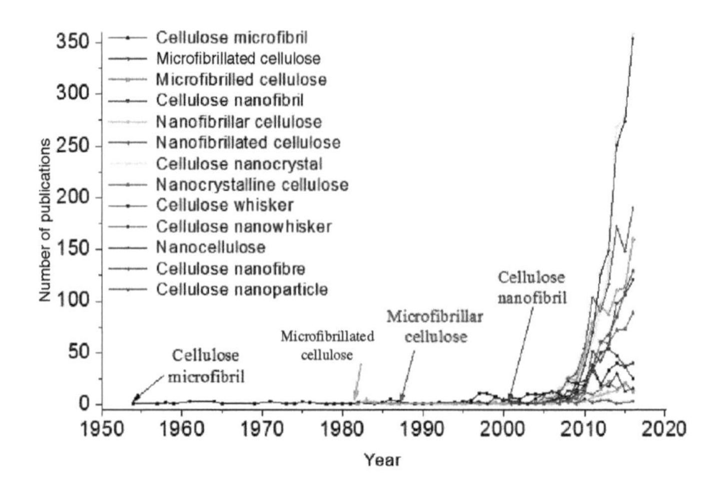

Figure 5.4 Publications about nanocellulose with various terminologies, 1954–2016.

hydrolysis in a little mild condition, Rånby and Bengt [2] first found 'rodlike particles'; unfortunately this discovery was soon forgotten within research on cellulose. In the 1980s, the term 'microfibrillated cellulose' became popular. Scientists from Japan [8–10] from Japan found that microfibrillated cellulose obtained from cellulose by the use of a refining process displayed very special physical and chemical properties, when compared with cellulose. In addition, they also found useful applications for microfibrillated cellulose in various industries, especially in the food industry. In the 1990s, the term 'cellulose whisker' became highly popular due to the discovery of its reinforcing effect on polymers [12, 13]. From the 2000s, various terms appeared as the boom in NC research continued. Researchers and governments around the world believed that this novel material would herald a bright future for humanity. Among the terminologies for NC, the terms 'nanocellulose' and 'cellulose nanocrystal(s)' have become increasingly popular in the last five years.

Some researchers define or classify these terminologies by the size of the cellulose materials [26]. Currently, standard terms and a definition of NC are under development by the ISO according to the length, width, aspect ratio, production methods and crystal structure (see Table 5.1). However, these standard terms seem inadequate for a full description of scope of terms of NC, e.g. it does not make a definition for cellulose nanowhisker.

5.2.2 NC sources

NC can be extracted from various cellulose resources. Until now, about 80 kinds of natural material are exploited for NC production, according to the published articles. Among these natural materials, plant fibres play a dominant role for the production of NC. As shown in Fig. 5.5, 84.51% of reported articles employed plant cellulose as the raw material in the production of NC. These plant fibres came mainly from: (1) wood fibre (eucalyptus [27–29], birch [30–32]); (2) seed fibre (cotton [33–36], kapok [37, 38]); (3) industrial product of cellulose fibre (Lyocell [39–41], microcrystalline

Table 5.1 Comparison of nanocellulose terms.

Terms	Size and aspect ratio	Crystal structure	Production method	Synonymous terms
Cellulose microfibril	Length 0.5–10 μm Width 10–100 nm	Contains both crystalline and amorphous regions	Mechanical treatment	Microfibrillar cellulose, Microfibrillated cellulose
Cellulose nanofibril	Width 10–100 nm Aspect ratio > 50	Contains both crystalline and amorphous regions	Mechanical treatment	Nanofibrillar cellulose, nanofibrillated cellulose
Cellulose nanocrystal	Width 3–10 nm 5<Aspect ratio <50	Contains pure crystalline structure	Chemical treatment	Nanocrystalline cellulose

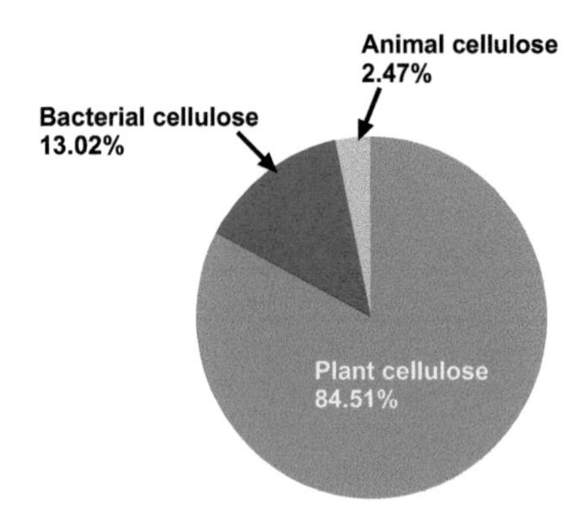

Figure 5.5 Proportion of resources for nanocellulose exploitation.

cellulose [42–44]); (4) Agro-waste (straw [45–47], shells [48–50], stalk [36, 51–53]); (5) industrial waste (waste paper [54–56], pulp residue []57–59], bagasse [60–62]); (6) stem fibre (bamboo [63–65], flax [66–68], hemp [69–71]); (7) leaf fibre [72–75]; (8) algae [76–78]; and (9) fruit fibre [79–81].

Wood fibre is the favourite material and 32.65% of reported articles employed wood fibre as the raw material (see Fig. 5.6). Seed fibre, especially cotton fibre, has the second highest proportion (up to 25.54%). These two kinds of fibre dominate the main resources, which reflects the fact that these two kinds of fibre can be obtained easily. Comfortingly, waste cellulose from agriculture and the industrial field is becoming more attractive to researchers, until now, 15.66% of reported articles describe research using the waste cellulose derived NC.

Due to the different chemical composition, various cell sizes and the structure of different raw materials, the final nanocellulose products present significant differences

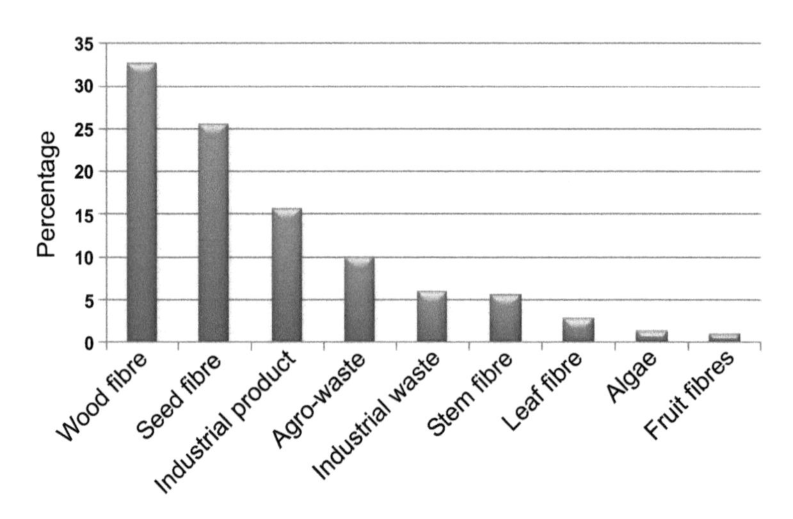

Figure 5.6 Proportion of plant-derived resources for nanocellulose exploitation

Table 5.2 Microstructure properties of nanocellulose from various sources.

Source		Length (nm)	Width (nm)	Aspect ratio	References
Wood fibre	Hardwood	26–129	1–3	29–30	[85]
	Softwood	97–331	1+-7	90–100	[86–88]
Seed fibre	Cotton	300–568	4–16	15–82	[89–91]
Industrial product	Lyocell	67.8	3.56–350	–	[92–94]
	MCC	136–769	3.9–33	14–31	[95–97]
Agro-waste	Wheat straw	84–800	4–25	7.5–20	[46, 98, 99]
	Rice straw	100–1000	1.8–6.9	20–30	[100]
	Bark	33–194	6–12	3.7–24.2	[101]
Industrial waste	Waste paper	5000	50	100	[102]
	Bagasse	143–353.4	3.5–16.6	15–70	[103–105]
Stem fibre	Bamboo	3–200	3–10	1–28	[88, 106]
	Flax	21–435	2–50	15.5	[107, 108]
	Kenaf	100–1400	7–84	10–50	[109, 110]
Leaf fibre	Banana	200–625	4–19	10–74	[111–113]
	Sisal	117–200	4–17	13–36	[88, 114, 115]
Algae		395–571	15.6–24.4	18.3–32.1	[76, 116]
Fruit fibre	Coir	40–90	11–19	4–5	[38]
	Oil palm empty fruit bunch	82.8–280	3.6–6.1	23–47	[80, 117–119]
Tunicate		400–2000	10–30	72–100	[120–123]
Bacterial cellulose		300	10–88	30–47	[124–126]

in their properties (Table 5.2), such as their geometries [26] and physical properties [82]. It is known that NC obtained from primary cell wall fibres is longer and thinner than that obtained from secondary cell wall fibres [83]. Table 5.2 shows: (1) the resources from tunicate and softwood fibre lead to higher aspect ratio NC; (2) resources from tunicate result in larger dimension NC; and (3) coir fibre is suitable for producing particle-size NC due to its shorter length. In addition, the various chemical compositions demand a variety of extraction processes to produce NC, e.g. non-wood plants generally contain less lignin than wood and therefore bleaching processes are less demanding [84].

5.3 Fabrication of NC

To some extent, plant fibre can be seen as a composite, which consists of cellulose, hemicellulose and lignin. As shown in Fig. 5.7, the NC fabrication process is a procedure to (1) remove the other chemical compositions, such as hemicellulose and lignin, with the aim of extracting cellulose; and (2) degrade cellulose to the nanoscale by using appropriate methods [127]. Prior to the degradation step, pretreatments such as alkali treatment and bleaching are required. The main goal of pretreatment is to remove certain amounts of lignin, hemicellulose, wax and oils, which cover the external surface of the fibre cell wall [82]. Alkali pretreatment [128] can also swell [129] and depolymerise the native cellulose structure, defibrillate the external cellulose fibrils and expose short-length crystalline cellulose [130]. Furthermore, bleaching treatment is required in order to remove the remaining lignin from the fibre [130]. According to the two-phase theory, cellulose contains amorphous and crystalline regions. It is relatively easy to degrade the amorphous phase of cellulose by mechanical, chemical or enzymatic methods; by contrast, it is quite difficult to break the crystalline region of cellulose due to its high crystalline degree.

Generally, the main methods to fabricate NC include: (1) the mechanical method; (2) the chemical method; and (3) the biological method. As aforementioned, the mechanical method gives rise to nanofibrillated cellulose, the chemical method produces cellulose nanocrystal, and the biological method is the main way to fabricate BC-based NC. The size, crystallinity index (CI) and yield of NC depends mainly on the methods of extraction. As shown in Table 5.3, the chemical method favours high CI of NC, while mechanical method favours of high yield of NC. For plant-derived cellulose, the biological treatment is always used as an assistant method of pretreatment for mechanical or chemical treatment; this assisting step consumes a lot of time but with less waste. These methods have similar mechanisms of extraction, however, namely (1) attack the defect or non-crystalline region of cellulose with chemical agents, fungus or mechanical force; and (2) the degradation of cellulose in the attack region.

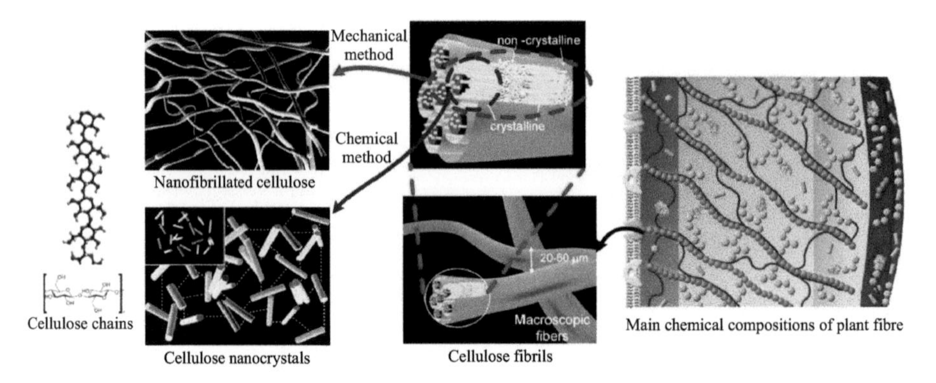

Figure 5.7 Fabrication of nanocellulose: from macroscale to nanoscale.

Table 5.3 Influence of source and fabrication method on final dimension, crystallinity index (CI) and yield of nanocellulose.

Source	Process	Dimension (nm)	CI (%)	Yield (%)	References
Eucalyptus (hardwood)	H_2SO_4 64%, 45 °C, 20 mins	5–10 (D), 100–300 (D)	86	29.58	[131]
	H_2SO_4 65%, 50 °C, 45 mins	7 (D), 100 (L)	89	17	[88]
	Pretreatment: ball mill; Sulphuric acid treatment: H_2SO_4 60%, 45 °C, 50 mins	15.1 (D), 240 (L)	56	64	[132]
	Pretreatment: ball mill; Mercerised treatment: NaOH 20%, 25 °C, 5 hours, Sulphuric acid treatment: H_2SO_4 60%, 45 °C, 50 mins	18.9 (D), 132 (L)	68	55	[132]
	Pretreatment: Cellulases, 50 °C, 48 hours; Mechanical treatment: domestic blender, 20,000 rpm, 30 secs	40–80 (D)	50	63	[133]
	Stirring 30 mins, 2000 rpm; Ultra-fine grinder 2507 rpm, 200 passes	18 (D)	–	60.3	[134]
	Pretreatment: TEMPO oxidation; Mechanical: high pressure homogeniser, 1000 bar	–	96	92	[135]
	Masuko MKCA6-2 grinder, 2500 rpm, 4 hours	70–90 (D)	78.3	–	[136]
Pine (softwood)	H_2SO_4 64%, 45 °C, 40 mins	3.4 (D), 166.7 (L)	93	45	[137]
	Stirring 30 mins, 2000 rpm; Ultra-fine grinder 2507 rpm, 200 passes	13 (D)	–	57	[134]
	Pretreatment: TEMPO oxidation; Mechanical: high pressure homogeniser, 1000 bar	–	88	86	[135]

(Continued)

Table 5.3 Influence of source and fabrication method on final dimension, crystallinity index (CI) and yield of nanocellulose.

Source	Process	Dimension (nm)	CI (%)	Yield (%)	References
Cotton fibre	H_2SO_4 65%, 63 °C, 30 mins	130 (D)	76	–	[138]
	H_2SO_4 64%, 45 °C, 30 mins	40 (D), 670 (L)	91.6	54.1	[139]
	H_2SO_4 60%, 45 °C, 3 hours	8 (D), 218 (L)	71.5	62.3	[140]
	Sulphuric acid pretreatment: H_2SO_4 64%, 45 °C, 1.5 hours; Ultrasonic treatment: 60–70 °C, 2 hours	20–30 (D), 200–300 (L)	80.3	70	[141
	Methanesulphonic acid, 3%; Ultrasonic processor, 10 mins	12.7 (D), 242 (L)	80	74.5	[142]
	Mixed acid (H_2SO_4 (98%), HCl (37%, 1/11 mol/mol); Ultrasonic processor, 55 °C, 7 hours	3–35 (D), 28–470 (L)	55.8	46.7	[143]
	Microfibrillation, 10-inch Masuko Supermasscolloider, 9 passes	5.9 (D), 150 (L)	88.2	47	[144]
Algae	H_2SO_4 64%, 45 °C, 45 mins	21.8 (D), 547.3 (L)	73	52.1	[145]
	H_2SO_4 64%, 50 °C, 50 mins	28.6 (D), 432 (L)	69.5	20.5	[116]
	Maleic acid (60%) 90 °C, 1 hour	–	97	37	[146]
	$NaClO_2$ (50%), 60 °C, 3 hours	–	95	53	[146]
	HBr, 80 °C, 3 hours	20 ± 4.4 (D)	94	42.2 ± 20	[76]
	Pretreatment: Cellulases, 50 °C, 48 hours; Mechanical treatment: domestic blender, 20,000 rpm, 30 secs	10–40 (D)	80	82	[133]

5.3.1 Chemical method

Using chemical agents to extract NC from cellulose resources is a traditional and popular method. In the literature, 43.54% employed this method to fabricate NC. The chemical extraction method mainly involves: (1) acid hydrolysis; (2) alkaline hydrolysis; and (3) chemo-mechanical or chem-enzymatic hydrolysis.

Nickerson *et al.* [1] first employed hydrochloric and sulphuric acid hydrolysis to produce cellulose crystallites from cellulose materials. However, this extraction method had not been used successfully to prepare a stable suspension of colloidal-sized cellulose nanocrystals until the 1950s by Rånby [6, 147]. Sulphuric acids have commonly been used to produce NC since the 1950s. Sulphuric acid treatment of cellulose results in scission of the glycosidic bonds, and thus causes a reduction in DP. The glycosidic bond scission occurs in three steps [148]: (1) protonation of glycosidic oxygen; (2) fission of the glycosidic bond with transfer of the positive charge to C-1; and (3) attack on the carbonium ion by water and re-formation of the hydronium ion. It has been found that the hydrolysis conditions, including reaction time, acid-to-pulp ratio and acid type, have a great influence on the dimensions and surface charge of the cellulose nanofibres [149]. Sulphuric acid provides a highly stable suspension with high negative charge (e.g. –69.7 mV for nanoscale cotton fibril [150]) due to the introduction of sulphate ester groups on the surface of the crystallites [151]. Hydrochloric acid has also been used for obtaining NC [152]: hydrochloric hydrolysis increases the thermal stability of cellulose nanostructures but chloride ions are easily removed by repeated rinsing with water and minimal surface charge on the cellulose nanofibres. One of the major challenges when using acid hydrolysis to fabricate NC is to increase the final yield after processing, typically around 30% (of initial weight), which potentially limits upscale production by this method [153].

5.3.2 Mechanical method

The fabrication of NC can be obtained through mechanical treatments. These methods mainly include: high-pressure homogenisation [154], ultrasonication [155–158], cryocrushing [159, 160] and grinding [161–163]. Among these methods, homogenisation seems to be the most popular [164]. The refining process is carried out prior to homogenisation because refining produces external fibrillation of the fibres by gradually peeling off the external cell wall layers (P and S1 layers) and exposing the S2 layer. It also causes internal fibrillation, which loosens the fibre wall, preparing the fibres for subsequent homogenisation treatment.

The passes through the homogeniser are considered as influencing the parameters for the resultant fibril dimensions and the physical properties of the fibril. For the influence of the number of passes through, it has been found that (1) additional passes contribute to an increase in surface area on the cellulose fibrils with a decrease in the diameter of the fibrils; (2) the aspect ratio of fibre bundles first increases with the increase in the number of passes through but, after a certain number, this value decreases [165]; and (3) higher passes give rise to a reduction in diameter distribution [149].

Mechanical treatment may result in longer and entangled nanoscale cellulose elements, leading to stronger networks and gels, but this treatment tends to damage the microfibril structure by reducing molar mass and the degree of crystallinity. In addition, a higher number of passes through a mechanical homogeniser increases the energy consumption for disintegration. Therefore, researchers use various pretreatments in order to overcome the above disadvantages. It has been found that combining pretreatments could lead to a controlled fibrillation drop to the nanoscale and a network of long and highly entangled cellulose elements. Among these combinations, enzyme-assisted [166–169] and chemo-mechanical techniques [170–173] are the main efficient means of pretreatment.

Enzymatic pretreatment can increase the reactivity and swelling of cellulosic fibres, due to degradation of the enzyme on cellulose chain. Enzymatic processes are generally carried out by three types of enzyme [167]: (1) endocellulase breaks the internal bonds to disrupt the crystalline structure of cellulose and expose individual cellulose polysaccharide chains; (2) exocellulase cleaves 2–4 units, such as cellobiose, from the ends of the exposed chains produced by endocellulase; and (3) cellobiase or β-glucosidase hydrolyses the exocellulase product into individual monosaccharides. This degradation contributes to the cellulosic fibre disintegration during homogenisation and reduces energy consumption. Enzymatic pretreatment was found to yield a higher average molar mass and larger aspect ratio than NC resulting from acidic pretreatment [166].

Chemical pretreatments typically promote fine production and fibre swelling, making defibrillation easier. The combination of chemical pretreatment with sonication is a commonly used chemo-mechanical technique. In 2004, Saito and Isogai [174] first reported the fabrication of NC with TEMPO-mediated oxidation. TEMPO-mediated oxidation produces easily fibrillated products; regulating the exposure time to a homogeniser can be used to control the nanofibril fraction within a fixed volume of oxidised fibres. In addition, researchers found that TEMPO-oxidised NC films had high optical transparency, flexibility, high-dimensional thermal stability and high oxygen barrier properties [175, 176]. This pretreatment introduces carboxylic acid groups in the C6 position of the glucose unit, which may also be utilised for surface modification. Small numbers of aldehyde groups are also introduced by the TEMPO pretreatment.

5.3.3 Drying of NC

To be precise, although chemical or mechanical methods are the way to fabricate NC suspension (gel), it is impractical and expensive to ship the material to customers in these forms (suspension or gel) due to the huge transportation costs [177]; furthermore, mixing with hydrophobic polymer would require the removal of water [178]. The drying process of NC is challenging work, which must overcome a number of significant issues, e.g. agglomeration and loss of key nanoscale characteristics [179] due to air–water interfaces, hydrogen bonding and van der Waals [180]. A successful drying process would allow NC to be transported in the dry or semi-dry state, with considerable cost benefits, while retaining the nanoscale characteristics. This subject has been investigated continuously since it was first reported in 2004 by Ishida *et al.* [181]. Five main drying methods have been employed to dry NC, i.e. freeze-drying [182–184], oven-drying [185–187], air-drying [188], spray-drying [189–191], supercritical drying [180, 192, 193].

Freeze-drying is the most frequently used method. This method creates highly networked structures of agglomerates with multi-scalar dimensions including nanoscale. It allows fast freezing and water evaporation, without overly affecting nanocellulose structure, but it is energy guzzling and time consuming.

Oven-drying is conducted by exposing the suspension in the oven under temperatures of 80–105 °C [180, 185, 186, 194] with circulating air. Oven-drying is the least promising method for the production of NC powder [195]. During oven-drying, forces resulting from the high temperatures and evaporation of water would drive the molecular contact of NC and cause agglomeration.

Air-drying is run as oven drying, with a slight difference in the temperature and humidity, and is always operated at room temperature and humidity [188].

Spray-drying is a suitable manufacturing process for drying NC suspensions: the particle sizes range from nano to micron in scale and are controllable, and the cost is low and the method is scalable [196]. The spray drying procedure comprises concentrating the liquid to the appropriate viscosity, pumping the liquid and dehydrating it by hot gas. It is a well-established technique that has been used as a standard industrial dehydration method in many areas, and it is the most suitable drying method for cellulose fibres that are used as reinforcement in a hydrophobic polymer matrix [186], because this procedure reduces the advanced irreversible hornification phenomenon, which occurs during the oven drying process.

Supercritical drying of the prepared suspension is conducted using a critical point dryer. Four steps are involved in the process [197]: (1) dehydration of the aqueous suspension with ethanol solutions; (2) replacement of ethanol with liquid CO_2; (3) the liquid CO_2 and cellulose mixture are pressurised and heated to supercritical condition; and (4) the liquid CO_2 is eliminated by decompression to the atmosphere. Supercritical drying creates highly networked structures of agglomerates with multi-scalar dimensions including the nanoscale [180].

5.4 Characterisation and properties of NC

5.4.1 Morphology and size of nanocellulose

As mentioned in Section 5.2.2, NC from different sources is of different geometrical dimensions. The precise shape and size of NC can be evidenced by microscopy (e.g. atomic force microscopy (AFM) [198], SEM [199], TEM [200]), spectroscopy (e.g. FTIR [201], nuclear magnetic resonance (NMR) [202], XRD [203]), and scattering techniques (e.g. DLS [204], small-angle neutron scattering (SANS) [205, 206], small-angle X-ray scattering (SAXS) [207, 208], and wide-angle X-ray scattering (WAXS) [209]). SEM, AFM and TEM are common analytical instruments for morphological analysis, although there are some slight differences within them. AFM and SEM can be used for structure and morphologies determination of cellulose whiskers and their nanocomposites. Field emission scanning electron microscopy (FEG-SEM) allows quick examination, giving an overview of the sample. However, the resolution is insufficient for obtaining detailed information. Conventional bright-field transmission electron microscopy (TEM) is capable of identifying individual NC, which enables the determination of their size and shape. AFM can be used to overestimate the width of the NC due to the tip-broadening effect. It has been found that the cross-section of the microfibrils observed by TEM is square, whereas the AFM topography of these microfibril surfaces shows a rounded profile due to convolution with the shape of the AFM tip [210].

Recently, combined analytical instrument and computer-assisted techniques have been developed. Xu *et al.* [211] first investigated the 3-D organisation of cellulose microfibrils in plastic resin-embedded, delignified cell walls of radiata pine early wood. By using IMOD software, they determined the diameter of microfibrils (around 3.2 nm). Meanwhile, they estimated the length of individual cellulose microfibrils. They tracked 90 slices out of 144 reconstructed slice images of 150 nm thick specimens. Within the 90 tracked slices, the length of the cellulose microfibrils was about 95 nm. With the image programme, Chinga-Carrasco and Syverud [212] assessed the multiscale structure of NFC films in detail and found that (1) the films made on the dynamic sheet former were exposed to shear forces during the fibril deposition on the forming wire; (2) the quantified local fibril orientations of the surface layers did not seem to reflect such differences; (3) the orientation anisotropy values were between 0.1 and 0.2 for all the samples; and (4) the surface structures had random orientations.

5.4.2 Mechanical properties of nanocellulose

By employing AFM [213–218] and spectroscopic techniques [219–222] researchers have revealed the mechanical properties of nanocellulose. The elastic modulus of nanocellulose has been measured in the axial direction (E_A) and transverse direction (E_T). The instruments for mechanical measurement and the results are summarised in Table 5.4.

Table 5.4 Mechanical properties of nanocellulose.

Techniques	Source	E_A (GPa)	E_T (GPa)	References
AFM	BC	78 ± 17	–	[213]
	Lyocell	92–104	–	[214, 215]
	Tunicate	TEMPO-oxidation 145.2 ± 31.3	–	[216]
		Acid hydrolysis 150.7 ± 28.8	–	
	Wood	–	18–50	[217]
	Tunicate	–	9 ± 3	[218]
XRD	Ramie	137	–	[219]
Raman spectroscopy	Tunicate	143	–	[220]
	Cotton	57–105	–	[221]
	BC	114	–	[222]

E_A: elastic modulus in axial direction; E_T: elastic modulus in transverse direction.

Axial elastic properties are typically measured using AFM atomic three-point bending of individual nanocellulose, XRD and in situ combination of tensile tests with Raman spectroscopy of thin mats of nanocellulose impregnated with epoxy (Table 5.4). It can be seen that the axial elastic modulus is 57–180 GPa, which is in agreement with the theoretical model, which gives a value of E_A = 124–155 GPa [223]. The transverse elastic properties have been measured using a combination of high-resolution AFM indentation and modelling, in which individual nanocellulose from wood and tunicate give values of E_T= 18–50 GPa and E_T= 9 ± 3 GPa, respectively. These results are similar to the those given by the theoretical model of E_T= 10–57 GPa [224, 225]. Note that there is a high probability of error with such measurements, associated with AFM sensitivity limits and with the assumptions in the model used to extract the mechanical properties.

5.4.3 Optical properties and liquid crystallinity of nanocellulose

In plant cell walls, cellulose microfibrils are arranged in a helicoidal pattern, which has been considered as an analog to a cholesteric order [226]. Stiff rod-like particles are known to be of liquid crystallinity [227]. Due to the high stiffness and aspect ratio, nanocellulose has a strong tendency to align along a vector direction under certain conditions, yet the crystals are readily dispersible, and lyotropic (in solution) behaviour is observed. This behaviour influences the other properties of nanocellulose significantly, such its polyelectrolytic nature and rheological behaviour.

When nanocelluloses are isolated from plant cell walls and placed in a dilute regime, the initial ordered domains are similar to tactoids. As the nanocellulose concentration is increased, the system becomes semi-diluted, whence rotation becomes inhibited. At a still higher concentration, an isotropic concentrated phase should be reached where rod motion is confined to small volumes. When a critical rod concentration is reached, the system becomes biphasic and some of the rods form an anisotropic phase in equilibrium with the isotropic phase. With increasing concentration in the biphasic region, the proportion of the anisotropic phase increases while the proportion of the isotropic phase decreases until the system becomes completely nematic liquid crystalline at a second critical concentration [228], and the chiral nematic orders can be retained after evaporation of the solvent. Marchessault *et al.* [229] first observed the birefringent character of nanocellulose suspension. The birefringent feature of nanocellulose suspension can be observed directly [230]. After drying, the nematic liquid crystal phase can be observed clearly with polarised microscopy [231] due to the nanocelluloses having high optical rotatory power reflecting a circularly polarised light in a limited wavelength band.

Various factors, such as size [232, 233], electrolyte [231, 234–236], dispersity, charge, and external stimuli, can affect the liquid crystallinity behaviour. Liquid crystallinity behaviour can also be influenced by the different process for preparing nanocellulose [237]. Due to the less separated single nanocellulose, the birefringence is somewhat weaker for sonicated suspension compared to homogenised and hydrolysed suspensions [237]. Different acid hydrolysis can also result in different liquid crystallinity behaviour, e.g. sulphuric acid and phosphoric acid derived crystals give a chiral nematic structure whereas hydrochloric acid derived crystals with post-treatment sulphonation gives a birefringent glassy phase [238, 239] that shows a cross-hatch pattern.

Because cellulose crystallites have a helical twist down the long axis, similar to a screw [239], when nanocellulose is submitted to a magnetic field, the long axes of nanocellulose become perpendicular to the magnetic field direction [240] and the distance between nematic planes along the cholesteric axis is shorter than that between the rods in a nematic plane [232]. Employing this feature, aligned nanofibres [241] and nanocellulose base composites (e.g. nanocellulose/PVA nanocomposite, nanocellulose/wood pulp composite [242, 243]) have been developed. The results showed that the mechanical properties along the perpendicular direction to the magnetic field were much stronger than that parallel to the magnetic field.

5.4.4 Polyelectrolytic nature of nanocellulose

A scattering technique was used to investigate cellulose microfibrils in 1961 and was pioneered by Marchessault *et al.* [244]. Scattering techniques were also used to investigate the polyelectrolytic behaviour of nanocellulose. These techniques include ultra-small-angle X-ray scattering (USAXS) [245], small-angle X-ray scattering (SAXS) [208], static light scattering (SLS) [246, 247] and dynamic light scattering (DLS) [246–248].

For the charged rods, the electrostatic interactions play a main role in the phase stability. It has been proved that the effect of the electrostatic interactions on the liquid crystal phase transition in solutions of rodlike polyelectrolytes can be characterised by two parameters, namely, the effective diameter and the twisting action. Additionally, excess of added monovalent electrolyte induces a decrease in the strength of the chiral interactions between the rods [231].

Furuta *et al.* [245] first investigated the ordering structure of nanocellulose with USAXS. The results showed the presence of a single broad scattering peak. The peak is shifted towards lower q values with the addition of simple electrolytes and disappeared at higher salt concentrations because of the disruption to the ordered arrangement. By using SLS and DLS, de Souza Lima *et al.* [247] revealed that the tunicate nanocellulose suspension is the strong electrostatic interaction and the long-range odder in this system.

5.4.5 Rheological properties of nanocellulose suspension

Marchessault and Morehead *et al.* [244] first investigated the rheological properties of nanocellulose suspension and found that the hydrodynamic properties of nanocellulose are related to the size and length distribution of the particles in suspension. Liquid crystals (LCs) may be divided into two subgroups: (1) lyotropic LCs, formed by mixing rigid rodlike molecules with a solvent; and (2) thermotropic LCs, formed by heating. Typically, the molecular weight, concentration and shear rate are the main factors that affect the rheological behaviour of the lyotropic LC suspension. Due to the effects of liquid crystallinity and its polyelectrolytic nature, the rheological behaviour of nanocellulose suspension is complicated. At low concentration, nanocellulose suspensions in the dilute regime are shear thinning. This behaviour increases when the concentration is raised and shows concentration dependence at low rates and very little concentration dependence at high rates [249]. However, at higher concentrations, where the suspensions are lyotropic, the shear rate vs viscosity plot obeys that for lyotropic LCs [250]. The effect of processing on the rheological behaviour has been demonstrated by the previous work [152], which found that the H_2SO_4-treated suspension showed no time dependence on viscosity, while the HCl-treated suspension was thixotropic at concentrations > 0.5 % (w/v) and anti-thixotropic at concentrations <0.3 %. Additionally, the effect of additives [251, 252], modification of nanocellulose [253–256] and post-treatment [257] on the rheological properties have been studied in recent years.

5.5 Application of NC

From middle 1990s, with the development of nanocellulose based composite [12, 13], the application of nanocellulose entered a new stage. In the last 26 years there has been extensive research into the application of nanocellulose. There have been more than 30 review articles and books describing preparation [258], morphology and structure [259, 260], properties [258, 260], modification [261], the application of nanocellulose [258, 262, 263] and patents relating to nanocellulose [264, 265]. In particular, nanocellulose nanocomposites [264, 266–271] have attracted much attention and these outstanding review reports have demonstrated the development of nanocellulose nanocomposite in recent years. Although most investigations concerning the application of nanocellulose are still at the laboratory stage, I am optimistic that, with the development of new technology, especially nanotechnology, nanocellulose will be used industrially and enter more industrial sectors. Due to their high purity, nanocelluloses from bacterial celluloses have been exploited commercially. Bacterial celluloses can be applied in areas where plant cellulose can hardly be used. Most applications are in the healthcare sector, e.g. skin transplants for sides, donor and receptor. With regard to nanocellulose from plant and tunicate cellulose, their application (at the laboratory scale) may be achieved by two methods: (1) without any modification; and (2) with some modifications.

5.5.1 Application of nanocellulose without modification

Nanocellulose without modification can be utilised as follows:

5.5.1.1 Nanopaper

Nanocellulose can be used as raw material for the preparation of nanopaper [272–277] or paper additives [278–280] in the paper industry. Using nanocellulose from wood cellulose, researchers have prepared high performance paper that almost matches steel in terms of strength. The foldable, low thermal expansion and optically transparent nanocellulose paper [281] is a perfect candidate for substrates for continuous roll-toroll processing in the future production of electronic devices, such as flexible displays, solar cells, e-papers and a myriad new flexible circuit technologies, and could replace the costly conventional batch processes based on glass substrates. Nakagaito and Yano [154] first fabricated high-content nanocellulose nanocomposite by mixing nanocellulose with phenol-formaldehyde (PF) resin. Henriksson *et al.* [272] first described the fabrication of nanopaper using solvent exchange techniques. This nanopaper sample showed a remarkable toughness (W_A = 15 MJ m^{-3}), strain-to-failure (10%), porosity (28%), Young's modulus (13.2 GPa) and tensile strength (214 MPa). Olsson *et al.* [273] demonstrated an interesting stiff magnetic nanopaper by compacting bacterial cellulose nanofibril aerogels. Recently, Berglund and his co-researchers [274, 276] developed a novel nanopaper through the addition of clay. This clay nanopaper has high strength (232 MPa) and modulus (13.4 GPa) and is characterised by substantial optical transparency (T_{600} of 42%), 200 mm diameter, and 21.9 nm surface smoothness. Meanwhile, clay nanopaper extends the property range of cellulose nanopaper, e.g. self-extinguishing characteristics and gas barrier properties.

5.5.1.2 Aerogel

Aerogels are highly porous and very lightweight materials that display a multitude of interesting properties, such as large specific surface area, extremely low thermal conductivity and sound propagation, and excellent shock absorption [282]. Aerogels can be prepared by replacing the liquid solvent in a gel with air without substantially altering the network structure or the volume of the gel body. The first aerogels were reported by Kistler in 1931 [283]. Aerogels based on nanofibrillated cellulose (NFC) may offer advantages due to its renewability. In 2001, Tan and co-workers [284] were the first to prepare cellulose aerogel. In 2004, Jin *et al.* [285] were the first to fabricate nanofibrillar cellulose aerogels with dissolution/regeneration of cellulose in aqueous calcium thiocyanate. Various drying methods (regular freeze-drying, rapid freeze-drying and solvent exchange drying) were investigated. It was found that solvent exchange drying gave highly porous aerogel composed of approximately 50 nm wide cellulose microfibrils and highest specific surface area (160–190 m^2 g^{-1}). Later, a new method for making nanocellulose aerogel was demonstrated – supercritical drying (SCD), developed by Hoepfner *et al.* [286]. Compared with freeze-drying, SCD is the most effective method for the synthesis of homo geneous aerogels. This method can result in much higher specific surface areas (250 m^2 g^{-1}) and bending strength (2 MPa). Recently, some novel nanocellulose-based aerogels have be described, e.g. nanocellulose/clay aerogel [287], and nanocellulose/titanium dioxide aerogel [288]. Bacterial cellulose has also been used as a raw material for making nanocellulose aerogel [273, 282, 289, 290]. Sehaqui *et al.* [291] introduced tert-butanol freeze-drying to make aerogel; due to the nature of tert-butanol, a lower extent of surface tension effects (capillary action) are displayed during the drying process. This resulted in aerogels with higher specific surface area (maximum of 284 m^2 g^{-1} for nanocellulose) compared with those prepared by conventional freeze-drying. More details of aerogels will be discussed in Chapter 10.

5.5.1.3 NC template

Due to the nanoscale, nanocellulose has been exploited as a template to make nanometals (e.g. nickel nanoparticles [292], porous titania [293], silver nanoparticles [294, 295] and gold nanoparticles [296]), nano-nonmetals [297] and nano-metal oxide [298, 299]. Nanocellulose releases electrons during the carbonisation process [300] and displays reducibility. Bearing this mechanism in mind, Yongsoon *et al.* [292] fabricated nickel nanoparticles by a thermal reduction process and prepared well-dispersed Ni nanocrystals on the carbonised nanocellulose around 5–12 nm in size. They also successfully synthesised selenium nanoparticles [297] using a thermal reduction process with nanocellulose. In 2010, Padalkar *et al.* [301] were the first to synthesize metallic nanoparticle chains by using a modified reductive deposition procedure involving cationic surfactant, cetyltrimethylammonium bromide (CTAB), which acts as a nanoparticle stabiliser, and a tunicate nanocellulose suspension. Ag nanoparticles are formed via the conventional reduction of $AgNO_3$, which is quite different from the thermal reduction process. This method resulted in the successful decoration of nanocellulose surfaces with Ag, Cu, Au, and Pt nanoparticles. Later in 2011, this research group successfully employed the same method to synthesise semiconductor [299].

5.5.1.4 Nanocomposites (hydrophilic polymer matrices)

Nanocomposites in general are two-phase materials in which one of the phases has at least one dimension in the nanometre range (1–100 nm). The advantages of nanocomposite materials are superior mechanical, barrier and thermal properties at low reinforcement levels, as well as their better recyclability, transparency and low weight [302, 303], when compared to conventional composites. Due to the high density of hydroxyl groups, nanocelluloses exhibit a strong hydrophilic surface. Nanocellulose without modification is always mixed with hydrophilic polymers for making nanocomposite. Solution casting is the main processing technique for nanocellulose (without modification) reinforced polymer matrix composites. Electrospinning has also been reported for making nanofibres [304, 305].

Nanocellulose has been used as reinforcing agent for nanocomposites. As Table 5.5 shows, the addition of nanocellulose significantly reinforces the tensile strength and modulus of the matrix, especially for polyvinyl alcohol (PVA) and poly(ethylene oxide) (PEO). With regards to strain, an interesting difference is displayed between natural polymers and water soluble polymers. The addition of nanocellulose seems to be able to facilitate an increase in strain for water soluble polymers. Due to the homogeneous and the strong hydrogen bonding interaction between the nanocellulose and matrix, nanocellulose content has a profound effect on the nanocomposite mechanical properties. Additionally, the amount of nanocellulose [306], the source of nanocellulose [307, 308] and the fabrication of nanocellulose [309] are considered as factors affecting the mechanical properties of nanocomposite.

The effect of nanocellulose on the water uptake of the composite is dependent on the matrix. For a matrix with lower water sensitivity (e.g. rubber), the addition of nanocellulose results in an increase in water uptake [310]. By contrast, for a matrix with stronger hydrophilicity (e.g. starch, soy protein), the formation of microfibril networks in the matrix prevents the swelling of the matrix and decreases water uptake. It has been proved that water sensitivity decreases linearly with the nanocellulose content [311–313]. Favier *et al.* [12, 13] were first to report the reinforcing effect of cellulose whiskers on a thermoplastic matrix. The addition of tunicin whiskers produced

Table 5.5 Mechanical properties of nanocellulose (without modification) reinforced nanocomposites.

		Strain (%)		Tensile strength (MPa)		Modulus (MPa)		
Type of matrix	Matrix	Pure matrix	Nano-composite	Pure matrix	Nano-composite	Pure matrix	Nano-composite	Ref.
Natural polymer	Starch	68.2 ± 3.1	57.3 ± 2.4 (5%)	3.9 ± 0.3	4.5 ± 0.2 (5%)	31.9 ± 5.1	34.5 ± 4.3 (5%)	[306]
	Amylo-pectin	3.8 ± 0.7	2.4 ± 0.8 (10%)	16.4 ± 6.6	15.3 ± 4.2 (10%)	683 ± 294	822 ± 279 (10%)	[314]
	Rubber	576 ± 36	849 ± 11 (6%)	0.56 ± 0.13	2.3 ± 0.4 (6%)	0.5 ± 0.14	1.0 ± 0.1 (6%)	[315]
	Chitosan	8.9	5.5–6.5	40	80–140	1.4	1.5–7.5	[168]
	Soy protein	29	2.5–11	16	32	0.5	1.19	[316]
Water soluble polymer	Polyvinyl alcohol	141.8	149.71 (5%)	32.38	33.07 (5%)	254.92	536.69 (5%)	[307]
	Poly (ethylene oxide)	176.4 ± 44.3	588 ± 102.5 (0.4%)	1.01 ± 0.15	1.74 ± 0.09 (0.4%)	32.7 ± 5.9	96.1 ± 10.7 (0.4%)	[304]

a spectacular improvement in the storage modulus of the matrix. These outstanding properties were ascribed to a mechanical percolation phenomenon, resulting from the interaction (via hydrogen bond forces) of cellulosic nanoparticles with the matrix.

The hydrogen bonding also gives rise to an increase in thermal stability [309, 310, 316–319] (Fig. 5.8 [318]). Moreover, nanocellulose/matrix interactions decrease with an increase in relative humidity (RH). The addition of nanocellulose affects the optical properties of the nanocellulose nanocomposite. For instance, Fig. 5.9 (chitosan/nanocellulose film is cited I reference [320] and soy protein is cited in reference [316]) shows that optical transmittance of the films decreases with an increase in cellulose whisker content, due to the occurrence of microphase separation caused by the aggregation of fillers within the matrix.

5.5.2 Application of nanocellulose with modification

As described above, nanocellulose possesses an abundance of hydroxyl groups on the surface, where chemical reactions take place. Among the three kinds of hydroxyl groups, the OH group on the sixth position acts as a primary alcohol, where most of the modification predominantly occurs [321]. Various chemical modifications of nanocellulose, such as acetylation [322–324], esterification [325–328], silylation [329, 330], oxidation [331–333] and grafting [334–336], have been reported. Most of these

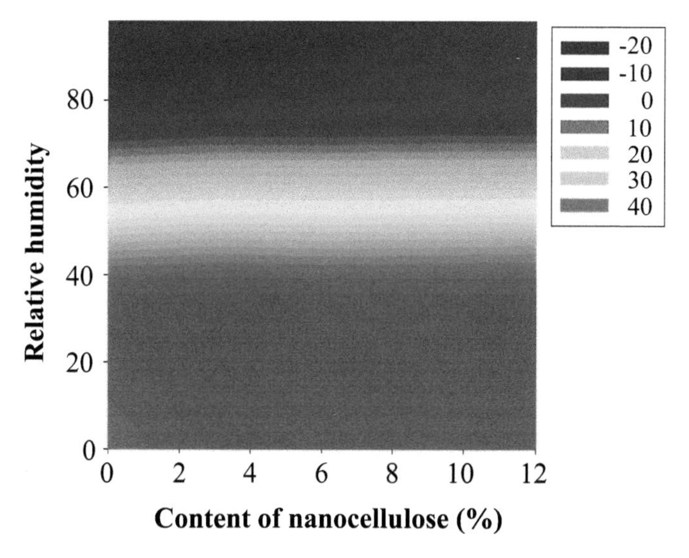

Figure 5.8 Effect of relative humidity and nanocellulose content on the glass transition temperature of PVA.

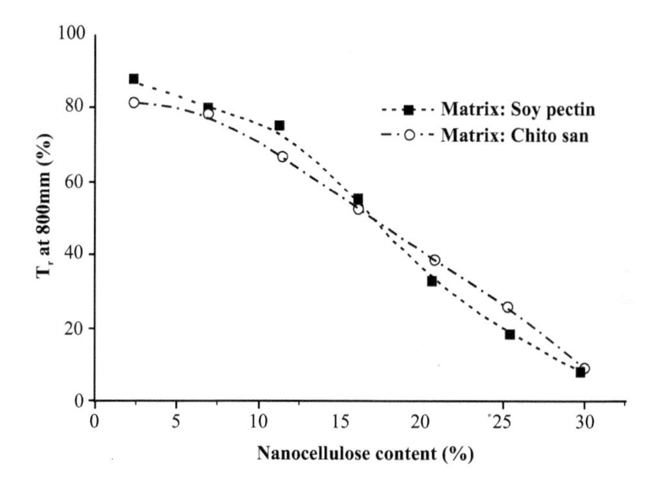

Figure 5.9 Dependence of nanocellulose content on optical transmittance (T_r) for soy pectin and chitosan matrix at 800 nm.

focused on the improvement of its dispersity and compatibility for different solvents or matrices that are suitable in the production of nanocomposites. These modifications open up a wide field for the application of nanocellulose. To date there is increased research focusing on the modification of nanocellulose due to the increasing potential of modified nanocellulose for application in various industrial sectors, such as stabilisers, nanocomposites, and so on.

5.5.2.1 Stabiliser

Since the beginning of the 20th century, it has been known that colloidal particles that are partially wetted by both aqueous and oil phases are capable of stabilising emulsions [337, 338]. The effectiveness of these solids as stabilisers depends on factors such as particle size, interparticle interactions and the wettability of the particles. In general, the formation of an oil-in-water (o/w) emulsion is favoured when the three-phase contact angle between the oil, the solid and the water is < 90°. When the angle is > 90°, the water-in-oil (w/o) emulsion will be stabilised. At the interface, the particles form rigid structures that can sterically inhibit the coalescence of emulsion droplets. Highly crystalline cellulosic materials also show this ability. In 1986, Oza and Frank [339] were the first to demonstrated the microcrystalline cellulose (MCC) stabilised emulsion. The authors proposed that, in addition to the 'active' role at the interface, the MCC particles form a three-dimensional network structure in the continuous phase, which provides further retardation of coalescence through entrapment of emulsion droplets. In 1997, Ougiya *et al.* [340] presented a study comparing o/w emulsions stabilised by bacterial cellulose (BC) with those stabilised by MCC and nanocellulose. These studies, describing the use of fibrillated cellulose materials as emulsion stabilisers, dealt with o/w emulsions. Due to the hydrophilic nature of cellulose, the fibrils and microfibrils are better wet by water than oil, and thus tend to stabilise water-continuous emulsions. In 2006 and 2007, Andresen *et al.* [341, 342] were the first to investigate the application of nanocellulose as a w/o emulsion stabiliser. They utilised chlorodimethyl isopropylsilane to modify the surface of nanocellulose. The resulting material has a contact angle > 90° [341, 342], and should consequently stabilise w/o emulsions [341, 342]. Recently, a number of investigations [343–346] about the fabrication of hydrophobised nanocellulose as a w/o or o/w stabiliser using novel methods have been reported consequently.

5.5.2.2 Nanocomposites (matrix is hydrophobic polymer)

It has been found that the addition of nanocellulose leads to superior mechanical properties for the polymer matrix (Table 5.6). Compounding extrusion, film stacking and casting-evaporation have been used for preparing nanocomposite. There have been a number of studies about the homogeneous dispersion of nanocellulose in the matrix. With the aim of overcoming this dispersion difficulty, roll-mill dispersion [347], various chemical pretreatments (Table 5.7) have been reported. Oksman and his co-worker [348] employed PVA, maleic anhydride, and acetic anhydride to modify nanocellulose for reinforcing a (polylactic acid) PLA matrix. As shown in Fig. 5.10, acetylation modification has a significant effect on the mechanical properties of nanocomposite. For instance, after maleic anhydride modification, the tensile strength and modulus increase to 90.46% and 34.48%, respectively.

Nanocellulose is an environmentally friendly material that could serve as a valuable renewable resource. New and emerging industrial extraction processes need to be optimised to achieve more efficient operations and this will require active research participation from both the academic and industrial sectors. The availability of materials based on nanocellulose remains limited, and more work is needed on the application of nanotechnology to turning nanocellulose into more valuable products. Increasing attention is being devoted to producing nanocellulose in larger

Table 5.6 Mechanical properties of nanocellulose reinforced hydrophobic polymer.

Matrix	Amount of nanocellulose (%)	Strain (%)	Tensile strength (MPa)	Modulus (GPa)	References
Polylactic acid	0	3.4 ± 0.4	58.9 ± 0.5	2.9 ± 0.6	[349]
	5	2.7 ± 0.1	71.2 ± 0.6	3.6 ± 0.7	
Polyurethane	0	751.4 ± 31.4	7.5 ± 1.0	8.2 ± 0.9	[350]
	0.5	1087 ± 61.9	26.9 ± 1.5	40.9 ± 3.3	
	1	994.2 ± 75.3	61.5 ± 4.8	42.4 ± 3.0	
	5	1110.3 ± 101.1	49.8 ± 6.6	44.9 ± 2.4	
Polyester	0	1.75	16.40	1.38	[351]
	1	1.86	16.97	1.58	
	2	1.89	17.81	1.74	
	3	1.53	18.02	1.90	
	4	1.80	21.69	2.51	
	5	2.50	22.64	2.59	

Table 5.7 Comparison of mechanical properties for unmodified and modified nanocellulose reinforced polylactic acid nanocomposite.

Modification	Amount of nanocellulose (%)	Strain (%)	Tensile strength (MPa)	Modulus (GPa)	References
Unmodified	0	3.4 ± 0.4	58.9 ± 0.5	2.9 ± 0.6	[349]
	5	2.7 ± 0.1	71.2 ± 0.6	3.6 ± 0.7	
PVA modification	0	3.4 ± 0.2	71.9 ± 2.0	3.31 ± 0.12	[348]
	5	2.4 ± 0.2	67.7 ± 0.8	3.71 ± 0.07	
Maleic anhydride modification	0	1.9 ± 0.2	40.9 ± 3.2	2.9 ± 0.1	[302]
	5	2.7 ± 0.5	77.9 ± 6.7	3.9 ± 0.3	
Acetic anhydride modification	0	2.06638	44.4431	0.91	[323]
	6	0.98534	71.615	1.05	

quantities, and to exploring various modification processes that enhance its properties, making it attractive for use in a wide range of industrial sectors. So far, most studies have focused on the mechanical and chiral nematic liquid properties of nanocellulose nanocomposites, but other research directions are also being explored. In nanocomposite systems, the homogeneous dispersion of nanocellulose in a polymer matrix is still a challenging problem, as aggregation or agglomeration of nanocellulose is commonly encountered.

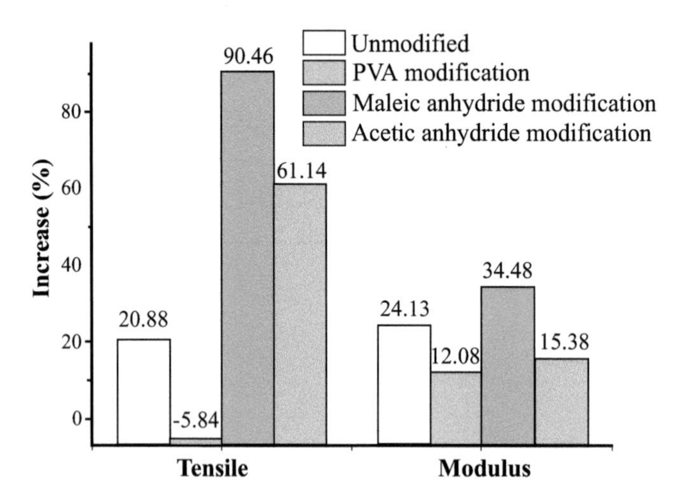

Figure 5.10 Comparison of tensile strength and modulus for unmodified and modified nanocellulose reinforced polylactic acid nanocomposite.

The appropriate modification of nanocellulose to impart functional characteristics to these nanomaterials is sorely needed if nanocellulose is to be successfully incorporated into a specific product system. It is anticipated that nanotechnology innovations in renewable resources, such as nanocellulose, will create a large market for future products based on nanocellulose.

5.6 Outlook

NC is an efficient bioadsorbent that combines high strength and flexibility with a high surface area and versatile surface chemistry. There is a wide range of facile hydrolysis and catalysed process routes that render nanocellulose into cellulose nanofibres and nanocrystals, with tailored surface properties. Due to its unique and exceptional renewability, biodegradability, abundance, mechanical and physiochemical properties and characteristics, the incorporation of a small amount of NC in the material matrix (polymer, ceramic and/or metal) enhances the mechanical strength of the latter by several order of magnitude. This can potentially broaden modern applications of NC in the future, e.g. structural materials, packaging materials, biomedical devices, coatings, drug delivery release, waste-water treatment, electronic devices, and automotive industry.

References

[1] Nickerson, R. F. & Habrle, J. A., Cellulose intercrystalline structure. *Industrial & Engineering Chemistry* **1947** 39 (11), 1507–1512.
[2] Rånby, B. G., Aqueous colloidal solutions of cellulose micelles. *Acta Chemica Scandinavica* **1949,** 3, 649–650.
[3] Rånby, B. G., Ribi, E., Über den Feinbau der Zellulose. *Experientia* **1950**, 6 (1), 12–14.

[4] Rånby, B. G., Fibrous macromolecular systems. Cellulose and muscle. The colloidal properties of cellulose micelles *Discussions of the Faraday Society* **1951**, 11, 158–&.
[5] Rånby, B. G., Fibrous macromolecular systems. Cellulose and muscle. The colloidal properties of cellulose micelles. *Discussions of the Faraday Society* **1951**, 11 (0), 158–164.
[6] Rånby, B. G., The cellulose micelles. *Tappi* **1952**, 35 (2), 53–58.
[7] Mukherjee, S. M. & Woods, H. J., X–ray and electron microscope studies of the degradation of cellulose by sulphuric acid. *Biochimica et Biophysica Acta* **1953**, 10, 499–511.
[8] Turbak, A. F., Snyder, F. W. & Sandberg, K. R. Food products containing microfibrillated cellulose. US 4341807 A, 1982.
[9] Turbak, A. F., Snyder, F. W. & Sandberg, K. R., Microfibrillated cellulose, a new cellulose product: Properties, uses, and commercial potential. In A. Sarko (ed.), *Proceedings of the Ninth Cellulose Conference: Symposium on Cellulose and Wood as Future Chemical Feedstocks and Sources of Energy, and General Papers. Volume 37* (pp. 815–827). Applied Polymer Symposia series, Wiley: New York, **1983**.
[10] Herrick, F. W., Casebier, R. L., Hamilton, J. K. & Sandberg, K. R., Microfibrillated cellulose: Morphology and accessibility. In A. Sarko (ed.), *Proceedings of the Ninth Cellulose Conference: Symposium on Cellulose and Wood as Future Chemical Feedstocks and Sources of Energy, and General Papers* (pp. 797–813). Applied Polymer Symposia series. Wiley: New York, **1983**.
[11] Boldizar, A., Klason, C., Kubát, J., Näslund, P. & Sáha, P., Prehydrolyzed cellulose as reinforcing filler for thermoplastics. *Int. J. Polym. Mater. Po.* **1987**, 11 (4), 229–262.
[12] Favier, V., Canova, G. R., Cavaille, J. Y., Chanzy, H., Dufresne, A. & Gauthier, C., Nanocomposite materials from latex and cellulose whiskers. *Polym. Advan. Technol.* **1995**, 6 (5), 351–355.
[13] Favier, V., Chanzy, H. & Cavaille, J. Y., Polymer nanocomposites reinforced by cellulose whiskers. *Macromolecules* **1995**, 28 (18), 6365–6367.
[14] Kaushik, M. & Moores, A., Review: Nanocelluloses as versatile supports for metal nanoparticles and their applications in catalysis. *Green Chem.* **2016**, 18 (3), 622–637.
[15] Jonoobi, M., Oladi, R., Davoudpour, Y., Oksman, K., Dufresne, A., Hamzeh, Y. & Davoodi, R., Different preparation methods and properties of nanostructured cellulose from various natural resources and residues: A review. *Cellulose* **2015**, 22 (2), 935–969.
[16] Osong, S. H., Norgren, S. & Engstrand, P., Processing of wood-based microfibrillated cellulose and nanofibrillated cellulose, and applications relating to papermaking: A review. *Cellulose* **2016**, 23 (1), 93–123.
[17] Nickerson, R., Hydrolysis and catalytic oxidation of cellulosic materials: A method for continuous estimation of free glucose. *Industrial & Engineering Chemistry Analytical Edition* **1941**, 13 (6), 423–426.
[18] Nickerson, R. F., Hydrolysis and catalytic oxidation of cellulosic materials. *Industrial & Engineering Chemistry* **1941**, 33 (8), 1022–1027.
[19] Nickerson, R. F. & Leape, C. B., Distribution of pectic acid in cotton fibers. *Industrial & Engineering Chemistry* **1941**, 33 (1), 83–86.
[20] Nickerson, R. F., Structure, properties, and utilization of cotton-influences of common agencies on behavior. *Industrial & Engineering Chemistry* **1942**, 34 (10), 1149–1154.
[21] Nickerson, R. F. & Habrle, J. A., Hydrolysis and catalytic oxidation of cellulosic materials. *Industrial & Engineering Chemistry* **1945**, 37 (11), 1115–1118.
[22] Nickerson, R. F. & Habrle, J. A., Hydrolysis and catalytic oxidation of cellulosic materials. *Industrial & Engineering Chemistry* **1946**, 38 (3), 299–301.
[23] Conrad, C. C. & Scroggie, A. G., Chemical characterization of rayon yarns and cellulosic raw materials. *Industrial & Engineering Chemistry* **1945**, 37 (6), 592–598.

[24] Battista, O. A. & Coppick, S., Hydrolysis of native versus regenerated cellulose structures. *Text. Res. J.* **1947**, 17 (8), 419–422.
[25] Hermans, P. H. & Weidinger, A., Change in crystallinity upon heterogeneous acid hydrolysis of cellulose fibers. *Journal of Polymer Science* **1949**, 4 (3), 317–322.
[26] Klemm, D., Kramer, F., Moritz, S., Lindstrom, T., Ankerfors, M., Gray, D. & Dorris, A., Nanocelluloses: A new family of nature-based materials. *Angew Chem. Int. Edit.* **2011**, 50 (24), 5438–5466.
[27] Zanini, M., Lavoratti, A., Zimmermann, M. V. G., Galiotto, D., Matana, F., Baldasso, C. & Zattera, A. J., Aerogel preparation from short cellulose nanofiber of the Eucalyptus species. *J. Cell. Plast.* **2017**, 53 (5), 503–512.
[28] Scatolino, M. V., Bufalino, L., Mendes, L. M., Guimaraes, M. & Tonoli, G. H. D., Impact of nanofibrillation degree of eucalyptus and Amazonian hardwood sawdust on physical properties of cellulose nanofibril films. *Wood Sci. Technol.* **2017**, 51 (5), 1095–1115.
[29] Vallejos, M. E., Felissia, F. E., Area, M. C., Ehman, N. V., Tarres, Q. & Mutje, P., Nanofibrillated cellulose (CNF) from eucalyptus sawdust as a dry strength agent of unrefined eucalyptus handsheets. *Carbohyd. Polym.* **2016**, 139, 99–105.
[30] Hellstrom, P., Heijnesson-Hulten, A., Paulsson, M., Hakansson, H. & Germgard, U., A comparative study of enzymatic and Fenton pretreatment applied to a birch kraft pulp used for MFC production in a pilot scale high-pressure homogenizer. *Tappi J.* **2016**, 15 (6), 375–381.
[31] Suopajärvi, T., Liimatainen, H. & Niinimäki, J., Morphological analyses of some micro and nanofibrils from birch and wheat straw sources. *J. Wood Chem. Technol.* **2015**, 35 (2), 102–112.
[32] Ponni, R., Paakkonen, T., Nuopponen, M., Pere, J. & Vuorinen, T., Alkali treatment of birch kraft pulp to enhance its TEMPO catalyzed oxidation with hypochlorite. *Cellulose* **2014**, 21 (4), 2859–2869.
[33] Thambiraj, S. & Shankaran, D. R., Preparation and physicochemical characterization of cellulose nanocrystals from industrial waste cotton. *Appl. Surf. Sci.* **2017**, 412, 405–416.
[34] Oikonomou, E. K., Mousseau, F., Christov, N., Cristobal, G., Vacher, A., Airiau, M., Bourgaux, C., Heux, L. & Berret, J. F., Fabric softener–cellulose nanocrystal interaction: A model for assessing surfactant deposition on cotton. *J. Phys. Chem. B* **2017**, 121 (10), 2299–2307.
[35] Boonmahitthisud, A., Nakajima, L., Nguyen, K. D. & Kobayashi, T., Composite effect of silica nanoparticle on the mechanical properties of cellulose-based hydrogels derived from cottonseed hulls. *J. Appl. Polym. Sci.* **2017**, 134 (10).
[36] Shamskar, K. R., Heidari, H. & Rashidi, A., Preparation and evaluation of nanocrystalline cellulose aerogels from raw cotton and cotton stalk. *Ind. Crop Prod.* **2016**, 93, 203–211.
[37] Zhang, J. P., Li, B. C., Li, L. X. & Wang, A. Q., Ultralight, compressible and multifunctional carbon aerogels based on natural tubular cellulose. *J. Mater. Chem. A* **2016**, 4 (6), 2069–2074.
[38] Deepa, B., Abraham, E., Cordeiro, N., Mozetic, M., Mathew, A., Oksman, K., Faria, M., Thomas, S. & Pothan, L., Utilization of various lignocellulosic biomass for the production of nanocellulose: a comparative study. *Cellulose* **2015**, 1–16.
[39] Babar, A. A., Peerzada, M. H., Jhatial, A. K. & Bughio, N. U., Pad ultrasonic batch dyeing of causticized lyocell fabric with reactive dyes. *Ultrason. Sonochem.* **2017**, 34, 993–999.
[40] Nicosia, A., Keppler, T., Muller, F. A., Vazquez, B., Ravegnani, F., Monticelli, P. & Belosi, F., Cellulose acetate nanofiber electrospun on nylon substrate as novel composite matrix for efficient, heat-resistant, air filters. *Chem. Eng. Sci.* **2016**, 153, 284–294.
[41] Yan, C. F., Yu, H. Y. & Yao, J. M., One-step extraction and functionalization of cellulose nanospheres from lyocell fibers with cellulose II crystal structure. *Cellulose* **2015**, 22 (6), 3773–3788.

[42] Zimmermann, M. V. G., da Silva, M. P., Zattera, A. J. & Santana, R. M. C., Effect of nanocellulose fibers and acetylated nanocellulose fibers on properties of poly(ethylene-co-vinyl acetate) foams. *J. Appl. Polym. Sci.* **2017**, 134 (17).

[43] Tummala, G. K., Joffre, T., Rojas, R., Persson, C. & Mihranyan, A., Strain-induced stiffening of nanocellulose-reinforced poly(vinyl alcohol) hydrogels mimicking collagenous soft tissues. *Soft Matter* **2017**, 13 (21), 3936–3945.

[44] Kumar, A., Rao, K. M., Kwon, S. E., Lee, Y. N. & Han, S. S., Xanthan gum/bioactive silica glass hybrid scaffolds reinforced with cellulose nanocrystals: Morphological, mechanical and in vitro cytocompatibility study. *Mater. Lett.* **2017**, 193, 274–278.

[45] Hassan, E. A., Hassan & M. L., Rice straw nanofibrillated cellulose films with antimicrobial properties via supramolecular route. *Ind. Crop Prod.* **2016**, 93, 142–151.

[46] Pereira, P. H. F., Waldron, K. W., Wilson, D. R., Cunha, A. P., de Brito, E. S., Rodrigues, T. H. S., Rosa, M. F. & Azeredo, H. M. C., Wheat straw hemicelluloses added with cellulose nanocrystals and citric acid. Effect on film physical properties. *Carbohyd. Polym.* **2017**, 164, 317–324.

[47] Espinosa, E., Tarres, Q., Delgado–Aguilar, M., Gonzalez, I., Mutje, P. & Rodriguez, A., Suitability of wheat straw semichemical pulp for the fabrication of lignocellulosic nanofibres and their application to papermaking slurries. *Cellulose* **2016**, 23 (1), 837–852.

[48] Movva, M. & Kommineni, R., Extraction of cellulose from pistachio shell and physical and mechanical characterisation of cellulose-based nanocomposites. *Materials Research Express* **2017**, 4 (4).

[49] Frone, A. N., Chiulan, I., Panaitescu, D. M., Nicolae, C. A., Ghiurea, M. & Galan, A. M., Isolation of cellulose nanocrystals from plum seed shells, structural and morphological characterization. *Mater. Lett.* **2017**, 194, 160–163.

[50] Bano, S. & Negi, Y. S., Studies on cellulose nanocrystals isolated from groundnut shells. *Carbohyd. Polym.* **2017**, 157, 1041–1049.

[51] Zhou, L., He, H., Jiang, C., Ma, L. & Yu, P., Cellulose nanocrystals from cotton stalk for reinforcement of poly(vinyl alcohol) composites. *Cell. Chem. Technol.* **2017**, 51 (1–2), 109–119.

[52] Theng, D., El Mansouri, N. E., Arbat, G., Ngo, B., Delgado–Aguilar, M., Pelach, M. A., Fullana–i–Palmer, P. & Mutje, P., Fiberboards made from corn stalk thermomechanical pulp and kraft lignin as a green adhesive. *Bioresources* **2017**, 12 (2), 2379–2393.

[53] Huang, S. W., Zhou, L., Li, M. C., Wu, Q. L. & Zhou, D. G., Cellulose nanocrystals &CNCs) from corn stalk: Activation energy analysis. *Materials* **2017**, 10 (1).

[54] Sazali, N., Salleh, W. N. W. & Ismail, A. F., Carbon tubular membranes from nanocrystalline cellulose blended with P84 co-polyimide for H-2 and He separation. *Int. J. Hydrogen Energ.* **2017**, 42 (15), 9952–9957.

[55] Putro, J. N., Santoso, S. P., Ismadji, S., Ju & Y. H., Investigation of heavy metal adsorption in binary system by nanocrystalline cellulose – Bentonite nanocomposite: Improvement on extended Langmuir isotherm model. *Micropor. Mesopor. Mat.* **2017**, 246, 166–177.

[56] Laitinen, O., Suopajarvi, T., Osterberg, M. & Liimatainen, H., Hydrophobic, superabsorbing aerogels from choline chloride-based deep eutectic solvent pretreated and silylated cellulose nanofibrils for selective oil removal. *ACS. Appl. Mater. Inter.* **2017**, 9 (29), 25029–25037.

[57] Sehaqui, H., Michen, B., Marty, E., Schaufelberger, L. & Zimmermann, T., Functional cellulose nanofiber filters with enhanced flux for the removal of humic acid by adsorption. *ACS. Sustain. Chem. Eng.* **2016**, 4 (9), 4582–4590.

[58] Sehaqui, H., Mautner, A., de Larraya, U. P., Pfenninger, N., Tingaut, P. & Zimmermann, T., Cationic cellulose nanofibers from waste pulp residues and their nitrate, fluoride, sulphate and phosphate adsorption properties. *Carbohyd. Polym.* **2016**, 135, 334–340.

[59] Ruiz-Palomero, C., Soriano, M. L. & Valcarcel, M., Sulfonated nanocellulose for the efficient dispersive micro solid-phase extraction and determination of silver nanoparticles in food products. *J. Chromatogr. A.* **2016**, 1428, 352–358.

[60] Uribe, B. E. B. & Tarpani, J. R., Interphase analysis of hierarchical composites via transmission electron microscopy. *Compos. Interface* **2017**, 24 (9), 849–859.

[61] Grande, R., Trovatti, E., Carvalho, A. J. F. & Gandini, A., Continuous microfiber drawing by interfacial charge complexation between anionic cellulose nanofibers and cationic chitosan. *J. Mater. Chem. A* **2017**, 5 (25), 13098–13103.

[62] Santucci, B. S., Bras, J., Belgacem, M. N., Curvelo, A. A. D. & Pimenta, M. T. B., Evaluation of the effects of chemical composition and refining treatments on the properties of nanofibrillated cellulose films from sugarcane bagasse. *Ind. Crop Prod.* **2016**, 91, 238–248.

[63] Singla, R., Soni, S., Kulurkar, P. M., Kumari, A., Mahesh, S., Patial, V., Padwad, Y. S. & Yadav, S. K., In situ functionalized nanobiocomposites dressings of bamboo cellulose nanocrystals and silver nanoparticles for accelerated wound healing. *Carbohyd. Polym.* **2017**, 155, 152–162.

[64] Razalli, R. L., Abdi, M. M., Tahir, P. M., Moradbak, A., Sulaiman, Y. & Heng, L. Y., Polyaniline-modified nanocellulose prepared from Semantan bamboo by chemical polymerization: Preparation and characterization. *RSC. Adv.* **2017**, 7 (41), 25191–25198.

[65] Saurabh, C. K., Mustapha, A., Masri, M. M., Owolabi, A. F., Syakir, M. I., Dungani, R., Paridah, M. T., Jawaid, M. & Khalil, H. P. S. A., Isolation and characterization of cellulose nanofibers from Gigantochloa scortechinii as a reinforcement material. *J. Nanomater.* **2016**.

[66] Madsen, B., Asian, M. & Lilholt, H., Fractographic observations of the microstructural characteristics of flax fibre composites. *Compos. Sci. Technol.* **2016**, 123, 151–162.

[67] Mikshina, P. V., Petrova, A. A., Faizullin, D. A., Zuev, Y. F. & Gorshkova, T. A., Tissue-specific rhamnogalacturonan I forms the gel with hyperelastic properties. *Biochemistry (Moscow)* **2015**, 80 (7), 915–924.

[68] Keryvin, V., Lan, M., Bourmaud, A., Parenteau, T., Charleux, L. & Baley, C., Analysis of flax fibres viscoelastic behaviour at micro and nano scales. *Composites Part A: Applied Science and Manufacturing* **2015**, 68 (0), 219–225.

[69] Sutka, A., Antsov, M., Jarvekulg, M., Visnapuu, M., Heinmaa, I., Maeorg, U., Vlassov, S. & Sutka, A., Mechanical properties of individual fiber segments of electrospun lignocellulose-reinforced poly(vinyl alcohol). *J. Appl. Polym. Sci.* **2017**, 134 (2).

[70] Pacaphol, K., Aht-Ong, D., The influences of silanes on interfacial adhesion and surface properties of nanocellulose film coating on glass and aluminum substrates. *Surf. Coat. Tech.* **2017**, 320, 70–81.

[71] Merkel, K., Rydarowski, H., Kazimierczak, J. & Bloda, A., Processing and characterization of reinforced polyethylene composites made with lignocellulosic fibres isolated from waste plant biomass such as hemp. *Composites Part B – Eng.* **2014**, 67, 138–144.

[72] Trifol, J., Sillard, C., Plackett, D., Szabo, P., Bras, J. & Daugaard, A. E., Chemically extracted nanocellulose from sisal fibres by a simple and industrially relevant process. *Cellulose* **2017**, 24 (1), 107–118.

[73] Santana, J. S., do Rosario, J. M., Pola, C. C., Otoni, C. G., Soares, N. D. F., Camilloto, G. P. & Cruz, R. S., Cassava starch-based nanocomposites reinforced with cellulose nanofibers extracted from sisal. *J. Appl. Polym. Sci.* **2017**, 134 (12).

[74] Echegaray, M., Mondragon, G., Martin, L., Gonzalez, A., Pena-Rodriguez, C. & Arbelaiz, A., Physicochemical and mechanical properties of gelatin reinforced with nanocellulose and montmorillonite. *Journal of Renewable Materials* **2016**, 4 (3), 206–214.

[75] Chaker, A., Mutje, P., Vilaseca, F. & Boufi, S., Reinforcing potential of nanofibrillated cellulose from nonwoody plants. *Polym. Composite.* **2013**, 34 (12), 1999–2007.

[76] Sucaldito, M. R. & Camacho, D. H., Characteristics of unique HBr-hydrolyzed cellulose nanocrystals from freshwater green algae (Cladophora rupestris) and its reinforcement in starch-based film. *Carbohyd. Polym.* **2017**, 169, 315–323.
[77] Ruan, C. Q., Stromme, M., Mihranyan, A. & Lindh, J., Favored surface-limited oxidation of cellulose with Oxone (R) in water. *RSC. Adv.* **2017**, 7 (64), 40600–40607.
[78] Gustafsson, S. & Mihranyan, A., Strategies for tailoring the pore-size distribution of virus retention filter papers. *ACS. Appl. Mater. Inter.* **2016**, 8 (22), 13759–13767.
[79] Solikhin, A., Hadi, Y. S., Massijaya, M. Y. & Nikmatin, S., Morphological, chemical, and thermal characteristics of nanofibrillated cellulose isolated using chemo–mechanical methods. *Makara Journal of Science* **2017**, 21 (2), 59–68.
[80] Lamaming, J., Hashim, R., Leh, C. P. & Sulaiman, O., Properties of cellulose nanocrystals from oil palm trunk isolated by total chlorine free method. *Carbohyd. Polym.* **2017**, 156, 409–416.
[81] Dasan, Y. K., Bhat, A. H. & Faiz, A., Development and material properties of poly(lactic acid)/poly(3-hydroxybutyrate-co-3-hydroxyvalerate)-based nanocrystalline cellulose nanocomposites. *J. Appl. Polym. Sci.* **2017**, 134 (5).
[82] Mondal, S., Preparation, properties and applications of nanocellulosic materials. *Carbohyd. Polym.* **2017**, 163, 301–316.
[83] Habibi, Y., Key advances in the chemical modification of nanocelluloses. *Chem. Soc. Rev.* **2014**, 43 (5), 1519–1542.
[84] Siro, I. & Plackett, D., Microfibrillated cellulose and new nanocomposite materials: a review. *Cellulose* **2010**, 17 (3), 459–494.
[85] Ojala, J., Sirvio, J. A. & Liimatainen, H., Nanoparticle emulsifiers based on bifunctionalized cellulose nanocrystals as marine diesel oil–water emulsion stabilizers. *Chem. Eng. J.* **2016**, 288, 312–320.
[86] Hosseinidoust, Z., Alam, M. N., Sim, G., Tufenkji, N. & van de Ven, T. G. M., Cellulose nanocrystals with tunable surface charge for nanomedicine. *Nanoscale* **2015**, 7 (40), 16647–16657.
[87] Salajkova, M., Berglund, L. A. & Zhou, Q., Hydrophobic cellulose nanocrystals modified with quaternary ammonium salts. *Journal of Materials Chemistry* **2012**, 22 (37), 19798–19805.
[88] Brito, B. S. L., Pereira, F. V., Putaux, J. L. & Jean, B., Preparation, morphology and structure of cellulose nanocrystals from bamboo fibers. *Cellulose* **2012**, 19 (5), 1527–1536.
[89] Moraes, A. D., de Goes, T. S., Hausen, M., Morais, J. P. S., Rosa, M. D., de Menezes, A., Mattoso, L. H. C. & Leite, F. D., Morphological characterization of cellulose nanocrystals by atomic force microscopy. *Materia – Brazil* **2016**, 21 (2), 532–540.
[90] Herrera, N., Salaberria, A. M., Mathew, A. P. & Oksman, K., Plasticized polylactic acid nanocomposite films with cellulose and chitin nanocrystals prepared using extrusion and compression molding with two cooling rates: Effects on mechanical, thermal and optical properties. *Composites Part A: Applied Science and Manufacturing* **2016**, 83, 89–97.
[91] Vestena, M., Gross, I. P., Muller, C. M. O. & Pires, A. T. N., Nanocomposite of poly(lactic acid)/cellulose nanocrystals: Effect of CNC content on the polymer crystallization kinetics. *J. Brazil. Chem. Soc.* **2016**, 27 (5), 905–911.
[92] Nindiyasari, F., Griesshaber, E., Zimmermann, T., Manian, A. P., Randow, C., Zehbe, R., Fernandez–Diaz, L., Ziegler, A., Fleck, C. & Schmahl, W. W., Characterization and mechanical properties investigation of the cellulose/gypsum composite. *J. Compos. Mater.* **2016**, 50 (5), 657–672.
[93] Cheng, M., Qin, Z. Y., Liu, Y. N., Qin, Y. F., Li, T., Chen, L. & Zhu, M. F., Efficient extraction of carboxylated spherical cellulose nanocrystals with narrow distribution

through hydrolysis of lyocell fibers by using ammonium persulfate as an oxidant. *J. Mater. Chem. A* **2014**, 2 (1), 251–258.

[94] Tanpichai, S., Sampson, W. W. & Eichhorn, S. J., Stress transfer in microfibrillated cellulose reinforced poly(vinyl alcohol) composites. *Composites Part A: Applied Science and Manufacturing* **2014**, 65 (0), 186–191.

[95] Wu, Q., Li, X. W., Fu, S. Y., Li, Q. & Wang, S. Q., Estimation of aspect ratio of cellulose nanocrystals by viscosity measurement: Influence of surface charge density and NaCl concentration. *Cellulose* **2017**, 24 (8), 3255–3264.

[96] Santamaria–Echart, A., Ugarte, L., Gonzalez, K., Martin, L., Irusta, L., Gonzalez, A., Corcuera, M. A. & Eceiza, A., The role of cellulose nanocrystals incorporation route in waterborne polyurethane for preparation of electrospun nanocomposites mats. *Carbohyd. Polym.* **2017**, 166, 146–155.

[97] Iskak, N. A. M., Julkapli, N. M. & Hamid, S. B. A., Understanding the effect of synthesis parameters on the catalytic ionic liquid hydrolysis process of cellulose nanocrystals. *Cellulose* **2017**, 24 (6), 2469–2481.

[98] Rafieian, F. & Simonsen, J., Fabrication and characterization of carboxylated cellulose nanocrystals reinforced glutenin nanocomposite. *Cellulose* **2014**, 1–14.

[99] Oun, A. A. & Rhim, J. W., Isolation of cellulose nanocrystals from grain straws and their use for the preparation of carboxymethyl cellulose-based nanocomposite films. *Carbohyd. Polym.* **2016**, 150, 187–200.

[100] Hu, S. X., Gu, J., Jiang, F. & Hsieh, Y. L., Holistic rice straw nanocellulose and hemicelluloses/lignin composite films. *ACS. Sustain. Chem. Eng.* **2016**, 4 (3), 728–737.

[101] Taflick, T., Schwendler, L. A., Rosa, S. M. L., Bica, C. I. D. & Nachtigall, S. M. B., Cellulose nanocrystals from acacia bark – Influence of solvent extraction. *Int. J. Biol. Macromol.* **2017**, 101, 553–561.

[102] Li, G. H., Tian, X. J., Xu, X. W., Zhou, C., Wu, J. Y., Li, Q., Zhang, L. Q., Yang, F. & Li, Y. F., Fabrication of robust and highly thermally conductive nanofibrillated cellulose/graphite nanoplatelets composite papers. *Compos. Sci. Technol.* **2017**, 138, 179–185.

[103] El-Wakil, N. A., Hassan, E. A., Abou-Zeid, R. E. & Dufresne, A., Development of wheat gluten/nanocellulose/titanium dioxide nanocomposites for active food packaging. *Carbohyd. Polym.* **2015** (0), 337–346.

[104] Gilfillan, W. N., Moghaddam, L. & Doherty, W. O. S., Preparation and characterization of composites from starch with sugarcane bagasse nanofibres. *Cellulose* **2014**, 21 (4), 2695–2712.

[105] Lam, N. T., Chollakup, R., Smitthipong, W., Nimchua, T. & Sukyai, P., Characterization of cellulose nanocrystals extracted from sugarcane bagasse for potential biomedical materials. *Sugar Tech.* **2017**, 19 (5), 539–552.

[106] Zhang, P. P., Tong, D. S., Lin, C. X., Yang, H. M., Zhong, Z. K., Yu, W. H., Wang, H. & Zhou, C. H., Effects of acid treatments on bamboo cellulose nanocrystals. *Asia–Pac. J. Chem. Eng.* **2014**, 9 (5), 686–695.

[107] Cao, X., Dong, H. & Li, C. M., New nanocomposite materials reinforced with flax cellulose nanocrystals in waterborne polyurethane. *Biomacromolecules* **2007**, 8 (3), 899–904.

[108] Qua, E. H. & Hornsby, P. R., Preparation and characterisation of nanocellulose reinforced polyamide – 6. *Plast. Rubber Compos.* **2011**, 40 (6–7), 300–306.

[109] Shi, J., Shi, S. Q., Barnes, H. M. & Pittman Jr, C. U., A chemical process for preparing cellulosic fibers hierarchically from kenaf bast fibers. *Bioresources* **2011**, 6 (1), 879–890.

[110] Kargarzadeh, H., Ahmad, I., Abdullah, I., Dufresne, A., Zainudin, S. Y. & Sheltami, R. M., Effects of hydrolysis conditions on the morphology, crystallinity, and thermal stability of cellulose nanocrystals extracted from kenaf bast fibers. *Cellulose* **2012**, 19 (3), 855–866.

[111] Cherian, B. M., Pothan, L. A., Nguyen-Chung, T., Mennig, G., Kottaisamy, M. & Thomas, S., A novel method for the synthesis of cellulose nanofibril whiskers from banana fibers and characterization. *J. Agr. Food Chem.* **2008**, 56 (14), 5617–5627.
[112] Silviya, E. K., Unnikrishnan, G., Varghese, S. & Guthrie, J. T., Surfactant effects on poly(ethylene-co-vinyl acetate)/cellulose composites. *Composites Part B: Engineering* **2013**, 47 (0), 137–144.
[113] Mueller, S., Weder, C. & Foster, E. J., Isolation of cellulose nanocrystals from pseudostems of banana plants. *RSC. Adv.* **2014**, 4 (2), 907–915.
[114] Wang, S. Q., Wei, C., Gong, Y. Y., Lv, J., Yu, C. B. & Yu, J. H., Cellulose nanofiber-assisted dispersion of cellulose nanocrystals@polyaniline in water and its conductive films. *RSC. Adv.* **2016**, 6 (12), 10168–10174.
[115] Johari, A. P., Mohanty, S., Kurmvanshi, S. K. & Nayak, S. K., Influence of different treated cellulose fibers on the mechanical and thermal properties of poly(lactic acid). *ACS. Sustain. Chem. Eng.* **2016**, 4 (3), 1619–1629.
[116] Hai, L. V., Son, H. N. & Seo, Y. B., Physical and bio-composite properties of nanocrystalline cellulose from wood, cotton linters, cattail, and red algae. *Cellulose* **2015**, 22 (3), 1789–1798.
[117] Benhamou, K., Dufresne, A., Magnin, A., Mortha, G. & Kaddami, H., Control of size and viscoelastic properties of nanofibrillated cellulose from palm tree by varying the TEMPO-mediated oxidation time. *Carbohyd. Polym.* **2014**, 99, 74–83.
[118] Ben Mabrouk, A., Kaddami, H., Magnin, A., Belgacem, M. N., Dufresne, A. & Boufi, S., Preparation of nanocomposite dispersions based on cellulose whiskers and acrylic copolymer by miniemulsion polymerization: Effect of the silane content. *Polymer Engineering & Science* **2011**, 51 (1), 62–70.
[119] Bendahou, A., Kaddami, H. & Dufresne, A., Investigation on the effect of cellulosic nanoparticles' morphology on the properties of natural rubber based nanocomposites. *Eur. Polym. J.* **2010**, 46 (4), 609–620.
[120] Le Goff, K. J., Gaillard, C., Garnier, C. & Aubry, T., Electrostatically driven modulation of the reinforcement of agarose hydrogels by cellulose nanowhiskers. *J. Appl. Polym. Sci.* **2016**, 133 (8).
[121] Nicharat, A., Sapkota, J., Weder, C. & Foster, E. J., Melt processing of polyamide 12 and cellulose nanocrystals nanocomposites. *J. Appl. Polym. Sci.* **2015**, 132 (45), 10.
[122] Sacui, I. A., Nieuwendaal, R. C., Burnett, D. J., Stranick, S. J., Jorfi, M., Weder, C., Foster, E. J., Olsson, R. T. & Gilman, J. W., Comparison of the properties of cellulose nanocrystals and cellulose nanofibrils isolated from bacteria, tunicate, and wood processed using acid, enzymatic, mechanical, and oxidative methods. *ACS. Appl. Mater. Inter.* **2014**, 6 (9), 6127–6138.
[123] Cheng, Q. Y., Ye, D. D., Chang, C. Y. & Zhang, L. N., Facile fabrication of superhydrophilic membranes consisted of fibrous tunicate cellulose nanocrystals for highly efficient oil/water separation. *J. Membrane Sci.* **2017**, 525, 1–8.
[124] Martinez-Sanz, M., Lopez-Rubio, A., Villano, M., Oliveira, C. S. S., Majone, M., Reis, M. & Lagaron, J. M., Production of bacterial nanobiocomposites of polyhydroxyalkanoates derived from waste and bacterial nanocellulose by the electrospinning enabling melt compounding method. *J. Appl. Polym. Sci.* **2016**, 133 (2).
[125] Martínez-Sanz, M., Vicente, A., Gontard, N., Lopez-Rubio, A. & Lagaron, J., On the extraction of cellulose nanowhiskers from food by-products and their comparative reinforcing effect on a polyhydroxybutyrate-co-valerate polymer. *Cellulose* **2015**, 22 (1), 535–551.
[126] Kalashnikova, I., Bizot, H., Bertoncini, P., Cathala, B. & Capron, I., Cellulosic nanorods of various aspect ratios for oil in water Pickering emulsions. *Soft Matter* **2013**, 9 (3), 952–959.

[127] Lee, H. V., Hamid, S. B. A. & Zain, S. K., Conversion of lignocellulosic biomass to nanocellulose: Structure and chemical process. *Scientific World Journal* **2014**.

[128] Han, J., Yue, Y., Wu, Q., Huang, C., Pan, H., Zhan, X., Mei, C. & Xu, X., Effects of nanocellulose on the structure and properties of poly(vinyl alcohol)–borax hybrid foams. *Cellulose* **2017**, 24 (10), 4433–4448.

[129] Kontturi, E. & Vuorinen, T., Indirect evidence of supramolecular changes within cellulose microfibrils of chemical pulp fibers upon drying. *Cellulose* **2009**, 16 (1), 65–74.

[130] Abraham, E., Deepa, B., Pothan, L. A., Jacob, M., Thomas, S., Cvelbar, U. & Anandjiwala, R., Extraction of nanocellulose fibrils from lignocellulosic fibres: A novel approach. *Carbohyd. Polym.* **2011**, 86 (4), 1468–1475.

[131] Neto, W. P. F., Silvério, H. A., Vieira, J. G., da Costa e Silva Alves, H., Pasquini, D., de Assunção, R. M. N. & Dantas, N. O., Preparation and characterization of nanocomposites of carboxymethyl cellulose reinforced with cellulose nanocrystals. *Macromolecular Symposia* **2012**, 319 (1), 93–98.

[132] Neto, W. P. F., Putaux, J. L., Mariano, M., Ogawa, Y., Otaguro, H., Pasquini, D. & Dufresne, A., Comprehensive morphological and structural investigation of cellulose I and II nanocrystals prepared by sulphuric acid hydrolysis. *RSC. Adv.* **2016**, 6 (79), 76017–76027.

[133] Xiang, Z. Y., Gao, W. H., Chen, L. H., Lan, W., Zhu, J. Y. & Runge, T., A comparison of cellulose nanofibrils produced from Cladophora glomerata algae and bleached eucalyptus pulp. *Cellulose* **2016**, 23 (1), 493–503.

[134] Wang, Q. Q. & Zhu, J. Y., Effects of mechanical fibrillation time by disk grinding on the properties of cellulose nanofibrils. *Tappi J.* **2016**, 15 (6), 419–423.

[135] Besbes, I., Vilar, M. R. & Boufi, S., Nanofibrillated cellulose from alfa, eucalyptus and pine fibres: Preparation, characteristics and reinforcing potential. *Carbohyd. Polym.* **2011**, 86 (3), 1198–1206.

[136] Lavoratti, A., Scienza, L. C. & Zattera, A. J., Dynamic-mechanical and thermomechanical properties of cellulose nanofiber/polyester resin composites. *Carbohyd. Polym.* **2016**, 136, 955–963.

[137] Moriana, R., Vilaplana, F. & Ek, M., Cellulose nanocrystals from forest residues as reinforcing agents for composites: A study from macro- to nano-dimensions. *Carbohyd. Polym.* **2016**, 139, 139–149.

[138] Azzam, F., Heux, L., Putaux, J.–L. & Jean, B., Preparation by grafting onto, characterization, and properties of thermally responsive polymer-decorated cellulose nanocrystals. *Biomacromolecules* **2010**, 11 (12), 3652–3659.

[139] Pirani, S. & Hashaikeh, R., Nanocrystalline cellulose extraction process and utilization of the byproduct for biofuels production. *Carbohyd. Polym.* **2013**, 93 (1), 357–363.

[140] Qi, W. H., Xu, H. N. & Zhang, L. F., The aggregation behavior of cellulose micro/nanoparticles in aqueous media. *RSC. Adv.* **2015**, 5 (12), 8770–8777.

[141] Kaushik, A. & Kaur, R., Thermoplastic starch nanocomposites reinforced with cellulose nanocrystals: Effect of plasticizer on properties. *Compos. Interface* **2016**, 23 (7), 701–717.

[142] Kunaver, M., Anzlovar, A. & Zagar, E., The fast and effective isolation of nanocellulose from selected cellulosic feedstocks. *Carbohyd. Polym.* **2016**, 148, 251–258.

[143] Wang, Z. H., Yao, Z. J., Zhou, J. T. & Zhang, Y., Reuse of waste cotton cloth for the extraction of cellulose nanocrystals. *Carbohyd. Polym.* **2017**, 157, 945–952.

[144] Farahbakhsh, N., Shahbeigi–Roodposhti, P., Sadeghifar, H., Venditti, R. A. & Jur, J. S., Effect of isolation method on reinforcing capability of recycled cotton nanomaterials in thermoplastic polymers. *J. Mater. Sci.* **2017**, 52 (9), 4997–5013.

[145] Chen, Y. W., Lee, H. V., Juan, J. C. & Phang, S. M., Production of new cellulose nanomaterial from red algae marine biomass Gelidium elegans. *Carbohyd. Polym.* **2016**, 151, 1210–1219.

[146] Guo, J. Q., Uddin, K. M. A., Mihhels, K., Fang, W. W., Laaksonen, P., Zhu, J. Y. & Rojas, O. J., Contribution of residual proteins to the thermomechanical performance of cellulosic nanofibrils isolated from green macroalgae. *ACS. Sustain. Chem. Eng.* **2017**, 5 (8), 6978–6985.

[147] Rånby, B. G., Physico-chemical investigations on animal cellulose (Tunicin). *Ark. Kemi.* **1952**, 4, 241–248.

[148] Grunert, M. Cellulose nanocrystals: Preparation, surface modification, and application in nanocomposites. 3047403, State University of New York College of Environmental Science and Forestry, United States, New York, 2002.

[149] Qua, E., Hornsby, P., Sharma, H. & Lyons, G., Preparation and characterisation of cellulose nanofibres. *J. Mat. Sci.* **2011**, 46 (18), 6029–6045.

[150] Satyamurthy, P., Jain, P., Balasubramanya, R. H. & Vigneshwaran, N., Preparation and characterization of cellulose nanowhiskers from cotton fibres by controlled microbial hydrolysis. *Carbohyd. Polym.* **2011**, 83 (1), 122–129.

[151] Svagan, A. Bio-inspired polysaccharide nanocomposites and foams. Doctoral thesis, Kungliga Tekniska högskolan, Stockholm, Sweden, 2006.

[152] Araki, J., Wada, M., Kuga, S. & Okano, T., Flow properties of microcrystalline cellulose suspension prepared by acid treatment of native cellulose. *Colloids and Surfaces A: Physicochemical and Engineering Aspects* **1998**, 142 (1), 75–82.

[153] Bondeson, D., Kvien, I. & Oksman, K., Strategies for preparation of cellulose whiskers from microcrystalline cellulose as reinforcement in nanocomposites. In K. Oksman & M. Sain (eds), *Cellulose Nanocomposites Processing, Characterization, and Properties. Volume 938* (pp. 10–25). American Chemical Society: **2006**.

[154] Nakagaito, A. N. & Yano, H., Novel high-strength biocomposites based on microfibrillated cellulose having nano-order-unit web-like network structure. *Applied Physics A: Materials Science & Processing* **2005**, 80 (1), 155–159.

[155] Cheng, Q., Wang, S., Rials, T. & Lee, S.–H., Physical and mechanical properties of polyvinyl alcohol and polypropylene composite materials reinforced with fibril aggregates isolated from regenerated cellulose fibers. *Cellulose* **2007**, 14 (6), 593–602.

[156] Johnson, R., Zink-Sharp, A., Renneckar, S. & Glasser, W., A new bio-based nanocomposite: Fibrillated TEMPO-oxidized celluloses in hydroxypropylcellulose matrix. *Cellulose* **2009**, 16 (2), 227–238.

[157] Tischer, P. C. S. F., Sierakowski, M. R., Westfahl, H. & Tischer, C. A., Nanostructural reorganization of bacterial cellulose by ultrasonic treatment. *Biomacromolecules* **2010**, 11 (5), 1217–1224.

[158] Pinjari, D. V. & Pandit, A. B., Cavitation milling of natural cellulose to nanofibrils. *Ultrason. Sonochem.* **2010**, 17 (5), 845–852.

[159] Chakraborty, A., Sain, M. & Kortschot, M., Cellulose microfibrils: A novel method of preparation using high shear refining and cryocrushing. *Holzforschung* **2005**, 59 (1), 102–107.

[160] Alemdar, A. & Sain, M., Isolation and characterization of nanofibers from agricultural residues-wheat straw and soy hulls. *Bioresource. Technol.* **2008**, 99 (6), 1664–1671.

[161] Ahmed, K. S., Vijayarangan, S. & Kumar, A., Low velocity impact damage characterization of woven jute-glass fabric reinforced isothalic polyester hybrid composites. *J. Reinf. Plast. Compos.* **2007**, 26 (10), 959–976.

[162] Iwamoto, S., Nakagaito, A. N. & Yano, H., Nano-fibrillation of pulp fibers for the processing of transparent nanocomposites. *Applied Physics A: Materials Science & Processing* **2007**, 89 (2), 461–466.

[163] Iwamoto, S., Abe, K. & Yano, H., The effect of hemicelluloses on wood pulp nanofibrillation and nanofiber network characteristics. *Biomacromolecules* **2008**, 9 (3), 1022–1026.

[164] Zimmermann, T., Bordeanu, N. & Strub, E., Properties of nanofibrillated cellulose from different raw materials and its reinforcement potential. *Carbohyd. Polym.* **2010**, 79 (4), 1086–1093.

[165] Lee, S. Y., Chun, S. J., Kang, I. A. & Park, J. Y., Preparation of cellulose nanofibrils by high-pressure homogenizer and cellulose-based composite films. *J. Ind. Eng. Chem.* **2009**, 15 (1), 50–55.

[166] Henriksson, M., Henriksson, G., Berglund, L. A. & Lindstrom, T., An environmentally friendly method for enzyme-assisted preparation of microfibrillated cellulose (MFC) nanofibers. *Eur. Polym. J.* **2007**, 43 (8), 3434–3441.

[167] Yoo, S. & Hsieh, J. S., Enzyme-assisted preparation of fibrillated cellulose fibers and its effect on physical and mechanical properties of paper sheet composites. *Industrial & Engineering Chemistry Research* **2010**, 49 (5), 2161–2168.

[168] Hassan, M., Hassan, E. & Oksman, K., Effect of pretreatment of bagasse fibers on the properties of chitosan/microfibrillated cellulose nanocomposites. *J. Mat. Sci.* **2011**, 46 (6), 1732–1740.

[169] Paakko, M., Ankerfors, M., Kosonen, H., Nykanen, A., Ahola, S., Osterberg, M., Ruokolainen, J., Laine, J., Larsson, P. T., Ikkala, O. & Lindstrom, T., Enzymatic hydrolysis combined with mechanical shearing and high-pressure homogenization for nanoscale cellulose fibrils and strong gels. *Biomacromolecules* **2007**, 8 (6), 1934–1941.

[170] Alemdar, A. & Sain, M., Biocomposites from wheat straw nanofibers: Morphology, thermal and mechanical properties. *Compos. Sci. Technol.* **2008**, 68 (2), 557–565.

[171] Chen, W., Yu, H., Liu, Y., Hai, Y., Zhang, M. & Chen, P., Isolation and characterization of cellulose nanofibers from four plant cellulose fibers using a chemical-ultrasonic process. *Cellulose* **2011**, 18 (2), 433–442.

[172] Kaushik, A. & Singh, M., Isolation and characterization of cellulose nanofibrils from wheat straw using steam explosion coupled with high shear homogenization. *Carbohydr. Res.* **2011**, 346 (1), 76–85.

[173] Hirota, M., Tamura, N., Saito, T. & Isogai, A., Water dispersion of cellulose II nanocrystals prepared by TEMPO-mediated oxidation of mercerized cellulose at pH 4.8. *Cellulose* **2010**, 17 (2), 279–288.

[174] Saito, T. & Isogai, A., TEMPO-mediated oxidation of native cellulose. The effect of oxidation conditions on chemical and crystal structures of the water-insoluble fractions. *Biomacromolecules* **2004**, 5 (5), 1983–1989.

[175] Fukuzumi, H., Saito, T., Iwata, T., Kumamoto, Y. & Isogai, A., Transparent and high gas barrier films of cellulose nanofibers prepared by TEMPO-mediated oxidation. *Biomacromolecules* **2008**, 10 (1), 162–165.

[176] Iwamoto, S., Kai, W., Isogai, T., Saito, T., Isogai, A. & Iwata, T., Comparison study of TEMPO-analogous compounds on oxidation efficiency of wood cellulose for preparation of cellulose nanofibrils. *Polym. Degrad. Stab.* **2010**, 95 (8), 1394–1398.

[177] O'Connor, B., Berry, R. & Goguen, R., Commercialization of cellulose nanocrystal (NCC™) production: A business case focusing on the importance of proactive EHS management. In M. S. Hull & D. M. Bowman (eds), *Nanotechnology Environmental Health and Safety (Second Edition)* (pp. 225–246). William Andrew Publishing: Oxford, **2014**.

[178] Hellrup, J., Nordstrom, J. & Mahlin, D., Powder compression mechanics of spray-dried lactose nanocomposites. *Int. J. Pharmaceut.* **2017**, 518 (1–2), 1–10.

[179] Plackett, D. & Iotti, M., Preparation of nanofibrillated cellulose and cellulose whiskers. In *Biopolymer Nanocomposites: Processing, Properties, and Applications*, 2013, pp. 309–338.

[180] Peng, Y. C., Gardner, D. J. & Han, Y. S., Drying cellulose nanofibrils: In search of a suitable method. *Cellulose* **2012**, 19 (1), 91–102.

[181] Ishida, O., Kim, D. Y., Kuga, S., Nishiyama, Y. & Brown, R. M., Microfibrillar carbon from native cellulose. *Cellulose* **2004**, 11 (3–4), 475–480.

[182] Yuan, H. H., Nishiyama, Y., Wada, M. & Kuga, S., Surface acylation of cellulose whiskers by drying aqueous emulsion. *Biomacromolecules* **2006**, 7 (3), 696–700.

[183] Eyholzer, C., Bordeanu, N., Lopez-Suevos, F., Rentsch, D., Zimmermann, T. & Oksman, K., Preparation and characterization of water-redispersible nanofibrillated cellulose in powder form. *Cellulose* **2010**, 17 (1), 19–30.

[184] Zepic, V., Poljansek, I., Oven, P., Skapin, A. S. & Hancic, A., Effect of drying pretreatment on the acetylation of nanofibrillated cellulose. *Bioresources* **2015**, 10 (4), 8148–8167.

[185] Quievy, N., Jacquet, N., Sclavons, M., Deroanne, C., Paquot, M. & Devaux, J., Influence of homogenization and drying on the thermal stability of microfibrillated cellulose. *Polym. Degrad. Stabil.* **2010**, 95 (3), 306–314.

[186] Nechita, P. & Panaitescu, D. M., Improving the dispersibility of cellulose microfibrillated structures in polymer matrix by controlling drying conditions and chemical surface modifications. *Cell. Chem. Technol.* **2013**, 47 (9–10), 711–719.

[187] Cervin, N. T., Johanson, E., Larsson, P. A. & Wagberg, L., Strong, water-durable, and wet-resilient cellulose nanofibril-stabilized foams from oven drying. *ACS. Appl. Mater. Inter.* **2016**, 8 (18), 11682–11689.

[188] Ramanen, P., Penttila, P. A., Svedstrom, K., Maunu, S. L. & Serimaa, R., The effect of drying method on the properties and nanoscale structure of cellulose whiskers. *Cellulose* **2012**, 19 (3), 901–912.

[189] Vartiainen, J., Pohler, T., Sirola, K., Pylkkanen, L., Alenius, H., Hokkinen, J., Tapper, U., Lahtinen, P., Kapanen, A., Putkisto, K., Hiekkataipale, P., Eronen, P., Ruokolainen, J. & Laukkanen, A., Health and environmental safety aspects of friction grinding and spray drying of microfibrillated cellulose. *Cellulose* **2011**, 18 (3), 775–786.

[190] Peng, Y. C., Han, Y. S. & Gardner, D. J., Spray-drying cellulose nanofibrils: Effect of drying process parameters on particle morphology and size distribution. *Wood Fiber Sci.* **2012**, 44 (4), 448–461.

[191] Zhong, T. H., Oporto, G. S., Peng, Y. C., Xie, X. F. & Gardner, D. J., Drying cellulose-based materials containing copper nanoparticles. *Cellulose* **2015**, 22 (4), 2665–2681.

[192] Peng, Y. C., Gardner, D. J., Han, Y., Cai, Z. Y. & Tshabalala, M. A., Influence of drying method on the surface energy of cellulose nanofibrils determined by inverse gas chromatography. *J. Colloid. Interf. Sci.* **2013**, 405, 85–95.

[193] Peng, Y. C., Gardner, D. J., Han, Y., Kiziltas, A., Cai, Z. Y. & Tshabalala, M. A., Influence of drying method on the material properties of nanocellulose I: Thermostability and crystallinity. *Cellulose* **2013**, 20 (5), 2379–2392.

[194] Butchosa, N. & Zhou, Q., Water redispersible cellulose nanofibrils adsorbed with carboxymethyl cellulose. *Cellulose* **2014**, 1–10.

[195] Liang, G. F., He, L. M., Cheng, H. Y., Zhang, C., Li, X. R., Fujita, S., Zhang, B., Arai, M. & Zhao, F. Y., ZSM-5-supported multiply-twinned nickel particles: Formation, surface properties, and high catalytic performance in hydrolytic hydrogenation of cellulose. *J. Catal.* **2015**, 325, 79–86.

[196] Brinchi, L., Cotana, F., Fortunati, E. & Kenny, J. M., Production of nanocrystalline cellulose from lignocellulosic biomass: Technology and applications. *Carbohyd. Polym.* **2013**, 94 (1), 154–169.

[197] Khalil, H. P. S. A., Davoudpour, Y., Aprilia, N. A. S., Mustapha, A., Hossain, S., Islam, N. & Dungani, R., Nanocellulose-based polymer nanocomposite: Isolation, characterization

and applications. In V. K. T. Akur (ed.), *Nanocellulose Polymer Nanocomposites* (pp. 273–309). Wiley: New York, **2014**.

[198] Li, R., Fei, J., Cai, Y., Li, Y., Feng, J. & Yao, J., Cellulose whiskers extracted from mulberry: A novel biomass production. *Carbohydr. Polym.* **2009**, 76 (1), 94–99.

[199] Abe, K., Iwamoto, S. & Yano, H., Obtaining cellulose nanofibers with a uniform width of 15 nm from wood. *Biomacromolecules* **2007**, 8 (10), 3276–3278.

[200] Zuluaga, R., Putaux, J.-L., Restrepo, A., Mondragon, I. & Ganan, P., Cellulose microfibrils from banana farming residues: isolation and characterization. *Cellulose* **2007**, 14 (6), 585–592.

[201] Horikawa, Y., Itoh, T. & Sugiyama, J., Preferential uniplanar orientation of cellulose microfibrils reinvestigated by the FTIR technique. *Cellulose* **2006**, 13 (3), 309–316.

[202] Duchesne, I., Hult, E., Molin, U., Daniel, G., Iversen, T. & Lennholm, H., The influence of hemicellulose on fibril aggregation of kraft pulp fibres as revealed by FE-SEM and CP/MAS 13C-NMR. *Cellulose* **2001**, 8 (2), 103–111.

[203] Peura, M., Müller, M., Vainio, U., Sarén, M.-P., Saranpää, P. & Serimaa, R., X-ray microdiffraction reveals the orientation of cellulose microfibrils and the size of cellulose crystallites in single Norway spruce tracheids. *Trees – Structure and Function* **2008**, 22 (1), 49–61.

[204] De Souza Lima, M. M., Wong, J. T., Paillet, M., Borsali, R. & Pecora, R., Translational and rotational dynamics of rodlike cellulose whiskers. *Langmuir* **2003**, 19 (1), 24–29.

[205] Terech, P., Chazeau, L. & Cavaille, J. Y., A small-angle scattering study of cellulose whiskers in aqueous suspensions. *Macromolecules* **1999**, 32 (6), 1872–1875.

[206] Bonini, C., Heux, L., Cavaillé, J.–Y., Lindner, P., Dewhurst, C. & Terech, P., Rodlike cellulose whiskers coated with surfactant: A small-angle neutron scattering characterization. *Langmuir* **2002**, 18 (8), 3311–3314.

[207] Yu, H., Liu, R. G., Shen, D. W., Wu, Z. H. & Huang, Y., Arrangement of cellulose microfibrils in the wheat straw cell wall. *Carbohyd. Polym.* **2008**, 72 (1), 122–127.

[208] Ebeling, T., Paillet, M., Borsali, R., Diat, O., Dufresne, A., Cavaillé, J. Y. & Chanzy, H., Shear-induced orientation phenomena in suspensions of cellulose microcrystals, revealed by small angle X-ray scattering. *Langmuir* **1999**, 15 (19), 6123–6126.

[209] Elazzouzi-Hafraoui, S., Nishiyama, Y., Putaux, J. L., Heux, L., Dubreuil, F. & Rochas, C., The shape and size distribution of crystalline nanoparticles prepared by acid hydrolysis of native cellulose. *Biomacromolecules* **2008**, 9 (1), 57–65.

[210] Hanley, S. J., Giasson, J., Revol, J.-F. & Gray, D. G., Atomic force microscopy of cellulose microfibrils: comparison with transmission electron microscopy. *Polymer.* **1992**, 33 (21), 4639–4642.

[211] Xu, P., Donaldson, L. A., Gergely, Z. R. & Staehelin, L. A., Dual-axis electron tomography: A new approach for investigating the spatial organization of wood cellulose microfibrils. *Wood Sci. Technol.* **2007**, 41 (2), 101–116.

[212] Chinga-Carrasco, G. & Syverud, K., Computer-assisted quantification of the multi-scale structure of films made of nanofibrillated cellulose. *J. Nanopart. Res.* **2010**, 12 (3), 841–851.

[213] Guhados, G., Wan, W. & Hutter, J. L., Measurement of the elastic modulus of single bacterial cellulose fibers using atomic force microscopy. *Langmuir* **2005**, 21 (14), 6642–6646.

[214] Cheng, Q. & Wang, S., A method for testing the elastic modulus of single cellulose fibrils via atomic force microscopy. *Composites Part A: Applied Science and Manufacturing* **2008**, 39 (12), 1838–1843.

[215] Cheng, Q., Wang, S. & Harper, D. P., Effects of process and source on elastic modulus of single cellulose fibrils evaluated by atomic force microscopy. *Composites Part A: Applied Science and Manufacturing* **2009**, 40 (5), 583–588.

[216] Iwamoto, S., Kai, W. H., Isogai, A. & Iwata, T., Elastic modulus of single cellulose microfibrils from tunicate measured by atomic force microscopy. *Biomacromolecules* **2009**, 10 (9), 2571–2576.

[217] Lahiji, R. R., Xu, X., Reifenberger, R., Raman, A., Rudie, A. & Moon, R. J., Atomic force microscopy characterization of cellulose nanocrystals. *Langmuir* **2010**, 26 (6), 4480–4488.

[218] Michael, T. P., András, V., John, D., Natalia, F., Bin, M., Ryan, W., Arvind, R., Robert, J. M., Ronald, S., Theodore, H. W. & James, B., Development of the metrology and imaging of cellulose nanocrystals. *Meas. Sci. Technol.* **2011**, 22 (2), 024005.

[219] Sakurada, I., Nukushina, Y. & Ito, T., Experimental determination of the elastic modulus of crystalline regions in oriented polymers. *Journal of Polymer Science* **1962**, 57 (165), 651–660.

[220] Šturcová, A., Davies, G. R. & Eichhorn, S. J., Elastic modulus and stress-transfer properties of tunicate cellulose whiskers. *Biomacromolecules* **2005**, 6 (2), 1055–1061.

[221] Rusli, R. & Eichhorn, S. J., Determination of the stiffness of cellulose nanowhiskers and the fiber–matrix interface in a nanocomposite using Raman spectroscopy. *Appl. Phys. Lett.* **2008**, 93 (3), 033111-033111-3.

[222] Hsieh, Y. C., Yano, H., Nogi, M. & Eichhorn, S., An estimation of the Young's modulus of bacterial cellulose filaments. *Cellulose* **2008**, 15 (4), 507–513.

[223] Tanaka, F. & Iwata, T., Estimation of the elastic modulus of cellulose crystal by molecular mechanics simulation. *Cellulose* **2006**, 13 (5), 509–517.

[224] Jaswon, M. A., Gillis, P. P. & Mark, R. E., The elastic constants of crystalline native cellulose. *Proceedings of the Royal Society of London. Series A. Mathematical and Physical Sciences* **1968**, 306 (1486), 389–412.

[225] Tashiro, K. & Kobayashi, M., Theoretical evaluation of three-dimensional elastic constants of native and regenerated celluloses: Role of hydrogen bonds. *Polymer.* **1991**, 32 (8), 1516–1526.

[226] Reis, D., Vian, B., Chanzy, H. & Roland, J.-C., Liquid crystal-type assembly of native cellulose-glucuronoxylans extracted from plant cell wall. *Biol. Cell.* **1991**, 73 (2–3), 173–178.

[227] Oldenbourg, R., Wen, X., Meyer, R. B. & Caspar, D. L. D., Orientational distribution function in nematic tobacco-mosaic-virus liquid crystals measured by X-ray diffraction. *Phys. Rev. Lett.* **1988**, 61 (16), 1851–1854.

[228] Davis, V. A., Ericson, L. M., Parra–Vasquez, A. N. G., Fan, H., Wang, Y., Prieto, V., Longoria, J. A., Ramesh, S., Saini, R. K., Kittrell, C., Billups, W. E., Adams, W. W., Hauge, R. H., Smalley, R. E. & Pasquali, M., Phase behavior and rheology of SWNTs in superacids. *Macromolecules* **2003**, 37 (1), 154–160.

[229] Marchessault, R. H., Morehead, F. F. & Walter, N. M., Liquid crystal systems from fibrillar polysaccharides. *Nature* **1959**, 184 (4686), 632–633.

[230] Siqueira, G., Abdillahi, H., Bras, J. & Dufresne, A., High reinforcing capability cellulose nanocrystals extracted from *Syngonanthus nitens* (Capim Dourado). *Cellulose* **2010**, 17 (2), 289–298.

[231] Dong, X. M., Kimura, T., Revol, J.-F. & Gray, D. G., Effects of ionic strength on the isotropic-chiral nematic phase transition of suspensions of cellulose crystallites. *Langmuir* **1996**, 12 (8), 2076–2082.

[232] Orts, W. J., Godbout, L., Marchessault, R. H. & Revol, J.-F., Enhanced ordering of liquid crystalline suspensions of cellulose microfibrils: A small angle neutron scattering study. *Macromolecules* **1998**, 31 (17), 5717–5725.

[233] Wang, N., Ding, E. & Cheng, R. S., Preparation and liquid crystalline properties of spherical cellulose nanocrystals. *Langmuir* **2008**, 24 (1), 5–8.

[234] Araki, J., Wada, M. & Kuga, S., Steric stabilization of a cellulose microcrystal suspension by poly(ethylene glycol) grafting. *Langmuir* **2001**, 17 (1), 21–27.

[235] Beck-Candanedo, S., Viet, D. & Gray, D. G., Induced phase separation in cellulose nanocrystal suspensions containing ionic dye species. *Cellulose* **2006**, 13 (6), 629–635.

[236] Beck-Candanedo, S., Viet, D. & Gray, D. G., Induced phase separation in low ionic-strength cellulose nanocrystal suspensions containing high molecular-weight blue dextrans. *Langmuir* **2006**, 22, 8690–8695.

[237] Oksman, K., Etang, J. A., Mathew, A. P. & Jonoobi, M., Cellulose nanowhiskers separated from a bio-residue from wood bioethanol production. *Biomass and Bioenergy* **2011**, 35 (1), 146–152.

[238] Araki, J., Wada, M., Kuga, S. & Okano, T., Birefringent glassy phase of a cellulose microcrystal suspension. *Langmuir* **2000**, 16 (6), 2413–2415.

[239] Fleming, K., Gray, D. G. & Matthews, S., Cellulose crystallites. *Chemistry – A European Journal* **2001**, 7 (9), 1831–1836.

[240] Sugiyama, J., Chanzy, H. & Maret, G., Orientation of cellulose microcrystals by strong magnetic fields. *Macromolecules* **1992**, 25 (16), 4232–4234.

[241] Kim, J., Chen, Y., Kang, K. S., Park, Y. B. & Schwartz, M., Magnetic field effect for cellulose nanofiber alignment. *J. Appl. Phys.* **2008**, 104 (9).

[242] Kvien, I. & Oksman, K., Orientation of cellulose nanowhiskers in polyvinyl alcohol. *Applied Physics A: Materials Science & Processing* **2007**, 87 (4), 641–643.

[243] Li, D., Liu, Z., Al–Haik, M., Tehrani, M., Murray, F., Tannenbaum, R. & Garmestani, H., Magnetic alignment of cellulose nanowhiskers in an all–cellulose composite. *Polym. Bull.* **2010**, 65 (6), 635–642.

[244] Marchessault, R. H., Morehead, F. F. & Koch, M. J., Some hydrodynamic properties of neutral suspensions of cellulose crystallites as related to size and shape. *Journal of Colloid Science* **1961**, 16 (4), 327–344.

[245] Furuta, T., Yamahara, E., Konishi, T. & Ise, N., Ordering in aqueous cellulose hydrolysate dispersions: An ultra-small-angle X-ray scattering study. *Macromolecules* **1996**, 29 (27), 8994–8995.

[246] Lu, X., Hu, Z. & Gao, J., Synthesis and light scattering study of hydroxypropyl cellulose microgels. *Macromolecules* **2000**, 33 (23), 8698–8702.

[247] de Souza Lima, M. M. & Borsali, R., Static and dynamic light scattering from polyelectrolyte microcrystal cellulose. *Langmuir* **2002**, 18 (4), 992–996.

[248] Bica, C. I. D., Borsali, R., Geissler, E. & Rochas, C., Dynamics of cellulose whiskers in agarose gels. 1. Polarized dynamic light scattering. *Macromolecules* **2001**, 34 (15), 5275–5279.

[249] Bercea, M. & Navard, P., Shear dynamics of aqueous suspensions of cellulose whiskers. *Macromolecules* **2000**, 33 (16), 6011–6016.

[250] Han, C. D., *Rheology and Processing of Polymeric Materials: Polymer Rheology. Volume 1.* Oxford University Press: New York, **2007**.

[251] Lowys, M. P., Desbrières, J. & Rinaudo, M., Rheological characterization of cellulosic microfibril suspensions. Role of polymeric additives. *Food Hydrocolloid* **2001**, 15 (1), 25–32.

[252] Agoda-Tandjawa, G., Durand, S., Gaillard, C., Garnier, C. & Doublier, J. L., Rheological behaviour and microstructure of microfibrillated cellulose suspensions/low-methoxyl pectin mixed systems. Effect of calcium ions. *Carbohyd. Polym.* **2012**, 87 (2), 1045–1057.

[253] Lewandowski, Z., Application of a linear synthetic polymer to improve the properties of cellulose fibers made by the NMMO process. *J. Appl. Polym. Sci.* **2002**, 83 (13), 2762– 2773.

[254] Gousse, C., Chanzy, H., Cerrada, M. L. & Fleury, E., Surface silylation of cellulose microfibrils: Preparation and rheological properties. *Polymer.* **2004**, 45 (5), 1569–1575.
[255] Lasseuguette, E., Roux, D. & Nishiyama, Y., Rheological properties of microfibrillar suspension of TEMPO-oxidized pulp. *Cellulose* **2008**, 15 (3), 425–433.
[256] Karppinen, A., Vesterinen, A.-H., Saarinen, T., Pietikäinen, P. & Seppälä, J., Effect of cationic polymethacrylates on the rheology and flocculation of microfibrillated cellulose. *Cellulose* **2011**, 18 (6), 1381–1390.
[257] Agoda-Tandjawa, G., Durand, S., Berot, S., Blassel, C., Gaillard, C., Garnier, C. & Doublier, J. L., Rheological characterization of microfibrillated cellulose suspensions after freezing. *Carbohyd. Polym.* **2010**, 80 (3), 677–686.
[258] Azizi Samir, M. A. S., Alloin, F. & Dufresne, A., Review of recent research into cellulosic whiskers, their properties and their application in nanocomposite field. *Biomacromolecules* **2005**, 6 (2), 612–626.
[259] Putaux, J.-L., Morphology and structure of crystalline polysaccharides: Some recent studies. *Macromolecular Symposia* **2005**, 229 (1), 66–71.
[260] Nishiyama, Y., Structure and properties of the cellulose microfibril. *Journal of Wood Science* **2009**, 55 (4), 241–249.
[261] Stenstad, P., Andresen, M., Tanem, B. & Stenius, P., Chemical surface modifications of microfibrillated cellulose. *Cellulose* **2008**, 15 (1), 35–45.
[262] Czaja, W. K., Young, D. J., Kawecki, M. & Brown, R. M., The future prospects of microbial cellulose in biomedical applications. *Biomacromolecules* **2006**, 8 (1), 1–12.
[263] Kohler, R. & Nebel, K., Cellulose-nanocomposites: Towards high performance composite materials. *Macromolecular Symposia* **2006**, 244 (1), 97–106.
[264] Siró, I. & Plackett, D., Microfibrillated cellulose and new nanocomposite materials: A review. *Cellulose* **2010**, 17 (3), 459–494.
[265] Duran, N., Paula Lemes, A. B. & Seabra, A., Review of cellulose nanocrystals patents: Preparation, composites and general applications *Recent Patents on Nanotechnology,* **2012**, 6 (1), 16–28.
[266] Hubbe, M. A., Rojas, O. J., Lucia, L. A. & Sain, M., Cellulosic nanocomposites: A review. *Bioresources* **2008**, 3 (3), 929–980.
[267] Dufresne, A. & Belgacem, M. N., Cellulose-reinforced composites: From micro- to nanoscale. *Polımeros: Ciência e Tecnologia* **2010**, 20, 1–10.
[268] Eichhorn, S., Dufresne, A., Aranguren, M., Marcovich, N., Capadona, J., Rowan, S., Weder, C., Thielemans, W., Roman, M., Renneckar, S., Gindl, W., Veigel, S., Keckes, J., Yano, H., Abe, K., Nogi, M., Nakagaito, A., Mangalam, A., Simonsen, J., Benight, A., Bismarck, A., Berglund, L. & Peijs, T., Review: Current international research into cellulose nanofibres and nanocomposites. *J. Mat. Sci.* **2010**, 45 (1), 1–33.
[269] Siqueira, G., Bras, J. & Dufresne, A., Cellulosic bionanocomposites: A review of preparation, properties and applications. *Polymers.* **2010**, 2 (4), 728–765.
[270] Moon, R. J., Martini, A., Nairn, J., Simonsen, J. & Youngblood, J., Cellulose nanomaterials review: Structure, properties and nanocomposites. *Chem. Soc. Rev.* **2011**, 40 (7), 3941–3994.
[271] Susheel, K., Alain, D., Bibin Mathew, C., Luc, A., James, N. & Elias, N., Cellulose-based bio- and nanocomposites: A review. *International Journal of Polymer Science* 2011, 2011.
[272] Henriksson, M., Berglund, L. A., Isaksson, P., Lindström, T. & Nishino, T., Cellulose nanopaper structures of high toughness. *Biomacromolecules* **2008**, 9 (6), 1579–1585.
[273] Olsson, R. T., Samir, M., Salazar–Alvarez, G., Belova, L., Strom, V., Berglund, L. A., Ikkala, O., Nogues, J. & Gedde, U. W., Making flexible magnetic aerogels and stiff magnetic nanopaper using cellulose nanofibrils as templates. *Nat. Nanotechnol.* **2010**, 5 (8), 584–588.

[274] Sehaqui, H., Liu, A., Zhou, Q. & Berglund, L. A., Fast preparation procedure for large, flat cellulose and cellulose/inorganic nanopaper structures. *Biomacromolecules* **2010**, 11 (9), 2195–2198.

[275] Chun, S.-J., Lee, S.-Y., Doh, G.-H., Lee, S. & Kim, J. H., Preparation of ultrastrength nanopapers using cellulose nanofibrils. *J. Ind. Eng. Chem.* **2011**, 17 (3), 521–526.

[276] Liu, A., Walther, A., Ikkala, O., Belova, L. & Berglund, L. A., Clay nanopaper with tough cellulose nanofiber matrix for fire retardancy and gas barrier functions. *Biomacromolecules* **2011**, 12 (3), 633–641.

[277] Sehaqui, H., Zhou, Q., Ikkala, O. & Berglund, L. A., Strong and tough cellulose nanopaper with high specific surface area and porosity. *Biomacromolecules* **2011**, 12 (10), 3638–3644.

[278] Ahola, S., Österberg, M. & Laine, J., Cellulose nanofibrils – adsorption with poly(amideamine) epichlorohydrin studied by QCM-D and application as a paper strength additive. *Cellulose* **2008**, 15 (2), 303–314.

[279] Hult, E.-L., Iotti, M. & Lenes, M., Efficient approach to high barrier packaging using microfibrillar cellulose and shellac. *Cellulose* **2010**, 17 (3), 575–586.

[280] Taipale, T., Osterberg, M., Nykanen, A., Ruokolainen, J. & Laine, J., Effect of microfibrillated cellulose and fines on the drainage of kraft pulp suspension and paper strength. *Cellulose* **2010**, 17 (5), 1005–1020.

[281] Nogi, M., Iwamoto, S., Nakagaito, A. N. & Yano, H., Optically transparent nanofiber paper. *Adv. Mater.* **2009**, 21 (16), 1595–1598.

[282] Liebner, F., Haimer, E., Wendland, M., Neouze, M.-A., Schlufter, K., Miethe, P., Heinze, T., Potthast, A. & Rosenau, T., Aerogels from unaltered bacterial cellulose: Application of scCO2 drying for the preparation of shaped, ultra-lightweight cellulosic aerogels. *Macromol. Biosci.* **2010**, 10 (4), 349–352.

[283] Kistler, S. S., Coherent expanded aerogels and jellies. *Nature* **1931**, 127, 741.

[284] Tan, C., Fung, B. M., Newman, J. K. & Vu, C., Organic aerogels with very high impact strength. *Adv. Mater.* **2001**, 13 (9), 644–646.

[285] Jin, H., Nishiyama, Y., Wada, M. & Kuga, S., Nanofibrillar cellulose aerogels. *Colloids Surf., A* **2004**, 240 (1–3), 63–67.

[286] Hoepfner, S., Ratke, L. & Milow, B., Synthesis and characterisation of nanofibrillar cellulose aerogels. *Cellulose* **2008**, 15 (1), 121–129.

[287] Gawryla, M. D., van den Berg, O., Weder, C. & Schiraldi, D. A., Clay aerogel/cellulose whisker nanocomposites: a nanoscale wattle and daub. *Journal of Materials Chemistry* **2009**, 19 (15), 2118–2124.

[288] Kettunen, M., Silvennoinen, R. J., Houbenov, N., Nykänen, A., Ruokolainen, J., Sainio, J., Pore, V., Kemell, M., Ankerfors, M., Lindström, T., Ritala, M., Ras, R. H. A. & Ikkala, O., Photoswitchable superabsorbency based on nanocellulose aerogels. *Adv. Funct. Mater.* **2011**, 21 (3), 510–517.

[289] Frensemeier, M., Koplin, C., Jaeger, R., Kramer, F. & Klemm, D., Mechanical properties of bacterially synthesized nanocellulose hydrogels. *Macromolecular Symposia* **2010**, 294 (2), 38–44.

[290] Haimer, E., Wendland, M., Schlufter, K., Frankenfeld, K., Miethe, P., Potthast, A., Rosenau, T. & Liebner, F., Loading of bacterial cellulose aerogels with bioactive compounds by antisolvent precipitation with supercritical carbon dioxide. *Macromolecular Symposia* **2010**, 294 (2), 64–74.

[291] Sehaqui, H., Zhou, Q. & Berglund, L. A., High-porosity aerogels of high specific surface area prepared from nanofibrillated cellulose (NFC). *Compos. Sci. Technol.* **2011**, 71 (13), 1593–1599.

[292] Shin, Y., Bae, I.–T., Arey, B. W. & Exarhos, G. J., Simple preparation and stabilization of nickel nanocrystals on cellulose nanocrystal. *Mater. Lett.* **2007**, 61 (14–15), 3215–3217.

[293] Shin, Y. & Exarhos, G. J., Template synthesis of porous titania using cellulose nanocrystals. *Mater. Lett.* **2007**, 61 (11–12), 2594–2597.

[294] Ifuku, S., Tsuji, M., Morimoto, M., Saimoto, H. & Yano, H., Synthesis of silver nanoparticles templated by TEMPO-mediated oxidized bacterial cellulose nanofibers. *Biomacromolecules* **2009**, 10 (9), 2714–2717.

[295] Drogat, N., Granet, R., Sol, V., Memmi, A., Saad, N., Klein Koerkamp, C., Bressollier & P., Krausz, P., Antimicrobial silver nanoparticles generated on cellulose nanocrystals. *J. Nanopart. Res.* **2010**, 1–6.

[296] Gruber, S., Taylor, R. N. K., Scheel, H., Greil, P. & Zollfrank, C., Cellulose-biotemplated silica nanowires coated with a dense gold nanoparticle layer. *Mater. Chem. Phys.* **2011**, 129 (1–2), 19–22.

[297] Shin, Y., Blackwood, J. M., Bae, I.-T., Arey, B. W. & Exarhos, G. J., Synthesis and stabilization of selenium nanoparticles on cellulose nanocrystal. *Mater. Lett.* **2007**, 61 (21), 4297–4300.

[298] Zhou, Y., Ding, E. Y. & Li, W. D., Synthesis of TiO_2 nanocubes induced by cellulose nanocrystal (CNC) at low temperature. *Mater. Lett.* **2007**, 61 (28), 5050–5052.

[299] Padalkar, S., Capadona, J., Rowan, S., Weder, C., Moon, R. & Stanciu, L., Self-assembly and alignment of semiconductor nanoparticles on cellulose nanocrystals. *J. Mat. Sci.* **2011**, 46 (17), 5672–5679.

[300] Hage, J. L. T., Reuter, M. A., Schuiling, R. D. & Ramtahalsing, I. S., Reduction of copper with cellulose in an autoclave, an alternative to electrolysis? *Miner. Eng.* **1999**, 12 (4), 393–404.

[301] Padalkar, S., Capadona, J. R., Rowan, S. J., Weder, C., Won, Y. H., Stanciu, L. A. & Moon, R. J., Natural biopolymers: Novel templates for the synthesis of nanostructures. *Langmuir* **2010**, 26 (11), 8497–502.

[302] Oksman, K., Mathew, A. P., Bondeson, D. & Kvien, I., Manufacturing process of cellulose whiskers/polylactic acid nanocomposites. *Compos. Sci. Technol.* **2006**, 66 (15), 2776–2784.

[303] Sorrentino, A., Gorrasi, G. & Vittoria, V., Potential perspectives of bio-nanocomposites for food packaging applications. *Trends in Food Science &, Technology* **2007**, 18 (2), 84–95.

[304] Park, W.-I., Kang, M., Kim, H. S. & Jin, H. J., Electrospinning of poly(ethylene oxide) with bacterial cellulose whiskers. *Macromolecular Symposia* **2007**, 249–250 (1), 289–294.

[305] Frenot, A., Henriksson, M. W. & Walkenström, P., Electrospinning of cellulose-based nanofibers. *J. Appl. Polym. Sci.* **2007**, 103 (3), 1473–1482.

[306] Cao, X. D., Chen, Y., Chang, P. R., Stumborg, M. & Huneault, M. A., Green composites reinforced with hemp nanocrystals in plasticized starch. *J. Appl. Polym. Sci.* **2008**, 109 (6), 3804–3810.

[307] Qua, E. H., Hornsby, P. R., Sharma, H. S. S., Lyons, G. & McCall, R. D., Preparation and characterization of poly(vinyl alcohol) nanocomposites made from cellulose nanofibers. *J. Appl. Polym. Sci.* **2009**, 113 (4), 2238–2247.

[308] Bhatnagar, A. & Sain, M., Processing of cellulose nanofiber-reinforced composites. *J. Reinf. Plast. Comp.* **2005**, 24 (12), 1259–1268.

[309] Lee, S. Y., Mohan, D. J., Kang, I. A., Doh, G. H., Lee, S. & Han, S. O., Nanocellulose reinforced PVA composite films: Effects of acid treatment and filler loading. *Fiber Polym.* **2009**, 10 (1), 77–82.

[310] Bras, J., Hassan, M. L., Bruzesse, C., Hassan, E. A., El–Wakil, N. A. & Dufresne, A., Mechanical, barrier, and biodegradability properties of bagasse cellulose whiskers reinforced natural rubber nanocomposites. *Ind. Crops Prod.* **2010**, 32 (3), 627–633.

[311] Dufresne, A. & Vignon, M. R., Improvement of starch film performances using cellulose microfibrils. *Macromolecules* **1998**, 31 (8), 2693–2696.

[312] Dufresne, A., Dupeyre, D. & Vignon, M. R., Cellulose microfibrils from potato tuber cells: Processing and characterization of starch–cellulose microfibril composites. *J. Appl. Polym. Sci.* **2000**, 76 (14), 2080–2092.

[313] Anglès, M. N. & Dufresne, A., Plasticized starch/tunicin whiskers nanocomposites. 1. Structural analysis. *Macromolecules* **2000**, 33 (22), 8344–8353.

[314] López-Rubio, A., Lagaron, J. M., Ankerfors, M., Lindström, T., Nordqvist, D., Mattozzi, A. & Hedenqvist, M. S., Enhanced film forming and film properties of amylopectin using micro-fibrillated cellulose. *Carbohyd. Polym.* **2007**, 68 (4), 718–727.

[315] Siqueira, G., Tapin-Lingua, S., Bras, J., Perez, D. d. S. & Dufresne, A., Mechanical properties of natural rubber nanocomposites reinforced with cellulosic nanoparticles obtained from combined mechanical shearing, and enzymatic and acid hydrolysis of sisal fibers. *Cellulose* **2011**, 18 (1), 57–65.

[316] Wang, Y., Cao, X. & Zhang, L., Effects of cellulose whiskers on properties of soy protein thermoplastics. *Macromol. Biosci.* **2006**, 6 (7), 524–531.

[317] Anglès, M. N., & Dufresne, A., Plasticized starch/tunicin whiskers nanocomposite materials. 2. Mechanical behavior. *Macromolecules* **2001**, 34 (9), 2921–2931.

[318] Roohani, M., Habibi, Y., Belgacem, N. M., Ebrahim, G., Karimi, A. N. & Dufresne, A., Cellulose whiskers reinforced polyvinyl alcohol copolymers nanocomposites. *Eur. Polym. J.* **2008**, 44 (8), 2489–2498.

[319] Fernandes, S. C. M., Freire, C. S. R., Silvestre, A. J. D., Pascoal Neto, C., Gandini, A., Berglund, L. A. & Salmén, L., Transparent chitosan films reinforced with a high content of nanofibrillated cellulose. *Carbohydr. Polym.* **2010**, 81 (2), 394–401.

[320] Li, Q., Zhou, J. P. & Zhang, L. N., Structure and properties of the nanocomposite films of chitosan reinforced with cellulose whiskers. *J. Polym. Sci. Pol. Phys.* **2009**, 47 (11), 1069–1077.

[321] Roy, D., Semsarilar, M., Guthrie, J. T. & Perrier, S., Cellulose modification by polymer grafting: A review. *Chem. Soc. Rev.* **2009**, 38 (7), 2046–2064.

[322] Çetin, N. S., Tingaut, P., Özmen, N., Henry, N., Harper, D., Dadmun, M. & Sèbe, G., Acetylation of cellulose nanowhiskers with vinyl acetate under moderate conditions. *Macromol. Biosci.* **2009**, 9 (10), 997–1003.

[323] Lin, N., Huang, J., Chang, P. R., Feng, J. & Yu, J., Surface acetylation of cellulose nanocrystal and its reinforcing function in poly(lactic acid). *Carbohyd. Polym.* **2011**, 83 (4), 1834–1842.

[324] Rodionova, G., Lenes, M., Eriksen, Ø. & Gregersen, Ø., Surface chemical modification of microfibrillated cellulose: Improvement of barrier properties for packaging applications. *Cellulose* **2011**, 18 (1), 127–134.

[325] Birgit, B. & John, R. D., Single-step method for the isolation and surface functionalization of cellulosic nanowhiskers. *Biomacromolecules* **2009**, 10 (2), 334–341.

[326] de Menezes, A. J., Siqueira, G., Curvelo, A. A. S. & Dufresne, A., Extrusion and characterization of functionalized cellulose whiskers reinforced polyethylene nanocomposites. *Polymer.* **2009**, 50 (19), 4552–4563.

[327] Lee, K. Y., Quero, F., Blaker, J., Hill, C., Eichhorn, S. & Bismarck, A., Surface only modification of bacterial cellulose nanofibres with organic acids. *Cellulose* **2011**, 18 (3), 595–605.

[328] Pahimanolis, N., Hippi, U., Johansson, L.–S., Saarinen, T., Houbenov, N., Ruokolainen, J. & Seppälä, J., Surface functionalization of nanofibrillated cellulose using click-chemistry approach in aqueous media. *Cellulose* **2011**, 18 (5), 1201–1212.

[329] Goussé, C., Chanzy, H., Excoffier, G., Soubeyrand, L. & Fleury, E., Stable suspensions of partially silylated cellulose whiskers dispersed in organic solvents. *Polymer.* **2002**, 43 (9), 2645–2651.

[330] Pei, A., Zhou, Q. & Berglund, L. A., Functionalized cellulose nanocrystals as biobased nucleation agents in poly(l-lactide) (PLLA) – Crystallization and mechanical property effects. *Compos. Sci. Technol.* **2010**, 70 (5), 815–821.

[331] Zhang, J., Jiang, N., Dang, Z., Elder, T. & Ragauskas, A., Oxidation and sulfonation of cellulosics. *Cellulose* **2008**, 15 (3), 489–496.

[332] Okita, Y., Fujisawa, S., Saito, T. & Isogai, A., TEMPO-oxidized cellulose nanofibrils dispersed in organic solvents. *Biomacromolecules* **2010**, 12 (2), 518–522.

[333] Nge, T., Nogi, M., Yano, H. & Sugiyama, J., Microstructure and mechanical properties of bacterial cellulose/chitosan porous scaffold. *Cellulose* **2010**, 17 (2), 349–363.

[334] Habibi, Y., Goffin, A. L., Schiltz, N., Duquesne, E., Dubois, P. & Dufresne, A., Bionanocomposites based on poly([varepsilon]-caprolactone)-grafted cellulose nanocrystals by ring-opening polymerization. *J. Mater. Chem.* **2008**, 18 (41), 5002–5010.

[335] Lasseuguette, E., Grafting onto microfibrils of native cellulose. *Cellulose* **2008**, 15 (4), 571–580.

[336] Zoppe, J. O., Österberg, M., Venditti, R. A., Laine, J. & Rojas, O. J., Surface interaction forces of cellulose nanocrystals grafted with thermoresponsive polymer brushes. *Biomacromolecules* **2011**, 12 (7), 2788–2796.

[337] Ramsden, W., Separation of solids in the surface-layers of solutions and 'suspensions' (observations on surface-membranes, bubbles, emulsions, and mechanical coagulation). *Proceedings of the Royal Society of London* **1903**, 72, 156–164.

[338] Briggs, T. R., Emulsions with finely divided solids. *Journal of Industrial & Engineering Chemistry* **1921**, 13 (11), 1008–1010.

[339] Oza, K. P. & Frank, S. G., Microcrystalline cellulose stabilized emulsions. *J. Disper. Sci. Technol.* **1986**, 7 (5), 543–561.

[340] Ougiya, H., Watanabe, K., Morinaga, Y. & Yoshinaga, F., Emulsion-stabilizing effect of bacterial cellulose. *Japan Society for Bioscience, Biotechnology, and Agrochemistry* **1997**, 61 (9), 1541–1545.

[341] Andresen, M., Johansson, L.–S., Tanem, B. S. & Stenius, P., Properties and characterization of hydrophobized microfibrillated cellulose. *Cellulose* **2006**, 13 (6), 665–677.

[342] Andresen, M. & Stenius, P., Water-in-oil emulsions stabilized by hydrophobized microfibrillated cellulose. *J. Dispersion Sci. Technol.* **2007**, 28 (6), 837–844.

[343] Lif, A., Stenstad, P., Syverud, K., Nydén, M. & Holmberg, K., Fischer–Tropsch diesel emulsions stabilised by microfibrillated cellulose and nonionic surfactants. *J. Colloid. Interface Sci.* **2010**, 352 (2), 585–592.

[344] Kalashnikova, I., Bizot, H., Cathala, B. & Capron, I., New Pickering emulsions stabilized by bacterial cellulose nanocrystals. *Langmuir* **2011**, 27 (12), 7471–7479.

[345] Xhanari, K., Syverud, K., Chinga-Carrasco, G., Paso, K. & Stenius, P., Structure of nanofibrillated cellulose layers at the o/w interface. *J. Colloid. Interface Sci.* **2011**, 356 (1), 58–62.

[346] Xhanari, K., Syverud, K. & Stenius, P., Emulsions stabilized by microfibrillated cellulose: The effect of hydrophobization, concentration and O/W ratio. *J. Disper. Sci. Technol.* **2011**, 32 (3), 447–452.

[347] Okubo, K., Fujii, T. & Thostenson, E. T., Multi-scale hybrid biocomposite: Processing and mechanical characterization of bamboo fiber reinforced PLA with microfibrillated cellulose. *Composites Part A: Applied Science and Manufacturing* **2009**, 40 (4), 469–475.

[348] Bondeson, D. & Oksman, K., Polylactic acid/cellulose whisker nanocomposites modified by polyvinyl alcohol. *Composites Part A: Applied Science and Manufacturing* **2007**, 38 (12), 2486–2492.

[349] Jonoobi, M., Harun, J., Mathew, A. P. & Oksman, K., Mechanical properties of cellulose nanofiber (CNF) reinforced polylactic acid (PLA) prepared by twin screw extrusion. *Compos. Sci. Technol.* **2010**, 70 (12), 1742–1747.

[350] Pei, A., Malho, J.-M., Ruokolainen, J., Zhou, Q. & Berglund, L. A., Strong nanocomposite reinforcement effects in polyurethane elastomer with low volume fraction of cellulose nanocrystals. *Macromolecules* **2011**, 44 (11), 4422–4427.

[351] Ten, E., Turtle, J., Bahr, D., Jiang, L. & Wolcott, M., Thermal and mechanical properties of poly(3-hydroxybutyrate-co-3-hydroxyvalerate)/cellulose nanowhiskers composites. *Polymer.* **2010**, 51 (12), 2652–2660.

Chapter 6
The interface and bonding mechanisms of plant fibre composites

Dr Yonghui Zhou and Professor Mizi Fan

This chapter focuses primarily on the interface and bonding mechanisms of plant fibre composites. It is organised in four main sections: an overview of the compatibility between the constituents of composites; modifications aimed at improving the compatibility and interfacial bonding of composites; the physical, mechanical and chemical bonding mechanisms; and the interface structure of composites. It provides a basis for further research and industrialisation of plant fibre composites.

6.1 Introduction

Plant fibre composite, as a significant branch of composite materials, has seen a rapid expansion over the last decade. This growth has largely been due to the advantageous features that plant fibres provide over inorganic fillers and/or reinforcements, i.e. abundance, environmental friendliness, biodegradability, nontoxicity, low cost and density, flexibility during processing, and high tensile strength and flexural modulus [1–6]. Compared with wood materials, plant fibre composite possesses better flexural and impact strength, higher moisture resistance, less shrinkage and improved weatherability. Regardless of these benchmark characteristics, however, optimisation of interfacial bonding between the plant fibres and the polymer matrix is one of the most indispensable procedures with respect to the optimal formulation of plant fibre composite [1].

The fibre–matrix interface is a reaction or diffusion zone in which two phases or components are physically, mechanically and/or chemically combined. Interfacial adhesion between the fibre and the matrix plays a fundamental role in terms of the factors that govern the mechanical characteristics of the composite [7]. The factors affecting the interfacial bonding between the fibre and the matrix are the mechanical interlocking, the molecular attractive forces and the chemical bonds. However, the naturally hydrophilic plant fibres are not inherently compatible with hydrophobic polymers. In addition to the pectin and waxy substances in plant fibre acting as a barrier to interlocking with the non-polar polymer matrix, the presence of plenty of hydroxyl groups hinders its operative reaction with the matrix [8–11]. Therefore, modification of the surface characteristics of the plant fibre and the hydrophobic polymer matrix is essential in order to formulate a reasonable composite with superior interfacial bonding and effective inherent stress transfer throughout the interface. Various approaches, including physical treatments (i.e. solvent extraction, heat treatment, corona and plasma treatments), physico-chemical treatments (i.e. laser, γ-ray and UV bombardment) [12] and chemical modifications, have been attempted for improving the compatibility and bonding between the lignocellulosic molecules and hydrocarbon-based polymers [13–14].

Increasing environmental awareness has brought substantial research and industrial investment into another class of plant fibre composites, in which the matrices – including starch, cellulose, chitin and chitosan, collagen, lignin, natural rubber, polyhydroxyalkanoate, poly(lactic acid) (PLA) and soy-based resins – are fully degradable and sustainable biopolymers. Unlike synthetic polymer-based composites, the similar polarities of both reinforcements and matrices impart the biodegradable composites better compatibility and interfacial adhesion [15]. However, some surface treatment of the fibres is specifically needed for the benefits of lowering moisture sensitivity, which generally leads to dimensional instability due to the fibre swelling, and hence loss of interface integrity.

6.2 Compatibility between constituents of plant fibre composites

6.2.1 Compatibility between hydrophilic and hydrophobic constituents

The main components of plant fibre include cellulose, hemicellulose, lignin, pectin, waxes and other low-molecule substances. Cellulose is the fundamental structural component, found in the form of slender rodlike crystalline microfibrils, aligned along the length of the fibre. It is a semicrystalline polysaccharide consisting of a linear chain of hundreds to thousands of β-(1-4)-glycosidic bonds linked D-glucopyranose with the presence of large numbers of hydroxyl groups. Hemicellulose is a lower molecular weight polysaccharide that functions as a cementing matrix between cellulose microfibrils, present along with cellulose in almost all plant cell walls. While cellulose is crystalline, strong, and resistant to hydrolysis, hemicellulose has a random, amorphous structure with little strength. Furthermore, it is hydrophilic and can be easily hydrolysed by dilute acids and bases. Lignin is a class of complex hydrocarbon polymer (cross-linked phenol polymers) that gives rigidity to plants. It is relatively hydrophobic and aromatic in nature. Pectin is a structural heteropolysaccharide contained in the primary cell walls of plants, giving the plants flexibility. Wax and water-soluble substances are used to protect fibre on the fibre surface [16–19]. The unique chemical structure makes the plant fibre hydrophilic in nature.

Although there are many advantages associated with the use of plant fibre as a reinforcement in polymer composites, the incorporation of plant fibre into hydrophobic and non-polar polymers leads to heterogeneous systems due to the lack of moderate compatibility and adhesion between the fibre and the matrix. These would also cause some problems in the composite processing and material performance, including poor moisture resistance, inferior fire resistance, limited processing temperatures, the formation of hydrogen bonds within the fibre itself, a tendency for the fibre to agglomerate into bundles, uneven distribution in the polar matrix during compound processing, and insufficient wetting of the fibre by the matrix which results in weak interfacial adhesion [16, 20–21].

By this token, various physical and chemical modifications have been tried in an attempt to decrease the hydrophilicity of plant fibre, enhance the wettability of the fibre by polymers and promote interfacial adhesion. Each approach has a different efficiency and uses a different mechanism for optimising the interfacial characteristics of the composite. Physical treatments, such as electric discharge, in general change the structural and surface properties of the fibre by introducing surface cross-linking, modifying the surface energy and/or generating reactive free radicals and groups, and thereby influence the mechanical bonding to the matrix. Chemical modification provides a means of permanently altering the nature of fibre cell walls by grafting polymers onto the fibres, cross-linking the fibre cell walls, or by using coupling agents [22]. Such modification strategies tackling with the compatibility and interfacial bonding of plant fibre composite is discussed in detail in the next section.

6.2.2 Compatibility between hydrophilic constituents (plant fibre biodegradable matrices)

Research on polymers obtained from different biorenewable resources, generally referred to as bio-based polymers and designed to be biodegradable, has increased substantially in recent times. It is well known that renewable resources such as bioproducts (e.g. cellulose or chitin, vegetable fats and oils, corn starch, etc.), bacteria, as well as non-renewable petroleum (e.g. aliphatic/aliphatic–aromatic copolymer), are the source of a variety of polymeric materials [15, 23–26]. Biodegradable polymers can accordingly be classified as naturally occurring or as synthetic based on their origins. Natural polymers are available in large quantities from renewable sources, i.e. cellulose, pectin, lignin, collagens, chitin/chitosan, starch, proteins, lipids, etc. Microbially synthesised polymers include polyhydroxyalkanoates (PHA) and bacterial cellulose, while chemically synthesised polymers are produced from conventional petroleum-based resource monomers, including polyacids, poly(vinyl alcohols) and polyesters [15, 27–29]. Biodegradable bioplastics can break down in either anaerobic or aerobic environments, depending on how they are manufactured, and thus are being envisaged as an alternative to their counterparts in olefin plastics. The incorporation of plant fibre into bioplastics to fabricate fully biodegradable composite materials has attracted attention for a variety of purposes, in particular for multifunctional applications, since many of these polymers, in addition to being biodegradable, also possess antimicrobial and antioxidant properties [30]. At this stage, the main impediment to the widespread use of bioplastic polymers is the high initial cost, currently from three to five times the cost of the extensively used resins such as PP, PVC, LDPE and HDPE [31].

Starch is not a true thermoplastic but it can be converted to thermoplastic starch (TPS) in the presence of plasticisers (i.e. water, glycerine, glycol, sorbitol, etc.) under high temperatures and shear by forming hydrogen bonds with the starch [32–33]. Compared to most synthetic plastics currently in use, the properties of TPS are unfortunately not compatible with its application as a thermoplastic, especially the highly water-soluble character and poor mechanical and thermal properties [34–35]. A good approach to tackling these drawbacks is the use of natural fibres as a reinforcement for TPS, including luffa fibre [33], sisal fibre [36], cotton fibre [37], sugarcane fibre [38], eucalyptus fibre [34], bagasse cellulose [39], wheat straw fibre [40] and winceyette fibre [32]. Starch is a naturally hydrophilic carbohydrate consisting of a large number of glucose units joined by glycosidic bonds. The similarities in chemical structure between TPS and plant fibre provide the resultant plant fibre/TPS composite with high compatibility and adhesion between the reinforcement and the TPS matrix. As a result, the inclusion of plant fibre not only leads to an improvement in the tensile strength and water and thermal resistance of the composite, but also hinders the re-crystallisation of starch, owing to the strong interaction between the fibre and the starch [32–35].

Polylactic acid (PLA) is one of the most promising biodegradable polymers and is derived from natural feedstock or agricultural products such as corn starch, rice, potatoes, sugar beet, etc. It has attracted great interest as a commodity polymer capable of replacing petrochemical polymers due to its similar mechanical properties to oil-derived plastics. However, the wider uptake of PLA is restricted by performance deficiencies, such as poor toughness arising from its inherent brittleness, and also its lower molecular weight and much higher production cost in comparison to conventional plastics [41–42]. In order to overcome its brittle nature, plant fibres have recently been incorporated into PLA matrix by researchers. Plant fibre/PLA composites containing less than 30 wt.% fibre have been shown to have increased tensile modulus and to have reduced tensile strength compared with PLA, and this has been attributed to factors that include the weak interfacial interaction between the hydrophobic PLA matrix and the hydrophilic cellulose fibres, and the lack of fibre dispersion due to a high degree of fibre agglomeration [43]. Various chemical methods of modifying the surface of the cellulosic fibres, including esterification, acetylation and cyanoethylation, along with coupling agents or compatibilisers, have been explored in an effort to improve the interaction and adhesion occurring at the interface between the PLA matrix and the fibres by inducing chemical bonding and enhancing compatibilisation [5, 43–46].

Polyhydroxyalkanoates (PHA) are a family of linear polyesters produced in nature by bacterial fermentation of sugar or lipids as intracellular carbon and energy storage granules. PHAs have a structure with the same three carbon backbones and distinct alkyl groups at the β or 3 positions. The main copolymers or homopolymers of the PHAs are poly(3-hydroxybutyrate) (PHB), poly(3-hydroxybutyrate-co-3-hydroxyvalerate) (PHBV), poly(3-hydroxybutyrate-co-hydroxyoctanoate) and poly(3-hydroxybutyrate-co-hydroxyhexanoate) [47]. As one of the most studied and promising alternatives to the current mainstream petrochemical plastics, PHA bioplastics have two major drawbacks: low thermal stability (they degrade at 200–250 °C) and brittleness. The addition of plant fibre to reinforce and toughen PHAs has been attempted. Percentage crystallinity of the formulated plant fibre/PHA composite was found to increase with the increase of fibril loading. The composites also exhibited higher thermal stability, higher glass transition temperature and higher heat distortion temperature than pure PHA. The composites could be developed further for various structural applications [48–51].

6.3 Modification of constituents of the composite

6.3.1 Physical treatments

Physical treatments of plant fibre alter the structure and surface properties of the fibres without the use of chemical agents, and thereby influence the mechanical bonding with the matrix in the composite. Radiation and discharge treatments, such as UV radiation, gamma radiation, corona and plasma treatments, are the most commonly used physical techniques in plant fibre composite with regards to improving the functional properties of the plant fibre, generally enabling significant physical and chemical changes as well as changes in the surface structure and surface energy of the plant fibre [52–55].

A typical treatment of gamma radiation is known to deposit energy on the plant fibre in the composite, and radicals are then produced on the cellulose chain by hydrogen and hydroxyl abstraction, ruptures of some carbon-carbon bonds, and chain scission. Simultaneously, peroxide radicals are generated when matrix polymers are irradiated in the presence of oxygen. These active sites in both the fibre and the matrix, produced by the gamma radiation, result in better bonding between the filler and polymer matrix, which consequently improves the mechanical strength of the composite [53, 56].

Corona treatment is a surface modification technique that uses a low-temperature corona discharge plasma to impart changes in the properties of a surface (e.g. surface energy). The corona plasma is generated by the application of high voltage to an electrode that has a sharp tip. The plasma forms at the tip. A linear array of electrodes is often used to create a curtain of corona plasma. The corona treatment of plant fibre causes the formation of high-energy electromagnetic fields close to the charged points, with consequent ionisation in their proximity. In the ionised region, the excited species (i.e. ions, radicals, amongst others) are present and the latter are active in surface modification, typically by the introduction of oxygen-containing groups in the molecular chain of the plant fibre [57]. Plastic is a man-made synthetic material that contains long homogeneous molecular chains which form a strong and uniform product. The chains of molecules are normally joined end-to-end, forming even longer chains, leaving only a few open chain ends and thus providing only a small number of bonding points at the surface. The small number of bonding points causes low adhesion and wettability, which is a problem in converting processes. A high-frequency charge provides both a more efficient and controllable method of increasing the adhesion and wettability of a plastic surface. During corona discharge treatment, electrons are accelerated to the surface of the plastic, resulting in the long chains to rupture, producing a multiplicity of open ends and forming free valences. The ozone from the electrical discharge creates oxygenation, which in turn forms new carbonyl groups with a higher surface energy. The result is an improvement in the chemical connection between the molecules in the plastic and the applied media/liquid. This surface treatment will not reduce or change the strength, neither will it change the appearance of the material. Such corona treatment has been widely applied in plant fibre composites, especially in composites involving polyolefins as the matrix [52, 58–59].

Plasma treatment is another useful physical technique for improving the surface properties of plant fibre and polymeric materials by utilising ingredients such as high-energy photons, electrons, ions, radicals and excited species. Generally, the modification of plant fibre by treatment in cold oxygen plasma obtained in a corona discharge under optimal operating conditions turns the fibre into a semi-active filler for the cellulosic compounds [60–62]. Thus, adhesion at the fibre/matrix interface increases with plasma treatment. The resultant impact is an improvement in the mechanical properties (i.e. flexural strength and modulus, tensile strength and modulus) of the composites. Nevertheless, the fibre may degrade over a longer time exposure due to the constant impact of particles on the surface, which inevitably weakens the interfacial adhesion [63–65]. UV radiation has also been employed to modify the surface characteristics of plant fibre composites [52, 66–67]. The UV treatment of single jute fibre and jute yarn was found to result in higher gains in polarity in comparison with those observed in relation to corona-treated counterparts. The polarity and yarn tenacity could be adjusted by increasing treatment time at a constant bulb-sample distance or alternatively decreasing the distance. For the benefits of improving the overall mechanical properties of plant fibre composites, an appropriate balance needs to be achieved between the increased polarity of the fibre surface and the decrease in fibre strength subsequent to excessive surface oxidation by UV radiation or corona discharge [52].

6.3.2 Chemical treatments

6.3.2.1 Alkaline treatment or mercerisation

Alkali treatment is one of the most commonly used chemical methods of modifying the cellulosic molecular structure of plant fibre when it is being used to reinforce thermoplastics and thermosets. This treatment, usually performed in NaOH aqueous solution, disrupts fibre clusters and forms amorphous at the expense of highly packed crystalline cellulose. The important modification occurring is the disruption of hydrogen bonding in the network structure. During the treatment, the sensitive hydroxyl groups (OH) are broken down, and thus react with water (H-OH) leaving the ionised reactive molecules to form alkoxide with NaOH (Scheme 1).

$$\text{Fibre} - \text{OH} + \text{NaOH} \rightarrow \text{Fibre} - \text{O} - \text{Na} + H_2O + \text{impurity} \qquad \text{(Scheme 1)}$$

As a result, the hydrophilic OH groups are reduced and the surface roughness of the fibre is increased. It also removes a certain portion of hemicellulose, lignin, waxes, and oils covering the external surface of the fibre cell wall, depolymerises cellulose and exposes the short-length crystallites [68–70]. Therefore, when the alkaline-treated plant fibre is used to reinforce polar polymer composites, in comparison with the composite filled with untreated plant fibre, the enhanced surface roughness and increased reactive sites exposed on the surface lead to better mechanical interlocking and adhesion with the matrix, both of which are in charge of the interfacial strength of the composite [71–72]. However, it should be noted that the superfluous alkali concentration will result in excess delignification of plant fibre, thus weakening or damaging the fibre being treated [68, 73].

The alkali treatment of plant fibre can also be carried out in combination with other treatments. Doan *et al.* reported on the surface structures of untreated jute fibre, alkali-treated jute fibre and alkali/organosilane/aqueous epoxy dispersions (ED) treated jute fibre [74]. The untreated jute fibre has a rather smooth surface due to the cementing made up of fats, waxes, lignin, pectin, and hemicellulose forming on the fibre surface [75]. The alkali treatment provided the fibre with a flaky or grooved surface by partially removing the cementing. The NaOH/(APS (3-aminopropyl-triethoxy-silane)+ED) treated fibre was covered by a sizing of varying thickness, forming a film on the fibre surface. By contrast, the NaOH/PAPS (3-phenyl-aminopropyl-trimethoxy-silane) treated fibre showed a much rougher surface since the sizing was less uniform in thickness or rather varying on a much higher scale, and appeared more flaky and less attached [74].

6.3.2.2 Acetylation treatment

Acetylation of plant fibre is known as an esterification method causing plasticisation of the fibre by introducing the acetyl functional group CH_3COO^-. The main purpose of the reaction is to substitute hydrophilic hydroxyl groups (OH) of the cell wall with the acetyl group CH_3COO^- from acetic anhydride ($CH_3COO{-}C{=}O{-}CH_3$), therefore rendering the fibre surface more hydrophobic (Scheme 2).

$$\text{Fibre} - OH + CH_3 - C(=O) - O - C\,(=O) - CH_3 \rightarrow \text{Fibre} - O - COCH_3 + CH_3COOH \qquad \text{(Scheme 2)}$$

The hydroxyl groups reacting are minor constituents of the fibre, i.e. hemicelluloses and lignin, and amorphous cellulose. This is because the hydroxyl groups in the crystalline regions of the fibre are closely packed with strong interlock bonding and are fairly inaccessible to chemical reagents [76]. In order to accelerate the reaction and maximise the degree of acetylation, an acid catalyst such as sulphuric acid and acetic acid are commonly used during the treatment [7, 76]. The esterification reaction not only stabilises the cell walls, particularly with regard to moisture uptake and consequent dimensional variation of the plant fibre, but it also provides the fibre rough surface tomography with less void contents, thereby allowing the adhesion of the fibre to the matrix to be improved [77–79]. It has been reported that the acetylation treatment of plant fibres resists up to 65% moisture absorption depending on the degree of acetylation [4, 76]. More importantly, in comparison to composites reinforced with untreated plant fibre, this esterification of plant fibre results in improved stress transfer efficiency at the interface and improved mechanical properties (tensile, flexural and impact properties) of its composites [80]. In addition, the enhanced hydrophobicity of the treated fibres is able to provide the composite with higher volume resistivity than that of untreated composites by reducing the dielectric constant of the composite [79].

6.3.2.3 Benzoylation treatment

Benzoylation treatment, an important transformation in organic synthesis, is another treatment aimed at decreasing the hydrophilicity of plant fibre. Prior to carrying out the reaction between the fibre and benzoyl groups ($C_6H_5C{=}O$), the plant fibre should be initially pretreated with NaOH aqueous solution in order to activate and expose the hydroxyl groups on the fibre surface. Thus, the fibre can be treated with benzoyl chloride, for which to be further substituted by benzoyl group, the reaction is given in Scheme 3 [73, 81–82]. This creates a more hydrophobic nature in the fibre and improves fibre matrix adhesion, thereby considerably increasing the strength and thermal stability of the composite [83].

$$\text{Fibre} - \text{O} - \text{Na} + \text{C}_6\text{H}_5 - \text{COCl} \rightarrow \text{Fibre} - \text{O} - \text{CO} - \text{C}_6\text{H}_5 + \text{NaCl} \qquad \text{(Scheme 3)}$$

6.3.2.4 Peroxide treatment

The interface property of plant fibre composite can also be improved by peroxide treatment. Peroxide is a chemical compound with the specific functional group ROOR containing the divalent ion bond O–O. In contrast to oxide ions, the oxygen atoms in the peroxide ion have an oxidation state of –1. Benzoyl peroxide and dicumyl peroxide are the most commonly used chemicals for plant fibre surface modification in the organic peroxides family. Both chemicals are highly reactive and are inclined to decompose to free radicals of the form $RO^{\cdot}$. The $RO^{\cdot}$ can then be grafted onto the cellulose macromolecules polymer chains by reacting with the hydrogen groups of the plant fibre and the polymer matrix (Scheme 4) [83, 81, 73]. As a result, good fibre matrix adhesion occurs along the interface of the composite [77, 84].

$$\begin{gathered} \text{RO} - \text{OR} \rightarrow 2\text{RO}^{\cdot}, \ \text{RO}^{\cdot} + \text{Polymer} \rightarrow \text{ROH} + \text{Polymer} \\ \text{RO}^{\cdot} + \text{Cellulose} \rightarrow \text{ROH} + \text{Cellulose}^{\cdot} \\ \text{Polymer}^{\cdot} + \text{Cellulose}^{\cdot} \rightarrow \text{Polymer} - \text{Cellulose} \end{gathered} \qquad \text{(Scheme 4)}$$

6.3.2.5 Silane treatment

Silane is an inorganic chemical compound with the chemical formula SiH_4. Silanes are used as coupling agents to modify the plant fibre surface. A typical silane coupling agent bears two reactive groups. One end of the silane agent with alkoxysilane groups is capable of reacting with hydroxyl-rich surfaces, namely wood or other plant fibres, whereas the other end is left to interact with the polymer matrix. Specifically, in the presence of moisture, the silane (hydrolysable alkoxy group) is able to react with water to form silanol, which is able to further react with the hydroxyl groups attached to the cellulose, hemicellulose and lignin molecules in the filler through an ether linkage with the removal of water [85]. The uptake of silane is very much dependent on a number of factors, including temperature, pH, hydrolysis time and the organo-functionality of silane [12]. The reaction scheme is given in Scheme 5.

$$\text{Fibre} - \text{OH} + \text{R} - \text{Si(OH)}_3 \rightarrow \text{Fibre} - \text{O} - \text{Si(OH)}_2 - \text{R} \qquad \text{(Scheme 5)}$$

On the other hand, the polysulphide bonds in silane molecules are able to react with the polymer matrix [85]. Therefore, the hydrocarbon chains provided by the application of silane restrain the swelling of the fibre by creating an entangled/cross-linked network due to covalent bonding between the matrix and the fibre [68, 86]. In addition, the introduced hydrocarbon chains are assumed to affect the wettability of the fibres and to improve the chemical affinity of the polymer matrix, thus enhancing the interfacial adhesion between the fibre and the matrix [83, 87].

6.3.2.6 Maleated coupling agents

The use of maleated coupling agents has proven to be an extremely efficient means of improving interfacial interactions between plant fibres and polymer matrices. The maleic anhydride groups react with the hydroxyl groups of the plant fibres and remove them from the fibre cells reducing hydrophilic tendency. The reaction scheme is given in Scheme 6.

H_2C–C(=O)–O–C(=O)–HC (maleic anhydride ring on polymer chain) + Fibre-OH ⟶ H_2C–C(=O)–O–Fibre / HC–C(=O)–OH (Scheme 6)

Moreover, the maleated coupler forms a C–C bond to the polymer chain with the matrix [1]. The strongest adhesion can be achieved when the covalent bonds are formed at the interface between the plant fibres and the coupling agent as well as molecular entanglement between the coupling agent and the polymer matrix [46]. The reaction mechanism of coupling agent with the plant fibre and the matrix can be explained as the activation of the copolymer by heating before the fibre treatment and then the esterification of the cellulose fibre. This treatment increases the surface energy of the plant fibre to a level much closer to that of the matrix, and thus results in better wettability and enhances the interfacial adhesion between the filler and the matrix [68].

6.4 Bonding mechanisms

In plant fibre composites, the main constituents are the reinforcing fibres and the polymer matrix phase. The properties and performance of the composites rely on three main parameters: matrix, reinforcement, and interface. The interface region between the fibre and the matrix has been recognised as playing a predominant role in governing the global material behaviour. Interfaces in composites, often considered as an intermediate region formed due to the bonding of the fibre and matrix, are in fact a zone of compositional, structural, and property gradients, typically varying in width from a single atom layer to micrometres. There is a close relationship between the processes that occur at the atomic, microscopic, and macroscopic levels at the interface. In fact, knowledge of the sequences of events occurring at these different levels is extremely important in understanding the nature of interfacial phenomena. The interface region controls the stress transfer efficiency between the fibre and the matrix and it is primarily dependent on the level of interfacial adhesion. A reasonable interfacial strength ensures the maximum stress level and can be maintained across the interface and from fibre to matrix without disruption. The efficiency of the load transfer is determined by the molecular interaction at the interface, along with the thickness and strength of the interfacial region formed [88]. The fibre–matrix interfacial bonding mechanisms in general include interdiffusion, electrostatic adhesion, chemical reactions and mechanical interlocking (see Fig. 6.1). Together these mechanisms are responsible for adhesion and usually one of them plays a dominant role.

Interdiffusion happens due to intimate intermolecular interactions between the molecules of the fibre substrate and the resin, which result from van der Waals forces or hydrogen bonding [47]. In fact, there are two stages involved in this adhesion mechanism, i.e. adsorption and diffusion. In the first stage, two constituents, fibre and matrix, should have intimate contact, which is in turn governed by two actions including spreading and

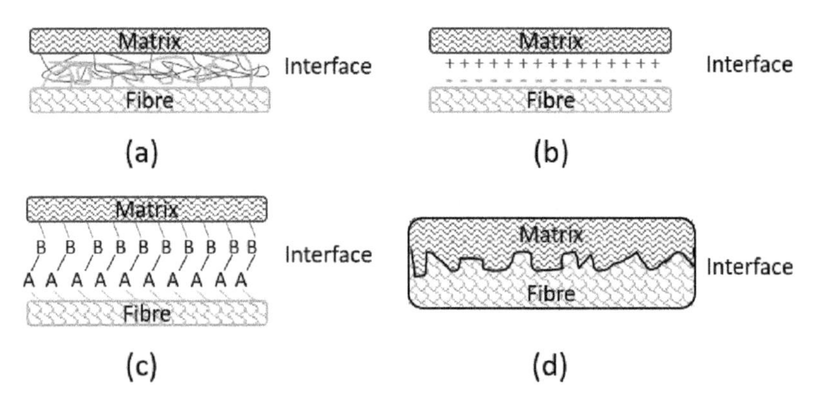

Figure 6.1 Fibre–matrix interfacial bonding mechanisms: (a) molecular entanglement following interdiffusion; (b) electrostatic adhesion; (c) chemical bonding; and (d) mechanical interlocking.

penetration. Once good wetting occurs, permanent adhesion is developed through molecular attraction, e.g. covalent, electrostatic, metallic and van der Waals. On the other hand, good wetting between the substrates leads to the interdiffusion of molecules of both fibre and matrix. The extent and degree of diffusion depends primarily on the chemical compatibility of the two constituents and the penetrability of the substrate [89].

Electrostatic adhesion is attributed to the creation of opposite charges (anionic and cationic) on the interacting surfaces of the fibre and the polymer matrix; thus, an interface consisting of two layers of opposites charges is formed, which accounts for the adhesion of the two constituents of the composite.

Chemisorption occurs when chemical bonds, including metallic, atomic and ionic bonds, are created between the substances as a result of chemical reactions [47]. Available physical and chemical bonds depend on the surface chemistry of the substrate and are sometimes collectively described as thermodynamic adhesion.

The mechanical interlocking phenomenon explains the adhesion when a matrix penetrates into the peaks, holes, valleys and crevices or other irregularities of the substrate, and mechanically locks to it [47]. It happens at a millimetre- and micron-length scale, and diffusion entanglement within the cell-wall pores of the fibre occurs on the nanoscale. Adhesion relying on mechanical interlocking can occur over larger length scales on the contact area. Adhesion, capillary forces, and other interaction factors can be ignored in most microscopic devices but at nanometre scales they often dominate the behaviour of the bonding quality. A flow of polymer resin proceeds into the interconnected network of lumens and open pores of the natural fibre, with flow moving primarily along the path of least resistance. The adhesive occupies the free volume within the cell wall and therefore inhibits shrinking and swelling. The adhesive penetration of the fibre occurs at two or more scales. There is micro-penetration, which occurs through the cell lumens and pits. Additionally, there is nano-penetration, which occurs in the cell wall. Macro-penetration of adhesive through process-induced cracks is also worth considering. Penetration of adhesive at any scale will impact bonding quality. Permeability is also a fibre-related factor controlling resin penetration. Permeability varies according to surface characteristics and direction e.g. tangential, radial and longitudinal. This mechanical interlocking mechanism is often used in polymer composites by etching the polymer surface to increase the surface roughness, thereby increasing the contact area for adhesive penetration and the mechanical interlocking of the substrate [89]. On the other hand, an increase in mechanical interlocking gives rise to the enhancement of other bonding systems/mechanisms.

6.5 Interface structure

6.5.1 Morphology

SEM is the most commonly used technique for investigating both fibre–matrix interactions at fracture surfaces and polymer distributions in plant fibre composites. It allows observation of monomer-impregnated samples directly and after cure to composites to yield information on the interaction of the polymer formed with the fibre components [1]. Migneault [90] used SEM to investigate the variations in wetting at the fibre–matrix interface of composites, among the different fibres employed being aspen wood, spruce bark and spruce wood fibres (Fig. 6.2). It was found that aspen fibres are completely wetted (noted A) whereas spruce and bark fibres are not in close contact with HDPE (noted C). The SEM micrographs also showed variations in interfacial adhesion and mechanical interlocking at the fibre–matrix interface. Aspen

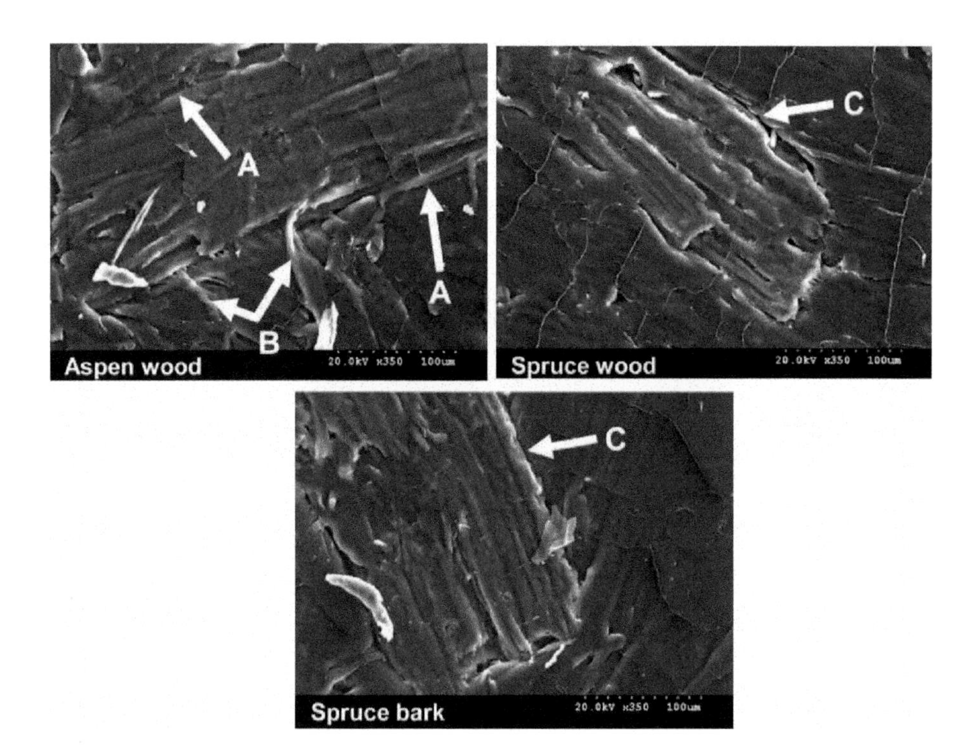

Figure 6.2 SEM images of the surface of WPC samples made with selected fibres. A = close contact/good wetting; B = macro-fibrils; C = no close contact) [90].

fibres presented macro-fibrils at the surface interlocking with the polymer matrix, thus increasing fibre reinforcement (noted B). The wetting and interlocking phenomena suggest a superior stress transfer in the case of aspen wood fibres, which explained the better performance of the aspen fibre reinforced composite than the other fibre-reinforced composites.

Pinto [91] examined the effect of silane treatment and Z-axis reinforcement on the morphological structure of jute-epoxy laminated composites by the use of SEM. The results are shown in Fig. 6.3. The fibres in the untreated plain weave sample (Fig. 6.3-I) appeared largely intact and undisturbed from their aligned position, denoting that the fibre–matrix interface failed before enough stress could be transferred to the fibres to break or pull them out. The untreated unidirectional sample (Fig. 6.3-III) showed a similar fracture, with the features (A) fractured matrix and (B) intact and aligned fibres to the untreated plain weave sample. The difference in this case is the dominance of the clean matrix fracture without any visible fibres, suggesting that the delamination resistance of this sample was primarily governed by matrix failure. Fig. 6.3-II shows the significant changes in the fracture surface of the silane-treated sample compared with the untreated counterpart, i.e. (A) pulled-out and fractured fibres and (B) matrix fragments clinging to the surface. While the matrix fracture is still present, the fibre pull-out and

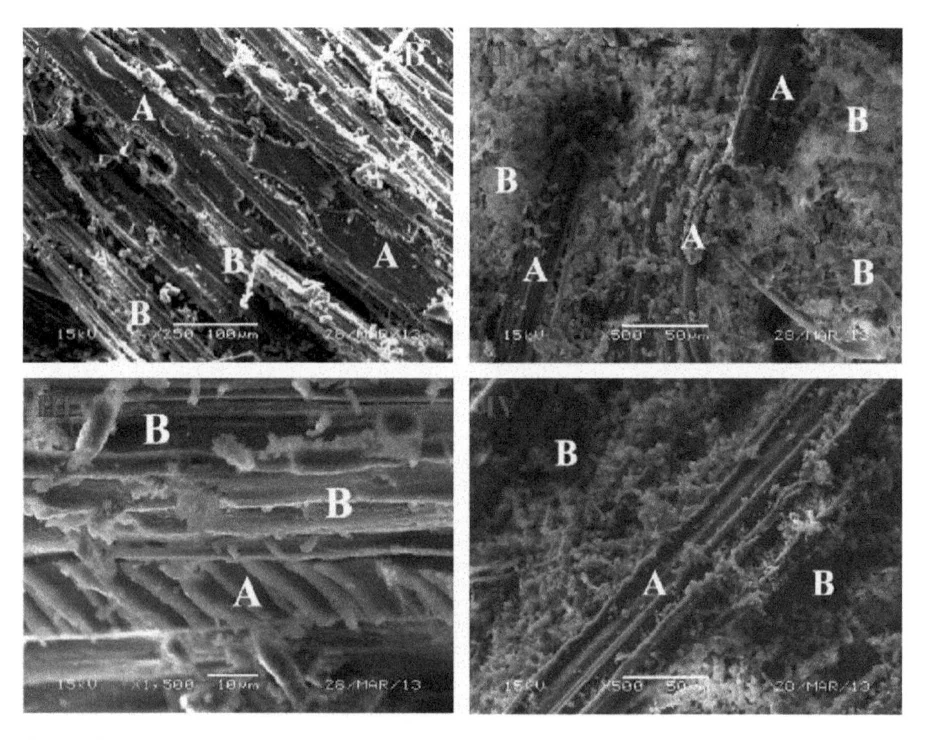

Figure 6.3 SEM images of the failure surfaces of (I) untreated and (II) silane-treated plain weave jute-epoxy composites; and (III) untreated and (IV) silane-treated unidirectional jute-epoxy composites [91].

breakage are seen to contribute to the sample's fracture resistance, implying that the interface between the fibre and the matrix is strong enough to support the stress transfer to the fibre to avoid the failure of the interface. The improved stress transfer from the matrix to the fibre, which led the fibre to be pulled-out prior to debonding, can also be seen on the surface of the silane-treated unidirectional composite, with the presence of more removed fibres from the epoxy matrix (A, Fig. 6.3-IV) and a great many matrix fragments attached to the surface (B, Fig. 6.3-IV).

6.5.2 Interfacial bonding capacity

6.5.2.1 Micromechanical measurements of interface shear capacity

Interpretation of fibre–matrix adhesion is of special significance for the successful design and proper utilisation of plant fibre composites. There are several micromechanical testing methods for measuring fibre–matrix interfacial adhesion: examples include the single fibre fragmentation test (SFFT), the single fibre pull-out test (SFPT), and the microbond test (MT). A detailed summary of these tests can be found in Kim & Mai [57].

The single fibre fragmentation test was originally developed from the early work of Kelly and Tyson [92], who investigated brittle tungsten fibres that broke into multiple segments in a metal matrix composite. In the fragmentation test (Fig. 6.4), a single fibre is totally encapsulated in a chosen polymer matrix, which in turn is loaded under tension. This experiment is done under a light microscope in order to observe the fragmentation process in situ. The fibre inside the resin breaks into increasingly smaller fragments at locations where the fibre's axial stress reaches its tensile strength. This requires a resin system with a sufficiently higher strain-to-failure than that of the fibre. When the fibre breaks, the tensile stress at the fracture location reduces to zero. Due to the constant shear in the matrix, the tensile stress in the fibre increases roughly

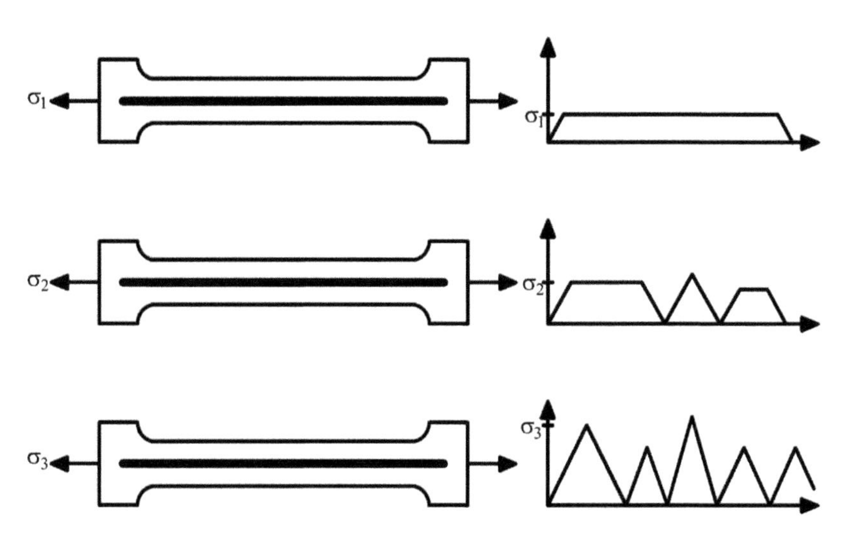

Figure 6.4 Schematic representation of the single fibre fragmentation test [93].

linearly at its ends to a plateau in longer fragments. The higher the axial strain, the more that fractures will be caused in the fibre, but at some level the number of fragments will become constant as the fragment length is too short to transfer sufficient stress into the fibre to cause further breakage [93]. The shortest fibre length that can break on application of stress is defined as the critical fibre length, l_c. The average interfacial shear strength τ, with regards to the fibre strength σ, and the fibre diameter d, can be estimated from a simple force–balance expression for a constant interfacial shear stress: $\tau = \sigma_f d/2l_c$.

In the single fibre pull-out test (Fig. 6.5), the fibre is embedded in a block of matrix. The free end is gripped and an increasing load is applied as the fibre is pulled out of the matrix while the load and displacement are measured [94]. At the first stage of pull-out loading, induced shear stresses along the fibre do not exceed the bond strength between the fibre and the matrix. Once the force required to pull the fibre out of the block is determined, the corresponding interfacial shear strength can be calculated. The maximum load, F, measured before the detachment of the fibre from the matrix is related to the average value of the fibre–matrix shear strength, τ, through the equation $F = \tau\pi dl$, where πd is the fibre circumference and l is the embedded fibre length [94].

The other test method of interfacial shear strength, the microbond test, is considered a modified single fibre-pull out test. It consists of first applying a small amount of resin to the fibre surface in the form of a droplet that forms concentrically around the fibre in the shape of an ellipsoid, and then applying a shearing force to pull the fibre out of the droplet or vice versa; thus the bead is restrained by opposing knife edges (Fig. 6.6) and stripped off, and the applied load and the blade displacement are recorded. Assuming that the interface is in a uniform state of shear stress, the average shear stress can be calculated by dividing the maximum measured force of debonding by the embedded area of fibre, i.e. $\tau = F/\pi dl$. The bond strength values can be used to investigate the dependence of composite performance on the energy-absorbing characteristics of the interface and to establish the extent to which the fibre surface treatment can alter bonding [1].

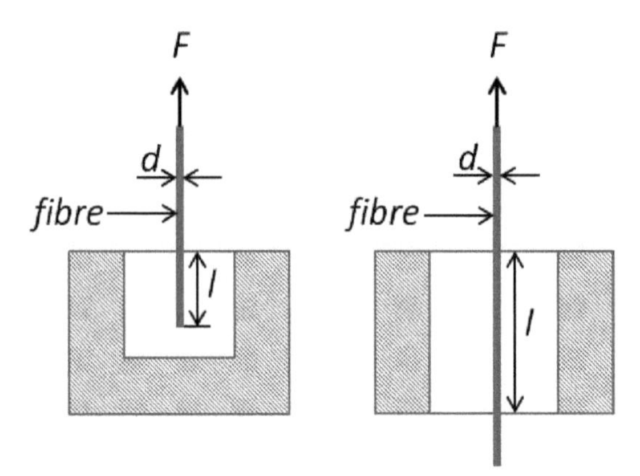

Figure 6.5 Schematic representation of the single fibre pull-out test.

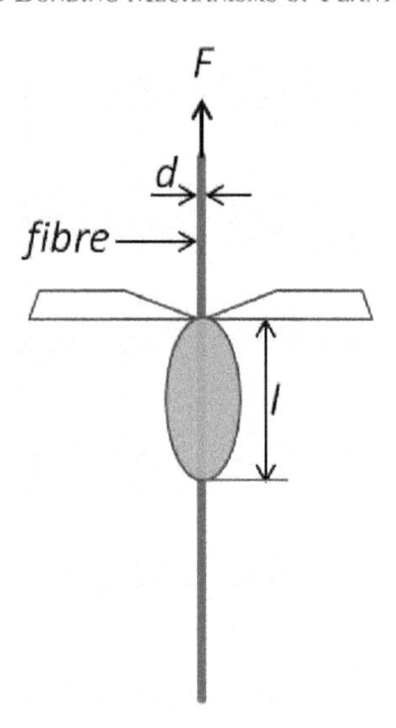

Figure 6.6 Schematic representation of the microbond test.

6.5.2.2 Spectroscopic measurements of interface bonding capacity

6.5.2.2.1 Fourier transform infrared spectroscopy (FTIR)

FTIR offers quantitative and qualitative analysis for organic and inorganic materials. It identifies the chemical bonds in a molecule by producing an infrared absorption spectrum. The resulting spectra produce a profile for the sample, a distinctive molecular fingerprint that can be used to screen and scan samples for many different components. This effective analytical instrument for detecting functional groups and characterising chemical bonding information has been extensively used in composite materials.

Lu *et al.* [95] investigated the effects of alkali soaking and silane coupling (KH560) modification of bamboo cellulose fibres (BCF) and the maleic anhydride (MA) grafting of poly(L-lactic acid) (PLLA) on the improved interfacial property of cellulose/PLLA composites. FTIR was employed to study the chemical structure of virgin cellulose and the changes after the pretreatments, as well as the chemical structure of PLLA and cellulose/PLLA composites. The results are shown in Fig. 6.7. The OH stretching vibration peak of virgin cellulose at 3421 cm^{-1} was shifted to 3415 cm^{-1} after the alkali treatment, which was ascribed to the disturbance of the hydrogen bond interaction that linked the cellulose and the impurities. The KH560-treated fibre demonstrated that new chemical bonds had been formed after the treatment, showing a peak at 1117 cm^{-1}, corresponding to the Si-O-C stretching vibration, and a peak at 802 cm^{-1} related to Si-C stretching in its FTIR spectrum. This was because the OH groups on silanol hydrolysed from KH560 condensed with the OH groups on cellulose, thus the $CH_2CH(O)CH_2O(CH_2)_3SiO^-$ group had been grafted onto the cellulosic molecules. The improved interfacial adhesion between the cellulosic fibre and the matrix after the modifications can also be confirmed by FTIR analysis

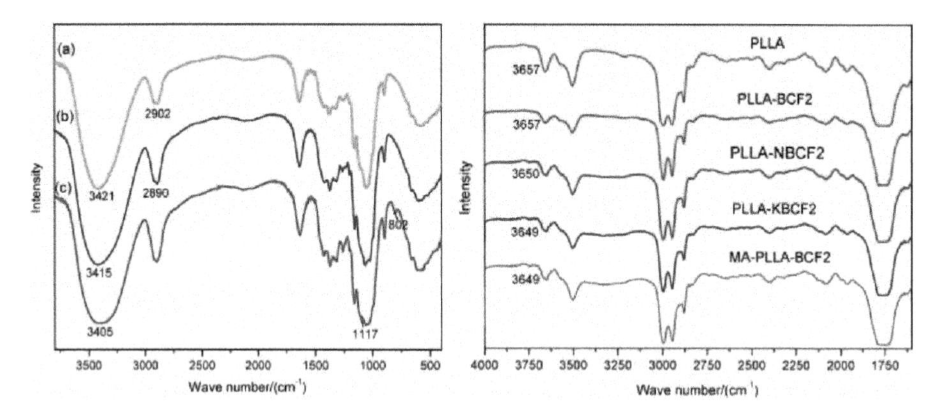

Figure 6.7 FTIR spectra of (a) BCF; (b) NaOH treated BCF (NBCR); (c) KH560 treated BCF (KBCF); PLLA and cellulose/PLLA composites [95].

of the composites, i.e. the filling of the untreated BCF in the composite did not lead to changes in the absorption peaks owing to the poor interaction between the fibre and the matrix, while after the three modifications, the OH stretching vibration at 3657 cm^{-1} had shifted to around 3650 cm^{-1}, indicating superior interaction between the BCFs and PLLA.

6.5.2.2.2 Nuclear magnetic resonance spectroscopy (NMR)

NMR is an analytical chemistry technique used in quality control and research for determining the composition and purity of a sample as well as its molecular structure. The principle behind NMR is that many nuclei have spin and all nuclei are electrically charged. An energy transfer from the base energy to a higher energy level (generally a single energy gap) happens when an external magnetic field is applied. The energy transfer takes place at a wavelength that corresponds to radio frequencies and when the spin returns to its base level, energy is emitted at the same frequency. The signal that matches this transfer is measured in a variety of ways and processed in order to yield an NMR spectrum for the nucleus concerned.

NMR has become one of the most significant techniques for determining the surface characteristics of plant fibre composites. Tavares *et al.* [96]studied polyurethane/natural fibre (sisal fibre, and sugarcane waste fibre, SWF) composites focusing on interaction, homogeneity and compatibility between composite components, using solid-state NMR employing magic angle spinning (MAS) and cross-polarisation/magic angle spinning (CPMAS) techniques. In comparison with fibres, the PU/SCF and PU/sisal composites changed the chemical shifts of C anomeric to low frequency, and their proton values $T_1^H\rho$ were higher than those of fibres. In addition, the values determined for CH-OH and CH_2-OH were increased for PU/SCF compared with SCF, and decreased for PU/sisal compared with sisal. This behaviour suggested that the PU/SCF had a better interaction between the matrix and the filler. Furthermore, since the relaxation parameter values increased, the modifications in molecular packing and fibre chains ordination would lead the PU to act as a plasticiser as well.

6.5.2.2.3 X-ray photoelectron spectroscopy (XPS)

XPS is a well-established technique for analysing the surface chemistry of a material, which provides valuable quantitative and chemical state information from the surface of the material being studied. XPS spectra are obtained by irradiating a solid surface with a beam of X-rays while simultaneously measuring the kinetic energy and electrons that are emitted from the top 1–10 nm of the material being analysed. A photoelectron spectrum is recorded by counting ejected electrons over a range of electron kinetic energies. Peaks appear in the spectrum from atoms that emit electrons of a particular characteristic energy. The energies and intensities of the photoelectron peaks enable the identification and quantification of all surface elements (except hydrogen). XPS has been used by researchers to scrutinise qualitatively and quantitatively the surface chemistry of treated and untreated plant fibres and their composites [97–99]. Apart from information it offers about the effect of surface treatments, this technique might also help indirectly in establishing mechanisms of interfacial behaviour, such as identifying boundary layers [94].

6.6 Conclusion and future trends

Plant fibre composites have increasingly found plenty of engineering applications in recent years, especially in North America, Europe, China and Japan. The optimised interface of the composites governing the overall property contributes to the reasonable formulation of the composites and thus the successful commercialisation. Understanding the interface and bonding mechanisms of the composites is the key issue that research efforts must address in order to maximise their application. The development and/or selection of efficient and cost-effective surface modifiers or dispersion aids or coupling agents, for both fibres and matrices, is the first critical area to be considered for future development. Current trends favour treatment that not only provides superb tailoring but also has a minimal impact on economics. In addition, more attention should be paid to the nanoscale interfacial characteristics of composites in order to figure out the stress transfer efficiency, interfacial penetration, interaction or adhesion at the nanoscale. Although the application of nanoscience in plant fibre composites (e.g. nanocellulose-reinforced composite) is currently at the infant laboratory stage, some of the plant fibre composites will inevitably be replaced by nanocomposites in the near future. Therefore, further research effort should be devoted to the study of the interface and bonding of biobased nanocomposites. This will help to bridge the gap between the scientific challenges and industrial production.

References

[1] George J, Sreekala M S Thomas S, 2001. A review on interface modification and characterization of natural fiber reinforced plastic composites. *Polymer Engineering & Science*, 41 (9), pp. 1471–1485.

[2] Mukhopadhyay S, Deopura B L & Alagiruswamy R, 2003. Interface Behavior in Polypropylene Composites. *Journal of Thermoplastic Composite Materials*, 16 (6), pp. 479–495.

[3] Lu J Z, Wu Q & McNabb H S, 2000. Chemical coupling in wood fiber and polymer composites: A review of coupling agents and treatments. *Wood and Fiber Science*, 32 (1), pp. 88–104.

[4] Bledzki A K & Gassan J, 1999. Composites reinforced with cellulose based fibres. *Progress in Polymer Science*, 24 (2), pp. 221–274.

[5] Abdul Khalil H P & Ismail H, 2000. Effect of acetylation and coupling agent treatments upon biological degradation of plant fibre reinforced polyester composites. *Polymer Testing*, 20 (1), pp. 65–75.

[6] Zadorecki P & Michell A J, 1989. Future prospects for wood cellulose as reinforcement in organic polymer composites. *Polymer Composites*, 10 (2), pp. 69–77.

[7] Kabir M M, 2012. Chemical treatments on plant-based natural fibre reinforced polymer composites: An overview. *Composites Part B: Engineering*, 43 (7), pp. 2883–2892.

[8] Bledzki A K, Gassan J & Theis S, 1998. Wood-filled thermoplastic composites. *Mechanics of Composite Materials*, 34 (6), pp. 563–568.

[9] Cantero G, 2003. Effects of fibre treatment on wettability and mechanical behaviour of flax/polypropylene composites. *Composites Science and Technology*, 63 (9), pp. 1247–1254.

[10] Kazayawoko M, Balatinecz J J & Matuana L M, Surface modification and adhesion mechanisms in woodfiber-polypropylene composites. *Journal of Materials Science*, 34 (24), pp. 6189–6199.

[11] Raj R G, 1989. Compounding of cellulose fibers with polypropylene: Effect of fiber treatment on dispersion in the polymer matirx. *Journal of Applied Polymer Science*, 38 (11), pp. 1987–1996.

[12] John M J & Anandjiwala RD, 2008. Recent developments in chemical modification and characterization of natural fiber-reinforced composites. *Polymer Composites*, 29 (2), pp. 187–207.

[13] Glasser W G, 1999. Fiber-reinforced cellulosic thermoplastic composites. *Journal of Applied Polymer Science*, 73 (7), pp. 1329–1340.

[14] Saheb D N & Jog J P, 1999. Natural fiber polymer composites: A review. *Advances in Polymer Technology*, 18 (4), pp. 351–363.

[15] Satyanarayana K G, Arizaga G G C & Wypych F, 2009. Biodegradable composites based on lignocellulosic fibers: An overview. *Progress in Polymer Science*, 34 (9), pp. 982–1021.

[16] Azwa Z N, 2013. A review on the degradability of polymeric composites based on natural fibres. *Materials and Design*, 47, pp. 424–442.

[17] Wong K J, Yousif B F & Low K O, 2010. The effects of alkali treatment on the interfacial adhesion of bamboo fibres. *Proceedings of the Institution of Mechanical Engineers, Part L: Journal of Materials Design and Applications*, 224 (3), pp. 139–148.

[18] Summerscales J, 2010. A review of bast fibres and their composites. Part 1 – Fibres as reinforcements. *Composites Part A: Applied Science and Manufacturing*, 41 (10), pp. 1329–1335.

[19] John M & Thomas S, 2008. Biofibres and biocomposites. *Carbohydrate Polymers*, 71 (3), pp. 343–364.

[20] Araújo J R, Waldman W R & De Paoli M A, 2008. Thermal properties of high density polyethylene composites with natural fibres: Coupling agent effect. *Polymer Degradation and Stability*, 93 (10), pp. 1770–1775.

[21] Dittenber D B & GangaRao H V S, 2012. Critical review of recent publications on use of natural composites in infrastructure. *Composites Part A: Applied Science and Manufacturing*, 43 (8), pp. 1419–1429.

[22] Xie Y, 2010. Silane coupling agents used for natural fiber/polymer composites: A review. *Composites Part A: Applied Science and Manufacturing*, 41 (7), pp. 806–819.

[23] Heredia A, 2003. Biophysical and biochemical characteristics of cutin, a plant barrier biopolymer. *Biochimica et Biophysica Acta (BBA) - General Subjects*, 1620 (1–3), pp. 1–7.

[24] Gáspár M, 2005. Reducing water absorption in compostable starch-based plastics. *Polymer Degradation and Stability*, 90 (3), pp. 563–569.
[25] Kurita K, 2001. Controlled functionalization of the polysaccharide chitin. *Progress in Polymer Science*, 26 (9), pp. 1921–1971.
[26] Wollerdorfer M & Bader H, 1998. Influence of natural fibres on the mechanical properties of biodegradable polymers. *Industrial Crops and Products*, 8 (2), pp. 105–112.
[27] Deshmukh A P, Simpson A J & Hatcher P G, 2003. Evidence for cross-linking in tomato cutin using HR-MAS NMR spectroscopy. *Phytochemistry*, 64 (6), pp. 1163–1170.
[28] Xu C P, 2003. Optimization of submerged culture conditions for mycelial growth and exo-biopolymer production by Paecilomyces tenuipes C240. *Process Biochemistry*, 38 (7), pp. 1025–1030.
[29] Matas A J, Cuartero J & Heredia A, 2004. Phase transitions in the biopolyester cutin isolated from tomato fruit cuticles. *Thermochimica Acta*, 409 (2), pp. 165–168.
[30] Thakur V K & Thakur M K, 2015. *Eco-friendly Polymer Nanocomposites*, New Delhi: Springer India.
[31] Yan L, Chouw N & Jayaraman K, 2014. Flax fibre and its composites: A review. *Composites Part B: Engineering*, 56 (January 2014), pp. 296–317.
[32] Ma X, Yu J & Kennedy J F, 2005. Studies on the properties of natural fibers-reinforced thermoplastic starch composites. *Carbohydrate Polymers*, 62 (1), pp. 19–24.
[33] Kaewtatip K & Thongmee J, 2012. Studies on the structure and properties of thermoplastic starch/luffa fiber composites. *Materials & Design*, 40 (September 2012), pp. 314–318.
[34] Culvelo A A S, Carvalho A J F & Angelli J A M, 2001. Thermoplastic starch-cellulosic fibers composites: Preliminary results. *Carbohydrate Polymers*, 45 (2), pp. 183–188
[35] Gironès J, 2012. Natural fiber-reinforced thermoplastic starch composites obtained by melt processing. *Composites Science and Technology*, 72 (7), pp. 858–863.
[36] Alvarez V A & Vázquez A, 2004. Thermal degradation of cellulose derivatives/starch blends and sisal fibre biocomposites. *Polymer Degradation and Stability*, 84 (1), pp. 13–21.
[37] Prachayawarakorn J, Sangnitidej P & Boonpasith P, 2010. Properties of thermoplastic rice starch composites reinforced by cotton fiber or low-density polyethylene. *Carbohydrate Polymers*, 81 (2), pp. 425–433.
[38] Guimarães J L, 2010. Studies of the processing and characterization of corn starch and its composites with banana and sugarcane fibers from Brazil. *Carbohydrate Polymers*, 80 (1), pp. 130–138.
[39] Teixeira E, 2009. Cassava bagasse cellulose nanofibrils reinforced thermoplastic cassava starch. *Carbohydrate Polymers*, 78 (3), pp. 422–431.
[40] Kaushik A, Singh M & Verma G, 2010. Green nanocomposites based on thermoplastic starch and steam exploded cellulose nanofibrils from wheat straw. *Carbohydrate Polymers*, 82 (2), pp. 337–345.
[41] Kovacevic Z, Bischof S & Fan M, 2015. The influence of Spartium junceum L. fibres modified with montmorrilonite nanoclay on the thermal properties of PLA biocomposites. *Composites Part B: Engineering*, 78 (September 2015), pp. 122–130.
[42] Wang Y N, Weng Y X & Wang L, 2014. Characterization of interfacial compatibility of polylactic acid and bamboo flour (PLA/BF) in biocomposites. *Polymer Testing*, 36 (June 2014), pp. 119–125.
[43] Masuelli M A M, 2013. *Fiber Reinforced Polymers: The Technology Applied for Concrete Repair*. Croatia: Intech.
[44] Sain M, Suhara P, Law S & Bouilloux A, 2005. Interface modification and mechanical properties of natural fiber–polyolefin composite products. *Journal of Reinforced Plastics and Composites*, 24 (2), pp. 121–130.

[45] Mohanty A K, Misra M & Drzal L T, 2001. Surface modifications of natural fibers and performance of the resulting biocomposites: An overview. *Composite Interfaces*, 8 (5), pp. 313–343.

[46] Huda M S, 2006. Wood-fiber-reinforced poly(lactic acid) composites: Evaluation of the physicomechanical and morphological properties. *Journal of Applied Polymer Science*, 102 (5), pp. 4856–4869.

[47] Liu D, 2012. Bamboo fiber and its reinforced composites: Structure and properties. *Cellulose*, 19 (5), pp. 1449–1480.

[48] Wong S, Shanks R & Hodzic A, 2002. Properties of Poly(3-hydroxybutyric acid) composites with flax fibres modified by plasticiser absorption. *Macromolecular Materials and Engineering*, 287 (10), pp. 647–655.

[49] Coats E R, 2008. Production of natural fiber reinforced thermoplastic composites through the use of polyhydroxybutyrate-rich biomass. *Bioresource Technology*, 99 (7), pp. 2680–2686.

[50] Bhardwaj R, 2006. Renewable resource-based green composites from recycled cellulose fiber and poly(3-hydroxybutyrate-co-3-hydroxyvalerate) bioplastic. *Biomacromolecules*, 7 (6), pp. 2044–2051.

[51] Krishnaprasad R, 2009. Mechanical and thermal properties of bamboo microfibril reinforced polyhydroxybutyrate Biocomposites. *Journal of Polymers and the Environment*, 17 (2), pp. 109–114.

[52] Gassan J & Gutowski V S, 2000. Effects of corona discharge and UV treatment on the properties of jute-fibre expoxy composites. *Composites Science and Technology*, 60 (15), pp. 2857–2863.

[53] Li R, 2015. Effect of gamma irradiation on the properties of basalt fiber reinforced epoxy resin matrix composite. *Journal of Nuclear Materials*, 466 (November 2015), pp. 100–107.

[54] Zaman H U, 2009. A comparative study between gamma and UV radiation of jute fabrics/polypropylene composites: Effect of starch. *Journal of Reinforced Plastics and Composites*, 29 (13), pp. 1930–1939.

[55] Khan R A, 2009. Effect of gamma radiation on the performance of jute fabrics-reinforced polypropylene composites. *Radiation Physics and Chemistry*, 78 (11), pp. 986–993.

[56] Khan M, 2008. Effect of gamma radiation on the physico-mechanical and electrical properties of jute fiber-reinforced polypropylene composites. *Journal of Reinforced Plastics and Composites*, 28 (13), pp. 1651–1660.

[57] Kim J K & Mai Y W, 1998. *Engineered Interfaces in Fiber Reinforced Composites*, Amsterdam: Elsevier.

[58] Ragoubi M, 2010. Impact of corona treated hemp fibres onto mechanical properties of polypropylene composites made thereof. *Industrial Crops and Products*, 31 (2), pp. 344–349.

[59] Ragoubi M, 2012. Effect of corona discharge treatment on mechanical and thermal properties of composites based on miscanthus fibres and polylactic acid or polypropylene matrix. *Composites Part A: Applied Science and Manufacturing*, 43 (4), pp. 675–685.

[60] Vladkova T G, 2006. Wood flour: New filler for the rubber processing industry. IV. Cure characteristics and mechanical properties of natural rubber compounds filled by non-modified or corona treated wood flour. *Journal of Applied Polymer Science*, 101 (1), pp. 651–658.

[61] Vladkova T G, Dineff P D & Gospodinova D N, 2004a, Wood flour: A new filler for the rubber processing industry. II. Cure characteristics and mechanical properties of NBR compounds filled with corona-treated wood flour. *Journal of Applied Polymer Science*, 91 (2), pp. 883–889.

[62] Vladkova T G, Dineff P D & Gospodinova D N, 2004b. Wood flour: A new filler for the rubber processing industry. III. Cure characteristics and mechanical properties of nitrile butadiene rubber compounds filled by wood flour in the presence of phenol-formaldehyde resin. *Journal of Applied Polymer Science*, 92 (1), pp. 95–101.

[63] Oporto G S, 2009. Forced air plasma treatment (FAPT) of hybrid wood plastic composite (WPC)–fiber reinforced plastic (FRP) surfaces. *Composite Interfaces*, 16 (7–9), pp. 847–867.
[64] Sarikanat M, 2014. The effect of argon and air plasma treatment of flax fiber on mechanical properties of reinforced polyester composite. *Journal of Industrial Textiles*, 0 (00), pp. 1–16.
[65] Morales J, 2006. Plasma modification of cellulose fibers for composite materials. *Journal of Applied Polymer Science*, 101 (6), pp. 3821–3828.
[66] Selden R, Nystrom B & Langstrom R, 2004. UV aging of poly(propylene)/wood-fiber composites. *Polymer Composites*, 25 (5), pp. 543–553.
[67] Zaman H U, Khan M A & Khan R A, 2009. Improvement of mechanical properties of jute fibers-polyethylene/polypropylene composites: Effect of green dye and UV radiation. *Polymer-Plastics Technology and Engineering*, 48 (August 2015), pp. 1130–1138.
[68] Li X, Tabil L G & Panigrahi S, 2007. Chemical treatments of natural fiber for use in natural fiber-reinforced composites: A review. *Journal of Polymers and the Environment*, 15 (1), pp. 25–33.
[69] Mohanty A K, Khan M A & Hinrichsen G, 2000. Surface modification of jute and its influence on performance of biodegradable jute-fabric/biopol composites. *Composites Science and Technology*, 60 (7), pp. 1115–1124.
[70] Ahmed K, Nizami S S & Riza N Z, 2014. Reinforcement of natural rubber hybrid composites based on marble sludge/silica and marble sludge/rice husk derived silica. *Journal of advanced research*, 5 (2), pp. 165–173.
[71] Ray D, 2001. Effect of alkali treated jute fibres on composite properties. *Bulletin of Materials Science*, 24 (2), pp. 129–135.
[72] Ouajai S & Shanks R A, 2005. Composition, structure and thermal degradation of hemp cellulose after chemical treatments. *Polymer Degradation and Stability*, 89 (2), pp. 327–335.
[73] Wang B, 2007. Pre-treatment of flax fibers for use in rotationally molded biocomposites. *Journal of Reinforced Plastics and Composites*, 26 (5), pp. 447–463.
[74] Doan T T L, Brodowsky H & Mäder E, 2012. Jute fibre/epoxy composites: Surface properties and interfacial adhesion. *Composites Science and Technology*, 72 (10), pp. 1160–1166.
[75] Mwaikambo L Y & Ansell M P, 2002. Chemical modification of hemp, sisal, jute, and kapok fibers by alkalization. *Journal of Applied Polymer Science*, 84 (12), pp. 2222–2234.
[76] Bledzki A K, 2008. The effects of acetylation on properties of flax fibre and its polypropylene composites. *Express Polymer Letters*, 2 (6), pp. 413–422.
[77] Sreekala M S, Kumaran M G & Thomas S, 2002. Water sorption in oil palm fiber reinforced phenol formaldehyde composites. *Composites Part A: Applied Science and Manufacturing*, 33 (6), pp. 763–777.
[78] Zurina M, Ismail H & Bakar A A, 2004. Rice husk powder-filled polystyrene/styrene butadiene rubber blends. *Journal of Applied Polymer Science*, 92 (5), pp. 3320–3332.
[79] Haseena A P, Unnikrishnan G & Kalaprasad G, 2007. Dielectric properties of short sisal/coir hybrid fibre reinforced natural rubber composites. *Composite Interfaces*, 14 (7-9), pp. 763–786.
[80] Joseph S, Koshy P & Thomas S, 2005. The role of interfacial interactions on the mechanical properties of banana fibre reinforced phenol formaldehyde composites. *Composite Interfaces*, 12 (6), pp. 581–600.
[81] Joseph K, Thomas S & Pavithran C. 1996. Effect of chemical treatment on the tensile properties of short sisal fibre-reinforced polyethylene composites. *Polymer*, 37 (23), pp. 5139–5149.
[82] Joseph P V, 2003. The thermal and crystallisation studies of short sisal fibre reinforced polypropylene composites. *Composites Part A: Applied Science and Manufacturing*, 34 (3), pp. 253–266.

[83] Kalaprasad G, 2004. Effect of fibre length and chemical modifications on the tensile properties of intimately mixed short sisal/glass hybrid fibre reinforced low density polyethylene composites. *Polymer International*, 53 (11), pp. 1624–1638.

[84] Sreekala M S, 2000. Oil palm fibre reinforced phenol formaldehyde composites: Influence of fibre surface modifications on the mechanical performance. *Applied Composite Materials*, 7 (5–6), pp. 295–329.

[85] Zhou Y, 2015. Lignocellulosic fibre mediated rubber composites: An overview. *Composites Part B: Engineering*, 76 (July 2015), pp. 180–191.

[86] Kalia S, Kaith B S & Kaur I, 2009. Pretreatments of natural fibers and their application as reinforcing material in polymer composites: A review. *Polymer Engineering & Science*, 49 (7), pp. 1253–1272.

[87] Mohanty S, 2004. Effect of MAPP as coupling agent on the performance of sisal-pp composites. *Journal of Reinforced Plastics and Composites*, 23 (6), pp. 2047–2063.

[88] Drzal L T & Madhukar M, 1993. Fibre–matrix adhesion and its relationship to composite mechanical properties. *Journal of Materials Science*, 28 (3), pp. 569–610.

[89] Kim J K & Pal K, 2010. *Recent Advances in the Processing of Wood-Plastic Composites*, Berlin: Springer Science.

[90] Migneault S, 2015. Effects of wood fiber surface chemistry on strength of wood–plastic composites. *Applied Surface Science*, 343 (July 2015), pp. 11–18.

[91] Pinto M A, 2013. Effect of surface treatment and Z-axis reinforcement on the interlaminar fracture of jute/epoxy laminated composites. *Engineering Fracture Mechanics*, 114 (December 2013), pp. 104–114.

[92] Kelly A & Tyson W R, 1965. Tensile properties of fibre-reinforced metals: Copper/tungsten and copper/molybdenum. *Journal of the Mechanics and Physics of Solids*, 13 (6), pp. 329–350.

[93] Feih S, 2004. *Testing Procedure for the Single Fiber Fragmentation Test*, Roskilde, Denmark: Technical University of Denmark.

[94] Pickering K, 2008. *Properties and Performance of Natural-Fibre Composites*, Cambridge: WoodHead Publishing.

[95] Lu T, 2014. Effects of modifications of bamboo cellulose fibers on the improved mechanical properties of cellulose reinforced poly(lactic acid) composites. *Composites Part B: Engineering*, 62 (June 2014), pp. 191–197.

[96] Tavares M I B, Mothe C G & Araujo C R, 2002. Solid-state nuclear magnetic resonance study of polyurethane/natural fibers composites. *Journal of Applied Polymer Science*, 85 (7), pp. 1465–1468.

[97] Tran L Q N, 2013. Understanding the interfacial compatibility and adhesion of natural coir fibre thermoplastic composites. *Composites Science and Technology*, 80 (May 2013), pp. 23–30.

[98] Gironès J, 2007. Effect of silane coupling agents on the properties of pine fibers/polypropylene composites. *Journal of Applied Polymer Science*, 103 (6), pp. 3706–3717.

[99] Kodal M, Topuk Z D & Ozkoc G, 2015. Dual effect of chemical modification and polymer precoating of flax fibers on the properties of short flax fiber/poly(lactic acid) composites. *Journal of Applied Poly*mer Science, 132 (48), pp. (1–13).

Chapter 7
Functional (nano) cellulose film

Dr Shan Lin and Dr Yonghui Zhou

This chapter focuses on advanced cellulosic product, cellulose and/or nanocellulose films. The chapter begins with an outline of the raw materials. There follows a detailed formulation of both aqueous and non-aqueous solvent dissolution and the fabrication of cellulose flat sheet film and cellulose hollow fibre membrane. The performance of a variety of cellulose films and their advanced applications is then discussed. Among the many products investigated is a range of cellulose separation membranes, cellulose based medical films, cellulose packaging films and cellulose photovoltaic films.

7.1 Introduction

Cellulose and/or nanocellulose film has attracted great attention among researchers and engineers in recent years, mainly due to the advantageous features that cellulose provides over synthetic polymers, i.e. abundance of raw material, low cost, environmentally friendly, renewability, biodegradability and non-toxicity. The development of new cellulose solvents, such as N-methyl morpholine-N-oxide (NMMO), dissolved alkaline systems, ionic liquids, and lithium chloride/dimethylacetamide (LiCl/DMAc), has also facilitated a speedy growth of cellulose films for various applications in separation, medical, packaging and optoelectronic industrial sectors.

7.2 Raw materials of cellulose film

Cellulose, the most abundant natural polymer on earth, is mainly obtained from plant fibres including wood, bamboo, grass, cotton and others. Cellulose is a linear polysaccharide consisting of D-glucose units (the so-called anhydroglucose units (AGU)) linked by β-1,4-glycosidic bonds, with the chemical formula $(C_6H_{10}O_5)_n$. The polymer unit contains three hydroxyl groups, namely the secondary hydroxyls at the C-2, C-3 positions and the primary hydroxyl at the C-6 position, which makes cellulose extremely hydrophilic. Regenerated cellulose (RC) refers to a class of materials manufactured by the conversion of natural cellulose to a soluble cellulosic derivative and subsequent regeneration, typically forming either a fibre (via polymer spinning) or a film (via polymer casting). Compared to natural cellulose, RC is less crystalline with lower molecular weight and entanglement.

Cellulose film can be divided into two categories: phase inversion film and composite film. To date, most commercial films are produced via the phase inversion method, mainly because of its simplicity, flexible production scale as well as the low cost of production [1, 2]. Phase inversion is a process of phase separation whereby the initial homogeneous polymer solution is transformed from a liquid to a solid state.

The transformation can be accomplished in several ways[3], namely (1) the immersion gel method (Liquid-Solid (L-S) method); (2) the solvent evaporation method; (3) water vapour induced phase separation; and (4) thermally induced phase separation (TIPS). The L-S method is the most commonly employed among these techniques. Selection of an appropriate solvent system is vitally important in the fabrication of cellulose film since it determines the structure and performance of the ultimate material. Great achievements in the dissolution of cellulose have been witnessed in the last decade, which will be discussed in detail in the next section.

7.3 Formulation of cellulose film

7.3.1 Dissolution methods

Cellulose molecules are difficult to dissolve in common solvents due to the compact structure of cellulose segment, which restricts the industrial application of cellulose. Traditional cellulose films are usually prepared by viscose or copper ammonia, but unfortunately the process is very complex with high material and energy consumption [4, 5], resulting in serious environmental pollution. Some new cellulose solvents including NMMO, alkaline systems, ionic liquids and LiCl/DMAc, have been developed in recent years. Cellulose solvents can be categorised into derivative solvents and non-derivative solvents; the latter can be further divided into non-aqueous and aqueous solvent systems.

7.3.1.1 Non-aqueous solvent dissolution

7.3.1.1(a) Amine oxide system

NMMO, a highly explosive solid chemical at room temperature, is the strongest and the most commonly used cellulose solvent in amine oxide system [6], although it needs to be molten under certain conditions. The dissolution of cellulose in NMMO hydrate (13.3% water) is usually carried out in a sealed vacuum apparatus at around 100°C as shown in Fig. 7.1 [7]. The mechanism of dissolution consists of a set of complex hydrogen bond formations and ionic interactions [8, 9]. Although industrial lyocell fibres (known as 'the 21st century green fibre') could be produced by the use of NMMO through a novel fibre-spinning process, some problems still exist, such as the

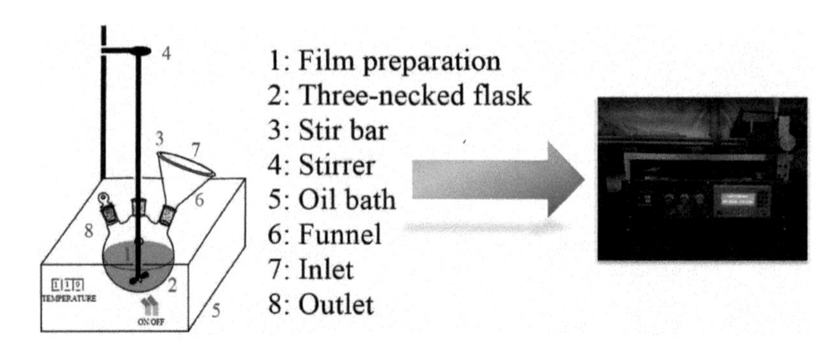

Figure 7.1 Diagram of cellulose dissolution [7].

high price of the solvent, harsh dissolving conditions, and difficulty and complexity in controlling the by-products, fibrillation and recovery rate [10].

7.3.1.1(b) LiCl/DMAc system

The content of LiCl in LiCl/DMAc solvent for dissolving cellulose is 5–9 wt%, such that the solubility of cellulose can reach 15–17 wt% at 150 °C [11]. The dissolution mechanism is that Li^+ forms complexes with the carbonyl of the DMAc, whilst Cl^- forms hydrogen bonds with the hydroxyl groups of the cellulose [12], thus undermining the association of the cellulose molecules. The main advantages of this method are that a high degree of cellulose can become soluble [13] and the dissolved cellulose is stable at room temperature, which enables its further processing. However, LiCl is expensive and difficult to recover, and it is therefore still mainly confined to laboratory research.

7.3.1.1(c) Ionic liquids

1-butyl-3-methyl-imidazole hydrochloride ([BMIM]Cl) is the earliest ionic liquid used to dissolve cellulose. Cellulose is soluble in ionic liquids containing strong hydrogen bond accepting ions (e.g. Cl^-). The dissolution of cellulose in ionic liquids starts from the dissociation of the ionic liquids into Cl^- and free positive ions at critical temperatures, which thus form hydrogen bonds with the hydroxyl groups of the cellulose, weakening the original hydrogen bonds within and between the cellulose molecules, and eventually undermining the network of hydrogen bonds of the cellulose molecules [14, 15]. Prior to achieving large-scale industrialisation of the dissolution of cellulose in ionic liquids, the production cost of ionic liquids will need to be reduced and a more efficient recovery approach will need to be developed.

7.3.1.2 Aqueous solvent dissolution

7.3.1.2(a) Inorganic salt solution

Aqueous cellulose solvent, a solution of inorganic salt mixture, is widely used to regenerate cellulose. The most common aqueous solvents are copper ammonium hydroxide solution (Cuam) and copper acetate diammonium hydroxide solution (Cuen). RC membrane obtained from Cuam solution is a high-quality haemodialysis membrane [16]. Even cellulose with a low molecular weight (e.g. 200 DP) can be dissolved in about 10% NaOH aqueous solution. Changing the pH of the solution can also regenerate cellulose.

7.3.1.2(b) Alkaline solvent system

A new cellulose solvent system (alkaline system) has recently been developed, including NaOH/urea [17–19], LiOH/urea [20, 21], NaOH/thiourea [22–24] and NaOH/urea/thiourea solutions [24]. When the NaOH/urea (7 wt% / 12 wt%) aqueous solution was precooled to –12 °C within 5 minutes under high-speed stirring, cellulose (average molecular weight (M_w) of 1.2×10^5) could be rapidly dissolved. Cellulose macromolecules and small solvent molecules form a stable network of hydrogen bonds at low temperatures, enabling the cellulose molecules to disperse into an aqueous solution as a transparent cellulose solution [26–30]. The solution can be stored for 1 week at 0–5 °C. This low temperature dissolution, beyond the traditional heating dissolution methods, is a 'green chemistry' process that does not use volatiles; however, the cryogenic temperatures should be strictly controlled.

7.3.1.2(c) Molten inorganic salt hydrates

Using molten inorganic salt hydrate as a new solvent of cellulose has attracted more and more attention [31]. Molten inorganic salt hydrate with the formula of $LiX \cdot H_2O$ ($X^- = I^-$, NO_3^-, CH_3COO^-, ClO_4^-) is capable of dissolving cellulose with DP values up to 1500 [32, 33], wherein $LiClO_4 \cdot 3H_2{}^{O}$ is a very effective solvent in which a transparent cellulose solution can be achieved in a few minutes due to the strong interaction between cellulose and the hydrated Li^+ ions. However, Li-containing compound is quite expensive, which is a hindrance to its further extensive application.

Cellulose swelling and dissolution in zinc chloride ($ZnCl_2$) aqueous solution has also been widely studied. Letters *et al.* [34] found that cellulose could be swollen in 55 wt% $ZnCl_2$ aqueous solution and was soluble in 63 wt% aqueous solution. Xu and Chen [35] studied the dissolution of cellulose in $ZnCl_2$ aqueous solution and the preparation of cellulose fibres. Zinc/cellulose composite was precipitated and regenerated in ethanol to obtain cellulose II [36]. The solubility of cellulose in $ZnCl_2$ aqueous was studied by Fischer *et al.* and Leipner *et al.* [37, 38]. Xiong and Zhao [39] found that cellulose could be dissolved in more than 65.0% (w/w) $ZnCl_2$ aqueous solution. Lu and Shen [40] found that bacterial cellulose dissolved in an aqueous solution of $ZnCl_2$ ($ZnCl_2 \cdot 3H_2{}^{O}$) could reach a maximum concentration of 5.5 wt%, and that regenerated bacterial cellulose spun fibre could be made by wet spinning with the cellulose solution. Bamboo dissolving pulp was dissolved in $ZnCl_2$ (4 wt% cellulose, 80 °C and 25 minutes dissolution time) to obtain cellulose solution [41]. Bamboo cellulose film was cast at 40 °C using pure water as the coagulation bath at room temperature. $ZnCl_2$ acted as a porogen during film formation and the wet film has a uniform porous reticulated structure.

7.3.2 Preparation of cellulose film

7.3.2.1 Preparation of cellulose flat sheet film

Cellulose solution is cast on a suitable substrate and then immersed in a non-solvent coagulating bath during the L-S procedure. Phase separation of cellulose is much more complex than that of amorphous polymers, such as polyethersulfone, due to the co-existing crystalline and amorphous regions in cellulose. There are numerous studies of cellulose films prepared by the L-S method, focussing on the relationship between the structure or morphology and the properties [42–44]. Membranes with higher water flux will usually be obtained in stronger solvent for the faster precipitation rate and less film shrinkage, while membranes with much shorter macrovoids beneath the skin layer will be obtained in weaker solvent, leading to higher rejection and much better mechanical properties.

During the phase inversion process a cellulose film formation is governed by two mechanisms, known as liquid–liquid demixing and crystallisation. The type of coagulation bath is one of the key elements in determining the sequence of liquid–liquid demixing and crystallisation. When NMMO is used as the solvent, a uniform symmetric sponge structure may be observed, with NMMO solution as a coagulant and the water flux increasing as the concentration of NMMO is increased; a porous structure may be observed by using alcohol solution as a coagulation medium, with the water flux increasing with increasing concentration of alcohol [45, 46]. Acting as a strong non-solvent, the water coagulation bath often leads to rapid liquid–liquid demixing and consequently asymmetric films consisting of finger-like voids may be formed [45–47]. Meanwhile, adding solvent to the coagulation bath may induce postponed liquid–liquid demixing and the formation of films with sponge-like structure. The addition of solvent accelerating liquid–liquid demixing helps to form a porous structure.

Different coagulation bath temperatures can result in the formation of different film structures. A high-temperature coagulation bath may result in the formation of a cellulose film with finger-like structures, while a low temperature leads to sponge-like structures [48, 49]. Zhang *et al.* studied the effect of coagulation bath temperatures on cellulose membrane morphology [48, 49]. By varying the coagulation bath temperature from 17 °C to 44 °C, the morphology of cellulose membranes changed from non-porous dense structures to finger-like structures. A similar conclusion was drawn by other researchers from their studies of the effect of coagulation bath temperatures on the morphology of cellulose membranes [46, 50].

Adding a non-solvent additive to the casting solution is one of the methods for improving the film morphology as well as the performance. Additives can be used as porogen to increase the viscosity of the solution or to accelerate the phase inversion process. Low-molecular inorganic chemicals as pore formers have attracted significant research attention recently. For example, the influence of ZnO powder on the formation of cellulose membrane was investigated by Lu and Wu [51]. The porosity of the membrane reached around 0.90 after the addition of a small amount of ZnO powder and the pore size of the membrane was adjustable.

7.3.2.2 Preparation of cellulose hollow fibre membrane

The preparation and application of cellulose hollow fibre membrane are limited in laboratory studies [52, 53] since the formation of hollow fibre membranes is much more complex than it is for flat sheet membranes. Although some of the information gained from the study of flat sheet membranes may be useful, similar conditions cannot simply be applied, because phase inversion takes place at both the inner and outer surfaces of hollow fibre membranes. Raw material composition, rheology and coagulation medium should be well controlled in order to obtain the desired performance. The combined effect of these factors may influence the ultimate membrane morphology and performance of cellulose hollow fibre membranes [54]. Increasing the distance of air gap is one of the methods that affect the outer skin formation of hollow fibre membranes [55, 56]. The water permeation flux of cellulose hollow fibre membranes tends to decrease with the increase in the air gap. The extrusion rate of polymer dope is important as it affects the ultimate membrane dimensions, especially the outer diameter and wall thickness. The formation of large pores should be prevented in the process of fabricating cellulose hollow fibre membranes, because these pores will undermine the mechanical properties of the membranes, especially under high pressure. The drying methods can also affect the morphology of the membranes. It is suggested that the ethanol-hexane exchange method is more suitable than natural drying, which tends to cause the largest shrinkage, making the membranes dense [52]. Cellulose hollow fibre ultrafiltration (UF) membranes for oil–water separation were fabricated from cellulose/NMMO·H_2O/polyethylene glycol (PEG400) (mass fraction ratio of 8/88/4) dope by Li *et al.* [57]. Mao *et al.* [58] reported that cellulose/polysulfone (PSF) bilayer hollow fibre membranes for dehydration of iso-propanol (IPA) and separation of carbon dioxide.

There are other methods of fabricating cellulose flat sheet or hollow fibre membranes, which are highly dependent on their applications. For example, from the separation point of view, the first task is to control the pore size of the membranes depending on the different fabricating methods [59].

7.4 Performance and application of cellulose film

7.4.1 Cellulose separation membrane

7.4.1.1 Performance of cellulose separation membrane

It's generally considered that RC membranes are highly hydrophilic and easy to clean, with strong stain resistance. Characterisation of the membranes can be divided into two categories: (1) determination of the structural parameters, such as pore size and distribution, thickness of the skin layer and porosity; and (2) measurement of penetration parameters such as rejection and separation factors, infiltration rate, etc.

7.4.1.1(a) Determination of structural parameters

Determination of pore size. Pore structure parameters include average pore size and pore size distribution. The cylindrical pores are straight holes through the membrane wall and the aperture can be directly observed by an electron microscopy, while the bending pores are determined indirectly using the bubble pressure method.

Visual measurement. The membrane is usually dried by the low-temperature freeze-drying method and can be directly observed by an electron microscope[50, 60]. Scanning electron microscopy (SEM) mainly uses secondary electron signal imaging to observe the surface morphology of the samples with a resolution of up to 10 nm. The membranes are dehydrated, sequentially frozen and quenched off in liquid nitrogen before they are sputtered with gold (thickness of about 20 nm). The pore size and distribution of the microfiltration membrane can be accurately characterised by SEM, which in general ranges from 0.1 μm to 1 μm. However, the pore size of the skin layer in an ultrafiltration membrane is too small to be accurately determined by SEM.

The morphology of cellulose membranes regenerated from $ZnCl_2$ solution was scrutinised using SEM [41]. It was found that the pores on the top surface of the membrane were larger than those on the bottom surface. This was because the top surface of the membrane was in direct contact with the water coagulation bath, which led to rapid liquid–liquid demixing, while the bottom surface was close to the glass.

Corn starch and cellulose were dissolved in the ionic liquid 1-ethyl-3-methylimidazolium acetate (EMIMAc) solvent and regenerated in a water and ethanol coagulation bath [61]. The morphology of the resulting membranes was studied using PHILIPS XL30 scanning electron microscope provided in Eco mode with a backscattered electron (BSE) detector at an accelerating voltage of 15 kV. The results revealed that in an ethanol coagulation bath the corn starch was surrounded by a three-dimensional cellulose network, while in a water coagulation bath the starch portion was leached with the formation of holes and channels. The morphology of cellulose/carboxymethyl cellulose (CMC) and cellulose/polyvinyl alcohol (PVA) blend membranes was also reported by Ibrahim *et al.* [62].

Thiol-ene chemistry has been used to modulate pore size and the surface properties of RC membranes [63]. A series of thiol PEGs (PEG-SH) (Mn, circa 2000, 6000 and 10,000 g/mol) was tethered onto vinylated surfaces to obtain membranes with a pore size ranging from 70 nm to 140nm, which were analysed by field emission electron microscopy (FSEM). It was observed that a novel layer of hydrogel was developed on the membrane surface, especially the cellulose-grafted PEG2000 membrane. The grafted hydrogel layer of cellulose-grafted PEG2000 membrane was thicker than those of the other two membranes, indicating that the change in morphology was not closely related to the graft chain length but more closely related to graft density.

The pore size and porosity of cellulose film can also be obtained from examination of AFM (atomic force microscope, AFM) images of the cross-section. The advantage of this is that the membrane surface can be measured without pretreatment at atmospheric condition; however, it is quite difficult to analyse when the roughness is equal to the pore size. Pang *et al.* [64] dissolved different types of cellulose (pine, cotton, bamboo and microcrystalline cellulose) in the ionic liquid 1-ethyl-3-methylimidazolium acetate to prepare environmentally friendly cellulose membranes. AFM images revealed that the morphology of cotton cellulose membranes was more uniform and smoother, indicating a higher degree of molecular orientation, which might be due to the higher content of cellulose (about 88~96%). Study of the morphology of cellulose membranes fabricated from NMMO solution using AFM has also been reported [65].

According to the physical effects associated with the pores, the corresponding physical quantity, pore size and distribution can be obtained indirectly from the theoretical formula, which is called the physical aperture, including bubble pressure method, filtration rate method, molecular weight cut-off method and the measurement of porosity [60].

Bubble pressure method. This was first proposed by Bechhold in 1970. The infiltrated membrane is fixed in the test pool and the pores are filled with water. Nitrogen is purged from the bottom of the membrane with slowly rising pressure. The bubble will infiltrate the pore when its radius is equal to the pore radius; thus the maximum pore size of the membrane corresponds to the pressure when the first bubble appears, and the minimum aperture corresponds to the pressure when the maximum bubble appears, which may be calculated using the Laplace equation, given by Eq. 1:

$$\gamma = \frac{2\sigma\cos\theta}{p} \qquad \text{(Eq. 1)}$$

where, σ is the surface tension of the liquid, θ is the contact angle of the liquid with the membrane, and γ is the membrane pore radius. The pore size and its distribution of open pores are measured by bubble pressure method. The disadvantage of this method is that the results vary with different liquids; also, the boosting speed and pore length can affect the measured results.

Filtration rate method. Assuming that the pore size of a porous membrane is the same as that of a straight cylindrical pore and the membrane is infiltrated with liquid at a certain pressure, the flow rate increases linearly with the increasing pressure, which meets the conditions of the Hagen-Poiseuille equation. The method is mainly used for the characterisation of microfiltration and ultrafiltration membranes, though the pore size of microfiltration and ultrafiltration membranes is not very uniform. Novel microporous membranes were successfully prepared from cellulose in NaOH/thiourea aqueous solution by coagulating in ammonium sulphate [$(NH_4)_2SO_4$] aqueous solution with different concentrations for 1–20 minutes [60]. The mean pore diameters measured by flow rate method ($2r_e$) and water permeability ($2r_f$) decreased from their maximum values (i.e. $2r_e$ = 514 nm, $2r_f$ = 43.3 nm) to their lowest values (i.e. $2r_e$ = 312 nm, $2r_f$ = 26.6 nm) as the concentration of $(NH_4)_2SO_4$ increased from 1 wt% to 10 wt%.

Molecular weight cut-off method. The pore structure and rejection performance of ultrafiltration membranes is usually reflected by molecular weight cut-off (MWCO), which is generally between 10^3 and 10^5. First, a series of standard reagents with different molecular weights (e.g. protein or polymer) is prepared and the retention is measured separately. Then a cut-off molecular weight curve is plotted and relative molecular mass of the rejection efficiency about 90% (polyethylene glycol solution as the evaluation liquid) or 95 % (protein solution as the evaluation liquid) is selected. This is the so-called MWCO of the ultrafiltration membrane. It should be mentioned that the molecular weight distribution of the selected standard reagents should be narrow and the reagents should be water soluble and low-pollution for the membrane.

Measurement of porosity. Porosity refers to the constitution percentage of membrane pore volume compared to the total membrane volume. In the literature, porosity generally refers to the total porosity, which can be measured by the difference in quality of the wet and dry membrane, given by Eq. 2:

$$\varepsilon = \frac{m_2 - m_1}{\rho V} \qquad \text{(Eq. 2)}$$

where ε is the porosity, m_2 and m_1 are the quality of the wet and dry membrane respectively, ρ is the density of the infiltration liquid. and V is the volume of the dry membrane.

Surface analysis. The stretching vibration peaks of hydrogen donor (such as –OH and $–NH_2$) and hydrogen bond acceptor (such as –C=O) observed by Fourier transform infrared spectroscopy (FTIR) reflect the strength of the hydrogen bonds of the cellulose film (the stretching vibration shifts to low wave number for the existence of hydrogen bonds). The characteristic absorption peaks of pure cellulose membranes include: an O–H stretching vibration peak at 3406 cm^{-1}, a C=O hydroxyl absorption peak at 1063 cm^{-1}, a β-glycosidic linkage characteristic absorption peak at 898 cm^{-1}, and cellulose characteristic peaks at 2920 cm^{-1}, 1645 cm^{-1} and 1376 cm^{-1} [66, 67]. Song *et al.* [68] dissolved 8 wt% cellulose in LiCl/DMAc and then alkyl ketene dimer (AKD) was added to react with the dissolved cellulose in a homogeneous medium, by which means composite membranes were fabricated by the phase inversion process. The characteristics of the β-keto ester absorption peak at 1703 cm^{-1} confirmed the reaction between the hydroxyl groups and the lactone ring. Chitosan and cellulose were dissolved simultaneously in $ZnCl_2$ solution and 1-ethyl-3-methylimidazolium acetate (EmimAc) respectively, and chitosan/cellulose blend membranes were obtained by Lin *et al.* [66] and Zhou *et al.* [67]. FTIR analysis showed that the two components in the blend membranes had good compatibility due to the hydrogen bonds between chitosan and cellulose molecules, which was consistent with other findings [69, 70]. In addition, hydrogen bonds were also found between cellulose and collagen in their blend membranes [71, 72]. ^{1}H NMR can also be used to determine the formation of hydrogen bonds as well as the strength of cellulose film. The formation of hydrogen bonds results in electron density averaging, thus producing chemical shifts of –OH, $-NH_2$ and other proton towards downfield.

X-ray photoelectron spectroscopy (XPS) analysis can also be used for the surface analysis of cellulose film. In XPS analysis, the sample is irradiated with X-rays, and the molecular or atomic electrons become free electrons from the atoms, namely photonics, and then the relationship between the photoelectron energy distribution and its strength is measured. The chemical composition of low-dose radiation modified cellulose membrane (commercially, dense RC flat membrane) has been determined by XPS [73]. Gamma irradiation caused a surface cleaning through the removal of an outermost carbonaceous contamination layer and partial loss of crystallinity of the cellulose matrix detected by X-ray diffraction (XRD) and FTIR, and there were new fracture propagation directions upon irradiation revealed by AFM. Porous cellulose and dense chitosan membranes were bombarded with argon and nitrogen-ion beams using 30 keV and 120 keV energy levels with a fluency of 5×10^{14} ions/cm^2 [74]. A dense structure was created without affecting the OH functional groups.

7.4.1.1(b) Permeability parameter

Permeation performance of cellulose membranes can be represented by the permeation rate, which is defined as a unit time permeation amount of per unit membrane area, i.e. the flux *J*, given by Eq. 3:

$$J = \frac{Q}{St} \qquad \text{(Eq. 3)}$$

where *Q* is the permeation flow, *S* is the effective membrane area, and *t* is the time.

For a liquid separation membrane, the membrane should generally be soaked for at least 24 hours prior to being measured at 25 °C after preloading for 30 minutes [75, 76].

7.4.1.1(c) Membrane fouling

Membrane fouling usually results in a decline in flux and an increase of transmembrane pressure, which in turn reflect the membrane fouling and cleaning processes.

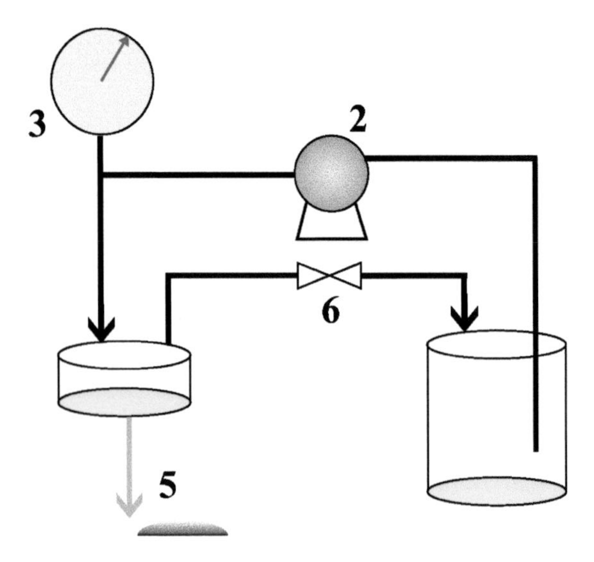

Figure 7.2 Film property evaluation instrument [7]. 1: Feed tank; 2: Pump; 3: Pressure gauge; 4: Membrane cell; 5: Permeate end; 6: Valve.

Membrane fouling can be characterised by SEM, energy dispersive X-ray spectroscopy (EDX), and FTIR.

The antifouling properties of microporous cellulose membranes modified through thio-ene chemistry were evaluated by cross-flow filtration of BSA solution [63]. The membrane was first pressurised with pure water under 1.0 bar, and the initial pure water flux was recorded as J_{wi}. Then the feed solution was changed to bovine serum albumin/phosphate-buffered saline (BSA/PBS) solution. The pure water flux of the resulting membrane J_m was recorded when the flux was steady. Thus, the membrane was rinsed with deionised water for 1 hour, and the pure water flux of the cleaned membrane J_{ww} was recorded. The water flux recovery (FR) was calculated from J_{wi} and J_{ww} using Eq. 4:

$$\mathrm{FR} = \frac{J_{ww}}{J_{wi}} \times 100\% \qquad \text{(Eq. 4)}$$

For irreversible fouling, the above equation can be written as Eq. 5:

$$\mathrm{R} = \frac{J_{ww}}{J_{wi}} = \left(1 - \frac{R_{ir}}{R_m + R_{ir}}\right) \times 100\% \qquad \text{(Eq. 5)}$$

where R_{ir} is the irreversible filtration resistance (m^{-1}) and R_m is the intrinsic membrane resistance (m^{-1}). R_m, R_{ir} and R_c can be calculated using Eq. 6, Eq. 7 and Eq. 8:

$$R_m = \frac{\Delta p}{\eta J_{wi}} \qquad \text{(Eq. 6)}$$

$$R_{ir} = \frac{\Delta p}{\eta J_{ww}} - R_m = \frac{\Delta p}{\eta J_{ww}} - \frac{\Delta p}{\eta J_{wi}} \quad \text{(Eq. 7)}$$

$$R_c = \frac{\Delta p}{\eta J_m} sm - R_m - R_{ir} \quad \text{(Eq. 8)}$$

where Δp is the transmembrane pressure (Pa), η is the feed viscosity (Pa·s) and R_c is the cake layer formation resistance (m^{-1}).

The antifouling properties of the membrane can be described quantitatively by the filtration resistance [63]. For pristine RC membrane, R_c is much larger than R_m, indicating that the fouling behaviour is mainly due to the cake layer formation. For cellulose-grafted polyethylene glycol (PEG) membranes, the greatly depressed R_c values were all lower than the R_m values, indicating a reduction in the cake layer formation. The R_m values of the cellulose-grafted PEG membranes were much higher than those of the RC membrane, mainly due to pore shrinkage. Higher density leads to lower R_c and R_{ir} values for the grafted PEG, while the length of the grafted chain had little effect on the average R_c and R_{ir} values.

7.4.1.1(d) Other parameters

The crystalline structure of cellulose II can be analysed by X-ray diffraction (XRD). The determination of the degree of crystallinity and the disorder parameter of the membranes by XRD analysis is also suitable for cellulose membranes [77]. The average crystallite size, D_{hlk}, can be estimated from wide-angle X-ray scattering (WAXS) interference using the Scherrer equation.

Mechanical properties. The tensile strength, elastic modulus, and elongation at break of membranes are generally measured using material testing machines [77]. The stability of membranes includes their chemical and thermal stability. Membranes that are chemically stable should not swell, dissolve or react with the feed solution. The thermal stability of the membranes is typically analysed by TGA or DTA [78, 79]. Homogenous cellulose/laponite aqueous composite films were prepared from the pre-cooling NaOH/urea aqueous systems [78]. The addition of laponite led to a great enhancement in tensile strength, up to 75.7%, due to the crosslinking effect of cellulose/laponite hybrids through intermolecular hydrogen bonds. However, the decomposition temperature of the composite films decreased from 330°C to 260 °C because the inorganic laponite catalysed the decomposition of cellulose [79, 80] with the decreasing degree of crystallinity of the films.

Hydrophilicity/hydrophobicity is an important surface characteristic parameter, which is closely related to the permeation and fouling status of the membranes, and it is usually characterised by the degree of swelling and the contact angle. Cellulose exhibits high hygroscopicity, arising from the interaction between its hydroxyl groups and water molecules. Microcrystalline cellulose was modified by crosslinking with toluene diisocyanate (TDI) in a DMAc/LiCl solvent system [81]. The hydrophilicity/hydrophobicity of the modified films was investigated by measurement of contact angle. It can be seen that the contact angles increased from 38.5° to 97.9° with increasing TDI content. The network of cellulose became dense and tight as more TDI reacted to the hydroxyl groups, and thus the surface of the films became more hydrophobic. Poly(N-isopropylacryamide) (PNIPAAm) and poly[(2-(diethyl-amino)ethyl methacrylate) (PDEAEMA) were grafted to the cellulose membrane by electron transfer for atom-transfer radical polymerisation [82]. Contact angles were determined by the pendant drop method with a water drop of 3 μL using an optical contact angle meter at room temperature and ambient humidity. The contact angles were measured on both sides of the drop by the ellipse-fitting calculation method. For the side grafted with PNIPAAm, the contact angle increased from 13.0° to 72.8° as the temperature increased from 20 °C to 50°C, exhibiting temperature-dependent changes. On the other hand, the contact angles of the PDEAEMA-grafted layer were 54.6° and 73.6° for pH values of 1.0 and 4.0 respectively and increased to 136.6° and 131.5° when the pH values increased to 7.0 and 10.0, indicating its pH-sensitive characteristics. The dual-responsive cellulose membrane might be applied in water treatment, separation, drug delivery, etc.

7.4.1.2 Application of cellulose separation membrane

The application of cellulose membranes includes microfiltration (MF), ultrafiltration (UF), nanofiltration (NF) and reverse osmosis (RO). Most microfiltration membranes are symmetrical and their pores are uniform (0.1–10 μm), generally with a high porosity – up to 80%. UF membranes are between microfiltration and nanofiltration with an MWCO from 1000 to 500,000, which may be related to the membrane pore size screening process. NF membranes are between reverse osmosis and ultrafiltration, and have a high retention for bivalent or multivalent ions with a molecular weight > 200. The desalination rate of the reverse osmosis membranes can reach 99.8–99.9% with an operating pressure of 1.5–10.5 MPa. The application of cellulose membranes covers nearly all industrial sectors, including environmental

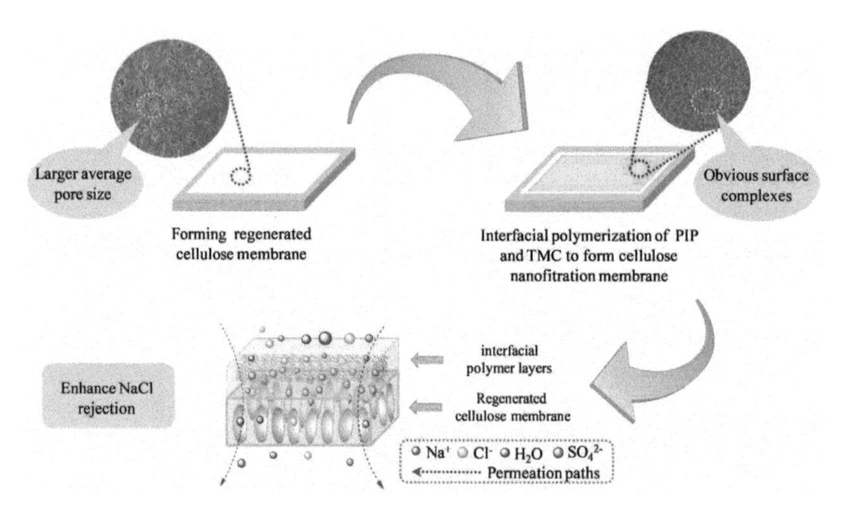

Figure 7.3 Process concept of filtration for the fabrication of cellulose-based nanofiltration membranes [7].

protection, electronics, energy, chemicals and biotechnology. Most cellulose membranes in water treatment research are UF membranes, applied to remove heavy metal ions, for oil and water separation, and for some industrial wastewater purification such as papermaking.

Chitin and cellulose were successfully dissolved in 7 wt% NaOH/12 wt% urea aqueous solution via a freezing/thawing method and their blend membranes were obtained by coagulating with 5 wt% Na_2SO_4. The results revealed that the blend membranes had a microporous structure, large surface area and an affinity to metal ions which enabled them to efficiently remove heavy metal ions (Hg^{2+}, Cu^{2+} and Pb^{2+}) from aqueous solution. The uptake capacity of the heavy metal ions increased with the increase in chitin content. The number of metal ions adsorbed onto the unit amount of the membrane was in the order $Hg^{2+} > Pb^{2+} > Cu^{2+}$, indicating a good adsorption of Hg^{2+} for the major adsorption mechanism, described as multi-interactions including complexation, electrostatic attraction, metal chelation and ionic exchange. This work provided a 'green' pathway for the removal of hazardous materials in wastewater [83].

Ultrafiltration can also be used to remove metal ions from aqueous solutions [84]. The polymer enhanced ultrafiltration (PEUF) technique uses a combination of polymers as the complexing agent for metal ions in ultrafiltration, because the pore sizes of membranes are not small enough to separate metal ions [85–87]. Water soluble polymers are commonly used as complexing agents via binding the metals to form macromolecular complexes in ultrafiltration, by which the complexed molecules are retained, leaving the noncomplexed ions going through the membrane [88]. Poly(vinyl alcohol) (PVA)/cellulose membranes were prepared by coating PVA mixture solutions on the filter paper and the application of these membranes in a batch-stirred ultrafiltration (metal removal from aqueous solutions) was studied [89]. The maximum retention of Fe(III) was 82% under a pressure of 60 psi and a pH of 3 with the presence of alginic acid as the complexing agent, while the maximum retention of Cu(II) was 44% under a pressure of 413.7 kPa and a pH of 7.

Dialdehyde groups can be introduced into a commercial RC ultrafiltration membrane by periodate oxidation [90] and can be further converted to nitrogen-containing derivatives through the Schiff base reaction with diethylenetriamine (DETA). The modified membrane was challenged with aqueous solution containing Pb(II) metal ions. It was found that increasing the feed pH and DETA concentration facilitated the removal of Pb^{2+}, while a higher concentration of Pb^{2+} was hard to be removed.

Cellulose/chitosan hybrid membranes were prepared by regenerating cellulose from chitosan acetic acid solution, and were much denser than the regenerated pure cellulose membranes with homogeneous porous structure [91]. The hybrid membranes exhibited a high rejection of Cu^{2+} at a pH of 5, due to the combination of nanofiltration and complexation of Cu^{2+} with amino groups of chitosan; the ion rejection varied accordingly as the pH of the feed solution changed.

Diaminobutane-based poly(propyleneimine) dendrimer functionalised with sixteen thiol groups, DAB-3-(SH)16, was successfully embedded in a swollen cellulosic support in order to achieve an easily handle engineered membrane [92]. The inclusion of dendrimer on one hand improved the elastic behaviour of the membrane, but on the other hand led to a significant reduction in the permeation of toxic heavy metals (Cd^{2+}, Hg^{2+} and Pb^{2+}). The dendrimer-modified membrane could possibly be applied in electrochemical water treatment.

Cellulose, chitin, and cellulose-chitin blends were regenerated from an ionic liquid and a barrier layer of nanofibrous high-flux UF composite membranes was prepared [93]. The permeation flux and rejection ratio of the membranes were investigated using an oil/water emulsion and a model of bilge water as the feed solution. The cellulose/chitin blend membranes exhibited 10 times higher permeation flux compared to a commercial UF membrane (PAN10, Sepro) with a similar rejection ratio after filtration over a time period of up to 100 hours.

The surfaces of low MWCO RC ultrafiltration membranes were modified by grafting poly(N-isopropylacrylamide) (PNIPAAm)-block-poly(oligoethylene glycol methacrylate) (PPEGMA) nanolayers using surface-initiated atom transfer radical polymerisation [94]. The modification changed the properties of membrane surface in order to limit foulant accumulation, providing an easy, chemical-free way to remove any attached foulants. The surface roughness of the modified membranes decreased slightly from 2.6 nm to 1.7 nm. The water flux of modified membranes was determined using synthetically produced water from an oil-in-water emulsion. The water flux of modified membranes was 13.8% higher than for unmodified membranes over a 40-hour cross-flow filtration run. The flux was fully recovered to initial levels for the modified membranes, while only around 81% was recovered for the unmodified membranes. The results indicate that PNIPAAm-b-PPEGMA modified ultrafiltration membranes could be applied to separate emulsified oils from synthetically produced water.

Membrane separation technology is applied more and more widely in the pulp and paper industry. The most commonly used is UF [95]. However, issues such as fouling and ineffective cleaning limit the further adoption of UF in pulp and paper mills [96, 97]. It has been suggested that hydrophilic membranes may reduce the fouling of the membrane due to its low fouling and good cleanability [98, 99]. RC membrane C30F, polyamide membrane PA50H, and polyethersulphone membrane PES50H were tested with ground wood mill circulation water from an integrated pulp and paper mill [100]. The results showed that the hydrophilic C30F (trade name UC030T) membranes fouled very little with the water flux close to 400 L/(m·h), while the hydrophobic PA50H and PES50H membranes fouled seriously and the restored water flux was only maintained for a short time after cleaning. The UF performances of the RC and polyethersulphone (PES) membranes were compared in chemithermomechanical pulp mill process water [101] originating from hardwood and softwood pulping. Although the hydrophobicity of the RC membranes increased remarkably for adsorptive fouling, the fouling was less than for PES membranes based on their flux recovery. Thus the hydrophilicity and morphology of the membranes had a clear influence on fouling. It was reported by Kallioinen that RC UF membrane C30F was particularly suitable for paper mill wastewater treatment [102], especially for acidic wastewater treatment [103]. The study also found that the performance of C30F under high pressure was stable and suitable for sulphite solution filtration under 550–750 KPa [104].

Cellulose-based membranes have emerged as an attractive alternative to non-biodegradable petrochemical materials. An important drawback, however, is that cellulose-based membranes are prone to biofouling [105–106]. Silver nanoparticles (AgNPs) encapped with polyacrylic acid were conjugated with chitosan/cellulose composite films to enhance the antimicrobial activity [107]. Compared to the chitosan/ cellulose composite films, the chitosan/cellulose–AgNPs composite films showed significantly improved antimicrobial activity. Contact active antibacterial RC membranes were prepared via a direct covalent linking of quaternary ammonium salts (QAS) and aminoalkyl groups [108]. The membranes showed strong antibacterial and bacteria anti-adhesion properties, with a bacteria killing ratio over 99.5%. Furthermore, the antibacterial property was robust across a wide range of pH and temperature conditions.

7.4.2 Cellulose-based medical films

Compared to synthetic polymer materials, natural polymer membranes have competent biocompatibility and blood compatibility. As high value-added biomedical materials, biomedical films have attracted great attention in the last decade. Cellulose medical films consist of cellulose films and cellulose composite films blended with other natural polymers, such as chitosan [67, 109–111], collagen [71, 72] or graft-cellulose membranes that have been surface modified [112]. Biomedical membranes can be applied in medical dressings [107], cell culture materials [112, 113] and tissue engineering materials [114], etc.

7.4.2.1 Properties of cellulose-based medical films

7.4.2.1(a) Moisturising and breathability

Moisturising of cellulose-based medical membranes is generally determined by the volume of absorption and retention liquid, whilst the breathability is determined by the permeability of water vapour transmission (WVTR)[115–117].

7.4.2.1(b) Antibacterial properties

The antibacterial properties of cellulose/chitosan blend films against *Escherichia coli (E. coli)* ATCC 25922 and *Staphylococcus aureus (S. aureus)* ATCC 25923 were tested by Zhou *et al.* [67]. All samples (6 mm-diameter films) were sterilised at 121°C for 20 minutes and then placed on the central part of Petri dishes that contained 20 mL of nutrient agar with 10^7 CFU/mL *E. coli* and *S. aureus*, respectively. Thus, dishes containing the blend films and bacteria were incubated at 37 °C for 24 hours and the inhibition zone was then measured. The results revealed that the cellulose films had no inhibitory effect against *E. coli* or *S. aureus*, while the chitosan/ cellulose blend films exhibited an effective antimicrobial ability against both *E. coli* and *S. aureus*. This demonstrates that chitosan provides the cellulose films with an antibacterial ability against both Gram-negative bacteria (*E. coli*) and Gram-positive bacteria (*S. aureus*). Similar results have also been reported by Shih *et al.* [110].

7.4.2.1(c) Biocompatibility

The biocompatibility of cellulose films can be determined by cell culture experiments [112]. Samples were cut into 6-mm diameter films and placed into 24-well polystyrene cell culture plates with cell adhesion molecules on the surface after sterilising under UV light for 2 hours. The culture medium (10% foetal bovine serum together with 1% penicillin–streptomycin solution, 10,000 u/ml) containing African green monkey kidney fibroblast cells (COS7) (cell density of 6×10^4) was transplanted onto surface of the films. The films were then placed in the incubator for 24 hours at 37°C with 5% CO_2 [112].

By contrast, COS7 cells were inoculated directly in culture wells without the samples. Cell morphology was observed with an optical microscope. A cytotoxicity test was carried out following the 3-(4,5-dimethylthiazol-2-yl)-2,5-diphenyltetra-zolium bromide (MTT) assay. The culture medium was removed after incubation for 24 hours. Fresh culture medium was added in each culture well (1 mL) with MTT (60 μL, 5 mg/mL) with the upper solution being carefully removed after 4 hours of being kept at 37°C. Then 1 mL of dimethyl sulfoxide (DMSO) was added to each well. The absorbance at 570 nm of the upper solution was measured by microplate reader, and thus the value of OD was obtained. Cell viability was calculated according to Eq. 9:

$$\text{Cell viability } (\%) = \frac{OD570_{(sample)}}{OD570_{(control)}} \qquad \text{(Eq. 9)}$$

where $OD570_{(control)}$ was obtained in the absence of the films and $OD570_{(sample)}$ was obtained in the presence of the films.

2, 3-dialdehyde cellulose (DARC) films were obtained from porous structured RC films oxidised by periodate oxidation, which were then reacted with collagen to obtain DARC/collagen composite films [113]. The cytotoxicity of the composite film was evaluated using NIH_3T_3 mice fibroblast cells. The DARC/collagen composite films showed good biocompatibility for scaffold material in tissue engineering.

7.4.2.1(d) Blood compatibility

Haemolysis, defined as the release of haemoglobin into plasma due to the damage of erythrocytes films, is directly related to the blood compatibility of materials [118].

Fresh anticoagulated blood from human volunteers (2 ml) was diluted with 2.5 ml of normal saline, which (0.2 ml) was then added into the test samples. The mixture was kept at 37 °C for 60 minutes and then centrifuged at 1500 rpm for 10minutes. The supernatant was transferred to a 96-well plate. The absorbance at 545 nm was measured by a BioTek™ synergy™ 2 Multi-Mode Microplate Reader. The mean value of three measurements was calculated. The blood (0.2 ml) was diluted in 10 ml of deionised water or normal saline as a positive control or negative control. The degree of haemolysis was calculated according to Eq. 10 [118]:

$$\text{Haemolysis} = \frac{D_t - D_{nc}}{D_{pc} - D_{nc}} \times 100\% \qquad \text{(Eq. 10)}$$

where D_t is the absorbance of the sample, D_{nc} is the negative control absorbance, and D_{pc} is the positive control absorbance.

Blood biocompatibility is generally considered to be relevant to protein adsorption and platelet adhesion, and thus the adsorbed protein may lead to thrombosis. *Wang et al.* [118] grafted a Zwitterionic brush onto the cellulose membrane via activator regeneration by the electron transfer (ARGET)–atom transfer radical polymerisation (ATRP) method for improving blood compatibility [118]. The platelet adhesion, haemolytic test and plasma protein adsorption showed the modified cellulose membrane having excellent blood compatibility featured on lower platelet adhesion and protein adsorption without causing haemolysis. A series of novel hydroxyl-capped phosphorylcholines (HOPC) (n = 5, 2a) was one-pot tethered onto a cellulose membrane with hexamethylene diisocyanate (HDI) as a coupling agent [119]. The results showed that cellulose membranes tethered with HOPC exhibit excellent haemocompatibility for low platelet adhesion and fibrinogen adsorption, as well as antibiofouling properties with bacterial adhesion resistance.

7.4.2.2 Application of cellulose-based medical films

7.4.2.2(a) Medical dressings

Chitin/chitosan is the second most abundant natural polymeric material, with excellent biocompatibility and blood compatibility. Cellulose and chitin or chitosan have good compatibility, and the use of their blend membranes for medical dressings has become a research hotspot in recent years [67, 109, 111, 120].

Chitosan/cellulose composite films can be made using trifluoroacetic acid as the solvent, and the composite films have antimicrobial activity against *E. coli* and *S. aureus* [109]. A series of chitosan/cellulose composite materials was prepared using ionic liquids. Cellulose and chitosan were simultaneously dissolved in 1-ethyl-3-methylimidazolium acetate (EmimAc) and cellulose/chitosan blend membranes were obtained by Zhou *et al.* [67]. The results revealed that the larger content of chitosan in the blend membranes, the better the antibacterial properties against *E. coli* and *S. aureus*, indicating that chitosan did not lose its high-efficiency, broad-spectrum antimicrobial performance during the dissolution and regeneration process.

Chitosan-crosslinked oxycellulose films were also prepared by Zhou *et al.* [120]. The crosslinking of chitosan with oxycellulose gave rise to excellent antibacterial ability against *E. coli* and *S. aureus* as shown in Fig. 7.1. The development of the chitosan/ cellulose or chitosan-crosslinked oxycellulose films has provided a basis for the possible development of various antimicrobial and biomedical products in the future. Chitin with a deacetylation degree of about 50% is water soluble and effective in promoting wound healing. Cellulose and water-soluble chitin were dissolved in a sodium hydroxide-urea-thiourea system and blend films were obtained [121]. The moisturising and breathability of the blend films reached around 32% and 55% respectively, and the films had antibacterial ability against *E. coli* and *S. aureus*. The antibacteriality of O-carboxymethyl chitosan/cellulose blend films fabricated in a sodium hydroxide-urea-thiourea system or NMMO has also been reported [122, 123].

Although there are numerous reports on the antibacterial properties of cellulose/ chitosan blend films, their medical application still remains at the laboratory research stage.

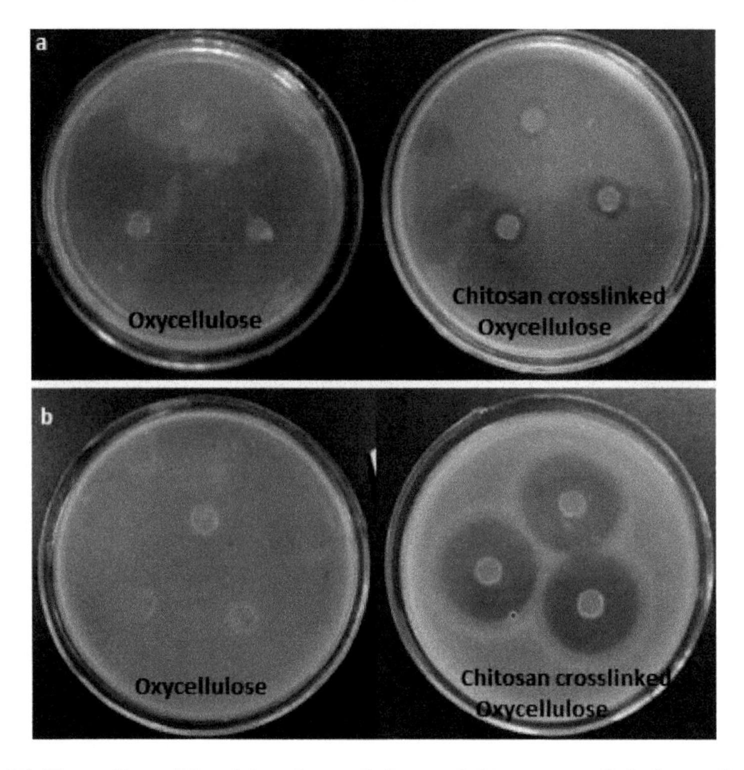

Figure 7.4 The antibacterial activity of oxycellulose and chitosan-crosslinked oxycellulose films against (a) *E. coli* and (b) *S. aureus* (b) [120].

7.4.2.2(b) Cell culture materials

Collagen is an important biomaterial as a form of structural protein in animal skin, bone and tendon. It can be applied widely in many biological or biomedical fields such as surgical sutures, wound healing and tissue engineering [124]. However, collagen has a number of important disadvantages, such as brittleness and poor water retention, which hinder its application to some extent. Cellulose/collagen films were successfully prepared through low-temperature dissolution of their blends by Zhang *et al.* [113], which enabled collagen to be used in a wet state because of its increased water resistance. A series of cellulose/collagen hydrolysate composite films was prepared in NaOH/urea aqueous solution, and thus crosslinked with genipin to further improve the mechanical properties of the films in the wet state [125]. The composite films exhibited good biocompatibility by proliferating COS7 cells on the surface, supporting cell adhesion and growth, and should have promising applications in the biomaterials field.

Porous structured RC films were first oxidised by periodate oxidation. Then collagen was immobilised on the DARC matrix via the Schiff base reaction between the NH_2 in the collagen and the CHO in the DARC backbone, and DARC/collagen composite films were obtained [113]. The composite films demonstrated a good equilibrium-swelling ratio, air permeability and water retention properties with a refined 3-dimensional (3D) network structure. The composite films also possessed favourable biocompatibility, which was evaluated using NIH_3T_3 mice fibroblast cells. They could therefore be used as scaffold material in tissue engineering.

Silk fibroin has been widely used in the biomedical and biotechnological fields for its excellent biocompatibility, biodegradability, relatively easy preparation technique and controllable process conditions. However, its poor mechanical properties (e.g. brittleness) have given rise to the introduction of cellulose to maximise its application [126, 127]. Silk fibroin/cellulose blend films were regenerated from hydrophilic ionic liquid, BmimCl by Shang *et al.* [126]. The results showed that the interactions between silk fibroin and cellulose in the blend films induced a conformation transition of the silk fibroin from random coil form or silk I to β-sheet structure. Furthermore, mouse fibroblast L929 cells showed significant adhesion and proliferation on the silk fibroin/cellulose blend films, which suggests that the treatment of ionic liquid and methanol may not affect the biocompatibility of silk fibroin/cellulose blend films [127].

Nanohydroxyapaptite (nHAP) particles were mixed with cellulose dissolved in pre-cooled 7 wt% NaOH/12 wt% urea aqueous solution. The blend solution was coagulated in Na_2SO_4 aqueous solution, and hybrid films were fabricated [128]. It was revealed that the nHAP with mean diameter of about 30 nm was uniformly dispersed and well immobilised in the nano- and micro- pores of the cellulose matrix due to the strong interaction between the inorganic particles and the organic phase. Furthermore, the results of a 293T cell viability assay showed that the HAP/cellulose films were safe, with excellent biocompatibility, indicating potential applications in the field of biomaterials.

Cellulose/soy protein isolate (SPI) hollow tube (CSC), CSC combined with Schwann cells (SCs) (CSSC) and CSSC combined with Pyrroloquinoline quinone (PQQ) (CSSPC), were constructed based on the composite membranes, respectively [114]. Three kinds of nerve guide conduits were applied to bridge and repair a sciatic nerve defect in rats, using autograft as control. With the comprehensive contributions from hollow CSC tube, SCs and PQQ, the CSSPC group had the strongest ability to repair and reconstruct the nerve structure and functions. It should, therefore, have great potential for nerve guide conduits in nerve tissue engineering.

7.4.2.2(c) Clinical detection materials

An RC membrane with peptide functionality used to selectively bind the influenza virus has been reported. The reaction kinetics of the heterogeneous two-phase reaction of epoxide-functionalised RC membrane with a model aliphatic amine, n-hexylamine, was further studied [129]. Cellulose affinity membranes for antibody purification were prepared using ionic liquid [BMIM]Cl and then the surface was functionalised with a robust biomimetic ligand [130]. The 10 wt% cellulose membrane showed enhanced morphological and mechanical properties. The modified membranes selectively bound and eluted human immunoglobulin (IgG). An integrated multilayer microfluidic device enabling raw sample pretreatment and immunoassay-based detection was reported by Chen *et al.* [131]. A polydimethylsiloxane (PDMS)-cellulose composite film was used to isolate plasma from raw samples based on the cross-flow principle.

Human IgG in rat whole blood samples and a raw urine sample could be detected by the device, which might function as a platform technology, enabling on-site detection of raw clinical samples.

7.4.3 Cellulose packaging films

Cellulose, as a raw material for new biodegradable packaging films, is of great significance for the establishment of ecological balance, environmental protection and economic development. Most current research on cellulose packaging films focuses on the films fabricated from cellulose-NMMO or cellulose-LiCl/DMAc systems. Compared to the viscose process, the performance of these films, for example, transparency, gas/water permeability, strength and chemical stability, has been greatly improved. The selection of the fabricating method is highly dependent on the application purpose. Moulding methods for formulating cellulose packaging films include the unidirectional casting method and the biaxial stretching method. The extrusion moulding method is extensively employed for obtaining excellent mechanical performance from packaging films.

7.4.3.1 Performance of cellulose packaging films

For the majority of foods, the most important functional performance of the packaging film is to prevent the migration of moisture and to prevent food deterioration due to water absorption or water loss, which can be assessed by measuring the film's permeability [132, 133]. The water content of food affects not only the sensory quality, microbial growth and reproduction, but also the chemical and enzymatic reactions caused by food spoilage. In addition to moisture migration, the migration of oxygen and carbon dioxide gas can seriously affect the quality of food and its storage stability. Therefore, the breathability of packaging films is vitally important [134–138]. The mechanical properties of cellulose packaging films are one of the main factors restricting their large-scale industrial application, which has gained great attention in recent years. In addition, cellulose packaging films are naturally transparent and antibacterial once modified.

Most research is focused upon improving the mechanical properties of cellulose packaging films, including process control, addition of inorganic particles and polymer enhancement.

Process control. Wood pulp was dissolved in LiCl/DMAc and biodegradable cellulose packaging films were fabricated by Zhang *et al.* [139]. The results revealed that the optimal concentration of cellulose was 7 wt% and compared with the coagulation bath concentration, the coagulation temperature had a greater impact on the mechanical properties of the films. The mechanical properties of cellulose films fabricated from different types of cellulose pulp dissolved in LiCl/DMAc were reported by Gao *et al.* [134]. It was found that the tensile strength and elongation at break of the membranes increased as the polymerisation degree of the same kind of cellulose pulp increased, and the mechanical strength of films made of softwood pulp was superior to that from hardwood pulp or cotton pulp.

Inorganic filler. The incorporation of inorganic nano-filler into polymer matrix can improve the properties of polymers. In the last few years, various nano-fillers, such as carbon nanotubes[140–142], nano-carbon black [143], graphite oxide [144], nano-hydroxyapatite [145, 146] and nano-montmorillonite/clay [147–149] have been used to enhance the RC's thermal and physical properties. The thermal stability of RC obtained from NMMO/water incorporated with montmorillonite (MMT) was significantly increased. However there were a number of drawbacks in the NMMO/water system, such as cellulose and solvent degradation side effects [150]. The RC/MMT nano-composite films were successfully prepared using the ionic liquid [BMIM]Cl [151]. The results confirmed that MMT improved the thermal stability of the regenerated cellulose. By adding 6 wt.% of MMT, the tensile strength and Young's modulus of the film were increased by 12% and 40%, respectively [151]. It was reported that there were hydrogen bonds present between the iron oxide and the hydroxyl groups of the RC membrane, and thus the tensile strength and elastic modulus of the iron oxide/cellulose were increased by around 39% and 57% compared to the virgin RC membrane [152]. It was found that nano-TiO_2 [153], nano-SiO_2 [154] and $CaCO_3$ [155] could also effectively enhance the mechanical properties of cellulose films.

Polymer enhancement. Cellulose and poly (vinylidene fluoride) (PVDF) were dissolved in DMAc/LiCl separately and the blend films were successfully prepared by coagulating their mixture solutions with water [135]. The elongation at break of the blend films, compared with that of virgin RC films, increased from 12 to 34%. The tensile strength was also improved, from 89 MPa to 106 MPa with the addition of PVDF up to 20 wt%. However, constantly increasing the content of PVDF (>20 wt%) undermined the mechanical properties of the films due to the appearance of a porous structure. It was suggested that cellulose could be miscible with polymers containing atoms of strong electronegativity.

Commercially viable 'green products' based on the use of natural resources for the matrices and reinforcements, are on the rise in a wide range of applications. It was reported that pretreated microcrystalline cellulose (PMCC) can be efficiently dissolved in the hydroxyl functionalised ionic liquid, 1-(2-hydroxylethyl)-3-methyl imidazolium chloride ([HeMIM]Cl), whereas nanocrystalline cellulose (NCC) can only be slightly dissolved in it. A green, all-cellulose nanocomposite film was easily prepared by adding NCC to the cellulose/ionic liquid solution. It was found that NCC acted as fillers for the RC matrix, thus improving the tensile strength, Young's modulus and the strain at failure of the cellulose films [156]. The tensile properties of cellulose matrix films reinforced by *Sterculia urens* short fibre were comparable to those of the conventional packaging materials, such as polypropylene, suggesting that the cellulose composite films have great potential in food packaging applications [157]. Cellulose microfibrils extracted from Agave fibres through chemical process were able to be dissolved in an ionic liquid, 1-allyl-3-methylimidazolium chloride (AmimCl) to a very large extent with a small amount of undissolved microfibrils; thus, regenerated cellulose composite films with self-reinforced microfibrils were prepared [158]. The average tensile strength, Young's modulus and elongation at break of the RC composite films were found to be 135 ± 8 MPa, 8150 ± 257 MPa and

3.2 ± 0.2%, respectively. The tensile strength and tensile modulus were higher than those of other biopolymer films, due to the covalent bonding between the soluble cellulose and the insoluble cellulose microfibrils. Recently, less expensive RC composite films using *Thespesia lampas* fibres as fillers have been reported and the influence of the addition of cellulose fibres on the properties was evaluated [159]. The results showed that the composite films possessed better tensile properties than conventional polyolefin polymers. Cellulose/starch/lignin composite films were prepared in AmimCl and coagulated with water [160]. Cellulose, starch and lignin had good affinity with each other, which gave rise to the excellent tensile strength without sacrificing the extensibility of the resulting films. The composite films also showed higher gas permeability values than RC films with a wider range of CO_2:O_2 permeation ratios (1.05–1.67).

The antibacterial property of cellulose packaging film is also very important, which is usually achieved by coating or blending with organic or inorganic antimicrobial preservatives [137, 161, 162]. The transparency of the films can be reflected by light transmittance through a haze meter [134]or UV–visible spectrophotometer [133, 156, 163].

7.4.3.2 Application of cellulose packaging film

It was proposed that NMMO cellulose packaging films could be used for packing tape and packing materials by Akzona in 1979. Cellulose packaging films used for agricultural wrapping tape were studied by the TITK laboratory in Germany and it was found that cellulose films had incomparable advantages compared to linen or polypropylene packaging films. Akozo Nobel in the Netherlands is Europe's largest copper ammonia dialysis membrane manufacturer, and they are currently exploring the production of cellulose packaging films using the NMMO process, which is expected to be applied to the largest area of application of traditional cellulose packaging films, i.e. artificial kidneys. Chicago Viskase, the world's largest supplier of cellulose casings, has carried out the industrial production of cellulose films.

Obtaining cellulose packaging films from NMMO is a green production process with no chemical reaction in it. Thus, the cellulose films have a more uniform thickness, higher mechanical strength and better water vapour permeability. Because they are completely naturally biodegradable, cellulose films can be used not only as pharmaceuticals or waste packaging materials but also as food packaging material due to their safety. In a word, cellulose packaging films offer great opportunities for the development of a green packaging industry.

7.4.4 Cellulose photovoltaic film

Cellulose has been re-discovered as a smart material due to its biocompatibility, easy modification, ecofriendliness, and low price. Cellulose or cellulose-based smart materials can be used in gas sensors, liquid sensors and strain sensors.

7.4.4.1 Properties of cellulose photovoltaic film

7.4.4.1(a) Gas sensing test

The chemical sensing ability of multi-walled carbon nanotubes (MWCNTs)/cellulose paper was investigated by Yun and Kim [164] in order to determine its suitability as a chemical vapour sensor. The chemical/electrical properties of MWCNTs/cellulose paper were investigated by measuring the electrical resistance between the electrodes of an interdigital transducer (IDT) using an inductance, capacitance and resistance (LCR) meter. The chemical vapour was provided by a bubbler system via injection of a carrier gas (argon).

7.4.4.1(b) Liquid sensing test

CNT–cellulose composite films were fabricated from a homogeneous dispersion of MWCNTs and cellulose in alkaline-urea aqueous solution and the response to liquid/water of the composite films was investigated [165]. U-shaped film samples coated with silver electrodes at both ends were used to determine the sample's resistance in distilled water and organic solvents (methanol, ethanol, and acetone). A measuring cycle consisted of immersion under certain temperatures controlled by a heating/cooling bath and with a drying step at 23°C and 50% relative humidity.

7.4.4.1(c) Humidity and temperature sensing test

Nanoscaled polypyrrole was introduced onto the surface of cellulose films by polymerisation-induced adsorption process and the application of cellulose films as a flexible humidity and temperature sensor was demonstrated [166]. An interdigital comb as electrode on cellulose/polypyrrole composite films via lift-off technique was used to evaluate the humidity and temperature-sensing behaviour. The electrical response of the composite films at different humidities and temperature levels was recorded using an LCR meter. The sensor was placed in an environmental chamber with controlled temperature and humidity and the humidity and temperature-sensitive properties were investigated.

7.4.4.1(d) Strain sensing test

RC films, termed as electro-active paper (EAPap), were prepared by dissolving cellulose fibres using LiCl/DMAc solvent system and curing the cellulose solution with deionised water and isopropyl alcohol solvent mixture [167]. Cellulose electro-active paper actuators were provided by coating both sides of the cellulose films with gold electrodes and the actuation durability was evaluated by measuring the bending displacement with time. The sample was placed in the environmental chamber with controlled temperature and humidity. The cellulose EAPap actuators produced a bending deformation as the electrical signal was sent out, which was recorded on the laser displacement sensor.

Cellulose films were prepared by dissolving cellulose in ionic liquid, and graphene nanoplatelets were loaded onto a cellulose matrix to improve the electro-active performance of the cellulose-based composite actuators [168]. The electro-active characteristics of the actuators were analysed through a set of direct current (DC) voltage experiments. The films were coated with gold electrodes and a DC voltage was applied. The motion of the actuators was recorded with image acquisition hardware.

7.4.4.1(e) Conductivity test

Transparent cellulose films were coated with indium tin oxide (ITO) nanoparticle solution and flexible and transparent cellulose/ITO layered films were fabricated [169].

The conductivity was measured using a 4-point probe station using the Au electrode as a driving electrode and Al electrode as the ground.

7.4.4.2 Applications of cellulose photovoltaic film

7.4.4.2(a) Cellulose sensors

Gas detection sensors can be fabricated from inorganic semiconductor metal oxides or organic conducting polymers; however, the former require a high working temperature and the latter have low thermal stability and a very long response time. Both chemically modified cellulose and cellulose-based composite with a conductive polymer can be used in gas sensors. The cellulose unit contains three hydroxyl groups (OH-group), which are highly active and can be chemically modified. MWCNTs covalently grafted cellulose film was fabricated by reacting imidazolides-MWCNTs with cellulose. It can be used for ammonia gas sensing and as a chemical vapour sensor [164].

Humidity sensors are widely used for environmental monitoring, industrial protection, as well as for human contact [170, 171]. Cellulose blended with organic conducting polymer enhanced not only the mechanical properties but also the sensing ability of materials. Nanocomposite based on cellulose and polypyrrole offered the unique property of cellulose combined with the electrical properties of polypyrrole, which can be capitalised upon to design flexible, biodegradable and low-cost humidity and temperature sensors [165]. Electrically conductive films composed of carbon nanotubes and cellulose matrix used as water sensors exhibited rapid response and high sensitivity, with a relative electrical resistance change of 5500–500% [165]. The sensitivity of cellulose–carbon nanotubes composites was mainly due to the hygroscopic swelling of the cellulose matrix.

Strain sensors can be used to monitor the state of buildings, bridges, and other critical infrastructure in real time in order to prevent any catastrophic failure [172]. The sensor is commonly fabricated as a thin film with active nanoparticles for providing electric charge under an external electric field. Cellulose has generally been used as a matrix for holding nanoparticles which can be both chemically modified and physically incorporated. Cellulose solution, prepared by dissolving cotton pulp in LiCl/DMAc solution reacting with MWNTs-imidazolides and MWNTs/cellulose EAPap, was obtained. It is found that the Young's modulus of the resulting film is remarkably enhanced compared to that of MWNTs blended electro-active paper, due to the covalent bonds [164]. The electrical conductivity of cellulose films was considerably increased with an increase in graphene nanoplatelets loading (0.10, 0.25, and 0.50 wt%) [168]. Graphene addition also enhanced the capability of an actuator being operated at higher excitation voltages of 3–7 V, whereas cellulose films could only be operated at an excitation voltage of 3 V. The greatest tip displacement was observed as 2.2 mm for cellulose/graphene (0.50 wt%) film under a DC voltage of 3 V, indicating a 267% increase in maximum tip displacement. Thus, graphene nanoplatelets loading resulted in better electro-active performance for cellulose-based composite actuators.

Conductive polymer polypyrrole and polyaniline were electro-deposited onto cellulose EAPap actuators by Kim *et al.* [173]. The results showed that the trilayer actuators were superior to their bilayer counterparts, and polyaniline-coated actuators gave rise to better performance than polypyrrole-based actuators in terms of bending displacement and response to humidity.

Cellulose itself has also been used as a sensor and actuator material [174, 175]. There are many advantages of cellulose EAPap, such as light weight, low cost, biodegradability, large deformation, low actuation voltage and low power consumption [176]. It is indicated that a 40:60 solvent mixture ratio of IPA and DI water as the coagulation bath for cellulose and LiCl/DMAc solution was optimal, which gave an improvement in surface roughness and Young's modulus as well as in the durability of the cellulose EAPap actuator [167]. It was found that cellulose EAPap provided promising potential as biodegradable and cheap piezoelectric polymer material. Strong shear electro-mechanical coupling was observed and the piezoelectric charge constant was in the range of 8–28.2 pC/N, which is similar to those for piezo polymer [177].

7.4.4.2(b) Other cellulose photovoltaic materials

It was reported that cellulose–ITO films with 15 wt% ITO nanoparticles showed outstanding electrical properties with a low resistivity to 5.0×10^3 and a high transmittance to 74% at 800 nm using a flexible and transparent cellulose substrate [178]. The coated ITO layer was securely attached to the surface of the cellulose through a simple spin-coating method without reducing the electrical characteristics. The cellulose–ITO films were eco-friendly, flexible, cheap and lightweight, with better electrical and optical properties, which could make them suitable for use as an optoelectrical material for flexible electronics and display devices. RC/graphene oxide nanosheets based multilayer films were successfully fabricated using a layer-by-layer assembly technique, with a thickness of about 400 nm per layer [178]. The conductivity of the film with 50 multilayers reached 1.3×10^4 S/m due to the formation of conductive networks throughout the insulating regenerated cellulose, as well as a 110.8% enhancement in the elastic modulus and a 262.5% increase in hardness, respectively. The conductive property of the multilayer film made it a promising candidate for biosensor and tissue engineering applications.

Gold nanoparticle thin film was deposited on one side of a cellulose membrane that was exposed to $AuCl_4$ ions which exhibited a broad surface plasmon resonance (SPR) peak at 529 nm [179]. It was observed that the membrane could be dried and rehydrated without any loss of gold nanoparticles and had the potential to be applied in the fabrication of gold thin films for semiconductor (solar cells) studies.

Furthermore, electro-deposited cellulose and cellulose composite films are attractive for the protection of sensor electrodes against interference and for controlling permeation in biosensors. Thin, porous cellulose films were readily electro-deposited onto boron-doped diamond electrode surfaces by Bonné *et al.* [180]. Ions partitioned into the electro-deposited cellulose films with relatively slow diffusion when immersed in aqueous electrolyte solution. It has been suggested that thin sensor films could be prepared by electro-deposition on various types of electrode surfaces.

In conclusion, the suitability of cellulose or cellulose-based materials as sensors is still being developed and there is a need for a marked improvement in the engineering properties together with higher efficiency in the sensor community [172].

7.5 Outlook

The advantageous features of cellulose compared with synthetic polymers provide great potential for the development of advanced cellulose and/or nanocellulose films. The development of new cellulose solvents, e.g. *N*-methyl morpholine-*N*-oxide (NMMO), dissolved alkaline systems, ionic liquids, and lithium chloride/ dimethylacetamide (LiCl/DMAc), have accelerated this development. Numerous products have been attempted, including a range of cellulose separation membranes, cellulose-based medical films, cellulose packaging films and cellulose photovoltaic films. However, many of these remain at the laboratory stage and much research and further innovation are necessary.

References

[1] Kesting R E, *Synthetic Polymeric Membranes: A Structural Perspective*, Wiley, New York, NY, 1985.

[2] Mulder M, *Basic Principles of Membrane Technology*, Kluwer Academic Publishers, Dordrecht, the Netherlands, 1991.

[3] Bottino A, Camera–Roda G, Capannelli G & Munari S, The formation of microporous polyvinylidene difluoride membranes by phase separation [J]. *Journal of Membrane Science*, 1991, 57 (1): 1–20.

[4] Inamoto M, Miyamoto I, Hongo T, Iwata M & Okajima K, Morphological formation of the regenerated cellulose membranes recovered from its cuprammonium solution using various coagulants [J]. *Polymer Journal*, 1996, 28 (6): 507–512.

[5] Hoenich N A, Cellulose for medical applications, *Bioresources*, 2006, 1 (2): 270–280.

[6] Michael M, Ibbett R N & Howarth O W. Interaction of cellulose with amine oxide solvents [J]. *Cellulose*, 2000, 7 (1): 21–23.

[7] Li S. Preparation and characterization of cellulose nanofiltration membranes with application in the depth treatment of drinking water. Master thesis of Fujian Agriculture and Forestry University, 2018.

[8] Fink H P, Weigel P, Purz H J & Ganster J. Structure formation of regenerated cellulose materials from NMMO– solutions, *Prog. Polym. Sci.* 26 (2001) 1473–1524.

[9] Lu A & Zhang L N. Advance in solvents of cellulose, [J]. *Acta Polymerica Sinica*, 2007, 10: 937–944.

[10] Li R. Basic scientific issues of spinning industrialization of cellulose dissolved at low temperature [D]. Doctoral thesis of Wuhan University, 2012.

[11] Terbojevich M, Cosani A, Conio G, Ciferri A & Bianchi E. Mesophase formation and chain rigidity in cellulose and derivatives. 3. Aggregation of cellulose in N, N–dimethylacetamide–lithium chloride[J]. *Macromolecules*, 1985, 18 (4): 640–646.

[12] Dawsey T R & Mccormick C L. The lithium chloride/dimethylacetamide solvent for cellulose–a literature–review. *J. Macromol. Sci. RMC*, 1990, C30: 405–440.

[13] Germain J S & Vincendon M. ^{1}H, ^{13}C and ^{7}Li nuclear magnetic resonance study of the lithium chloride-N,N-dimethylacetamide system [J]. *Organic Magnetic Resonance*, 1983, 21 (6): 371–375.

[14] Pinkert A, Marsh K N, Pang S S & Staiger M P. Ionic liquids and their interaction with cellulose. *Chem. Rev.* 2009, 109: 6712–6728.

[15] Swatloski R P, Spear S K, Holbrey J D & Rogers R D. Dissolution of cellulose with ionic liquids [J]. *Journal of the American Chemical Society*, 2002, 124 (18): 4974–4975.

[16] Hoenich N A, Woffindin C, Stamp S, Roberts S J & Turnbull J. Synthetically modified cellulose: An alternative to synthetic membranes for use in haemodialysis [J]. *Biomaterials*, 1997, 18 (19): 1299–1303.

[17] Zhou J P & Zhang L N. Solubility of cellulose in NaOH urea aqueous solution [J]. *Polymer Journal*, 2000, 32 (10): 866–870.

[18] Zhang L N, Ruan D & Zhou J P. Structure and properties of regenerated cellulose films prepared from cotton linters in NaOH/urea aqueous solution [J]. *Industrial & Engineering Chemistry Research*, 2001, 40 (25): 5923–5928.

[19] Qi H S, Chang C Y & Zhang L. Properties and applications of biodegradable transparent and photoluminescent cellulose films prepared via a green process [J]. *Green Chemistry*, 2009, 11 (2): 177–184.

[20] Weng L H, Zhang L N, Ruan D, Shi L H & Xu J. Thermal gelation of cellulose in a NaOH/thiourea aqueous solution [J]. *Langmuir* 2004, 20 (6): 2086–2093.

[21] Zeng J, Li R, Liu S L & Zhang L. Fiber-like TiO_2 nanomaterials with different crystallinity phases fabricated via a green pathway [J]. *ACS Applied Materials & Interfaces*, 2011, 3 (6): 2074–2079.

[22] Cai J, Liu Y T & Zhang L N. Dilute solution properties of cellulose in LiOH/urea aqueous system [J]. *Journal of Polymer Science Part B: Polymer Physics*, 2006, 44 (21): 3093–3101.

[23] Cai J, Zhang L N, Chang C Y, Cheng G Z, Chen X M & Chu B. Hydrogen-bond-induced inclusion complex in aqueous cellulose/LiOH/urea solution at low temperature [J]. *Chem. Phys. Chem.*, 2007, 8 (10): 1572–1579.

[24] Liu S L, Zeng J A, Tao D D & Zhang L N. Microfiltration performance of regenerated cellulose membrane prepared at low temperature for wastewater treatment [J]. *Cellulose*, 2010, 17 (6): 1159–1169.

[25] Zhou X D, Zhu P & Zhang L. Preparation and properties of soluble chitin/cellulose blend membranes, *Membrane Science and Technology*, 2009, 29 (2): 12–15.

[26] Cai J, Zhang L N & Liu S L. Dynamic self-assembly induced rapid dissolution of cellulose at low temperatures. *Macromolecules*, 2008, 41: 9345–9351.

[27] Cai J, Zhang L N, Chang C Y, Cheng G Z, Chen X M & Chu B. Hydrogen-bond-induced inclusion complex in aqueous cellulose/LiOH/urea solution at low temperature. *Chemphyschem*, 2007, 8: 1572–1579.

[28] Lu A, Liu Y T, Zhang L N & Potthast A. Investigation on metastable solution of cellulose dissolved in NaOH/urea aqueous system at low temperature. *J. Phys. Chem. B*, 2011, 115: 12801–12808.

[29] Lue A, Liu Y T, Zhang L & Potthas A. Light scattering study on the dynamic behavior of cellulose inclusion complex in LiOH/urea aqueous solution. *Polymer*, 2011, 52: 3857– 3864.

[30] Medronho B & Lindman B. Brief overview on cellulose dissolution/regeneration interactions and mechanisms, *Advances in Colloid and Interface Science*, 2015, 222: 502–508.

[31] Sen S, Martin J D & Argyropoulos D S. Review of cellulose non-derivatizing solvent interactions with emphasis on activity in inorganic molten salt hydrates. *ACS. Sustain. Chem. Eng.* 2013, 1: 858–870.

[32] Fischer S, Voigt W & Fischer K. The behavior of cellulose in hydrated melts of the composition $LiX{\cdot}nH_2O(X^-=I^-, NO_3^-, CH_3COO^-, ClO_4^-)$ [J]. *Cellulose*, 1999, 6 (3): 213– 219.

[33] Leipner H, Fischer S, Brendler E & Voigt W. Structural changes of cellulose dissolved in molten salt hydrates. *Macromolecular Chemistry and Physics*, 2000, 201 (15): 2041–2049.

[34] Letters K. Viskosimetrische Untersuchungen über die Reaktion von Cellulose mit konzentrierten Chlorzinklösungen. Kolloidzeitschrift, 1932, 58 (2): 229–235.

[35] Xu Q & Chen L F. Preparing cellulose fibre from zinc–cellulose complexes, *Textile Techn. Int.*, 1996, 40: 19–21.

[36] Grinshpan D D, Lushchik L G N, Tsygankova G, Voronkov V G, Irklei V M & Chegolya A S. Process of preparing hydrocellulose fibers and films from aqueous solutions of cellulose in zinc chloride. *Fibre Chemistry*, 1988, 20 (6): 365–369.

[37] Fischer S, Leipner H, Thümmler K, Brendler E & Peters J. Inorganic molten salts as solvents for cellulose. *Cellulose*, 2003, 10 (3): 227–236.

[38] Leipner H, Fischer S, Brendler E & Voigt W. Structural changes of cellulose dissolved in molten salt hydrates. *Macromolecular Chemistry and Physics*, 2000, 201 (15): 2041–2049.

[39] Xiong J, Ye J & Zhao X F. Solubility of cellulose in $ZnCl_2$ aqueous solution and structure of regenerated cellulose. *Journal of South China University of Technology*, 2010, 38 (2): 23–27.

[40] Lu X K & Shen X Y. Solubility of bacteria cellulose in zinc chloride aqueous solutions. *Carbohydrate Polymers*, 2011, 86 (1): 239–244.

[41] Lin S. Preparation of antibacterial cellulose membranes and their application in the depth treatment of water. Doctoral thesis of Fujian Agriculture and Forestry University, 2013.

[42] Abe Y & Mochizuki A. Hemodialysis membrane prepared from cellulose/N–methylmorpholine–N–oxide solution. I. Effect of membrane preparation conditions on its permeation characteristics. *Journal of Applied Polymer Science*, 2002, 84 (12): 2302–2307.

[43] Abe Y & Mochizuki A. Hemodialysis membrane prepared from cellulose/N-methylmorpholine-N-oxide solution. II. Comparative studies on the permeation characteristics of membranes prepared from N-methylmorpholine-N-oxide and cuprammonium solutions. *Journal of Applied Polymer Science*, 2003, 89 (2): 333–339.

[44] Zhang Y P, Shao H L, Wu C X & Hu X C. Formation and characterization of cellulose membranes from N-methylmorpholine-N-oxide solution. *Macromolecular Bioscienc*e, 2001, 1 (4): 141–148.

[45] Zhang Y P, Shao H L, Shen X Y & Hu X C. Study on cellulose UF membrane prepared by NMMO technology. *Journal of Henan Normal University*, 2000, 28 (4): 47–51.

[46] Lu Y C & Wu Y X. Influence of coagulation bath on morphology of cellulose membranes prepared by NMMO method. *Journal of Chemical Engineering of Chinese Universities*, 2007, 21 (3): 398–403.

[47] Kesting R E. Phase inversion membranes. *Materials Science of Synthetic Membranes*. Austin, TX: American Chemical Society, 1985: 131–164.

[48] Zhang Y P, Shao H L, Shen X Y & Hu X C. Effect of coagulation bath and casting solution on cellulose UF membranes. *Journal of China Textile University*, 2000, 26 (4): 90–92.

[49] Zhang Y P, Hu S Z, Shao H L, Shen X Y & Hu X C. Influence of coagulation bath on Lyocell membrane performance and surface topography. *Journal of Dong Hua University*, 2002, 28 (3): 1–6.

[50] Li R, Zhang L N & Xu M. Novel regenerated cellulose films prepared by coagulating with water: Structure and properties. *Carbohydrate Polymers*, 2012, 87 (1): 95–100.

[51] Lu Y C & Wu Y X. Pore structure control of cellulose membrane prepared using the NaOH/urea method, *J. Tsinghua Univ*. 2007, 47 (9): 1503–1506.

[52] Yamamoto K I, Ogawa T, Matsuda M, Iino A, Yakushiji T, Miyasaka T & Sakai K. Membrane potential and charge density of hollow fiber dialysis membranes. *Journal of Membrane Science*, 2010, 355 (1–2): 182–185.

[53] Jie X M, Cao Y M, Qin J J & Liu J H. Influence of drying method on morphology and properties of asymmetric cellulose hollow fiber membrane [J]. *Journal of Membrane Science*, 2005, 246 (2): 157–165.

[54] Sukitpaneenit P & Chung T G. Molecular elucidation of morphology and mechanical properties of PVDF hollow fiber membranes from aspects of phase inversion, crystallization and rheology. *Journal of Membrane Science*, 2009, 340 (1–2): 192–205.

[55] Khayet M. The effects of air gap length on the internal and external morphology of hollow fiber membranes. *Chemical Engineering Science*, 2003, 58 (14): 3091–3104.

[56] Ren J, Wang R, Zhang H Y, Li Z, Liang D T & Tay J H. Effect of PVDF dope rheology on the structure of hollow fiber membranes used for CO_2 capture. *Journal of Membrane Science*, 2006, 281 (1–2): 334–344.

[57] Li H J, Cao Y M, Qin J J, Jie X M, Wang T H & Liu J H. Development and characterization of anti–fouling cellulose hollow fiber UF membranes for oil–water separation. *Journal of Membrane Science*, 2006, 279 (1–2): 328–335.

[58] Mao Z M, Jie X M, Cao Y M, Wang L N, Li M & Yuan Q. Preparation of dual-layer cellulose/polysulfone hollow fiber membrane and its performance for isopropanol dehydration and CO_2 separation. *Separation and Purification Technology*, 2011, 77 (1): 179–184.

[59] Mulder M. *Basic Principles of Membrane Technology. Second edition* [M]. Netherlands: Kluwer Academic Publishers, 1996.

[60] Ruan D, Zhang L N, Mao Y, Zeng M & Li X B. Microporous membranes prepared from cellulose in NaOH/thiourea aqueous solution, *Journal of Membrane Science*, 2004, 241: 265–274.

[61] Liu W Q & Budtova T. Ionic liquid: A powerful solvent for homogeneous starch-cellulose mixing and making films with tuned morphology, *Polymer*, 2012, 53 (25): 5779–5787.

[62] Ibrahim M M, Koschella A, Kadry G & Heinze T. Evaluation of cellulose and carboxymethyl cellulose/poly(vinyl alcohol) membranes, *Carbohydrate Polymers*, 2013, 95 (1): 414–420.

[63] Yuan T, Meng J Q, Gong X Z, Zhang Y F & Xu M L. Modulating pore size and surface properties of cellulose microporous membrane via thio-ene chemistry, *Desalination*, 2013, 328: 58–66.

[64] Pang J H, Wu M, Zhang Q H, Tan X, Xu F, Zhang X M & Sun R C. Comparison of physical properties of regenerated cellulose films fabricated with different cellulose feedstocks in ionic liquid, *Carbohydrate Polymers*, 2015, 121: 71–78.

[65] Yokota S, Kitaoka T & Wariishi H. Surface morphology of cellulose films prepared by spin coating on silicon oxide substrates pretreated with cationic polyelectrolyte, *Applied Surface Science*, 2007, 253 (9): 4208–4214.

[66] Lin S, Chen L H, Huang L L, Cao S L, Luo X L, Liu K & Huang Z H. Preparation and characterization of chitosan/cellulose blend films using $ZnCl_2 \cdot 3H_2O$ as a solvent. *Bioresources*, 2012, 7: 5488–5499.

[67] Zhou Y H. Preparation and characterization of antibacterial cellulose/chitosan film. Master's thesis of Fujian Agriculture and Forestry University, 2013.

[68] Song X M, Chen F S & Liu F S. Preparation and characterization of alkyl ketene dimer (AKD) modified cellulose composite membrane, *Carbohydrate Polymers*, 2012, 88 (2): 417–421.

[69] Luo K, Yin J, Khutoryanskaya O V & Khutoryanskiy V V. Mucoadhesive and elastic films based on blends of chitosan and hydroxyethylcellulose, *Macromol. Biosci.* 2008, 8 (2): 184–192.

[70] Xu Y X, Kim K M, Hanna M A & Nag D. Chitosan-starch composite film: Preparation and characterization, *Ind. Crop. Prod.* 2005, 21 (2): 185–192.

[71] Zhang M, Ding C C, Huang L L, Chen L H & Yang H Y. Interactions of collagen and cellulose in their blends with 1-ethyl-3-methylimidazolium acetate as solvent, *Cellulose*, 2014, 21: 3311–3322.

[72] Zhang M, Ding C C, Chen L H & Huang L L. The preparation of cellulose/collagen composite films using 1-ethyl-3-methylimidazolium acetate as a solvent, *Bioresources*, 2014, 9 (1): 756–771.

[73] Vázquez M I, Heredia-Guerrero J A, Galán P, Benitez J J & Benavente J. Structural, chemical surface and transport modifications of regenerated cellulose dense membranes due to low-dose radiation. *Materials Chemistry and Physics*, 2011, 126 (3): 734–740.

[74] Wanichapichart P, Taweepreeda W, Choomgan P & Yu L D. Argon and nitrogen beams influencing membrane permeate fluxes and microbial growth. *Radiation Physics and Chemistry*, 2010, 79 (3): 214–218.

[75] Yang G, Xiong X P & Zhang L. Microporous formation of blend membranes from cellulose/konjac glucomannan in NaOH/thiourea aqueous solution. *Journal of Membrane Science*, 2002, 201 (1–2): 161–173.

[76] Chen Y, Zhang L, Gu J M & Liu J. Physical properties of microporous membranes prepared by hydrolyzing cellulose/soy protein blends [J]. *Journal of Membrane Science*, 2004, 241 (2): 393–402.

[77] Fink H P, Weigel P, Purz H J & Ganster J. Structure formation of regenerated cellulose materials from NMMO-solutions, *Prog. Polym. Sci.*, 2001, 26: 1473–1524.

[78] Yuan Z W, Fan Q R, Dai X N, Zhao C, Lv A J, Zhang J J, Xu G Y & Qin M H. Cross–linkage effect of cellulose/laponite hybrids in aqueous dispersions and solid films, *Carbohydrate Polymers*, 2014, 102: 431– 437.

[79] Zeng J, Li R, Liu S & Zhang L. Fiber-like TiO_2 nanomaterials with different crystallinity phases fabricated via a green pathway. *ACS Applied Materials & Interfaces*, 2011, 3: 2074–2079.

[80] Zhou J, Li R, Liu S, Li Q & Zhang L. Structure and magnetic properties of regenerated cellulose/Fe_3O_4 nanocomposite films. *Journal of Applied Polymer Science*, 2009, 111: 2477–2484.

[81] Qiu X Y, Tao S M, Ren X Q & Hu S W. Modified cellulose films with controlled permeatability and biodegradability by crosslinking with toluene diisocyanate under homogeneous conditions, *Carbohydrate Polymers*, 2012, 88: 1272–1280.

[82] Qiu X Y, Ren X Q& Hu S W. Fabrication of dual-responsive cellulose-based membrane via simplified surface-initiated ATRP. *Carbohydrate Polymers*, 2013, 92 (2): 1887–1895.

[83] Tang H, Chang C & Zhang L. Efficient adsorption of Hg^{2+} ions on chitin/cellulose composite membranes prepared via environmentally friendly pathway, *Chemical Engineering Journal*, 2011, 173: 689–697.

[84] Tavares C R, Vieira M, Petrus J C & Bortoletto F. Ultrafiltration/complexation process for metal removal from pulp E.C. and paper industry wastewater. *Desalination*, 2002, 144 (1–3): 261–265.

[85] Juang R S & Chen M N. Retention of copper (II)-EDTA chelates from dilute aqueous solutions by a polyelectrolyte-enhanced ultrafiltration process. *Journal of Membrane Science*, 1996, 119 (1): 25–37.

[86] Juang R S & Shiav R C. Metal removal from aqueous solutions using chitosan-enhanced membrane filtration. *Journal of Membrane Science*, 2000, 165 (2): 159–167.

[87] Kozlowski C A & Walkowiak W. Removal of chromium(VI) from aqueous solutions by polymers inclusion membranes. *Water Research*, 2002, 36 (19): 4870–4876.

[88] Yang L, Hsiao W W & Chen P. Chitosan–cellulose composite membrane for affinity purification of biopolymers and immunoadsorption. *Journal of Membrane Science*, 2002, 197 (1–2): 185–197.

[89] Çifci C & Kaya A. Preparation of poly(vinyl alcohol)/cellulose composite membranes for metal removal from aqueous solutions. *Desalination*, 2010, 253 (1–3): 175–179.

[90] Madaeni S S & Heidary F. Improving separation capability of regenerated cellulose ultrafiltration membrane by surface modification. *Applied Surface Science*, 2011, 257 (11): 4870–4876.
[91] Xiong X P, Duan J J, Zou W W, He X M & Zheng W. A ph-sensitive regenerated cellulose membrane. *Journal of Membrane Science*, 2010, 363 (1–2): 96–102.
[92] Algarra M, Vázquez M I, Alonso B, Casado C M, Casado J & Benavente J. Characterization of an engineered cellulose-based membrane by thiol dendrimer for heavy metals removal, *Chemical Engineering Journal*, 2014, 253: 472–477.
[93] Ma H, Hsiao B S & Chu B. Thin-film nanofibrous composite membranes containing cellulose or chitin barrier layers fabricated by ionic liquids. *Polymer*, 2011, 52 (12): 2594–2599.
[94] Wanderaa D, Wickramasingheb S R & Hussona S M. Modification and characterization of ultrafiltration membranes for treatment of produced water. *Journal of Membrane Science*, 2011, 373 (1–2): 178–188.
[95] Zhou Y T, Zhao H, Bai H L, Zhang L P & Tang H W. Papermaking effluent treatment: A new cellulose nanocrystalline/polysulfone composite membrane. *Procedings Environmental Sciences*, 2012, 16: 145–151.
[96] Mättäri M & Nyström M. Membranes in the pulp and paper industry. In A. K. Paddy, A. M. Sastre & S. S. H. Rizvi (eds), *Handbook of Membrane Separations: Chemical, Pharmaceutical, and Biotechnological Applications*, CRC Press, Boca Raton, Fl, 2009.
[97] Weis A, Bird M R, Nyström M & Wright C. The influence of morphology, hydrophobicity and charge upon the long-term performance of ultrafiltration membranes fouled with spent sulphite liquor [J]. *Desalination*, 2005, 175 (1): 73–85.
[98] Huuhilo T, Väsänen P, Nuortila-Jokinen J & Nyström M. Influence of shear on flux in membrane filtration of an integrated pulp and paper mill circulation water [J]. *Desalination*, 2001, 141 (3): 245–258.
[99] Maartens A, Jacobs E P & Swart P. UF of pulp and paper effluent: Membrane fouling – prevention and cleaning [J]. *Journal of Membrane Science*, 2002, 209 (1): 81–92.
[100] Huuhilo T, When P V, Nuortila-Jokinen J & Nyström M. Influence of shear on flux in membrane filtration of integrated pulp and paper mill circulation water. *Desalination*, 2001, 141 (3): 245–258.
[101] Puro L, Kallioinen M, Mänttäri M, Natarajan G, Cameron D C & Nyström M. Performance of RC and PES ultrafiltration membranes in the filtration of pulp mill process waters. *Desalination*, 2010, 264 (3): 249–255.
[102] Weis A, Bird M R, Nyström M & Wright C. The influence of morphology, hydrophobicity and charge upon the long-term performance of ultrafiltration membranes fouled with spent sulphite liquor. *Desalination*, 2005,175 (1): 73–85.
[103] Kallioinen M, Mänttäri M, Nyström M, Nuortila-Jokinen J, Nurminen P & Sutela T. Membrane evaluation for the treatment of acidic clear filtrate. *Desalination*, 2010, 250 (3): 1002–1004.
[104] Kallioinen M, Pekkarinen M, Mänttäri M, Nuortila-Jokinen J & Nyström M. Comparison of the performance of two different regenerated cellulose ultrafiltration membranes at high filtration pressure. *Journal of Membrane Science*, 2007, 294 (1–2): 93–102.
[105] Worthley C H, Constantopoulos K T, Ginic-Markovic M, Pillar R J, Matisons J G & Clarke S. Surface modification of commercial cellulose acetate membranes using surface-initiated polymerization of hydroxyethyl methacrylate to improve membrane surface biofouling resistance. *Journal of Membrane Science*, 2011, 385–386: 30–39.
[106] Liu C X, Zhang D R, He Y, Zhao X S & Bai R B. Modification of membrane surface for anti-biofouling performance: Effect of anti-adhesion and anti-bacteria approaches. *Journal of Membrane Science*, 2010, 346 (1): 121–130.

[107] Lin S, Chen L H, Huang L L, Cao S L, Luo X L & Liu K. Novel antimicrobial chitosan–cellulose composite films bioconjugated with silver nanoparticles, *Industrial Crops and Products*, 2015, 70: 395–403.

[108] Meng J Q, Zhang X, Ni L, Tang Z, Zhang Y F, Zhang Y J & Zhang W. Antibacterial cellulose membrane via one–step covalent immobilization of ammonium/amine groups, *Desalination*, 2015, 359: 156–166.

[109] Wu Y B, Yu S H & Mi F L. Preparation and characterization on mechanical and antibacterial properties of chitosan/cellulose blends. *Carbohydrate Polymers*, 2004, 57 (2): 435–440.

[110] Shih C M, Shieh Y T & Twu Y K. Preparation and characterization of cellulose/chitosan blend films, *Carbohydrate Polym*ers, 2009, 78: 169–174.

[111] Xiao W J, Chen Q & Wu Y. Dissolution and blending of chitosan using 1, 3-dimethylimidazolium chloride and 1-H-3-methylimidazolium chloride binary ionic liquid solvent. *Carbohydrate Polymers*, 2011, 83: 233–238.

[112] Pei Y. Construction, structure and properties of biomedical materials based on cellulose, [D]. Doctoral thesis, Wuhan University, 2013.

[113] Cheng Y M, Lu J T, Liu S L, Zhao P, Lu G Z & Chen J H. The preparation, characterization and evaluation of regenerated cellulose/collagen composite hydrogel films, *Carbohydrate Polymers*, 2014, 107: 57–64.

[114] Luo L H, Gan L, Liu Y M, Tian W Q, Tong Z, Wang X, Huselstein C & Chen Y. Construction of nerve guide conduits from cellulose/soy protein composite membranes combined with Schwann cells and pyrroloquinoline quinone for the repair of peripheral nerve defect, *Biochemical and Biophysical Research Communications*, 2015, 457: 507–513.

[115] ASTM Standard E96–00. Standard test methods for water vapor transmission of materials. Annual book of ASTM standards (Vol. 4.06) Philadelphia: ASTM, 2000.

[116] Ou S Y, Kwok K C & Kang Y J. Changes in vitro digestibility and availablelysine of soy protein isolate after formation of film. *Journal of Food Engineering*, 2004, 64: 301–305.

[117] Toméa L C, Goncalvesa C M B, Boaventurab M, Brandão L, Mendes A M, & Silvestre A J D. Preparation and evaluation of the barrier properties of cellophane films modified with fatty acids. *Carbohydrate Polymers*, 2011, 83: 836–842.

[118] Wang M, Yuan J, Huang X B, Cai X M, Li L & Shen J. Grafting of carboxybetaine brush onto cellulose membranes via surface-initiated ARGET–ATRP for improving blood compatibility, *Colloids and Surfaces B: Biointerfaces*, 2013,103: 52–58.

[119] Yuan J, Tong L X H, Yi B, Wang X, Shen J & Lin S C. Synthesis and one-pot tethering of hydroxyl-capped phosphorylcholine onto cellulose membrane for improving hemocompatibility and antibiofouling property, *Colloids and Surfaces B: Biointerfaces*, 2013, 111: 432–438.

[120] Zhou Y H, Fan M Z, Luo X L, Huang L L & Chen L H. Acidic ionic liquid catalyzed crosslinking of oxycellulose with chitosan for advanced biocomposites. *Carbohydrate Polymers*, 2014, 113:108–114.

[121] Zhou X D, Zhu P & Zhang L. Preparation and properties of soluble chitin/cellulose blend membranes, *Membrane Science and Technology*, 2009, 29 (2) :12–15.

[122] Zhang L, Yang L B, Sui S Y, Dong Z H & Lu P. Study on the preparation and performance of the antibacterial cellulose/chitosan derivative composite film, *Textile Auxiliaries*, 2010, 27 (6): 20–22.

[123] Zhuang X P & Liu X F. Blend films of O-carboxymethyl chitosan and cellulose in N-methylmorpholine, N-oxide monohydrate. *Journal of Applied Polymer Science*, 2006, 102 (5): 4601–4605.

[124] Liu X & Ma P X. Phase separation, pore structure, and properties of nanofibrous gelatin scaffolds. *Biomaterials*, 2009, 30 (25): 4094–4103.

[125] Pei Y, Yang J, Liu P, Xu M & Zhang X Z. Fabrication, properties and bioapplications of cellulose/collagen hydrolysate L. N. composite films, *Carbohydrate Polymers*, 2013, 92: 1752–1760.

[126] Shang S M, Zhu L & Fan J T. Physical properties of silk fibroin/cellulose blend films regenerated from the hydrophilic ionic liquid, *Carbohydrate Polymers*, 2011, 86: 462–468.

[127] Zhou L, Wang Q, Wen J C, Chen X & Shao Z Z. Preparation and characterization of transparent silk fibroin/cellulose blend films, *Polymer*, 2013, 54: 5035–5042.

[128] He M, Chang C Y, Peng N & Zhang L N. Structure and properties of hydroxyapatite/cellulose nanocomposite films, *Carbohydrate Polymers*, 2012, 87: 2512–2518.

[129] Koh Y P, Karim M N & Simon S L. Heterogeneous reaction kinetics of epoxide-functionalized regenerated cellulose membrane and aliphatic amine, *Thermochimica Acta*, 2012, 543: 18–23.

[130] Barroso T, Temtem M, Hussain A, Aguiar-Ricardo A & Roque A C. Preparation and characterization of a cellulose affinity membrane for human immunoglobulin G (IgG) purification, *Journal of Membrane Science*, 2010, 348: 224–230.

[131] Chen X, Zhang L L, Li H, Sun J H, Cai H Y & Cui D F. Development of a multilayer microfluidic device integrated with a PDMS–cellulose composite film for sample pretreatment and immunoassay, *Sensors and Actuators A*, 2013, 193: 54–58.

[132] Gao S S, Wang J Q & Jin Z W. Phase transformation conditions on the structure and properties of cellulose packaging films in LiCl/DMAc, *China Printing and Packaging Study*, 2010, 2: 433–433.

[133] Abdulkhani A, Marvast E H, Ashori A, Hamzeh Y & Karimi A N. Preparation of cellulose/polyvinyl alcohol biocomposite films using1-n-butyl-3-methylimidazolium chloride, *International Journal of Biological Macromolecules,* 2013, 62: 379–386.

[134] Gao S S, Wang J Q, Jin Z W & Zhang H. Influence of type and polymerization degree of pulp on structure and properties of cellulose packaging film, *Packaging Engineering*, 2011, 32(13): 8–10.

[135] Zhang X M, Feng J X, Liu X Q & Zhu J. Preparation and characterization of regenerated cellulose/poly(vinylidene fluoride) (PVDF) blend films, *Carbohydrate Polymers*, 2012, 89: 67–71.

[136] Mahmoudian S, Wahit M U, Ismail A F & Yussuf A A. Preparation of regenerated cellulose/montmorillonite nanocomposite films via ionic liquids, *Carbohydrate Polymers*, 2012, 88: 1251–1257.

[137] Zhang Y Y, Wang J Q. Research on the fresh–keeping effect of antibacterial cellulose film from NMMO process for pork packaging, *Packaging Engineering*, 2007, 28 (12): 20–22.

[138] Gao S S & Wang J Q. Preparation of cellulose packaging films with high oxygen permeability, *China Printing and Packaging Study*, 2012, 4 (6): 43–47.

[139] Zhang C F, Wang J Q, Zhang Y Y & Qi Y. Research on cellulose packaging films prepared by LiCl/DMAc process, *Packaging Engineering*, 2008, 29 (1): 21–23.

[140] Kim D H, Park S Y, Kim J & Park M. Preparation and properties of the single-walled carbon nanotube/cellulose nanocomposites using N-methylmorpholine-N-oxide monohydrate. *Journal of Applied Polymer Science*, 2010, 117: 3588–3594.

[141] Rahatekar S S, Rasheed A, Jain R, Zammarano M, Koziol K K & Windle A H. Solution spinning of cellulose carbon nanotube composites using room temperature ionic liquids. *Polymer*, 2009, 50: 4577–4583.

[142] Zhang H, Wang Z G, Zhang Z N, Wu J, Zhang J & He J S. Regenerated-cellulose/multiwalled-carbon-nanotube composite fibers with enhanced mechanical properties prepared with the ionic liquid 1-allyl-3-methylimidazolium chloride. *Advanced Materials*, 2007, 19, 698–704.

[143] Zhang H, Guo L, Shao H & Hu X. Nano-carbon black filled lyocell fiber as a precursor for carbon fiber. *Journal of Applied Polymer Science*, 2006, 99: 65–74.

[144] Han D, Yan L, Chen W, Li W & Bangal P R. Cellulose/graphite oxide composite films with improved mechanical properties over a wide range of temperature. *Carbohydrate Polymers*, 2011, 83: 966–972.

[145] Tsioptsias C & Panayiotou C. Preparation of cellulose-nanohydroxyapatite composite scaffolds from ionic liquid solutions. *Carbohydrate Polymers*, 2008, 74: 99–105.

[146] Zadegan S, Hosainalipour M, Rezaie H R, Ghassai H & Shokrgozar M A. Synthesis and biocompatibility evaluation of cellulose/hydroxyapatite nanocomposite scaffold in 1-n-allyl-3-methylimidazolium chloride. *Materials Science and Engineering*: C, 2011, 31: 954–961.

[147] Cerruti P, Ambrogi V, Postiglione A, Rychlý J, Matisová–Rychlá L & Carfagna C. Morphological and thermal properties of cellulose/montmorillonite nanocomposites. Biomacromolecules, 2008, 9: 3004–3013.

[148] Delhom C D, White–Ghoorahoo L A & Pang S S. Development and characterization of cellulose/clay nanocomposites. *Composites Part B: Engineering*, 2010, 41: 475–481.

[149] Lee J, Sun Q & Deng Y. Nanocomposites from regenerated cellulose and nanoclay. *Journal of Biobased Materials and Bioenergy*, 2008, 2: 162–168.

[150] Cerruti P, Ambrogi V, Postiglione A, Rychlý J, Matisová-Rychlá L & Carfagna C. Morphological and thermal properties of cellulose/montmorillonite nanocomposites. *Biomacromolecules*, 2008, 9: 3004–3013.

[151] Mahmoudian S, Wahit M U, Ismail A F & Yussuf A A. Preparation of regenerated cellulose/montmorillonite nanocomposite films via ionic liquids, *Carbohydrate Polymers*, 2012, 88: 1251–1257.

[152] Yadav M, Mun S, Hyun J & Kim J. Synthesis and characterization of iron oxide/cellulose nanocomposite film, *International Journal of Biological Macromolecules*, 2015, 74: 142–149.

[153] Gao S S & Wang J Q. Influence of modified TiO_2 on performance and structure of cellulose packaging films, *China Printing and Packaging Study*, 2012, 4 (2): 67–72.

[154] Wang J Q, Xu M, Jin Z W, Zhao M X & Zhang L. Study of morphological structure and property of SiO_2 nano-particle/cellulose packaging film, *Packaging Engineering*, 2009, 30 (9):1–4.

[155] Zhang L, Wang J Q & Xu M. Influence of inorganic filling material on cellulose packaging films prepared by NMMO method, *Packaging Engineering*, 2009, 30 (1): 62–64.

[156] Ma H, Zhou B, Li H, Li Y Q & Ou S Y. Green composite films composed of nanocrystalline cellulose and a cellulose matrix regenerated from functionalized ionic liquid solution, *Carbohydrate Polymers*, 2011, 84: 383–389.

[157] Jayaramudu J, Reddy G S M, Varaprasad K, Sadiku E R, Ray S S & Rajulu A V. Preparation and properties of biodegradable films from Sterculia urens short fiber/cellulose green composites, *Carbohydrate Polymers*, 2013, 93: 622–627.

[158] Reddy K O, Zhang J, Zhang J & Rajulu A V. Preparation and properties of self–reinforced cellulose composite films from Agave microfibrils using an ionic liquid, *Carbohydrate Polymers*, 2014, 114: 537–545.

[159] Ashok B, Reddy K O, Madhukar K, Cai J, Zhang L & Rajulu A V. Properties of cellulose/Thespesia lampas short fibers biocomposite films, *Carbohydrate Polymers*, 2015, 127: 110–115.

[160] Wu R L, Wang X L, Li F, Li H Z & Wang Y Z. Green composite films prepared from cellulose, starch and lignin in room-temperature ionic liquid, *Bioresource Technology*, 2009, 100: 2569–2574.

[161] Chen Z H, Gao S S, Li Z T, Yan P & Liu Y. Research on preservation packaging of fresh-cut Broccoli with modified nano ZnO/cellulose film, *China Printing and Packaging Study*, 2014, 6 (4): 112–116.
[162] Gao S S, Wang J Q, Liu B & Zhang L. Research on the fresh-keeping effect of cellulose film from LiCl/DMAc process to fresh–cut potato, *Packaging Engineering*, 2008, 29 (11): 16–18.
[163] Qi H, Cai J, Zhang L & Kuga S, Properties of films composed of cellulose nanowhiskers and a cellulose matrix regenerated from alkali/urea solution. *Biomacromolecules*, 2009, 10 (6): 1597–1602.
[164] Yun S & Kim J. Multi-walled carbon nanotubes-cellulose paper for a chemical vapor sensor. *Sens Actuators B: Chem*, 2010, 150: 308–313.
[165] Qi H S, Mäder E & Liu J W. Unique water sensors based on carbon nanotube–cellulose composites, *Sensors and Actuators B*, 2013, 185: 225–230
[166] Mahadeva S K, Yun S & Kim J. Flexible humidity and temperature sensor based on cellulose–polypyrrole nanocomposite. *Sens Actuators A: Phys*, 2011, 165: 194–199.
[167] Yun S, Chen Y, Nayak J N & Kim J. Effect of solvent mixture on properties and performance of electro-active paper made with regenerated cellulose, *Sensors and Actuators B*, 2008, 129: 652–658.
[168] Sen I, Seki Y, Sarikanat M, Cetin L, Gurses B O, Ozdemir O, Yilmaz O C, Sever K, Akar E & Mermer O. Electroactive behavior of graphene nanoplatelets loaded cellulose composite actuators, *Composites: Part B*, 2015, 69: 369–377.
[169] Khondoker M A H, Yang S Y, Mun S C & Kim J. Flexible and conductive ITO electrode made on cellulose film by spin-coating, *Synthetic Metals*, 2012, 162: 1972–1976.
[170] Rittersma Z M. Recent achievements in miniaturised humidity sensors–a review of transduction techniques. *Sens Actuators A: Phys*, 2002, 96: 196–210.
[171] Yeo T L, Sun T & Grattan K T V. Fibre-optic sensor technologies for humidity and moisture measurement. *Sens Actuators A: Phys*, 2008, 144: 280–295.
[172] Ummartyotin S & Manuspiya H. A critical review on cellulose: From fundamental to an approach on sensor technology, *Renewable and Sustainable Energy Reviews*, 2015, 41: 402–412.
[173] Kim J, Deshpande S D, Yun S & Li Q. A comparative study of conductive polypyrrole and polyaniline coatings on electro-active papers, *Polym. J.*, 2006, 38: 659–668.
[174] Kim H S, Li Y & Kim J. Electro-mechanical behavior and direct piezo electricity of cellulose electro-active paper. *Sens Actuators A: Phys*, 2008, 147: 304–309.
[175] Kim J, Yun S & Ounaies Z. Discovery of cellulose as a smart material, *Macromolecules*, 2006, 39: 4202–4206.
[176] Kim J, Song C S & Yun S. Cellulose based electro-active papers: Performance and environmental effects, *Smart Mater. Struct.*, 2006, 15: 719–723.
[177] Kim J & Seo Y B, Electro-active paper actuators, *Smart Mater. Struct.*, 2002, 11: 355–360.
[178] Sosibo N, Mdluli P, Mashazi P, Tshikhudo R, Skepu A, Vilakazi S & Nyokong T. Facile deposition of gold nanoparticle thin films on semi-permeable cellulose substrate, *Materials Letters*, 2012, 88: 132–135.
[179] Tang L, Li X, Du D & He C J. Fabrication of multilayer films from regenerated cellulose and grapheme oxide through layer-by-layer assembly, *Progress in Natural Science: Materials International*, 2012, 22 (4): 341–346.
[180] Bonné M J, Helton M, Edler K & Marken F. Electro-deposition of thin cellulose films at boron-doped diamond substrates, Electrochemistry Communications, 2007, 9 (1): 42–48.

Chapter 8
Cellulose-based textile materials

Professor Liulian Huang and Professor Fang Huang

This chapter introduces cellulose-based textiles, natural fibre based textiles, such as cotton, ramie and other natural fibre textiles, and regenerated cellulosic fibre textiles, such as viscose and Lyocell textiles. The chapter first overviews the sources of cellulose textiles for both natural and regenerated cellulosic textile fibres and then discusses the properties of natural cotton and ramie textile fibres. Finally, the chapter examines the production of regenerated textiles fibres, followed by the production, properties and application of regenerated cellulosic textiles, including viscose-based textiles and Lyocell fabrics.

8.1 Sources of cellulose-based textile fibres

Textile fibre is the basic raw material for textile production and has undergone many structural changes. In the early 18th century, the main sources for textiles were silk and natural plant fibres, which mainly came from bast fibres. Last century witnessed the gradual ascendancy of cotton in the development of textile history. Cotton is suitable for centralised planting, commensurate with high productivity and good serviceability. Advances in processing technologies have also facilitated cotton becoming the dominant textile fibre. Nevertheless, recent decades have also seen the development of regenerated cellulosic textiles, which are made from cellulose-rich regenerated fibres.

8.1.1 Natural cellulosic fibres

Natural fibre is any hair-like raw material directly obtained from animal, vegetable or mineral sources, and convertible to non-woven fabrics or after spinning into yarns and woven cloths. Natural fibre accounts for 50% of the fibres in the textile industry. Natural fibres can be classified into different species according to their origin: vegetable or cellulose-based fibres include cotton, flax and jute; animal or protein-based fibres include wool, mohair and silk; and mineral fibres include asbestos and basalt fibres.

The two most common natural textile fibres are cotton and ramie fibres.

8.1.1.1 Cotton textile fibre

The main component of cotton is cellulose, which accounts for 94% of the total mass. Cotton fibre is a soft and fluffy staple. Cotton fibres can be classified as medium-staple cotton, long-staple cotton and short-staple cotton fibres.

Medium-staple cotton, renamed upland cotton, mainly from *Gossypium hirsutum*, native to Central America, Mexico, the Caribbean and southern Florida (90% of world production). The density and the length of fibre are medium, normally around 2.12–1.5 dtex and 25–35 mm respectively. The strength is around 4.5 cN.

Long-staple cotton, renamed Sea Island cotton, mainly from *Gossypium barbadense*, native to tropical South America (8% of world production). The fibre is thin and long. Normally the length is more than 33 mm and the density is around 1.54–1.18 dtex. The strength is up to 4.5 cN.

Short-staple cotton, e.g. those from *Gossypium arboretum*, native to India and Pakistan and *Gossypium herbaceum*, native to southern Africa and the Arabian Peninsula. This kind of cotton fibre is no longer used because the fibre is too thick and short.

8.1.1.2 Ramie textile fibre

Ramie is a flowering plant in the nettle family, Urticaceae, native to eastern Asia. Ramie is one of the oldest fibre crops, having been used for at least six thousand years, and is principally used for fabric production. Ramie fibre is a bast fibre and the part used is the bark (phloem) of the ramie stalks. The extraction of the ramie fibre occurs in three stages. First, the cortex or bark is removed and this can be done by hand or machine. It is called decortication. Second, the cortex is scraped to remove most of the outer bark, the parenchyma in the bast layer and some of the gums and pectins. Finally, the cortex material is washed, dried and degummed to extract the spinnable fibres.

Ramie fibre is one of the strongest natural fibres and is known especially for its ability to hold shape, reduce wrinkling and introduce a silky lustre to the fabric's appearance. Ramie fibre is not as durable as other plant fibres and is therefore usually used as a blend with other fibres, such as cotton or wool fibres. It is similar to linen in absorbency, density and microscopic appearance, although it cannot be dyed as well as cotton fibre. Because of its high molecular crystallinity, ramie fibre is stiff and brittle, and will break if folded repeatedly; it lacks resiliency and is low in elasticity and potential elongation.

Ramie fibre is used to make such products as industrial sewing thread, packing materials, fishing nets and filter cloths. It is also made into fabrics for household furnishings (upholstery, canvas) and clothing, frequently in blends with other textile fibres – for instance, when mixed with wool, the shrinkage is greatly reduced compared to pure wool products. Shorter fibres and waste are used in paper manufacture. Ramie ribbon is used in fine bookbinding as a substitute for traditional linen tape. The main chemical components of ramie are cellulose, gelatin, hypoxylogen and wax.

8.1.1.3 Other plant textile fibres

There are a number of other bast and leaf textile fibres. Bast fibre is the fascicular bundle fibre of the bast zone from the stem of Gemini Plants, such as flax, jute, cannabis and abutilon fibres. Leaf fibre is the vascular bundle fibre of the sheath and blade area from the monocotyledons, such as sisal, Manila hemp and pineapple fibres.

Flax fibre has a higher strength, less stretch, and is soft and easy to weave. This kind of textile product keeps its shape easily and absorbs sweat. Jute, cannabis and kenaf fibres are short and rough, so their products are appropriate as packaging materials. Sisal and Manila hemp fibres are hard and rough; therefore, the fibres are also called the hard cellulose. Because the cell wall has already lignified, the fibre is long, has less stretch, high strength, is non-corrosive and has seawater immersion resistance. This fibre is widely used for rope on boats, in mining and in package materials.

8.1.2 Regenerated cellulose textile fibres

Regenerated cellulose fibres are converted from the fine short fibres that normally come from trees, and usually fine long fibres used in textiles and non-wovens. A regenerated fibre is one formed when a natural polymer or its chemical derivative is dissolved and extruded. The chemical nature of the natural polymer is either retained or regenerated after the fibre formation. Regenerated cellulose fibres remain unique among the mass-produced fibres because they are the only ones to use natural polymer directly. The total regenerated cellulose fibre was 5,080,000 tonnes in 2015 compared to only 3,641,000 tons in 2011 [1].

Originally, the word 'rayon' was applied to any cellulose-based manufactured fibre, including cellulose acetate fibres. However, the definition of rayon was clarified in 1951 and now rayon includes textiles fibres and filaments composed of regenerated cellulose, excluding acetate. In Europe the fibres are now generally known as viscose, with the term 'viscose rayon' being used whenever confusion between the fibre and the cellulose xanthate solution may occur.

8.1.2.1 Lyocell and Modal fibres

Regenerated fibres produced via the direct dissolution of cellulose in organic solvents are generically known as Lyocell fibres (commercial name Tencel®). This technique can cope with the difficulties of cellulose solution through an easy-to-dissolve cellulose derivative, e.g. xanthate, or a cellulose complex, such as cuprammonium. It must be noted that an ideal process, one that could dissolve the cellulose directly from ground wood, is still some way off, although significant progress has been made since the early 1980s.

Modal fibre comes from wood and can degrade naturally after use. The price of modal fibres is half that of Tencel fibres. It is the second generation of regenerated cellulose. Modal can be mixed with the other celluloses to achieve better quality. Compared to the cotton, Modal has better moisture absorption and air permeability.

8.1.2.2 Regenerated bamboo fibre

Bamboo fibres can be divided into regenerated and natural bamboo fibres. Regenerated bamboo fibre is the product of *Phyllostachys pubescens*, *Sinocalamus oldhami*, *Dendrocalamous affinis* after steam processing by hydrochloric acid hydrolysis, bleaching and wet spinning. Natural bamboo fibre is the product of *Phyllostachys pubescens* processed by natural biological agent. Bamboo textile is considered an environmentally friendly material. Bamboo fibre is similar in characteristics to viscose fibre, having good moisture absorbance, air permeability, drapability and dyeability; other features include antibiosis, deodorisation and being ultraviolet-proof.

8.1.2.3 Viscose fibre

Viscose fibre is made from a mixture of chitin, chitosan and callouses through a wet spinning process. Chitin can be found in the shells of shrimp, turtle and insert, and the cell walls of mushrooms and algae. The features of the viscose fibre are its bioactivity, biodegradability and biocompatibility. Viscose fibre has good moisture absorbance and retention. The textile can also absorb odours, which may be the result of the mixture of chitin and cotton.

8.2 Properties of natural cellulosic textile fibres

8.2.1 Properties of cotton textile fibres

Cotton textile fibre has good absorption ability, being able to absorb 20–30% moisture from the surrounding air, and hence can be heavily dyed through substantive staining, reduction staining and sulphurisation staining. Cotton fibre is lacking in mould resistance and is easily affected by the microbiology, causing mildew.

Washing shrinkage is 4–10%. Cotton fibres have good dilatability, are light and heat resistant, but do not have good heat and electrical conduction. The strength may reduce when the cotton is oxidised through exposure to air, and its colour may change to yellow when the temperature reaches 120°C and dark brown when the temperature reaches 250°C.

Cotton textile fibres are sensitive to acid and can be destroyed in organic or inorganic acids. However, they are stable in alkali, although the strength will be decreased in strong alkali. Organic acid will not damage the textile but inorganic acid can cause eyelets. It is worth noting that after chemical reaction with 20% alkali solution, cotton textile can be changed to mercerised cotton textile. Mercerised cotton is smooth with a good lustre.

8.2.1.1 Length and linear density

The length of medium cotton is 23–33 mm and for long-staple cotton it is 33–45 mm, depending on the process and quality of the cottons. The linear density is the quality of unit length, equal to the mass in grams of 1000 metres of fibre. Tex is the name of unit used to measure the linear density and can be expressed as follows:

$$T_d = 1000\ W/(L \times N) \tag{8.1}$$

where T is the average fibre linear density in dtex, W is the mass of bundle fibres in mg, L is the length of bundle fibres in mm and N is the number of fibres in the bundle.

The type of cotton and planting environment can affect the linear density. Long-staple cotton is normally thinner with a linear density of 1.11–1.43 dtex. Medium-staple cotton is thicker with a linear density of around 1.43–2.22 dtex. A ripen cotton normally has lower linear density and has better strength, which can be equal to the yarn of cotton cellulose. This type of cotton cellulose is used to produce low linear density yarn. It should be noted that a low linear density may be caused by things such as unripe and/or dead fibres, and this type of cotton cellulose will be twisted and/or broken, producing nap and short fibres.

8.2.1.2 Strength and elongation

Strength and elongation represent the strength of stretching or specific strength, the breaking length and the ratio of the breaking strength. The strength of stretching is the maximum external force when the fibre stretches to breaking point. The fracture strength ratio (*Pt* in N/tex) is the maximum external force (*P* in Newtons) that the unit linear density can bear and is used to compare the stretch of fibres of different fineness. The unit is Newton per tex. It can be calculated using the following equation:

$$Pt = P/\text{Ntex} \tag{8.2}$$

Strength is one of the prerequisites for cellulose as a textile material as the cellulose will be subject to an external force during processing. In addition, if the fibre strength is higher, the yarn strength will be higher. The breaking length of medium-staple cotton is 20–30 km, but for long-staple cotton it is higher. The ratio of the breaking and elongation of cotton cellulose is 3–7% and can be expressed as follows:

$$\varepsilon = (L_a - L_o)/L_o \times 100\ (\%) \tag{8.3}$$

where ε is the ratio of the breaking and elongation (%), L_a is the breaking length; L_o is the original length.

8.2.1.3 Chemical stability

The main ingredient of cotton textile fibres is cellulose, which is alkali resistant but not acid resistant. Acid can cause the lytic response of cellulose to break the large molecules. Although low concentration alkali solution does not damage cotton cellulose at normal temperature, it can make the cellulose swell. Thus, the treatment of cellulose textile fibres with a certain concentration of sodium hydroxide or liquid nitrogen can cause the cellulose swell in breadth and change the cross-section into a circle, and hence twisting disappears. The above changes lead cellulose to be looked like silk. Stretching the cellulose during the swelling may also change the internal structure and result in an increase in cellulose strength. The process is called mercerising.

8.2.2 Properties of ramie textile fibres

Linen cellulose fibres have good moisture absorbance and heat conducting properties. The textile fibres feel cool and smooth, and do not stick to the body with sweating. They are light, insect- and mould-resistant, and have a warm color. The textile fibres have good strength, are less prone to static electricity and cause no pollution. They adapt to body and skin excretion and secretion.

8.2.2.1 Length and linear density

The unit length of linen cellulose textile fibres is normally 50–120 mm for ramie, 17–25 mm for flax, 2–4 mm for jute and 2–6 mm for meson. The linear density is normally 0.91–0.4 tex for ramie and 0.29 tex for flax. The single cellulose width of jute is around 10–28 μm. Bundle fibre is the product after the prime processing. After carding, the bundle fibre changes to technical fibre, ready for spinning. The type of linen, degree of degumming degree and the carding number can affect the linear density of technical fibres.

8.2.2.2 Strength and elongation

Linen cellulose textile fibre normally has the greatest strength and least elongation among natural textile fibres. The broken strength of ramie is up to 40–50 km. The broken length is 2–4% for ramie, 3% for flax and 3% for jute. The elasticity of linen cellulose is poor. Cloth made from linen wrinkles easily.

8.2.2.3 Absorbing moisture

Linen cellulose textile fibres have strong absorbing capacity, especially in the case of jute. Moisture absorption can reach up to 14% when exposed to normal environmental conditions. Linen cellulosic textile fibres are also able to absorb and distribute moisture rapidly, so that textiles made from linen are comfortable in summer.

8.3 Preparation and properties of regenerated textile fibres

8.3.1 The production of rayon fibres

8.3.1.1 The preparation of rayon

Excepting the pulp material and the equipment, there is a generic procedure for textile fibre production, including (1) production of the rayon; (2) preparation before the spinning; (3) fibre formation; and (4) post-treatment of the fibre.

The pulp requires two chemical modifications for formulating rayon. The first one involves pulp reacting with alkali and changing to alkali cellulose; the second one is alkali cellulose reacting with carbon disulphide and changing to cellulose ester. Through these two reactions, strong polar sulphonic acid groups are added to the cellulose molecule, which undissolved directly in the alkali solution, and new products can dissolve in the solvent to make rayon. However, this viscose is rough and needs to be refined for spinning (Fig. 8.1).

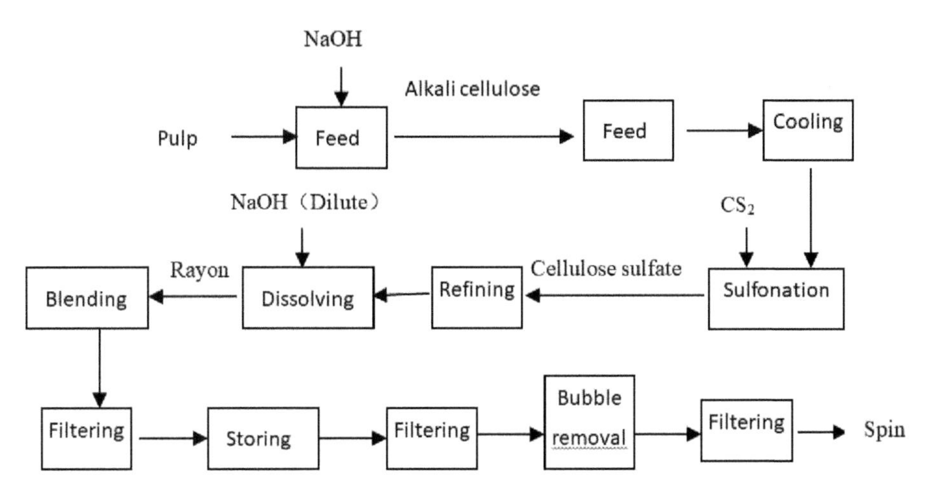

Figure 8.1 Diagram of rayon production.

Before processing, the different batch of the pulp needs to be mixed to ensure the consistent property of the viscose fibres. A certain concentration alkali solution is then added to produce the alkali cellulose, via the following reaction:

$$C_6H_9O_4 - OH + NaOH \rightarrow C_6H_9O_4 - Na + H_2O$$

The alkali cellulose is exposed to the air after breaking, and the molecular chain breaks. The average polymerisation is adjusted to achieve the necessary viscose degree according to the oxidisation. A high viscose degree may cause problems during the processing. The level of cellulose ageing is different depending on the different types of cellulose; some celluloses do not have the ageing process.

The alkali cellulose may react with carbon dioxide and the resultant product is cellulose sulphonate:

$$C_6H_9O_4 - Na + CS_2 \rightarrow C_6H_9O_4 - O - \overset{\overset{\Large S}{/\!/}}{C} - SNa$$

The filtering process is required to remove the solid or half-solute matter before spinning. The process can avoid blocking of the spinneret hole. Such blocking not only causes difficulties in spinning, but also affects the quality of products. Viscose cellulose needs to be filtered 3–4 times.

8.3.1.2 Spinning of rayon textile fibres

Cellulose fibre is formed from the rayon in the acidic coagulating bath. The cellulose fibres are stretched, cut, refined and dried to get the final product (Fig. 8.2).

During the spinning, the cellulose sulphonate is degraded into cellulose fibre and water. The spin process can be classified into three types, i.e. single spinning, twin spinning and poly-spinning.

Chemical reactions in spinning. The major components of rayon are cellulose sulphate, NaOH and water. It also contains minor components, such as CS_2 and Na_2CS_3. The major chemical process in spinning is the decomposition of cellulose sulphonate and acid-base neutralisation:

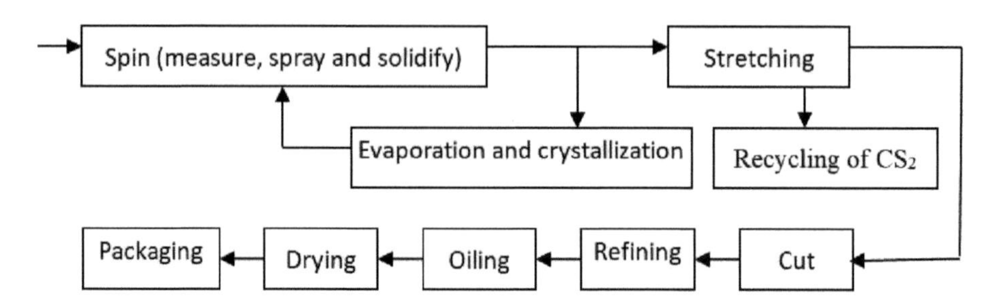

Figure 8.2 Spinning and post-treatment.

$$
\begin{array}{l}
/O\ (C_6H_9O_4)_n\ (OH)_{n-1} \\
C{=}S \qquad\qquad + H_2SO_4 \rightarrow NaHSO_4 + CS_2 \uparrow + (C_6H_{10}O_5)_n \\
\backslash SNa
\end{array}
$$

$$NaHSO_4 + NaOH \rightarrow Na_2SO_4 + H_2O$$

$$NaOH + H_2SO_4 \rightarrow NaSO_3 + 2H_2O$$

The side reactions are as follows:

$$Na_2CS_3 + H_2SO_4 \rightarrow Na_2SO_4 + CS_2 \uparrow H_2S \uparrow$$

$$Na_2S_x + H_2SO_4 \rightarrow Na_2SO_4 + H_2S \uparrow + (X - 1)\ S \downarrow$$

The decomposition of the cellulose sulphonate is the process of cellulose regeneration, which is influenced by H^- ions. A higher number of H^- ions will increase the decomposition.

Physical reactions in spinning. During the decomposition of the cellulose sulphonate, free hydroxyls are released to form new hydrogen bonds and eventually to form gels. There are sulphuric acids, sodium sulphate and zinc sulphate in the coagulation. The function of the sulphuric acid is to stimulate the decomposition of cellulose sulphonate and to neutralise the base. The function of the sodium sulphate is to stimulate the coagulation of rayon and reduce the decomposition. Zinc sulphate could improve the cellulose fibre formation and physical strength.

8.3.1.3 Post-treatment of rayon textile fibres

Post-treatment is to remove the impurities in the rayon. There are a number of procedures to be carried out after the spinning of the rayon fibres.

Cutting. Cutting is to cut long fibres into three categories, i.e. short fibres (38 mm), medium fibres (51–76 mm) and long fibres (76–102 mm). The cutter should be sharp and should be adjustable to different length requirements.

Refining. Refining includes washing, sulphur removal, bleaching and acid washing. The quality of the final product can be improved significantly. Water wash is carried out first to remove the salts, acids and a number of other impurities. In particular, metal ions should be removed. Temperature is important for an effective wash. Sulphur removal then follows. Chemicals, such as NaOH, Na_2S and Na_2SO_3, are normally used to remove the sulphurs. The mechanism of the process is as follows:

$$6NaOH + 4S \rightarrow 2Na_2S + NaSO_3 + 5H_2O$$

$$Na_2S + {}_xS \rightarrow Na_2S_{x+1}$$

The alkali concentration should be controlled under 7 g/L and below 45 °C. Na_2S may corrode equipment, while Na_2SO_3 shows less corrosion effect.

Bleaching is the third step of the post-treatment to improve the brightness of rayon. Chemicals, such as NaClO, H_2O_2, and $NaClO_2$, are applied to the rayon. The bleaching pH should be controlled between 8 and 10, concentration 0.5–1.0 g/L, and temperature 25–30 °C. H_2O_2 can be used in weak alkali medium, but its high price and instability have implications for its wide use. $NaClO_2$ can be used in acidic medium, but this may cause corrosion and involve high energy consumption.

Acid wash is used to remove some metal ions and impurities. HCl or H_2SO_4 can be used. HCl is the most commonly used in practice, with a concentration of 1–2 g/L and a temperature of 25–30 °C.

Oiling. Oiling is to improve the fibre friction and softness.

Drying. The drying temperature should be controlled at 110–120°C. The moisture content of the final product fibres is 8–11%.

8.3.2 Production and properties of Lyocell fibres

8.3.2.1 Production of Lyocell fibres

Lyocell fibre is produced from regenerated cellulose dissolved in N-methyl morpholine-N-oxide (NMMO) through wet spinning (Fig. 8.3). Lyocell fibre has the combined properties of natural and synthetic fibres, having good physical properties and being cost effective. It has high dry and wet strength. It is resistant to heat. Below 190°C, internal damage is only 7% in 30 s.

Lyocell production is similar to rayon production. It is dissolved in the NMMO to get the regenerated cellulose. The cellulose can be chelated with NNMO and then dissolved in NNMO [2]. After that, the Lyocell is obtained through the spin process. The dissolving process follows the following mechanism:

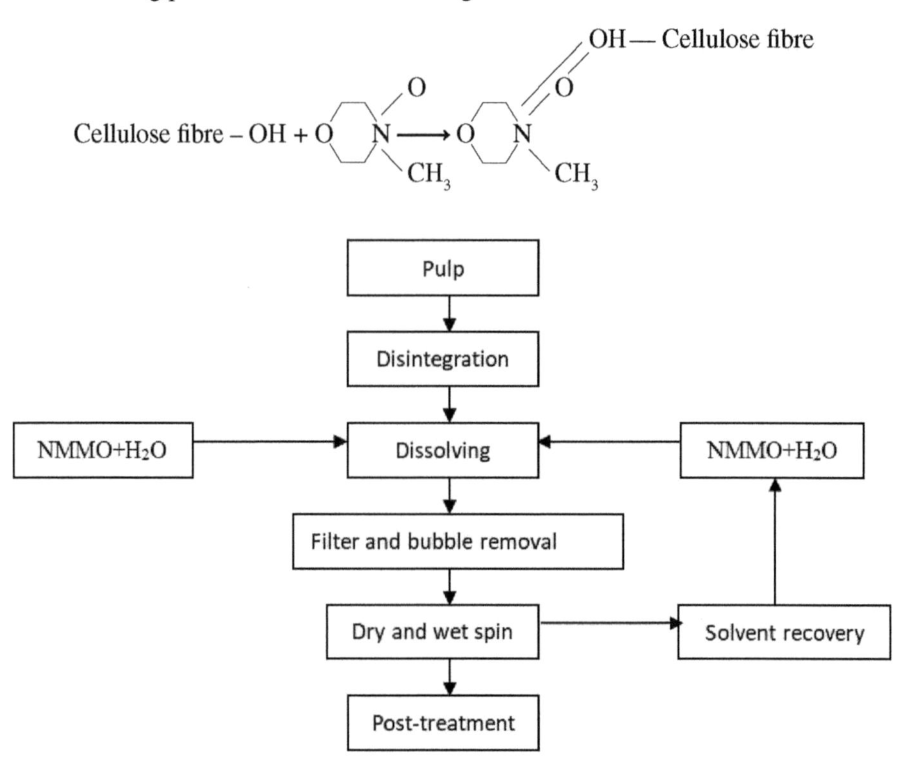

Figure 8.3 Diagram of Lyocell fibre production.

It should be noted that it is necessary to determine the chemical composition of pulps, such as α-cellulose, cellulose, lignin, ash, dirt, polysaccharides and pulp brightness. The solution temperature, dissolving time and solvent quantity are also prime parameters for the production. The NMMO solvent can be recycled through filtration, ion exchange and reduced pressure evaporation. The NMMO solvent recovery can be up to 97% through this process.

8.3.2.2 Properties of Lyocell fibres

Lyocell has similar properties to cotton and rayon. The molecular weight (DP) and crystallinity are between cotton and rayon (Table 8.1). The cross-section of Lyocell is round. The surface is smooth. The fibre is curl, but highly orientated. Lyocell fibre has crystalline and amorphous regions. The CI is 49% for Lyocell, 70% for cotton and 30% for rayon.

The wet/dry strength of Tencel® fibre is much higher than that of rayon fibre (Table 8.2). Lyocell fibre is able to absorb moisture of 11%, which is similar to rayon, but higher than cotton. The water retention of Lyocell fibre is higher than cotton but lower than rayon.

When Lyocell fibre gets wet, the fibre diameter increases by 27%. This makes the network of fibre denser (Fig 8.4) [3]. When the textile re-dries, the re-spaces between the fibres lead to good drapability.

Table 8.1 The molecular weight and crystallinity of Lyocell fibre.

	Cotton	Lyocell fibre	Rayon
DP	10 000	500–550	250–300
Crystallinity(%)	70	50	30

Table 8.2 Properties of Tencel® in comparison with other fibres.

Properties	Tencel® fibre	Common rayon	Modified rayon	High modulus rayon	Cotton	Polyester
Denier (den)	1.5	1.5	1.5	1.5	–	1.5
Dry strength (g/den)	4.8~5.0	2.4~2.5	2.6~3.1	4.1~4.3	2.4~2.9	4.8~6.0
Dry stretch (%)	14~16	20~25	20~25	13~15	7~9	25~35
Wet strength (g/den)	4.2~4.6	1.1~1.7	1.2~1.8	2.3~2.5	3.1~3.6	4.8~6.0
Wet stretch (%)	16~18	25~30	25~35	13~15	12~14	25~35
Water retention (%)	65	90	90	75	50	3

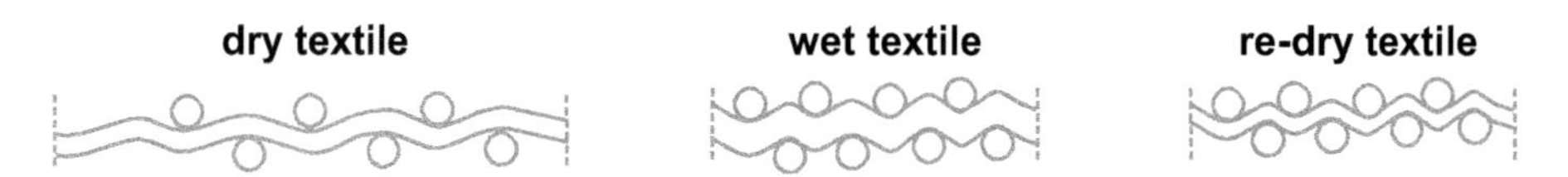

Figure 8.4 Change in network structure of Lyocell textile upon dry-wetting process.

Lyocell fibres can be used for the production of clothes with good handling properties and physical strength. Lyocell can also be used in many industrial sectors, such as non-woven cloth, filters, specialty papers and some medical-use fibres.

8.4 Production and properties of regenerated cellulose textile

8.4.1 Viscose fibre based textiles

8.4.1.1 Production of viscose fibre based textiles

Viscose is a cellulose textile fibre largely used for conventional and technical applications, obtained by cellulose regeneration from carbon disulphide solution [4]. Viscose can also be made into the common form of rayon, which is used for many types of textile production, including clothing. Viscose rayon has a silky appearance and feel, and also has an ability to breathe in a manner similar to cotton weaves. One of the more popular properties of viscose rayon is that the fabric tends to drape very well, which makes it ideal for use in simple curtains, as well as the perfect fabric for lining draperies. Viscose is also being used for furnishing fabrics, being a staple for towels and tablecloths, and it is being made into high-tenacity yarn for tyres.

Fig. 8.5 provides an overview of the complex processes that are used to produce technical textiles. Plaiting and knotting are used for manufacturing ropes (e.g. rope ladders, scramble nets, gym ropes and decorative rope) and nets (e.g. mosquito nets). Rope is being used in barrier/garden-decking, tree-felling, scaffolding, marquees, gymnasiums (battling ropes), timber, fencing, marine, construction, transportation, brewery and equestrian applications. In manufacturing it is being used, for example, in paper mills.

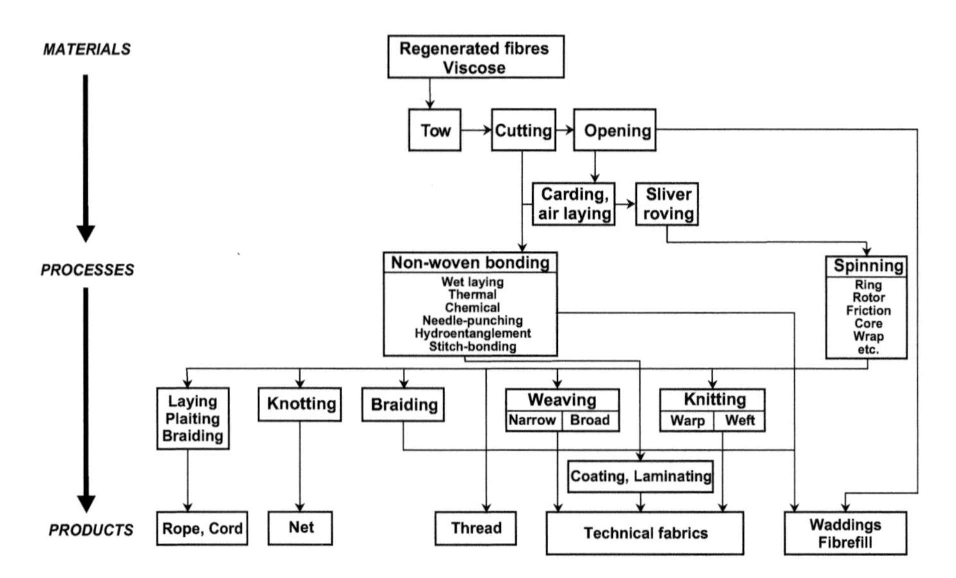

Figure 8.5 Flow diagram of textile processing.

8.4.1.2 Properties of viscose fibre based textiles

Viscose fibre has myriad merits, which makes it a popular fibre to work with. Its versatile allows blends very well with other fibres. Viscose fibre based textiles inherit many properties from viscose fibres, such as breathability, good drapery, colour retention, and excellent absorbent properties. The textiles are very smooth, relatively light, strong and robust, soft and comfortable, and inexpensive. The textile products do not trap body heat or build up static.

Thanks to these characteristics, a wide range of products has been developed in many industrial sectors. Nevertheless, some care should be taken during wearing and washing, as viscose textiles can shrink when washed, wrinkle easily and deteriorate with exposure to light. They are also susceptible to mildew.

8.4.2 Preparation and properties of Lyocell fibre based fabrics

8.4.2.1 Production of Lyocell textiles

Lyocell fabric is an eco-friendly fabric that represents a milestone in the development of environmentally sustainable textiles. Compared to cotton, Lyocell wrinkles less, is softer, more absorbent, and much more resistant to ripping. In addition, Lyocell is a high-tenacity cellulosic fibre, which gives high strength to the fabric. In terms of physical properties, Lyocell is more like cotton than rayon. Like other cellulosic fibres, it is breathable, absorbent, and very comfortable to wear. Lyocell is, in fact, more absorbent than cotton or silk, but slightly less absorbent than wool, linen or rayon.

Yarn Manufacturing. Once Lyocell fibre has been produced, either cut staple fibre or continuous tow, it is converted to yarns and fabrics by a range of conventional textile processes. The most common way of using Lyocell fibre is as cut staple, with 1.4 and 1.7 dtex fibres being cut to 38 mm and converted into a spun yarn using machinery as used for cotton fibres.

Lyocell is able to open easily with little nep formation. In sliver and roving, the fibres pack together, giving high cohesion and therefore requiring high drafting forces. Lyocell yields very regular yarns with high tensile strength and few imperfections. Lyocell blends well with other fibres, including cotton, viscose, linen, wool, silk, nylon and polyester. Lyocell adds strength to the yarn as well as enhancing the performance and aesthetic properties of the final fabric. Minimal carding power is required, as the fibre is very open. In drawing, sliver detectors may need to be re-set to adjust for the low bulk of Lyocell. In roving, the twist should be low to avoid too high cohesion. Optimisation is very important at this stage of the process.

Yarn steaming should be avoided wherever possible. Amongst other things, steaming cellulosic fibres affects fibre dye affinity, twist liveliness and splice strength. Dye affinity for cellulosic fibres reduces with increasing steam temperature, the influence of which on Lyocell fibre is greater than that for other cellulosic fibres, such as cotton and viscose. Steaming should be avoided, therefore, unless this can be well controlled. Twist liveliness can be reduced in other ways, such as by storing yarn on a ring tube for 16–24 hours in a high- humidity environment prior to winding.

Fabric manufacturing. Weaving of Lyocell fabrics can be successfully carried out on most conventional looms and in a wide range of structures. The structure needs to be carefully engineered with the dyeing/finishing route to develop the best performance and aesthetics. Very tight structure can cause problems in dyeing and tends to result in fabrics with poorer caring performance.

Dyeing and finishing of Lyocell. The dyeing and finishing of Lyocell fabrics are the key procedures. There are three characteristics of the fibres that can be manipulated to produce fabrics with attractive and differentiated aesthetics: namely, the ease of fibrillation, high nodules, and wet swelling characteristics. Fibrillation can yield the characteristic 'peach skin' effect to the surface touch of fabrics made from this fibre, but unwanted and uncontrolled fibrillation can also impair the quality; much dyeing and finishing development has focused on this aspect.

Lyocell is a cellulosic fibre and can be dyed with colours normally used for cotton. Compared with unmercerised cotton, Lyocell – except with a few reactive, vat and a number of direct dyes (pale shades) – can be dyed to a greater depth by exhaust techniques and, as such, many shades can be attained at lower cost, particularly with reactive dyes. The dyeing mechanisms for most reactive dyes are similar, i.e. the reactive dye is first exhausted on to the cellulose fibre using salt and then alkali is added to fix the dye.

Many modern reactive dyestuffs contain two or three reactive groups. A key discovery, made early in the development of Lyocell, was that these multifunctional dyestuffs can cross-link the fibre and prevent or inhibit the fibrillation of the fibre. The manipulation of this fibrillation is critical for the development of fabric aesthetics.

8.4.2.2 Properties of Lyocell textiles

The advantage of Lyocell textiles starts right from its manufacturing stage. Compared to the commonly used viscose fibres, Lyocell fibre involves fewer steps and chemicals in its manufacture. Solvent and water can be recycled in the Lyocell process. The Lyocell process also starts with pulp; the remaining manufacturing process as discussed is completely new. The environmentally friendly production, the use of renewable raw materials and its biodegradability make Lyocell superior to many other regenerated fibres. The production method is potentially more cost effective and faster than that used to make viscose rayon. The Lyocell process takes three hours to produce fibre compared with the 40 hours needed to make viscose rayon staple.

Lyocell fibres have the unique characteristics of soft and silky feeling, lustre and bulky touch. Fabrics made out of this fibre show very good drape and fluidity, something unexpected for a fabric of this weight. A rich look stands out as the hallmark of its aesthetics. Interestingly, Lyocell has high dry tenacity and modulus. It is the strongest cellulosic fibre when dry, even stronger than cotton or linen. It also retains much of its strength when wet. Its wet tenacity is higher than that of cotton and other cellulosic fibres. Compared to viscose, it is two times stronger when dry and three times when wet. The reason for this is the average degree of polymerisation and the number of crystalline zones being greater in Lyocell compared to that in conventional viscose rayon, high wet modulus (HWM) modal, or polynosic fibres.

Lyocell is not only an environmentally friendly fibre, but it also offers more desirable properties, such as a highly crystalline structure in which crystalline domains are continuously dispersed along the fibre axis, good wet strength, as well as excellent dry strength. This makes Lyocell water washable. Furthermore, Lyocell is able to take up more dyes and it shrinks less when wetted and dried than other cellulose fibres, such as cotton and viscose rayon.

8.4.2.3 Application of Lyocell textiles

Lyocell has good resilience. It does not wrinkle as badly as rayon, cotton or linen, and some wrinkles can fall out if the garment is hung in a warm moist area, such as a bathroom after a hot shower. A light pressing will renew the appearance if needed. Slight shrinkage is typical in Lyocell garments. Lyocell is more expensive to produce than cotton or rayon, but nevertheless it is used in many products for daily use. Lyocell staple is used in apparel items, such as denim, chino, underwear and other casual wear clothing and towels. Filament fibres are used in items that have a silkier appearance, such as women's clothing and men's dress shirts. Lyocell can be blended with a variety of other fibres, such as silk, cotton, rayon, polyester, linen, nylon, and wool.

The early stage of the commercialisation of Lyocell focused on the fashion textile apparel sector. However, Lyocell is now targeted equally at the industrial sector, with particular emphasis on the key non-wovens market of wipes, filters and feminine hygiene products. The key difference between traditional textile production and non-wovens production is the omission of the yarn stage from the production process. Non-wovens form fibres into a web, and fibre bonding or entangling impart integrity and control the function, hand feeling and appearance of the resultant non-woven substrate. Staple fibres are produced to suit carded dry laid, air laid and wet laid processes. Lyocell is also used in conveyer belt, specialty paper, medical dressing, surgical swabs, drapes, gowns, floppy disc liners, filtration cloth and lining materials.

8.5 Outlook

Cellulose-based textiles can be classified into two categories: natural cellulose-based textiles and synthetic cellulose-based textiles. Typical natural cellulose-based textiles, such as cotton and ramie, have excellent fibre strength and weaving properties. Recently developed synthetic cellulose-based textiles, such as viscose rayon and Lyocell, have properties similar to natural cellulose-based textiles and are widely applied in a diversity of domains.

Compared with traditional natural cellulose-based textiles, synthetic cellulose-based textiles have several advantages, such as good drapability and dyeing properties, good durability and runnability, and excellent hand feeling. As such they were widely applied not only in the textile industry, but also in the pharmaceutical and chemical engineering industries.

To date the use of synthetic cellulose-based textiles remains limited and the industry requires further advances in production technologies, i.e. simplification of the process involved and minimisation of pollution, eventually reducing costs accordingly. With the necessary technological advances in place, synthetic cellulose-based textiles have a potentially prosperous future.

References

[1] Wang D. Global fiber production in 2015 and trends in chemical fiber production. *Polyester Industry*, 2016, 29 (2): 55–56.

[2] Lei T, Wang J & Zhang C. Lyocell fiber – 21st century environmentally friendly fiber. *Textile Science Research*, 1998, 3: 11–14.

[3] Lin Z & Ke J. The history, processing, property and future trend of Lyocell fiber. *Textile Science Research*, 2000, 1: 29–33.

[4] Dall'Acqua L, Tonin C, Varesano A, Canetti M, Porzio W & Catellani M. Vapour phase polymerisation of pyrrole on cellulose-based textile substrates. *Synthetic Metals*, 2006, 156: 379–386.

Chapter 9
Cellulose-based functional detection materials

Dr Yudong Lu

Due to their abundance, high strength and stiffness, low weight and biodegradability, nanoscale cellulose fibre materials (e.g., microfibrillated cellulose and bacterial cellulose) serve as promising candidates for bio-nanocomposite production. Such new high-value materials are the subject of continuing research and are commercially interesting in terms of new products for the pulp and paper industry and the agricultural sector [1]. Cellulose has emerged as an attractive substrate for the production of economical, disposable, point-of-care (POC) analytical devices. The development of novel methods of (bio)activation is central to broadening the application space of cellulosic materials [2].

9.1 Cellulose derivative functional detection materials

Functional detection materials have been prepared with cellulose, cellulose acetate membrane and cellulose diacetate, and have found an increasingly wide utilisation in all fields, such as enzyme electrode immobilisation procedures [3–5].

Researcher prepared functional detection materials, such as fluorescence-based sensing schemes, glucose biosensors, H_2O_2 biosensors, used cellulose nanocrystals (CNCs) and poly(vinyl alcohol) (PVA), crystalline cellulose-poly pyrrole (NCC-PPY) film, cellulose nanocrystals and poly(N-isopropylacrylamide) hydrogels [6–9].

Mun *et al.* prepared an inexpensive, flexible and disposable cellulose ZnO hybrid film (CZHF) and its feasibility for a conductometric glucose biosensor was investigated. CZHF is fabricated by simply blending ZnO nanoparticles with cellulose solution prepared by dissolving cotton pulp with lithium chloride/N,N-dimethylacetamide solvent. The enzyme activity of the glucose biosensor increases as the ZnO weight ratio increases linearly. The CZHF can detect glucose in the range of 1–12 mM [10].

Functional detection materials were prepared using cellulose-based nanomaterials. An amperometric glucose biosensor was fabricated by the adsorption of glucose oxidase (GOx) to an Ag/PPy nanoparticle-ethyl cellulose composite material modified platinum electrode [11–12].

A highly responsive glucose biosensor was developed based on immobilisation of glucose oxidase in gold nanorod medium by cross-linking with glutaraldehyde [13]. A novel matrix, gold nanoparticles-bacterial cellulose nanofibres (Au-BC) nanocomposite was developed for enzyme immobilisation and biosensor fabrication due to its unique properties such as satisfying biocompatibility, good conductivity and extensive surface area [14]. A biosensor for analysis of diazinon pesticide was also fabricated.

The basic element of this biosensor was a gold electrode modified with an immobilised acetylcholinesterase enzyme layer formed by entrapment with glutaraldehyde cross-linked cellulose acetate [15].

Functional detection materials were prepared using cellulose-carbon nanotube composite materials. Young-Hoo Kim *et al.* have prepared BC-CNT composite electrodes by directly filtering CNTs through BC hydrogel. And glucose oxidase was immobilised on BC-CNT composite electrodes. Biocompatible electrodes have many potential applications in the biomedical field, such as biosensors, biofuel cells and bioelectronic devices. Wu *et al.* have prepared a conductive cellulose-multiwalled carbon nanotube (MWCNT) matrix with a porous structure and good biocompatibility. Glucose oxidase was encapsulated in this matrix and thereby immobilised on a glassy carbon surface. Sensors were prepared based on dialdehyde cellulose/carbon nanotube/ionic liquid nanocomposite, multifunctional carbon nanotube (CNT)-cellulose films, modified nanostructured cellulose by different concentration and volume of dispersed multi-walled carbon nanotube (MWCNT) and double-walled carbon nanotube (DWCNT) solutions [16–21].

9.2 Synthesis of SERS substrate through the HPMC template method

Surface-enhanced Raman scattering (SERS) is a useful analytical technique with many significant advantages for sensitive chemical analysis and interfacial studies [22]. The ability to induce SERS is dependent on two key factors: resonant surface plasmon excitation of a metal substrate and the close proximity of analyte molecules to the metal substrate surface [23]. Surface plasmons are the result of the collective excitation of conduction band electrons near coinage metal surfaces, namely gold, copper and silver [24]. A wide variety of substrates has been found to exhibit SERS: electrochemically modified electrodes [25], colloids, films [26] and, more recently, regular particle arrays [27].

Colloidal dispersions of metal nanoparticles are widely used as efficient SERS substrates, and a range of 'soft' templates, such as polyvinyl pyrrolidone [28], poly(acrylic acid), CTAB [29], Chitosan, starch and DNA [30], has been used to prepare metal nanoparticles. Increasing awareness of green chemistry has evoked interest in developing an eco-friendly approach to the synthesis of nanoparticles. Cellulose is the most important neutral polymer from renewable sources. The aqueous solution of hydroxypropyl methyl cellulose (HPMC) contains size-confined, nano-sized polls of intermolecular origin. The polyhydroxylated HPMC shows dynamic supramolecular association helped by intramolecular and intermolecular hydrogen bonds forming molecular level pools, which act as a template for nanoparticle growth.

9.2.1 Preparation of silver nanoparticles

A definite weight of HPMC (viscosity: 4000 cps, 2% (w/v) in water at 20 °C;) and a certain amount of sodium citrate were dissolved in 50 ml of distilled water under stirring. After complete dissolution, the temperature of the reaction medium was raised to the desired temperature (55–95°C). Then, 50 ml silver nitrate solution was added drop-wise to the HPMC solution, stirring continuously for 30 minutes. After the mixture had been stirred for a given time, the product was washed with water and centrifuged at 5000 rpm for 5 minutes, and the precipitate was dissolved in 10 ml of water.

9.2.2 Reaction mechanism for formation of silver nanoparticles

Previous reports (Laurent *et al.*, 2008; Raveendran, Fu, & Wallen, 2003) have disclosed that solutions of polymers can be used for the synthesis and stabilisation of nanoparticles. HPMC macromolecules consist of chemically modified cellulose chains containing reducing groups (Fig. 9.1).

For the synthesis of silver nanoparticles, the generally accepted mechanism suggests a two-step process. In the first step, a portion of silver ions in a solution is reduced by the available reducing groups of the HPMC. The atoms thus produced act as nucleation centres and catalyse the reduction of the remaining silver ions present in the bulk solution. Subsequently, the atoms coalesce leading to the formation of silver clusters. The surface silver ions are again reduced and in this way the aggregation process does not cease until high values of nuclearity have been attained, which results in larger particles.

9.2.3 Effect of the HPMC concentration

Different concentrations of HPMC were used to assist the growth of silver nanoparticles via the reduction of silver nitrate with sodium citrate at a fixed temperature of 75 °C. The UV–vis absorption spectra and SERS spectra of the so-produced silver nanoparticles were recorded in each case after a fixed duration of 150 minutes (Fig. 9.2).

Fig. 9.2a shows the UV–vis spectra of the silver colloid obtained using HPMC as the stabilising agent. The results reveal that there is a gradual increase in the absorption intensity, by increasing the HPMC concentration up to 0.15%, which could be ascribed to the number of silver ions being reduced by the HPMC and sodium citrate. The same absorption band shifts from 465 nm to 455 nm by increasing the HPMC concentrations from 0.05% to 0.15%. Further increase in the HPMC concentration results in a decrease in the absorption intensity, which could be ascribed to the high viscosity of HPMC presenting an impediment to the rate of reduction of silver ions by the sodium citrate.

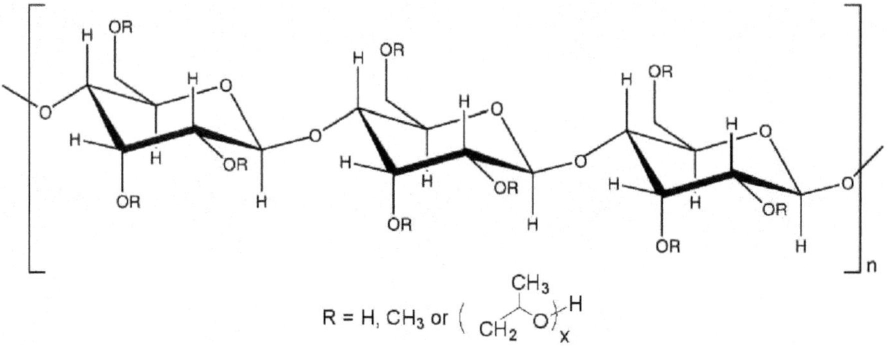

R = H, CH_3 or $(CH_2CH(CH_3)O)_xH$

Figure 9.1 The molecular structure of HPMC.

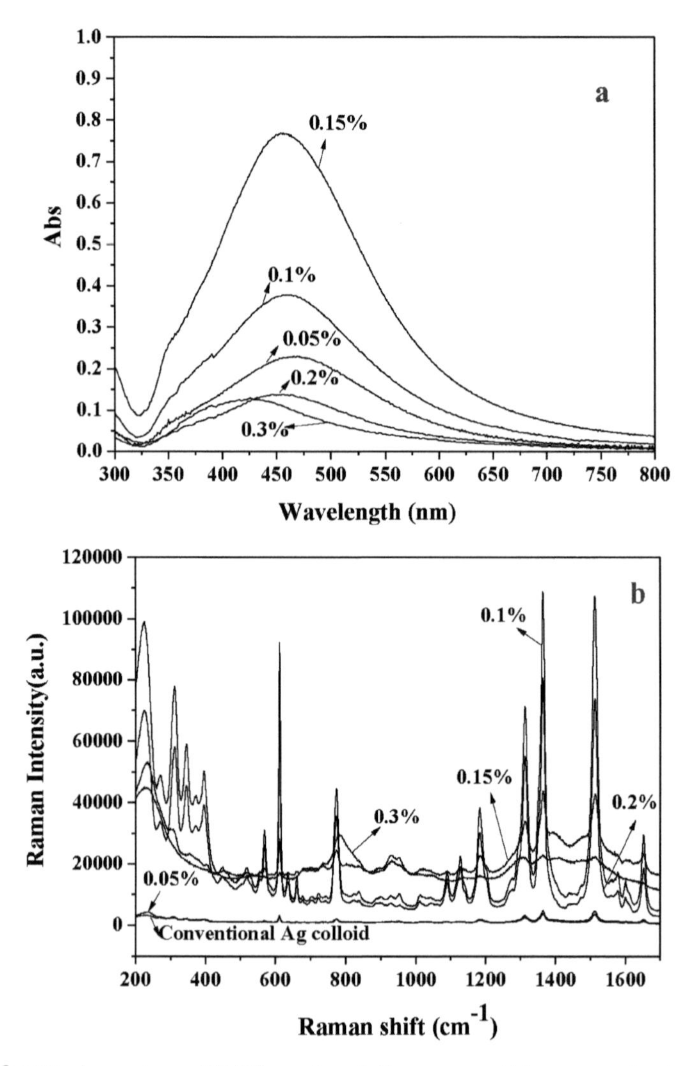

Figure 9.2 UV–vis spectra and SERS spectra of silver nanoparticles prepared using different HPMC concentrations. Reaction conditions: 50 ml of HPMC and 0.1% sodium citrate; 50 ml 10 mM silver nitrate; temperature, 75°C; duration, 150 minutes.

Fig. 9.2b shows the SERS spectra of silver nanoparticles prepared under different HPMC concentrations using Rhodamine 6G (R6G) (10^{-7} M) as a probe molecule. The results reveal that the silver nanoparticles prepared at 0.1% and 0.15% concentrations of HPMC have been shown to provide elegant SERS signals of R6G. Compared to conventional Ag colloid (Lee & Meisel, 1982), it provides a good SERS substrate with a relatively high enhancement. Differences in the relative intensities can be attributed to differences in the morphology of the silver particles due to different HPMC concentrations. The SERS performance of silver nanoparticles prepared under 0.1%

concentrations HPMC is superior to others, demonstrating the advantage of this new method of preparing Ag-based SERS substrates.

9.2.4 Effect of concentration of silver nitrate

0.1% HPMC was used to assist the growth of silver nanoparticles with different concentrations of silver nitrate, namely, 5 mM, 10 mM and 20 mM via the reduction with 0.1% sodium citrate at a fixed temperature of 75 °C. The UV–vis absorption spectra and SERS spectra of the so-produced silver nanoparticles were recorded in each case after a fixed duration of 150 minutes (Fig. 9.3).

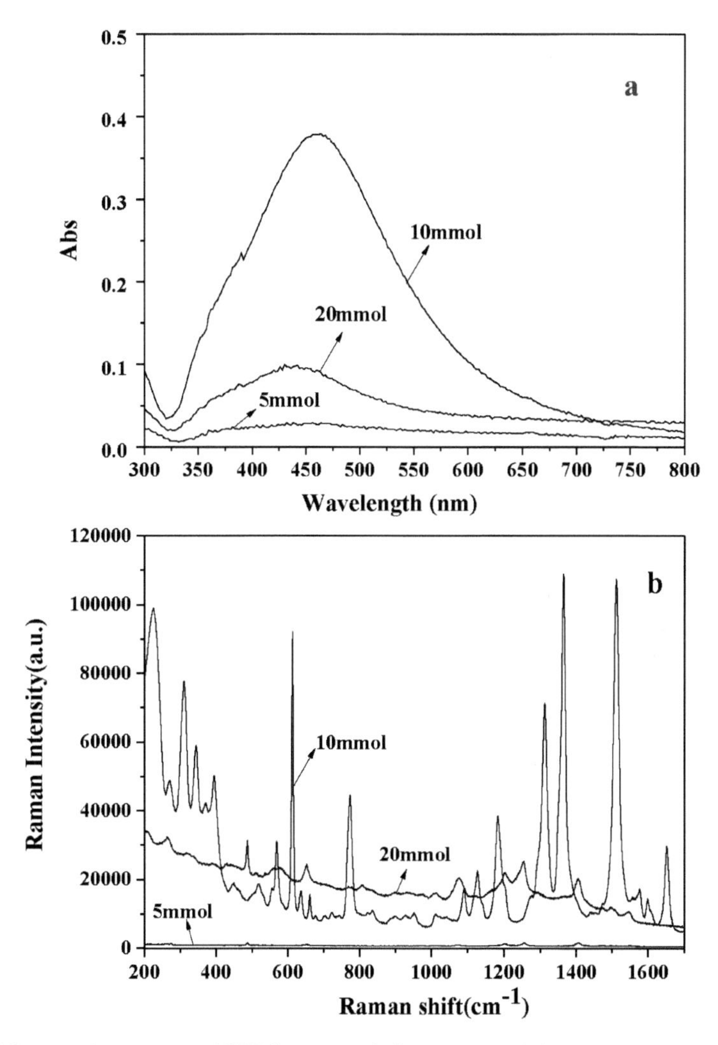

Figure 9.3 UV–vis spectra and SERS spectra of silver nanoparticles prepared using different silver nitrate concentrations. Reaction conditions: 50 ml of 0.1% HPMC and 0.1% sodium citrate; 50 ml silver nitrate; temperature, 75 °C; duration, 150 minutes.

The data in Fig. 9.3a shows that the absorption intensity of silver nanoparticles prepared with 10 mM silver nitrate is stronger than others. There is a gradual increase in the absorption intensity, by increasing the silver nitrate concentration up to 10 mM. Further increase in the silver nitrate concentration results in a decrease in the absorption intensity; the plasmon band is broad.

The data in Fig. 9.3b shows the SERS spectra of silver nanoparticles synthesised with different silver nitrate concentrations using R6G (10^{-7} M) as a probe molecule. The results show that silver nitrate concentrations play a crucial role in the performance of silver nanoparticles. The SERS performance of silver nanoparticles prepared with 10 mM silver nitrate is superior to others.

9.2.5 Effect of temperature

Fig. 9.4 shows the UV–vis spectra and SERS spectra of silver nanoparticles prepared at different temperatures.

At the low temperature (55 °C), the plasmon band is broad and a simple test for silver ions using NaCl solution indicates low conversion of silver ions to metallic silver nanoparticles at this duration. Increasing the temperature to 85°C, leads to outstanding enhancement in the plasmon intensity indicating that large numbers of silver ions are reduced and used for cluster formation. A further increase in temperature, to 95 °C, leads to significant decrement in the absorption intensity, which could be ascribed to the high viscosity of HPMC presenting an impediment to the rate of reduction of silver ions by the sodium citrate.

The data reveals that there is a gradual increase in the Raman intensity with a rise in reaction temperature up to 75°C (Fig. 9.4b). Increasing in the reaction temperature further, to 95 °C, leads to a decrease in the Raman intensity. It is clear from the data that preparation of silver nanoparticles at 75 °C represents the optimum condition.

9.2.6 Effect of reaction duration

Fig 9.5 shows the UV–vis spectra and SERS spectra of silver nanoparticles prepared at different reaction durations, while other parameters were kept constant.

The data in Fig.9.5a reveals that there is a gradual increase in the absorption intensity by increasing reaction duration up to 150 minutes. Further prolonging of the reaction duration, cause a decrease in the absorption intensity. The absorption band shifts from 460 nm to 471 nm, which indicates higher aggregation of silver nanoparticles.

The data in Fig. 9.5b reveals that the silver nanoparticles provide elegant SERS signals of R6G. The Raman signal can be amplified by adsorbing the R6G on the surface of metal nanoparticles with amplification as high as 10^5. The SERS performance of silver nanoparticles prepared at 150 minutes duration is superior to others. Based on the above, the optimum duration for preparation of silver nanoparticles is 150 minutes. These results are in agreement with the expected data obtained from the UV–vis spectra (Fig. 9.5a).

9.2.7 Effect of the reducing agent

Fig. 9.6 shows the UV–vis absorption spectra and the SERS spectra of silver nanoparticles prepared at different concentrations of sodium citrate, while other parameters were kept constant.

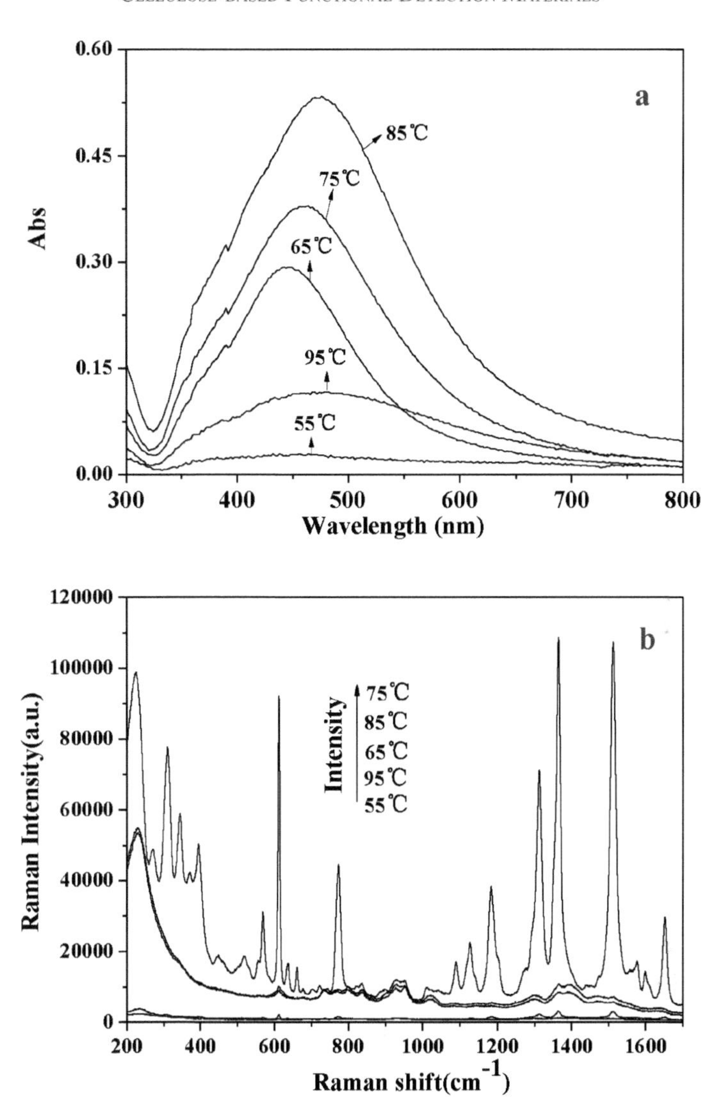

Figure 9.4 UV–vis spectra and SERS spectra of silver nanoparticles prepared using different temperatures. Reaction conditions: 50 ml of 0.1% HPMC and 0.1% sodium citrate; 50 ml 10 mM silver nitrate; duration, 150 minutes.

It is clear that there is a gradual increase in the absorption intensity, by increasing the sodium citrate concentration up to 0.1%. Further increase in the sodium citrate concentration leads to a decrease in the absorption intensity (Fig. 9.6a). Remarkable enhancements were found in the SERS spectrum of R6G adsorbed on the silver nanoparticles prepared at 0.1% concentrations of sodium citrate compared with those adsorbed on the other silver nanoparticles (Fig. 9.6b). The experiments indicated

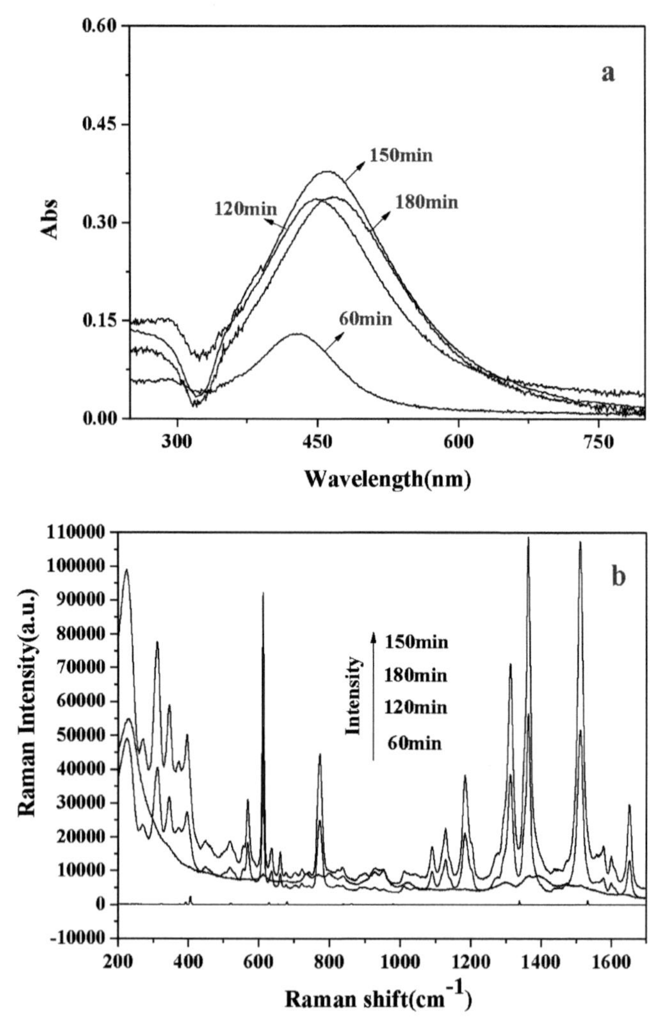

Figure 9.5 UV–vis spectra and SERS spectra of silver nanoparticles prepared at different durations. Reaction conditions: 50 ml of 0.1% HPMC and 0.1% sodium citrate; 50 ml 10 mM silver nitrate; temperature, 75°C.

that the concentration of sodium citrate played a significant role in the formation and growth of the silver nanoparticles.

Fig. 9.7 shows the TEM image of silver nanoparticles prepared at the optimum conditions of synthesis of SERS substrates: 0.1% HPMC, 10 mM silver nitrate at 75 °C, then reducing in 0.1% reducing agent at 150 minutes duration. As seen in Fig. 9.7, particles are mostly near-spherical in shape with an average size of 60–70 nm. The SERS enhancement of 6-months aged silver nanoparticles was successfully tested as well.

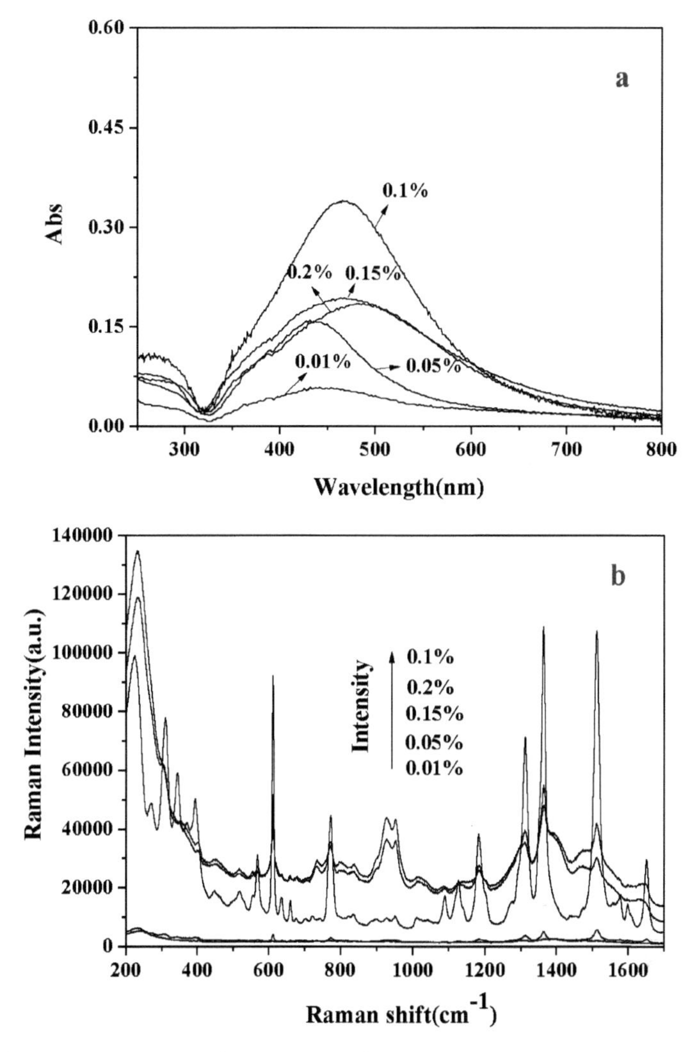

Figure 9.6 UV–vis spectra and SERS spectra of silver nanoparticles prepared using different sodium citrate concentrations. Reaction conditions: 50 ml of 0.1% HPMC; 50 ml 10 mM silver nitrate; temperature, 75°C; duration, 150 minutes.

9.3 Synthesis of SERS substrates through the CMC Template Method

Cellulose is one of the most abundant naturally occurring biopolymers and it is commonly found in the cell walls of plants and certain algae. Cellulose has three reactive hydroxyl groups per anhydroglucose repeating unit that form intermolecular and intramolecular hydrogen bonds. These bonds strongly influence the chemical reactivity and solubility of cellulose [31]. Carboxymethyl cellulose (CMC) macromolecules

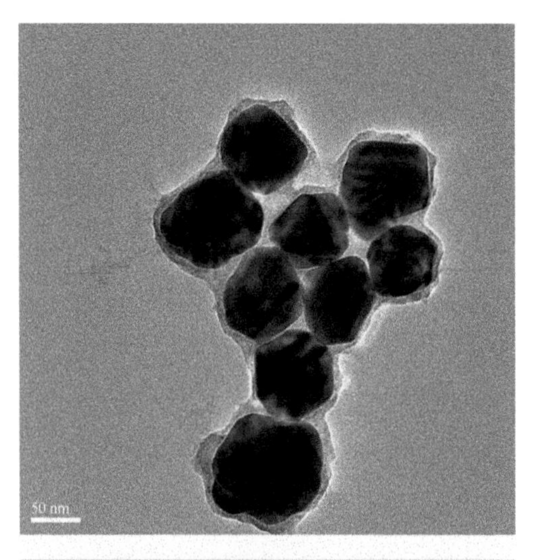

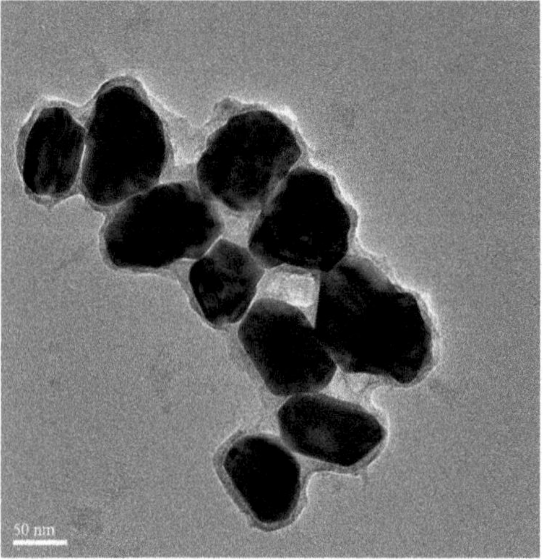

Figure 9.7 TEM image of silver nanoparticles prepared at the optimum conditions.

consist of chemically modified cellulose chains containing reducing groups and carboxyl groups. All these components with their anionic and reducing properties support the utilisation of CMC as a reducing and stabilising agent for the synthesis of silver nanoparticles. The negatively charged solubilised CMC facilitates the attraction of the positively charged silver cations to the polymeric chains followed by reduction with the existing reducing groups. It is well known that aqueous solution of CMC contains size-confined, nanosized polls of intermolecular origin [32]. The polyhydroxylated CMC shows dynamic supramolecular association helped by intramolecular and intermolecular hydrogen bond forming molecular level pools, which act as a template for nanoparticle growth.

9.3.1 Preparation of silver nanoparticles

A definite weight of CMC (average molecular weight: 2,50,000; viscosity: 2500–4500 mPa.s, 1% (w/v) in water at 25°C; degree of substitution of CMC: 0.80–0.95) and sodium citrate were dissolved in 50 ml of distilled water using a heated magnetic stirrer. After complete dissolution, the temperature of the reaction medium was raised to the desired temperature (60–90 °C). A certain amount of silver nitrate solution was then added drop-wise for 30 minutes. The reaction mixture was continuously stirred for different durations (0.5–3.5 hours). After the mixture was stirred for a given time, the product was washed and centrifuged at 5000 rpm for 5 minutes, and the precipitate was dissolved in 10 mL of water.

9.3.2 Reaction mechanism for the formation of silver nanoparticles

The molecular structure of CMC is shown in Fig. 9.8. CMC macromolecules consist of chemically modified cellulose chains containing reducing groups and carboxyl groups. The negatively charged solubilised CMC facilitates the attraction of the positively charged silver cations to the polymeric chains followed by reduction with the existing reducing groups.

For the synthesis of silver nanoparticles, the generally accepted mechanism suggests a two-step process, i.e. atom formation and then polymerisation of the atoms. In the first step, a portion of metal ions in a solution is reduced by the available reducing groups of the CMC. The atoms thus produced act as nucleation centres and catalyse the reduction of the remaining metal ions present in the bulk solution. Subsequently, the atoms coalesce leading to the formation of metal clusters. The surface ions are again reduced and in this way the aggregation process does not cease until high values of nuclearity are attained, which results in larger particles. The process is stabilised by the interaction with the polymer, so preventing further coalescence [33].

9.3.3 Effect of CMC concentration

Different concentrations of CMC, namely, 0.075%, 0.15%, 0.3%, 0.75% and 1.5% were used to assist the growth of silver nanoparticles via the reduction of silver nitrate with sodium citrate at a fixed temperature of 75°C. The UV–vis absorption spectra of

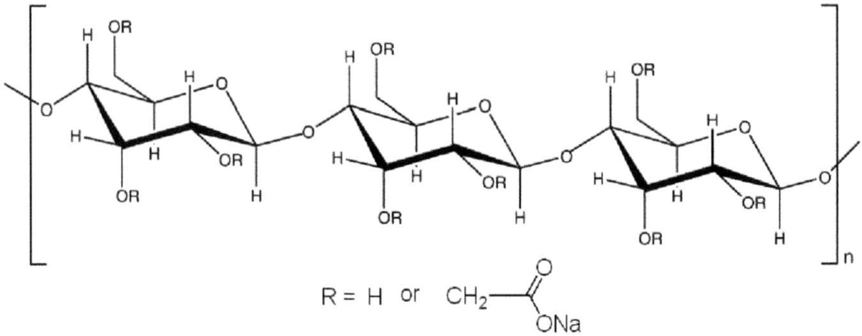

Figure 9.8 Molecular structure of carboxymethyl cellulose.

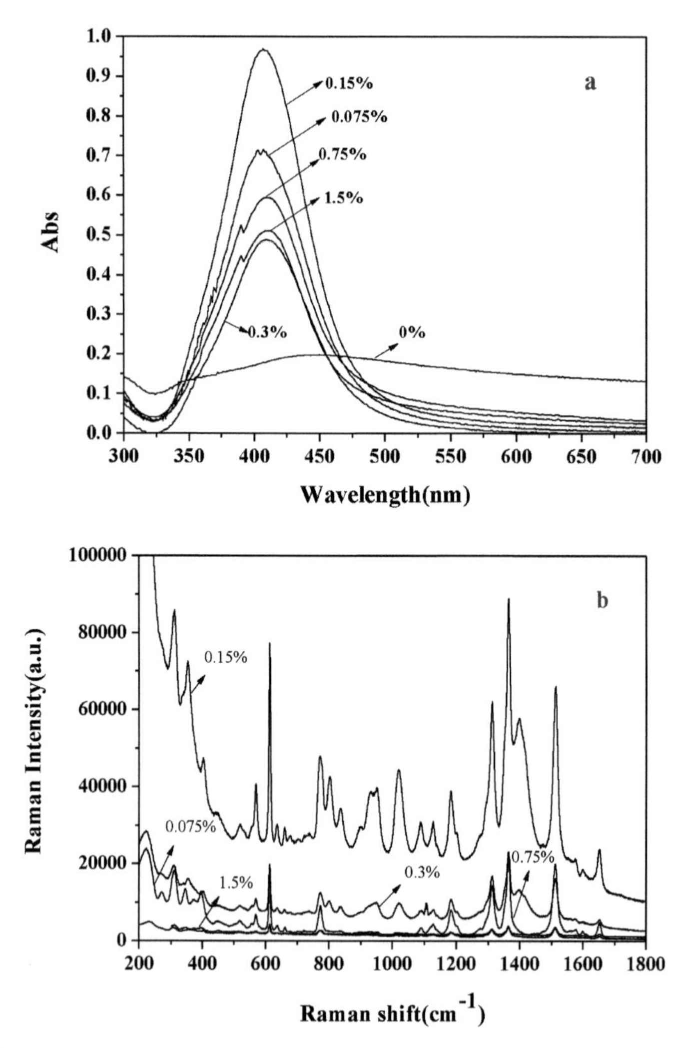

Figure 9.9 UV–vis spectra and SERS spectra of silver nanoparticles prepared using different CMC concentrations. Reaction conditions: 50 ml of CMC and 0.8% sodium citrate; 50 ml 25 mM $AgNO_3$; temperature, 75 °C; duration, 3 hours.

the so-produced silver nanoparticles were recorded in each case after a fixed duration of 3 hours (Fig. 9.9).

Fig. 9.9a shows the UV–vis spectra of the silver colloid obtained using CMC as the stabilising agent. The results reveal that regardless of the CMC concentration used, similar plasmon bands are formed at a wavelength 405 nm with the formation of the ideal bell shape that is characteristic for the formation of silver nanoparticles.

The surface-enhanced Stokes Raman signal is proportional to the Raman cross-section of the adsorbed molecule, the excitation laser intensity, and the number of molecules that are involved in the SERS process. The SERS effect occurs because of the very strong electromagnetic fields and field gradients available in the so-called 'hot spots' of the fractal colloidal silver cluster. Therefore, the molecules involved in the SERS effect are predominantly those adsorbed on aggregates that are favourable for surface plasmon resonances.

Fig. 9.9b shows the SERS spectra of silver nanoparticles prepared under different CMC concentrations using Rhodamine 6G (R6G) as a probe molecule. The results reveal that the silver nanoparticles have been shown to provide elegant SERS signals of R6G. Differences in the relative intensities can be attributed to differences in the morphology of the silver particles due to different CMC concentrations. The SERS performance of silver nanoparticles prepared under 0.15% concentrations CMC is superior to others, demonstrating the advantage of this new method of preparing Ag-based SERS substrates.

9.3.4 Concentration of silver nitrate

0.15% CMC was used to assist the growth of silver nanoparticles with different concentrations of silver nitrate, namely, 20 mM, 25 mM and 30 mM, via the reduction of silver nitrate with sodium citrate at a fixed temperature of 75 °C. The UV–vis absorption spectra and SERS spectra of the so-produced silver nanoparticles were recorded in each case after a fixed duration of 3 hours (Fig. 9.10).

The data in Fig. 9.10a shows the UV–vis spectra of the silver colloid obtained using CMC as the stabilising agent at different silver nitrate concentrations. The results reveal that similar plasmon bands are formed at a wavelength 405 nm with the formation of the ideal bell shape that is characteristic of the formation of silver nanoparticles. The same absorption band shifts from 415 nm to 403 nm upon increasing the silver nitrate concentration from 20 mM to 30 mM.

The data in Fig. 9.10b shows the SERS spectra of silver nanoparticles synthesised with different $AgNO_3$ concentrations using R6G as a probe molecule. The results reveal that there are small differences in the relative intensities due to different silver nitrate concentrations. The SERS performance of silver nanoparticles prepared under 25 mM silver nitrate is superior to others.

9.3.5 Effect of temperature

Fig. 9.11 shows the UV–vis spectra of silver nanoparticle prepared at different temperatures. It is clear from the data that preparation of silver nanoparticles at 75 °C represents the optimum condition. At the low temperature (60 °C) the plasmon band is broad and a simple test for silver ions using NaCl solution indicates low conversion of silver ions to metallic silver nanoparticles at this duration. Increasing the temperature to 75 °C, leads to outstanding enhancement in the plasmon intensity, indicating that large numbers of silver ions are reduced and used for cluster formation. Further increasing the temperature up to 90 °C, leads to significant decrement in the absorption intensity, which indicates less stability and higher aggregation of silver nanoparticles.

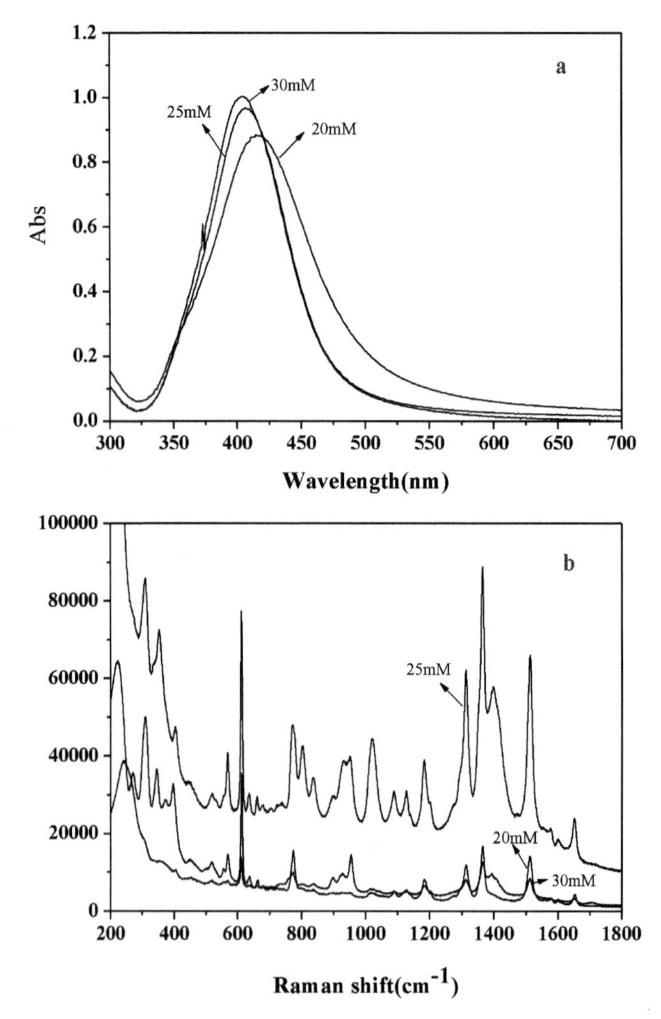

Figure 9.10 UV–vis spectra and SERS spectra of silver nanoparticles prepared using different silver nitrate concentrations. Reaction conditions: 50 ml of 0.15% CMC and 0.8% sodium citrate; 50 ml $AgNO_3$; temperature, 75 °C; duration, 3 hours.

Fig. 9.12 shows a TEM image of silver nanoparticles prepared at different temperatures. The silver nanoparticles prepared at 60 °C were too small; it was difficult to get the precipitation of the silver nanoparticles using the centrifuging method, which indicated that the silver nanoparticles prepared at 60°C could not be used as SERS substrates. The silver nanoparticles prepared at 90°C were too large to be used as SERS substrates. Based on the above, the optimum temperature for preparation of silver nanoparticles SERS substrates is 75°C. It is obvious from the TEM image in Fig. 9.12b that most of the silver nanoparticles are spherical, with diameters ranging from 40 nm to 60 nm.

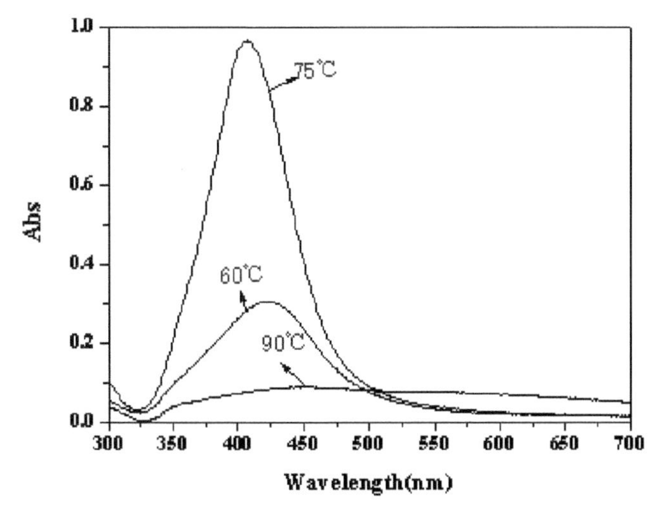

Figure 9.11 UV–vis spectra of silver nanoparticles prepared at different temperatures. Reaction conditions: 50 ml of 0.15% CMC and 0.8% sodium citrate; 50 ml 25 mM $AgNO_3$; duration, 3 hours.

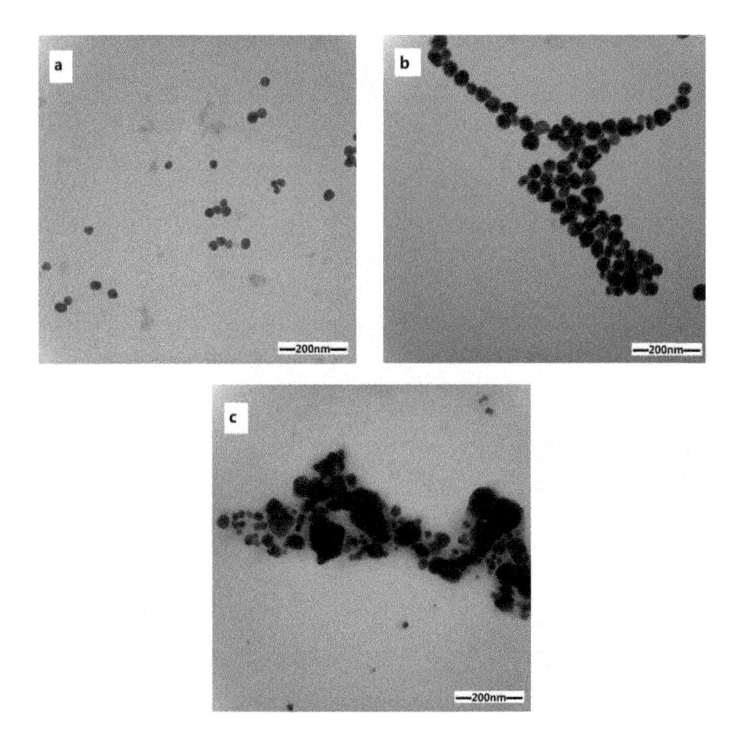

Figure 9.12 TEM of silver nanoparticles prepared at different temperatures (a) 60 °C; (b) 75°C; (c) 90°C. Reaction conditions: 50 ml of 0.15% CMC and 0.8% sodium citrate; 50 ml 25 mM $AgNO_3$; duration, 3 hours.

9.3.6 Effect of reaction duration

Preparation of silver nanoparticles was carried out at 75 °C with 0.15% CMC, and samples from the reaction medium were withdrawn at different time intervals, namely, 1, 1.5, 2, 2.5, 3 and 3.5 hours, for recording the UV–vis absorption spectra and the SERS spectra of the formed silver nanoparticles at these time intervals (Fig. 9.13).

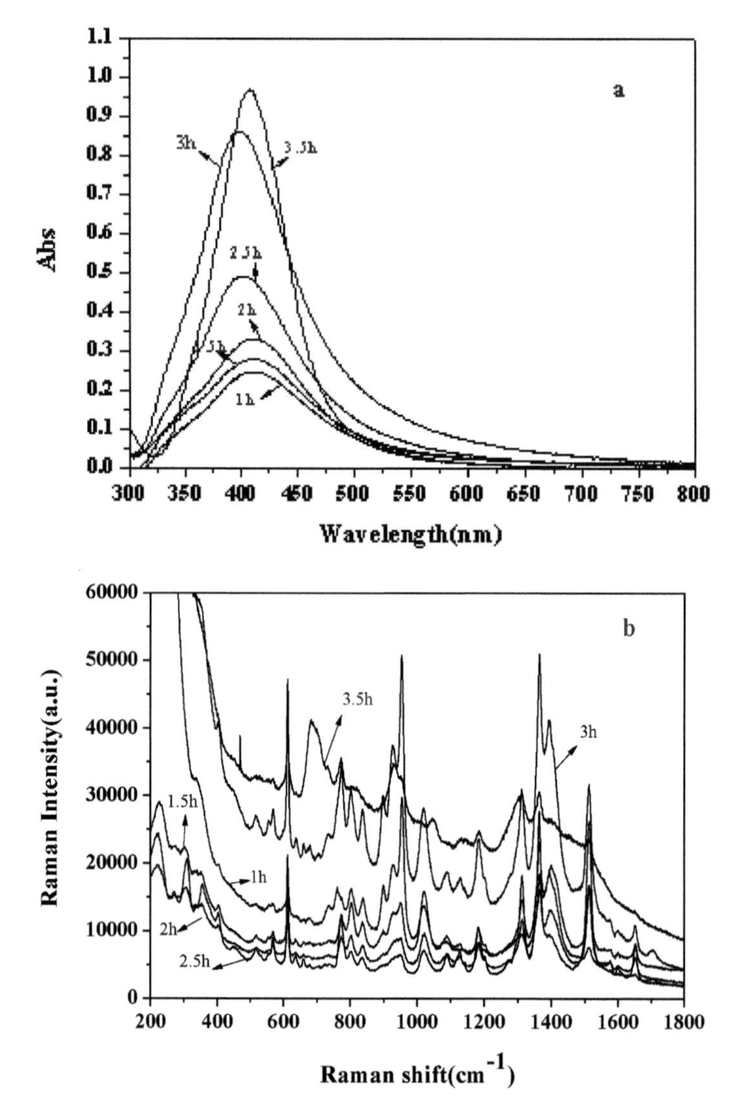

Figure 9.13 UV–vis spectra and SERS spectra of silver nanoparticles prepared at different durations. Reaction conditions: 50 ml of 0.15% CMC and 0.8% sodium citrate; 50 ml 25 mM $AgNO_3$; temperature, 75°C.

The data in Fig. 9.13a suggests several important findings, which can be presented as follows:

(i) at the early stage reaction duration (before 2 h) the plasmon band is broad and a simple test for silver ions using NaCl solution indicates low conversion of silver ions to metallic silver nanoparticles;
(ii) prolonging the reaction duration to 2.5 hours leads to outstanding enhancement in the plasmon intensity, indicating that large numbers of silver ions are reduced and used for cluster formation;
(iii) further prolonging the reaction duration to 3 hours, the peak corresponding to silver nanoparticles becomes more sharp and acquires an ideal bell shape. No marked improvement in the peak was noticed upon prolonging the reaction duration to 3.5 hours, and the absorption band shifts from 418 nm to 431 nm, which indicates higher aggregation of silver nanoparticles.

The data in Fig. 9.13b reveals that the SERS performance of silver nanoparticles prepared at 3 hours duration is superior to others. Based on the above, the optimum duration for preparation of silver nanoparticles is 3 hours. These results are in agreement with the expected data obtained from the UV–vis spectra(Fig. 9.13a).

9.3.7 Effect of the reducing agent

Fig. 9.14 shows the UV–vis absorption spectra and the SERS spectra of silver nanoparticles prepared at different concentrations of sodium citrate, while other parameters were kept constant.

The experiments indicated that the concentration of sodium citrate played a significant role in the formation and growth of the silver nanoparticles. It is clear that there is a gradual increase in the absorption intensity, by increasing the sodium citrate concentration up to 0.8%, which could be ascribed to numbers of silver ions being reduced and used for cluster formation. A further increase in the sodium citrate concentration led to a decrease in the absorption intensity.

To demonstrate the enhancement of Raman spectral intensity for the silver nanoparticles substrate, SERS spectra of R6G on silver nanoparticles are shown in Fig. 9.14b.

We found remarkable enhancement in the SERS spectrum of R6G adsorbed on the silver nanoparticles prepared at 0.8% concentrations of sodium citrate compared with those adsorbed on the other silver nanoparticles. Compared to the conventional Ag colloid, it provides a good SERS substrate with a relatively high enhancement.

9.4 Surface-enhanced Raman scattering study of silver nanoparticles prepared using MC as a template

In this work, a simple and effective approach to the aqueous phase synthesis of crystalline silver nanoparticles was employed, based on the reduction of silver ions by trisodium citrate in the presence of polymeric stabiliser, hydroxylpropyl methyl cellulose (MC). The surface-enhanced Raman scattering of the silver nanoparticles was also investigated.

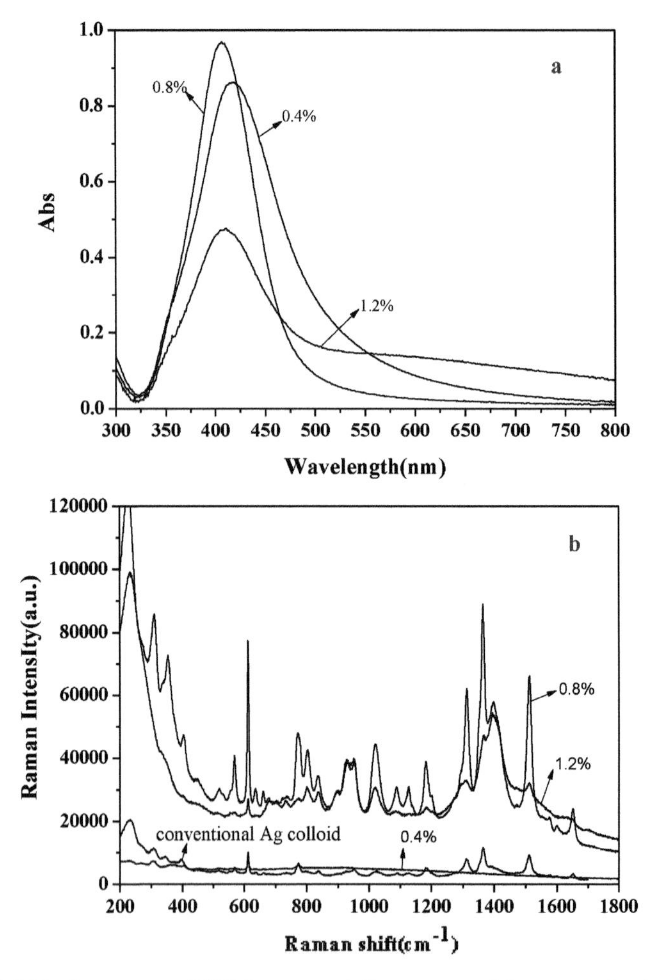

Figure 9.14 UV–vis spectra and SERS spectra of silver nanoparticles prepared using different sodium citrate concentrations. Reaction conditions: 50 ml of 0.15% CMC; 50 ml 25mM $AgNO_3$; temperature, 75 °C; duration, 3 hours.

9.4.1 Preparation of nanoparticles

A definite weight of MC (average molecular weight: 320,000–350,000; degree of polymerisation: 1.5–2.0; viscosity: 4000 cps, 2% (w/v) in water at 20°C;) was dissolved in 50 ml of distilled water under stirring. After complete dissolution, the temperature of the reaction medium was raised to 75 °C. Then, 50 ml of silver ammonia chloride solution (5 mM, 10 mM, 20mM or 30 mM) was added drop-wise to the MC solution (MC concentrations were 0.1%, 0.15%, 0.2%, 0.3% or 0.5%) maintaining continuous stirring for 30 minutes. After the mixture had been stirred for a given time (60, 90, 120 or 150 minutes), the product was washed with water and centrifuged at 5000 rpm for 5 minutes, and the precipitate was dispersed in 10 ml of water.

9.4.2 Effect of MC concentration

Different concentrations (0.1%, 0.15%, 0.2%, 0.3% and 0.5%) of MC were used to assist the growth of silver nanoparticles via the reduction of silver ammonia chloride (10 mM) with sodium citrate (0.1%) at a fixed temperature of 75 °C.

The effect of different concentrations of MC on the shape, size and dispersity of Ag nanoparticles is truly noticeable from the TEM images (Fig. 9.15).

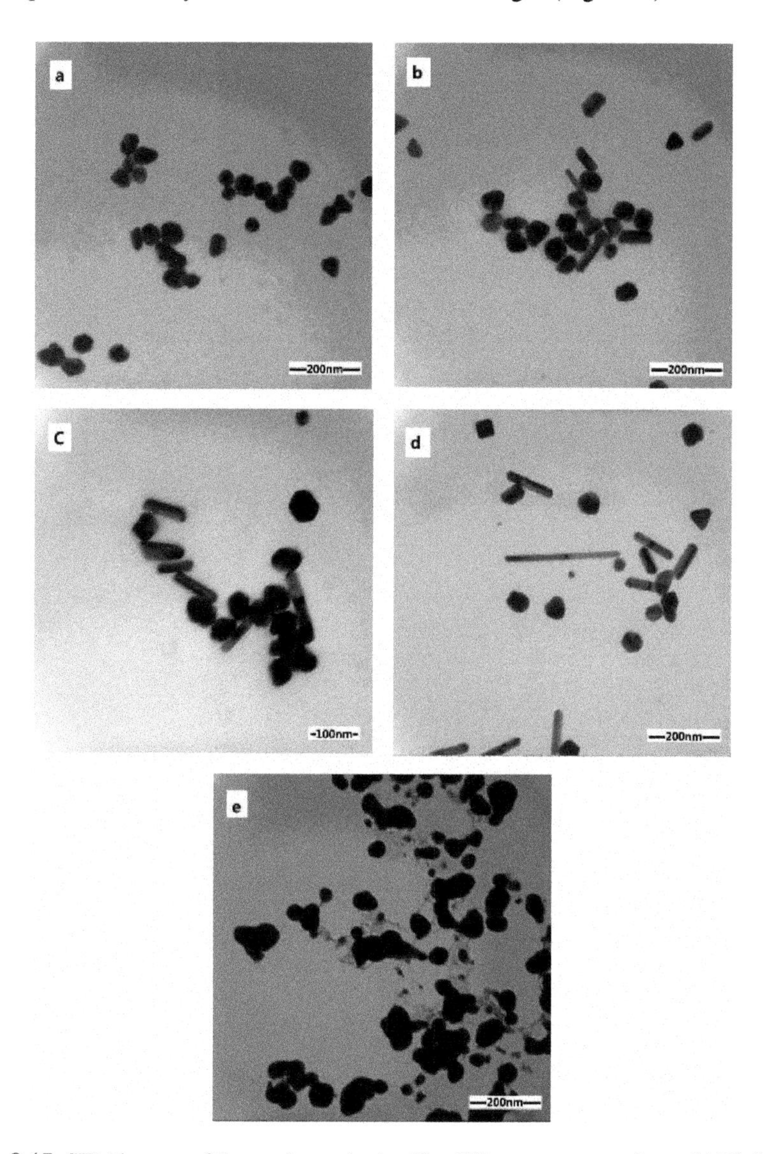

Figure 9.15 SEM images of the products obtained by different concentrations of MC (10 mM silver ammonia chloride, 0.1% sodium citrate): (a) 0.1% MC; (b) 0.15% MC; (c) 0.2% MC; (d) 0.3% MC; (e) 0.5% MC.

As seen in Fig.9.15, the concentration of MC (from 0.1% to 0.5%) has a significant effect on the reduction efficiency, particle shape and particle size (Fig. 9.15a–e). When the concentration of MC is 0.1% (Fig. 9.15a), particles are mostly spherical and triangular in shape, with an average size of 50 nm. When the concentration of MC is 0.15% (Fig. 9.15b), particles are mostly spherical and rodlike in shape, with an average size of 60 nm. When the concentration of MC is 0.2% (Fig. 9.15c), particles are mostly spherical and rodlike in shape, with an average size of 80 nm. When the concentration of MC is 0.3% (Fig. 9.15d), particles are mostly rodlike, spherical, triangular and quadrangular in shape, with an average size of 75 nm. When the concentration of MC is 0.5% (Fig. 9.15e), particles are mostly spherical in shape, with size ranging from 20 nm to 100 nm.

UV–vis absorption spectra and SERS spectra of the produced silver nanoparticles were recorded in each case after a fixed duration of 120 min (Fig. 9.16). Fig. 9.16a shows UV–vis spectra of silver colloid obtained using MC as a stabilising agent. It is clear that the concentration of MC has a great effect on the absorption peak.

Results reveal that the absorption intensity increases initially with the increase in the MC concentration, and then reached a maximum when the MC concentration was 0.3%. A further increase in the MC concentration results in a decrease in the absorption intensity. This could be ascribed to the high viscosity of the MC presenting an impediment to the rate of reduction of silver ions by sodium citrate and indicate a tendency towards aggregation. This can be identified in the SEM image in Fig. 9.16e.

Fig. 9.16b shows SERS spectra of silver nanoparticles prepared under different concentrations of MC using Rhodamine 6G (R6G) (10^{-7} M) as a probe molecule. It can be seen that the silver nanoparticles provide elegant SERS signals of R6G. SERS performance of silver nanoparticles prepared under 0.2% concentrations of MC is superior to others. The Raman signal can be amplified by adsorbing R6G on the surface of metal nanoparticles and amplification as high as 10^5.

Metallic nanostructures have a significant effect on the SERS enhancement. It is well known that Raman enhancement is associated with plasmonic 'hot spots', which occur near the contact point of metallic nanoparticles or nanogaps between two or more particles [34]. The 'rough surface' can act as 'hot sites' for surface plasma.

From a comparison of the Raman spectra of R6G (10^{-7} M) in Fig. 9.16b, the Raman enhancement of silver nanoparticles prepared under 0.2% concentration of MC is seen to be better than others. This can be identified by the SEM image in Fig. 9.15c. As shown in Fig. 9.15c, particles are mostly spherical and rodlike in shape, which easily appear as having a "rough surface", such as edges, corners, and protuberances, etc. [35].

9.4.3 Concentration of silver ammonia chloride

0.2% MC was used to assist the growth of silver nanoparticles with different concentrations of silver ammonia chloride, namely, 5 mM, 10 mM, 20 mM and 30 mM via reduction with sodium citrate (0.1%) at a fixed temperature of 75 °C.

The effect of different concentrations of silver ammonia chloride on the shape, size and dispersity of Ag nanoparticles is truly noticeable from the TEM images (Fig. 9.17).

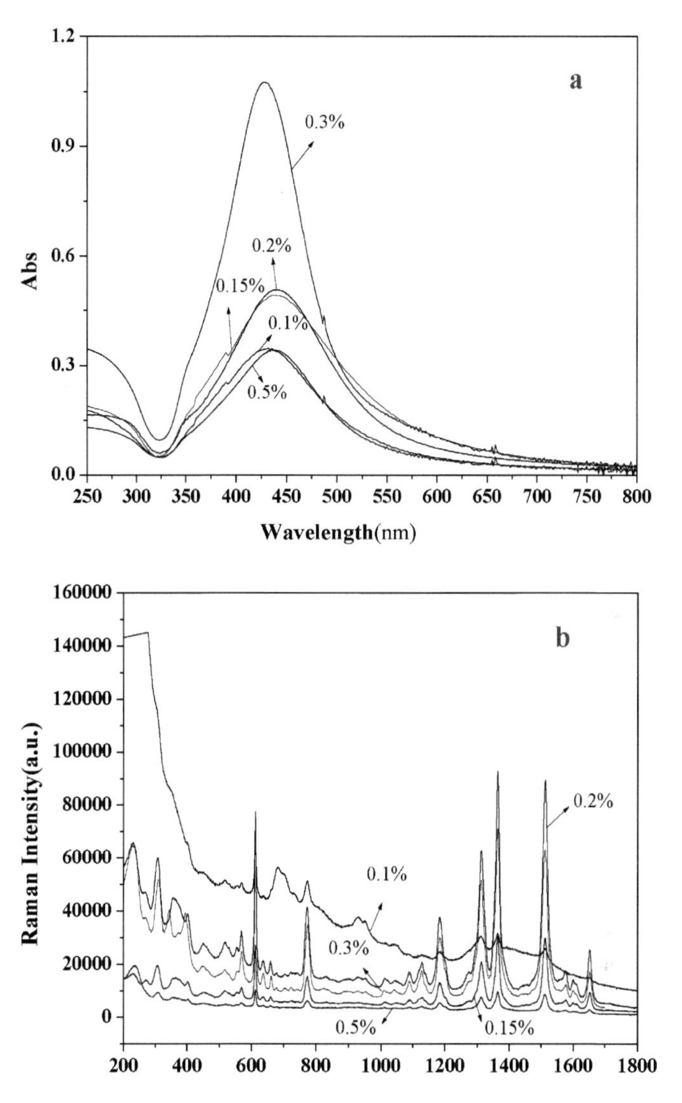

Figure 9.16 UV–vis spectra and SERS spectra of silver nanoparticles prepared using different concentrations of MC (10 mM silver ammonia chloride, 0.1% sodium citrate): (a) 0.1% MC; (b) 0.15% MC; (c) 0.2% MC; (d) 0.3% MC; (e) 0.5% MC.

As seen in Fig. 9.17a–d, when the concentration of silver ammonia chloride is 5 mM, particles are mostly spherical in shape, with size ranges of 60 nm. With an increase in the concentration of silver ammonia chloride, particles are mostly rodlike and spherical in shape, with an average size of 75 nm. A further increase in the concentration of silver ammonia chloride to 30 mM, gives particles that are mostly spherical in shape, with size ranging from 20 nm to 110 nm.

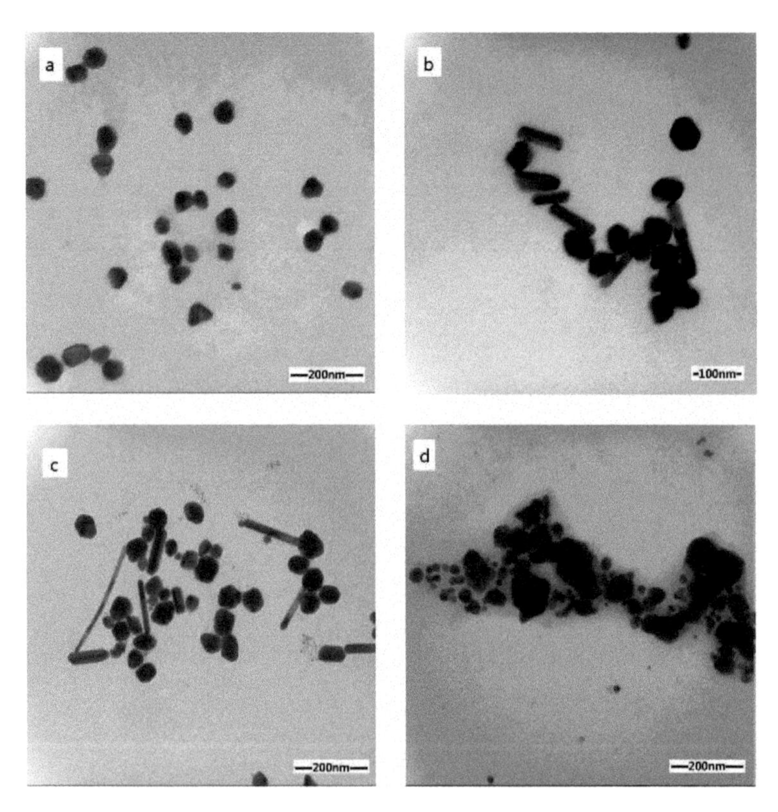

Figure 9.17 SEM images of the products obtained by different concentrations of silver ammonia chloride (0.2% MC, 0.1% sodium citrate,): (a) 5 mM; (b) 10 mM; (c) 20 mM; (d) 30 mM.

UV–vis absorption spectra and SERS spectra of the so-produced silver nanoparticles were recorded in each case after a fixed duration of 120 minutes (Fig. 9.18).

The data in Fig. 9.18a shows UV–vis spectra of the silver colloid obtained using MC as a stabilising agent at different concentrations of silver ammonia chloride. The results reveal that similar plasmon bands are formed at a wavelength of 440 nm, with the formation of the ideal bell shape that is characteristic of the formation of silver nanoparticles. The absorption band shifts from 433 nm to 441 nm with an increase in the silver ammonia chloride concentration from 5 mM to 10 mM. This indicated that silver nanoparticles became bigger, which can be identified in the SEM image in Fig. 9.17.

There is a gradual increase in absorption intensity, by increasing silver ammonia chloride concentration up to 10 mM. A further increase in the silver ammonia chloride concentration, resulted in a decrease in the absorption intensity; the plasmon band is broaden and a simple test for silver ions using NaCl solution indicates low conversion of silver ions to metallic silver nanoparticles. When the silver ammonia chloride concentration was increased to 30 mM, the absorption intensity decreased and exhibited

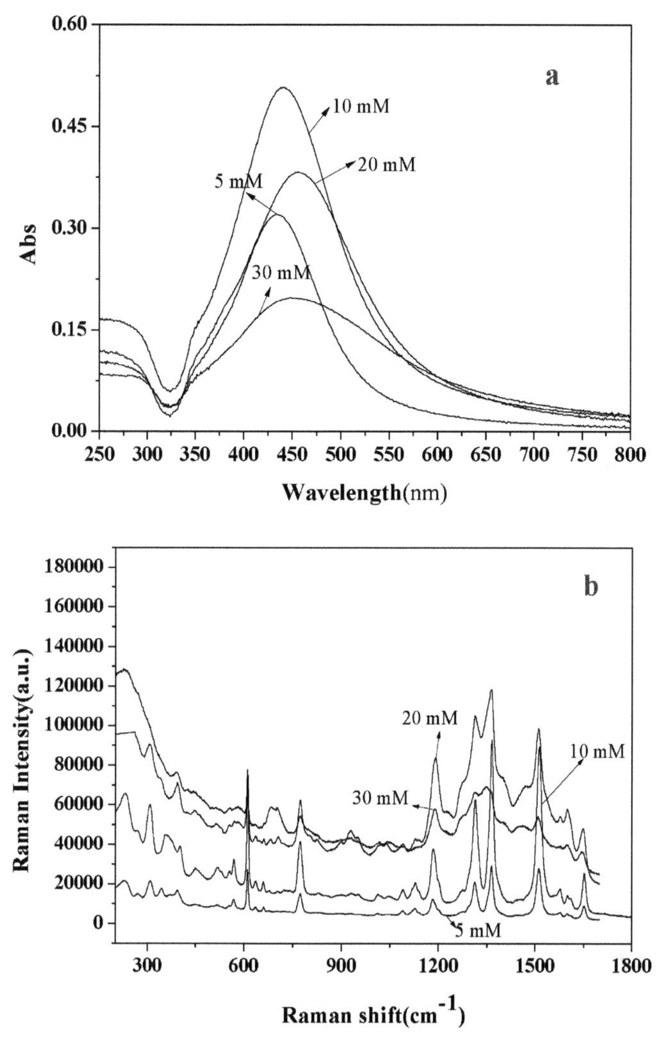

Figure 9.18 UV–vis spectra and SERS spectra of silver nanoparticles prepared using different concentrations of silver ammonia chloride. Reaction conditions: 50 ml of 0.2% MC and 0.1% sodium citrate; 50 ml silver ammonia chloride; temperature, 75 °C; duration, 120 minutes.

a broad band, which indicated the formation of silver nanoparticles with higher aggregation. These results are in agreement with the SEM image in Fig. 9.17d.

The data in Fig. 9.18b shows SERS spectra of silver nanoparticles synthesised with different silver ammonia chloride concentrations using R6G (10^{-7} M) as a probe molecule. The results reveal that silver ammonia chloride concentrations have a significant effect on enhancing the performance of silver nanoparticles. Silver nanoparticles prepared under 20 mM silver ammonia chloride provide SERS signals of R6G that are stronger than others.

Although the surface of silver nanoparticles prepared under 20 mM silver ammonia chloride appeared 'rougher' and the Raman signal stronger than others, the baseline is not as good as silver nanoparticles prepared under 10 mM. Considering the size and morphology of nanoparticles, the optimum silver ammonia chloride concentrations was 10 mM.

9.4.4 Effect of the reducing agent

The effect of different concentrations of sodium citrate on the shape, size and dispersity of Ag nanoparticles is truly noticeable from the TEM images (Fig. 9.19). Fig. 9.19 reveals that the number of Ag nanoparticles increases by increasing the concentration of sodium citrate from 0.05% to 0.3%.

Fig. 9.20 shows UV–vis absorption spectra and SERS spectra of silver nanoparticle prepared at different concentrations of sodium citrate (0.05%, 0.1%, 0.2% or 0.3%), with 0.2% MC at a temperature of 75°C after a fixed duration of 120 minutes.

The experiments indicated that the concentration of sodium citrate played a significant role in the formation of the silver nanoparticles. It is clear that there is a gradual increase in the absorption intensity by increasing the sodium citrate concentration, which could be ascribed to numbers of silver ions being reduced and used for cluster formation. With an increase in the sodium citrate concentration, the absorption band

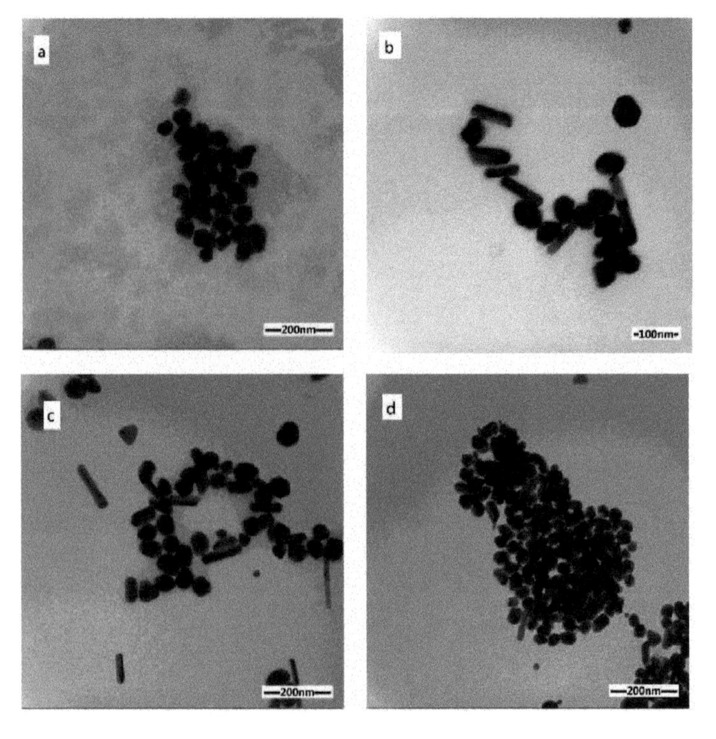

Figure 9.19 SEM images of the products obtained by different concentrations of sodium citrate (10 mM silver ammonia chloride, 0.2% MC): (a) 0.05% sodium citrate; (b) 0.1% sodium citrate; (c) 0.2% sodium citrate; (d) 0.3% sodium citrate.

shifts from 440 nm to 455 nm, which indicates the diameter of silver nanoparticles had become larger.

To demonstrate the enhancement of the Raman spectral intensity for the silver nanoparticles substrate, SERS spectra of R6G on silver nanoparticles are shown in Fig. 9.20b. We found remarkable enhancements in the SERS spectrum of R6G adsorbed on the silver nanoparticles prepared at 0.2% concentrations of sodium citrate compared with those adsorbed on the other silver nanoparticles.

The 'rough surface' of silver nanoparticles can act as 'hot sites' for surface plasma. The silver nanoparticles show stronger Raman enhancement with rougher surfaces of

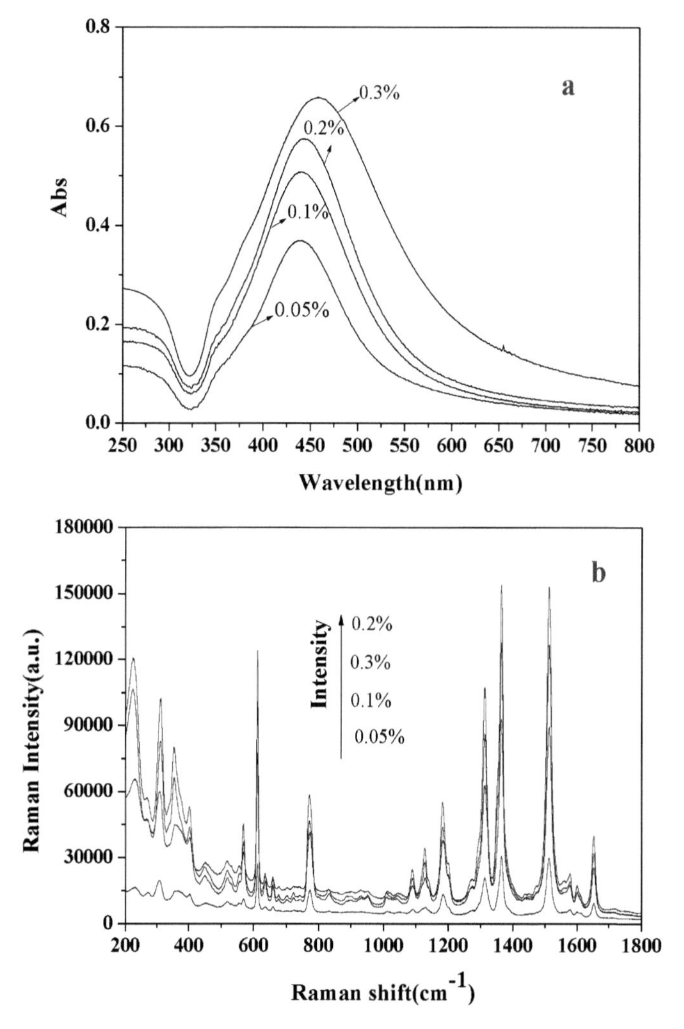

Figure 9.20 UV–vis spectra and SERS spectra of silver nanoparticles prepared using different concentrations of sodium citrate. Reaction conditions: 50 ml of 0.2% MC; 50 ml 10 mM silver ammonia chloride; temperature, 75°C; duration, 120 minutes.

silver nanoparticles. The results show that the density of silver nanoparticles also has a significant effect on the Raman enhancement. The optimum concentration of sodium citrate is 0.2%.

9.4.5 Effect of reaction duration

Preparation of silver nanoparticles was carried out at 75°C with 0.2% MC, and samples from the reaction medium were withdrawn at different time intervals, namely, 60, 90, 120, 150 and 180 minutes for recording the UV–vis absorption spectra and the SERS spectra of the formed silver nanoparticles at these time intervals (Fig. 9.21).

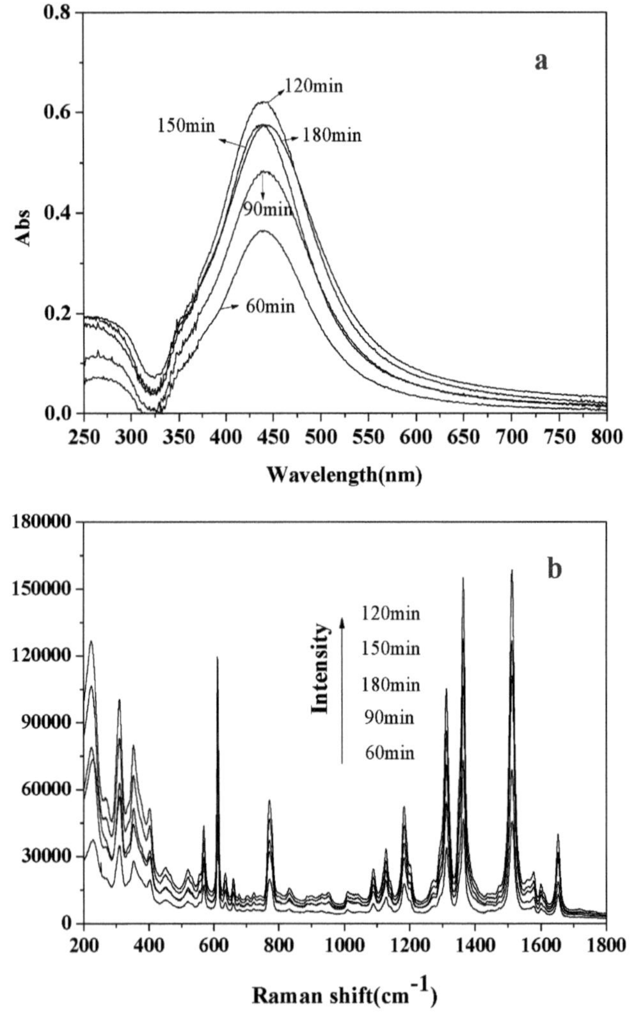

Figure 9.21 UV–vis spectra and SERS spectra of silver nanoparticles prepared at different durations. Reaction conditions: 50 ml of 0.2% MC and 0.2% sodium citrate; 50 ml 10 mM silver ammonia chloride; temperature, 75°C.

The data in Fig. 9.21a reveals that there is a gradual increase in the absorption intensity by increasing the reaction duration up to 120 minutes. When the reaction duration is 60 minutes, the plasmon band is broad, which indicates that the conversion rate of silver ions to metallic silver nanoparticles is low. With a reaction duration of more than 90 minutes, the plasmon intensity leads to outstanding enhancement, which indicates that large numbers of silver ions are being reduced and used. Further prolongation of the reaction duration, gave a decrease in the absorption intensity. An absorption band shift from 440 nm to 446 nm, indicates higher aggregation of silver nanoparticles.

The data in Fig. 9.21b reveals that the silver nanoparticles provide elegant SERS signals of R_6G. The Raman signal can be amplified by adsorbing the R_6G on the surface of metal nanoparticles and amplification as high as 10^5. The SERS performance of silver nanoparticles prepared at 120 minutes is superior to others. Based on the above, the optimum duration for the preparation of silver nanoparticles is 120 minutes. These results are in agreement with the expected data obtained from the UV–vis spectra (Fig. 9.21a).

9.5 Preparation of silver nanoparticles and their application to nasopharyngeal cancer detection

9.5.1 Surface-enhanced Raman scattering (SERS) spectra of blood plasma

Fig. 9.22 shows the SERS spectrum of a blood plasma sample from a patient with nasopharyngeal cancer, obtained by mixing the plasma with Ag colloid at a 1:1 proportion (A), the regular Raman spectrum of the same plasma sample without the silver

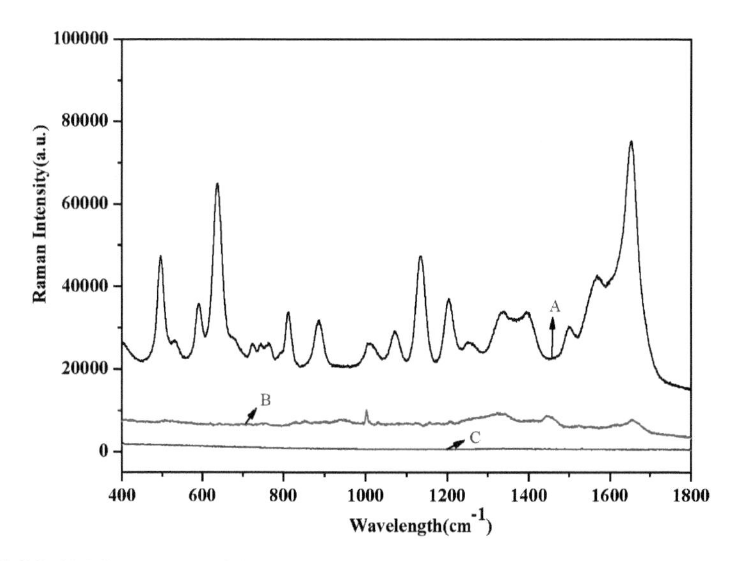

Figure 9.22 SERS spectrum of the blood plasma sample from a patient with nasopharyngeal cancer, obtained by mixing the plasma with Ag colloid at a 1:1 proportion. (A) the regular Raman spectrum of the same plasma sample without the silver sol (B) and the background Raman signal of the anticoagulant agent EDTA mixed with Ag colloid (C).

sol (B), and the background Raman signal of the anticoagulant agent EDTA mixed with Ag colloid (C).

The data in Fig. 9.22 reveals that the intensity of many dominant vibrational bands have been increased dramatically by SERS, indicating that there were strong interactions between the silver colloids and the blood plasma. Only a few Raman peaks could be observed in the native blood plasma without the addition of silver solution. Fig. 9.22c shows the background Raman signal of the anticoagulant with added Ag sol. We see no interference signal in the interested spectral range [36].

9.5.2 SERS spectra of normal subject blood plasma samples and nasopharyngeal cancer patient plasma samples

Fig. 9.23 shows a comparison of the normalised mean SERS spectra obtained from 22 normal subject blood plasma samples and 22 nasopharyngeal cancer patient plasma samples. A comparison was made of the mean spectrum for the normal blood plasma (black line, $n = 22$) versus that of the nasopharyngeal cancer (red line, $n = 22$). The green and black shadows represent the standard deviation of nasopharyngeal carcinoma patients and normal human plasma SERS spectra. The bottom curve was the difference spectrum.

There were spectral regions where the standard deviations did not overlap, and the differences were thus significant and reproducible.

The most obvious differences between the normal and nasopharyngeal cancer plasma can be found in the peaks at 495, 636, 1134 and 1653 cm^{-1}.

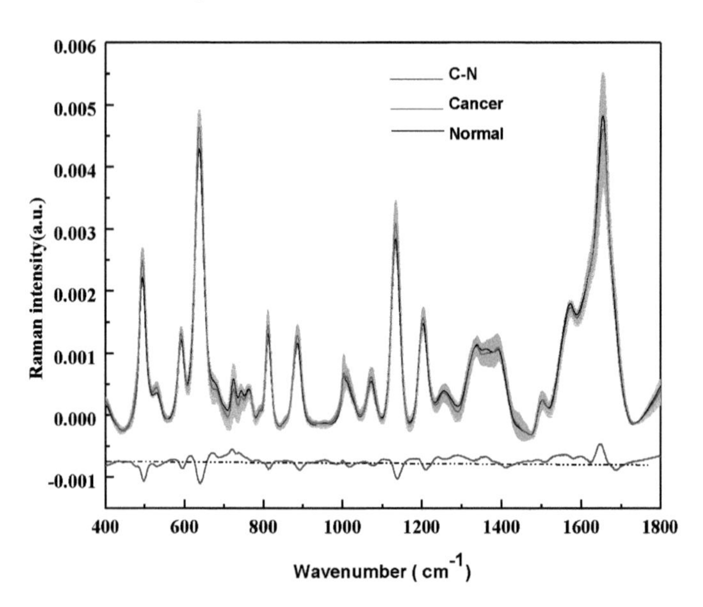

Figure 9.23 Comparison of the mean spectrum of normal blood plasma (black line, $n = 22$) versus that for nasopharyngeal cancer (red line, $n = 22$). The green and black shadows represent the standard deviation for nasopharyngeal carcinoma patients and normal human plasma SERS spectra. The bottom curve is the difference spectrum.

9.5.3 PCA combined with LDA of normal subject blood plasma samples and nasopharyngeal cancer patient plasma samples

Principal component analysis (PCA) combined with linear discriminate analysis (LDA) was performed to test the capability of plasma SERS spectra for differentiating between cancer and normal blood plasma SERS spectra.

9.5.3.1 Two-dimensional plot using scores of PC5 and PC6

An independent sample T-test on all the PC scores comparing normal and cancerous groups showed that there were three PCs (PC5, PC6 and PC8) that were most diagnostically significant for discriminating normal and cancerous groups.

Fig. 9.24 shows a comparison between normal and cancer groups by 2-D plot using scores of PC5 and PC6. The white square stands for the nasopharyngeal cancer group and the red circle stands for the healthy volunteer group. Fig. 9.24 shows there were obvious differences between the white squares and the red circle.

9.5.3.2 Three-dimensional mapping of the PCA result

In order to incorporate all significant SERS spectral features, LDA was used to generate diagnostic algorithms using the PC scores for the three most significant PCs (PC5, PC6 and PC8). Fig. 9.25 shows a 3-D scatter plot with PC5, PC6 and PC8 as the three axes of the PCA result of the nasopharyngeal cancer group (black square) and the healthy volunteer group (red circle) .

A probability distribution for the plasma SERS spectroscopy of normal people and gastric cancer patients analysed by PCA-LDA is shown in Fig. 9.26. The results show that the corresponding sensitivity and specificity are 90.9% and 97.5%, respectively.

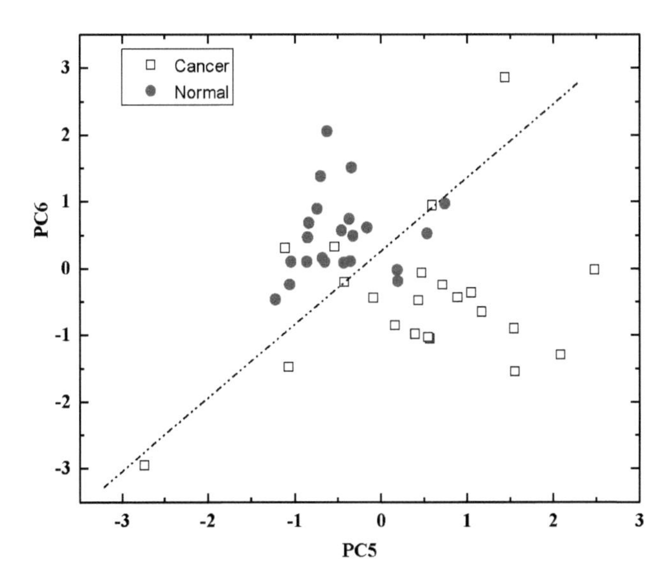

Figure 9.24 2-D plot using scores of PC5 and PC6.

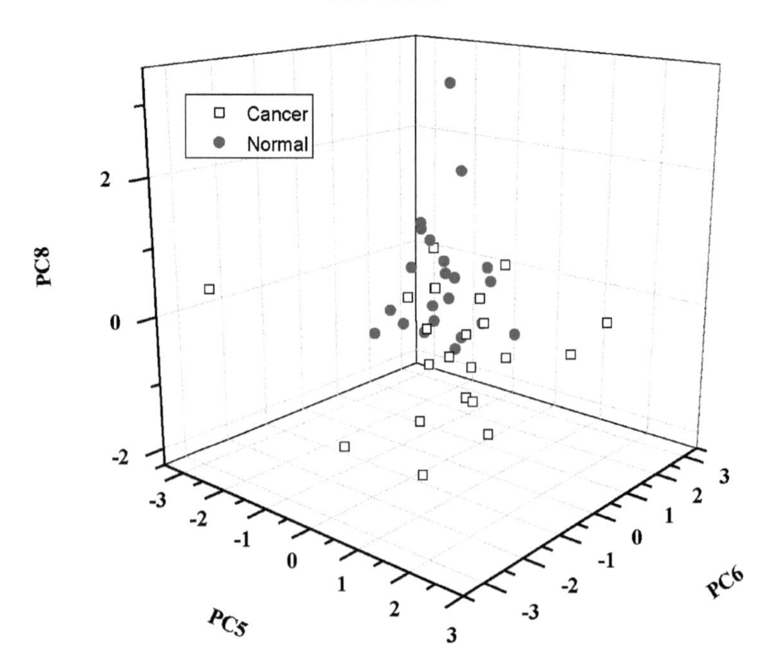

Figure 9.25 Three-dimensional mapping of the PCA result for the nasopharyngeal cancer group (black square) and the healthy volunteer group (red circle).

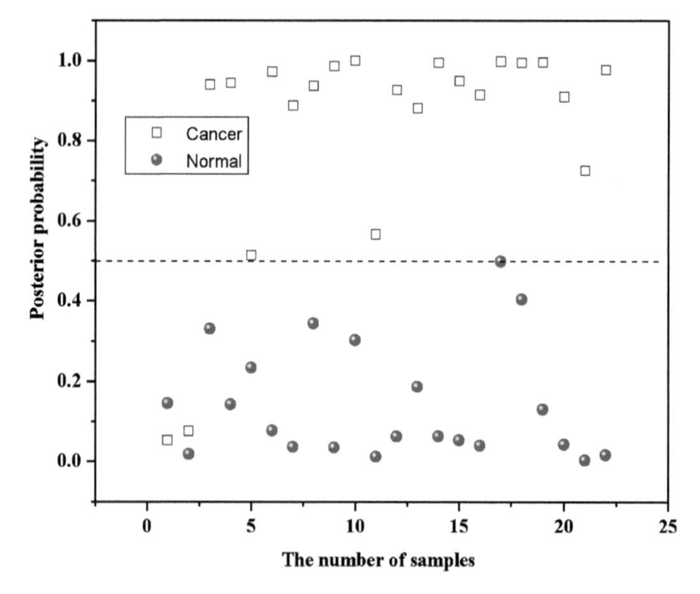

Figure 9.26 Probability distribution of plasma SERS spectroscopy of normal people and gastric cancer patients analysed by PCA-LDA. (The blue squares and red circles represent the nasopharyngeal carcinoma patient and normal human plasma SERS spectra. The probability of the corresponding diagnosis in line graph for $p = 0.5$.)

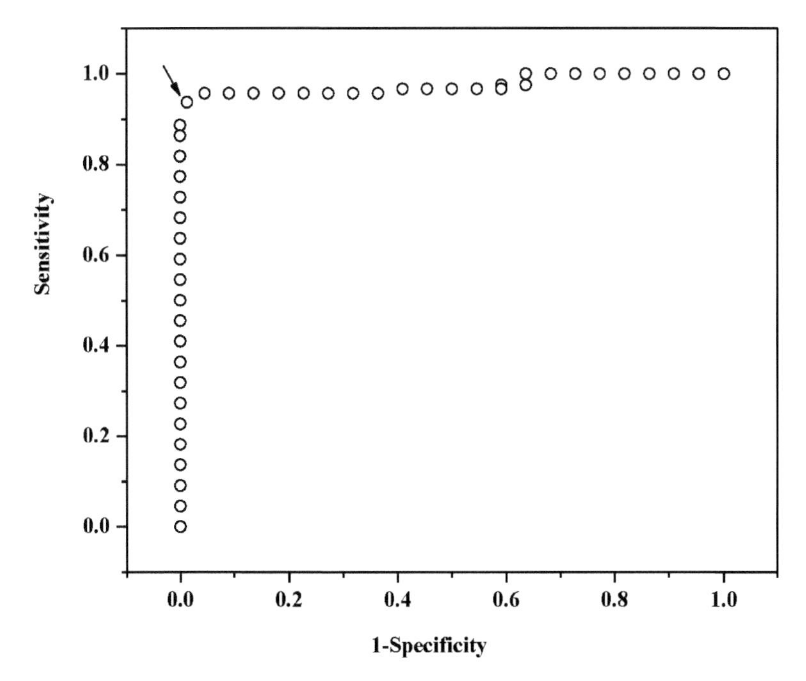

Figure 9.27 The receiver operating characteristic (ROC) curve of the discrimination result for using PCA-LDA-based SERS spectral classification with leave-one-out, cross-validation method. The integrated areas under the ROC curve are 0.987.

9.5.3.3 Receiver operating characteristic (ROC) curve

To further evaluate the performance of the PCA–LDA-based diagnostic algorithm for nasopharyngeal cancer diagnosis, the ROC curve (Fig. 9.27) was generated at different threshold levels. The area under the ROC curve was 0.987 (Fig. 9.27). The ROC curve of the discrimination result for using PCA-LDA-based SERS spectral classification with leave-one-out cross-validation method. The integrated areas under the ROC curve are 0.987. The corresponding sensitivity and specificity were 93.65% and 98.72%, respectively. The result shows that the blood plasma SERS spectra can be used for nasopharyngeal cancer detection with high sensitivity and specificity.

References

[1] Siró I & Plackett D. Microfibrillated cellulose and new nanocomposite materials: A review [J]. *Cellulose*, 2010, 17 (3): 459–494.

[2] Derikvand F, Yin D L T, Barrett R, *et al.* Cellulose-based biosensors for esterase detection [J]. *Analytical chemistry*, 2016, 88 (6): 2989–2993.

[3] Vaidya R & Wilkins E. Effect of interference on amperometric glucose biosensors with cellulose acetate membranes [J]. *Electroanalysis*, 1994, 6 (8): 677–682.

[4] Wang S, Ji X & Yuan Z. Study of cellulose acetate membrane-based glucose biosensors [J]. *Chinese Journal of Biotechnology*, 1994, 11 (3): 199–205.

[5] She-Ying D, Guang-Zhe G U, Zhu-Qing Y U, *et al.* Hydrogen peroxide biosensor based on cellulose diacetate-ionic liquid film immobilizing myoglobin [J]. *Chinese Journal of Analytical Chemistry*, 2011, 39 (9): 1358–1362.

[6] Schyrr B, Pasche S, Voirin G, *et al.* Biosensors based on porous cellulose nanocrystal–poly (vinyl alcohol) scaffolds [J]. *ACS Applied Materials & Interfaces*, 2014, 6 (15): 12674–12683.

[7] Derikvand F, Yin D L T, Barrett R, *et al.* Cellulose-based biosensors for esterase detection[J]. *Analytical chemistry*, 2016, 88 (6): 2989–2993.

[8] Lee A D H. Conductive nanocrystalline cellulose polymer composite film as a novel mediator in biosensor applications [D]. 2011.

[9] Berg D & Mugo S. Cellulose nanocrystals and poly (N-isopropylacrylamide) for hydrogen peroxide based biosensors [J]. *URSCA Proceedings*, 2015, 1.

[10] Mun S, Maniruzzaman M Ko H U, *et al.* Preparation and characterisation of cellulose ZnO hybrid film by blending method and its glucose biosensor application [J]. *Materials Technology*, 2015, 30 (supp.7): B150–B154.

[11] Ren X Z, Zhang P X, Liu J H, *et al.* A super highly sensitive glucose biosensor based on Ag/PPy nanoparticle-ethyl cellulose hybrid materials [C]. *Advanced Materials Research.* 2009, 58: 15–20.

[12] Singh V & Ahmad S. Synthesis and characterization of carboxymethyl cellulose-silver nanoparticle (AgNp)-silica hybrid for amylase immobilization [J]. *Cellulose*, 2012, 19 (5): 1759–1769.

[13] Ren X, Chen D, Meng X, *et al.* Amperometric glucose biosensor based on a gold nanorods/cellulose acetate composite film as immobilization matrix [J]. *Colloids and Surfaces B: Biointerfaces*, 2009, 72 (2): 188–192.

[14] Wang W, Zhang T J, Zhang D W, *et al.* Amperometric hydrogen peroxide biosensor based on the immobilization of heme proteins on gold nanoparticles–bacteria cellulose nanofibers nanocomposite [J]. *Talanta*, 2011, 84 (1): 71–77.

[15] Ramadhan L & Jahiding M. Analysis of diazinon pesticide using potentiometric biosensor based on enzyme immobilized cellulose acetate membrane in gold electrode [C]. IOP Conference Series: Materials Science and Engineering. IOP Publishing, 2016, 107 (1): 012–013.

[16] Kim Y H, Park S, Won K, *et al.* Bacterial cellulose–carbon nanotube composite as a biocompatible electrode for the direct electron transfer of glucose oxidase [J]. *Journal of Chemical Technology and Biotechnology*, 2013, 88 (6): 1067–1070.

[17] Wu X, Zhao F, Varcoe J R, *et al.* Direct electron transfer of glucose oxidase immobilized in an ionic liquid reconstituted cellulose–carbon nanotube matrix [J]. *Bioelectrochemistry*, 2009, 77 (1): 64–68.

[18] Shen G, Zhang X & Zhang S. A label-free electrochemical aptamer sensor based on dialdehyde cellulose/carbon nanotube/ionic liquid nanocomposite [J]. *Journal of the Electrochemical Society*, 2014, 161 (12): B256–B260.

[19] Rahatekar S S, Rasheed A, Jain R, *et al.* Solution spinning of cellulose carbon nanotube composites using room temperature ionic liquids [J]. *Polymer*, 2009, 50 (19): 4577–4583.

[20] Qi H, Liu J, Gao S, *et al.* Multifunctional films composed of carbon nanotubes and cellulose regenerated from alkaline–urea solution [J]. *Journal of Materials Chemistry A*, 2013, 1 (6): 2161–2168.

[21] Toomadj F, Farjana S, Sanz-Velasco A, *et al.* Strain sensitivity of carbon nanotubes modified cellulose [J]. *Procedia engineering*, 2011, 25: 1353–1356.

[22] Alvarez-Puebla R A, Dos Santos Jr D S, & Aroca R F. Surface-enhanced Raman scattering for ultrasensitive chemical analysis of 1 and 2-naphthalenethiols. *Analyst*, 2004. 129 (12): 1251–1256.

[23] Cui Shiqiang, Liu Yunchun, Yang Zhousheng, & Wei Xianwen. Construction of silver nanowires on DNA template by an electrochemical technique. *Materials & Design*, 2007, 28 (2): 722–725.

[24] Jiang Lizhong, Wu Zhanpeng, Wu Dezhen, Yang Wantai, & Jin Riguang. Controllable embedding of silver nanoparticles on silica nanospheres using poly (acrylic acid) as a soft template. *Nanotechnology*, 2007, 18 (18): 1–6.

[25] Ni Fan, Feng Helena, Gorton Lo, & Cotton Therese M. Electrochemical and SERS studies of chemically modified electrodes: Nile Blue A, a mediator for NADH oxidation. *Langmuir*, 1990, 6 (1): 66–73.

[26] Leopold N & Lendl B. A new method for fast preparation of highly surface-enhanced Raman scattering (SERS) active silver colloids at room temperature by reduction of silver nitrate with hydroxylamine hydrochloride. *The Journal of Physical Chemistry B*, 2003, 107 (24): 5723–5727.

[27] Qian Yong, Lu Shunbao, & Gao Fenglei. Synthesis of copper nanoparticles/carbon spheres and application as a surface-enhanced Raman scattering substrate. *Materials Letters*, 2012,

[28] Wang Hongshui, Qiao Xueliang, Chen Jianguo, Wang Xiaojian, & Ding Shiyuan. Mechanisms of PVP in the preparation of silver nanoparticles. *Materials Chemistry and Physics*, 2005, 94 (2): 449–453.

[29] Yu Ying, Du Fei-Peng, Yu Jimmy C, Zhuang Yuan-Yi & Wong Po-Keung. One-dimensional shape-controlled preparation of porous Cu^2 O nano-whiskers by using CTAB as a template. *Journal of Solid State Chemistry*, 2004, 177 (12): 4640–4647.

[30] Zhou Qun, Zhao Gui, Chao Yanwen, Li Yan, Wu Ying & Zheng Junwei. Charge-transfer induced surface-enhanced Raman scattering in silver nanoparticle assemblies. *The Journal of Physical Chemistry C*, 2007, 111 (5): 1951–1954.

[31] Sundar S T, Sain M M & Oksman, K. Characterization of microcrystalline cellulose and cellulose long fiber modified by iron salt [J]. *Carbohydrate Polymers*, 2010, 80 (1): 35–43.

[32] *Fu & Wallen S L.* Completely 'green' synthesis and stabilization of metal nanoparticles. *Journal of the American Chemical Society,* 2003, 125 (46): 13940–13941.

[33] Goia, D V. Preparation and formation mechanisms of uniform metallic particles in homogeneous solutions. *Journal of Materials Chemistry*, 14 (4): 451–458, 2004.

[34] Zhang J, Li X, Sun X & Li Y: Surface enhanced Raman scattering effects of silver colloids with different shapes. *The Journal of Physical Chemistry B*, 2005, 109: 12544–12548.

[35] Potara M, Maniu D & Astilean S. The synthesis of biocompatible and SERS-active gold nanoparticles using chitosan. *Nanotechnology*, 2009, 20 (31) 315602.

[36] Feng S, Chen R, Lin J, *et al.* Nasopharyngeal cancer detection based on blood plasma surface-enhanced Raman spectroscopy and multivariate analysis [J]. *Biosensors and Bioelectronics*, 2010, 25 (11): 2414–2419.

Chapter 10
Nanocellulose-based aerogels

Dr Omar Abo Madyan and Professor Mizi Fan

This chapter begins with a discussion of the raw materials and processes used for nanocellulose production and for generating various nanocelluloses with different properties. It then outlines both inorganic and organic matrices that have been reported for the preparation of nanocellulose-based aerogel composites, along with cross-linkers and other additives that have been shown to be effective in nanocellulose aerogel preparation. The chapter concludes with an account of the various manufacturing methods, performance and applications of nanocellulose aerogels.

10.1 Raw materials of nanocellulose aerogels

10.1.1 Nanocellulose

Nanocellulose is mainly characterised by its source, whether it is derived from trees or other plants to produce cellulose nanocrystals (CNCs) or cellulose nanofibrils (CNFs), or from certain bacteria and algae to produce bacterial nanocellulose (BC) [1]. Each group has a different production process as well as different properties and morphologies, as summarised in Table 10.1 [2–3]. CNCs are needle-like cellulose crystals of 10–20 nm in width and several nanometres in length (Fig. 10.1a), sourced from various biological sources (e.g. bleached wood pulp, cotton manila, tunicin and bacteria), often by strong acid hydrolysis (Table 10.1). Acid treatments remove the nanocellulose components and most amorphous cellulose, and produce highly purified cellulose crystals, making nanocellulose highly crystalline [4].

CNFs form long, flexible fibre networks with a fibril diameter similar to or larger than CNCs (Fig. 10.1b). CNFs can be produced through a combination of both mechanical treatment and chemical pre-treatment by TEMPO-mediated oxidation (2,6,6,-tetramethylpiperidine-1-oxyl radical) with a multi-pass high-pressure homogenisation to introduce carboxyl groups into the CNF to facilitate their dispersion, or through enzymatic hydrolysis or direct mechanical fibrillation [5]. CNFs contain both crystalline and amorphous regions [4].

Static and stirred culture are the two most common ways to produce BC through the use of microorganisms. Static culture results in the accumulation of a thick, leather-like white BC pellicle at the air-liquid interface, while for the stirred culture the cellulose is produced in the dispersed medium, forming irregular pellets or suspended fibres [6]. Nanocellulose produced from bacteria usually has a diameter of 2–4 nm, which then aggregates in the form of ribbon- shaped microfibrils of 80 × 4 nm. The overlapping and intertwisted cellulose ribbons form a non-woven mat with very high water content (Fig. 10.1c).

Table 10.1 Nanocellulose materials.

Type of nanocellulose	Synonyms	Typical sources	Formation and average size
Cellulose nanofibrils (CNFs)	• microfibrillated cellulose • nanofibrils cellulose • microfibrils cellulose • nanofibrillated	wood, sugar beet, potato tuber, hemp, flax	delamination of wood pulp by mechanical pressure before and/or after chemical or enzymatic treatment diameter: 5–60 nm length: several micrometres
Nanocrystalline cellulose (CNC)	• cellulose nanocrystals • crystallite cellulose • whiskers/ rodlike • cellulose microcrystals	wood, cotton, hemp, flax, wheat straw, mulberry bark, ramie, Avicel, tunicin, cellulose from algae and bacteria	acid hydrolysis of cellulose from many sources diameter: 5–70 nm length: 100–250 nm (from plant celluloses); 100 nm to several micrometres (from celluloses of tunicates, algae, bacteria)
Bacterial nanocellulose (BNC)	• bacterial cellulose • microbial cellulose • biocellulose	low molecular weight sugars and alcohols	bacterial synthesis diameter: 20–100 nm; different types of nanofibre networks

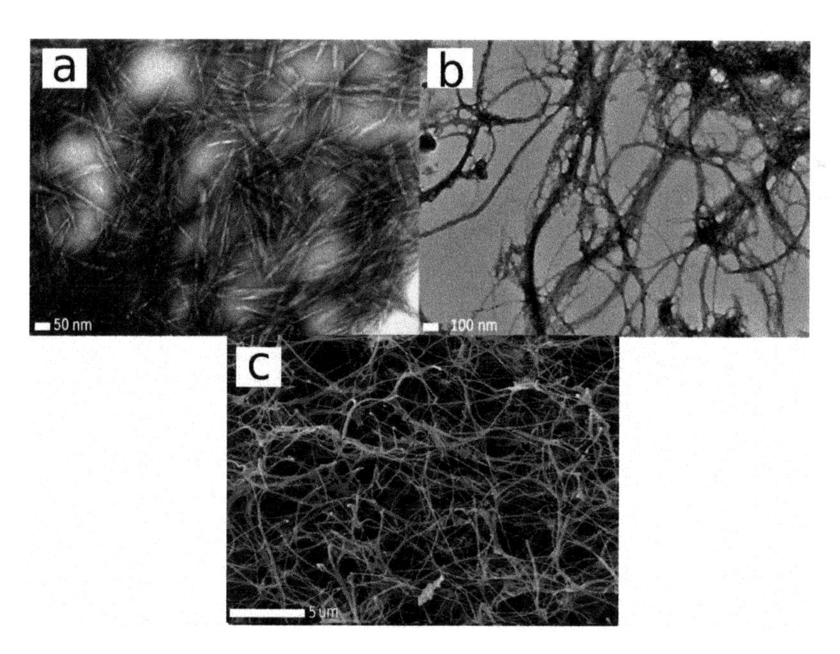

Figure 10.1 Geometry of nanocellulose: (a) CNC; (b) CNF; and (c) BC.

Nanocellulose is highly valuable for producing nanocomposites due to its beneficial attributes, such as its sustainability, abundance, good mechanical properties with a Young's modulus of 130 GPa, a density of around 1.5–1.6 g cm^{-3} and a specific Young's modulus of around 65 J g^{-1} for microfibrils and 85 J g^{-1} for nanocrystals [2]. It also has high tensile strength and stiffness, high flexibility, and good electrical and thermal properties. Nanocellulose is an environmentally friendly material [7–8].

10.1.2 Inorganic matrix

10.1.2.1 Alginate

Alginates are structural polysaccharides extracted from brown algae. At a molecular level alginates are composed of guluronic (G) and mannuronic (M) acid units, forming regions of M-blocks, G-blocks, and blocks of alternating sequence (MG-blocks). The relative proportion of these sequential organisations depends on the source [9]. Alginate can form a thermally stable and biocompatible hydrogel in the presence of di- or tri-cations. Alginate is of interest as a potential biopolymer film component because of its unique colloidal properties, which include thickening, stabilising, suspending, film-forming, gel-producing and emulsion-stabilising [10].

10.1.2.2 Clay

Clay layers can be formed from one tetrahedral sheet linked to one octahedral sheet, known as a one to one (1:1) layer, such as kaolinite. Clay systems have cations between each sheet to balance the surface charge on the layers. This structure requires no counter ions within the structure due to the octahedral sheets balancing the charge on the surface of the tetrahedral sheets. The clay layers system can also be formed from one octahedral sheet between two tetrahedral sheets, known as a two to one (2:1) layer, such as smectite micas, vermiculite and chlorites. These minerals have cations between each sheet to balance the surface charge on the layers. The best-known clay minerals of the 2:1 layer type are the smectite group, and are widely used in various branches of industry due to their high cation exchange capacity, swelling ability and high surface area [11–12]. Physical properties, such as particle size, shape, and distribution, can define the clay minerals and their application, alongside important properties that include surface chemistry, area and charge. The properties of clay materials, such as viscosity, absorption, plasticity, dry and fired strength, casting rate, permeability, and bond strength, are closely related to process used and to the end products.

10.1.2.3 Graphene oxide (GO)

Carbon atoms can be arranged in a single layer organised in a crystalline structure of hexagonal cells to form graphene. All graphite materials are based on graphene, such as graphite, which is a three-dimensional carbon-based material made up of millions of layers of graphene. Graphene oxide (GO) is also a derivative of graphene, characterised by a carbon monolayer sheet of graphene with the inclusion of sp^3 hybridised carbons whereby the surface is functionalised by oxidation [13]. Due to these functionalities, GO is easily dispersible in water and other organic solvents, making it a very useful material for electronic devices for thermal management, for transparent conductive electrodes and as a matrix to improve the properties of composites [14].

10.1.3 Organic matrix

10.1.3.1 Polyvinyl alcohols (PVA)

Polyvinyl alcohols (PVA) are synthetic polymers used in various commercial applications, such as medicine and food, due to their unique chemical and physical properties [15]. Combining PVA with other biodegradable polymers, such as polysaccharides and biodegradable synthetic polyesters, may result in resourceful biodegradable composites. Additionally, PVA has been used in the production of low-cost environmentally friendly compatible and versatile composites, ranging from sugar cane, starch, clay, carbon nanotubes, wood fibres, cement and organ ceramics. These composites are used in a variety of forms and industrial applications, including in the fibre and textiles industries for sizing and finishing, coatings, adhesives, emulsifiers, and colloidal stabilisers [16].

10.1.3.2 Poly(*N*-isopropylacrylamide) (PNIPAAm)

Thermo-responsive PNIPAAm has excellent swelling and mechanical properties, and most importantly exhibits a thermo-responsive, lower critical solution temperature (LCST) behaviour, at which inverse solubility upon heating and a critical transition from hydrophilic to hydrophobic takes place at a specific temperature, losing about 90% of its volume [17]. Since PNIPA expels its liquid content at a temperature near that of the human body, PNIPA has been investigated by many researchers for possible applications in tissue engineering and controlled drug delivery.

10.1.3.3 Soy protein

Soy protein is a by-product of the soy oil industry and currently utilized in applications, such as animal feed. Its valorisation as a food supplement is currently being pursued. Both the growing demand for soy oil and the concurrent anticipated availability of these inexpensive, residual SPs have stimulated interest in developing new functional materials. This is mainly because beyond its abundance and nutritional value, soy protein possesses valuable properties, such as biodegradability and biocompatibility. It also displays a multiplicity of chemical functionality, amphoteric behaviour and pH responsiveness [18–19].

10.1.3.4 Lignin

Lignin is a natural amorphous polymeric material, considered a waste material or by-product of pulping. Lignin is one of the most abundant and renewable materials. Lignin is an extremely complex three-dimensional polymer formed by radical coupling polymerisation of *p*-hydroxycinnamyl, coniferyl and sinapyl alcohols. These three lignin precursor monolignols give rise to the so-called *p*-hydroxyphenyl (H), guaiacyl (G) and syringyl (S) phenylpropanoid units, which manifest in different proportions in lignin from different groups of vascular plants, as well as in different plant tissues and cell wall layers [20]. However, lignin has yet to reach its full potential, and has not been fully utilised. Currently, it is mostly used as fuel to fire the pulping boilers. Current studies have been focused on exploring its value-added applications, such as enhancing the mechanical properties of polymeric composites due to their phenolic base structure. Lignin has recently been exploited as a cheap raw material in the production of organic aerogels [21].

10.1.3.5 Polylactic acid (PLA)

Poly(lactic acid) or polylactic acid or polylactide (PLA) is a biodegradable and bioactive thermoplastic aliphatic polyester derived from renewable resources, such as corn starch, cassava roots, chips or starch or sugarcane. PLA is an aliphatic polyester made up of lactic acid (2-hydroxy propionic acid) building blocks [22]. PLA is prepared by first condensing the lactic acid and removing water continuously, leading to low molecular weight PLA. This process has several problems, including the need for a high temperature, continuous removal of by-products and long reaction times. Current large-scale producers have progressed to ring-opening polymerisation of lactide (cyclic dimer of lactic acid) promoted by protic compounds as initiators with tin(II) octoate ($Sn(Oct)_2$) as a catalyst, to achieve high molecular weight PLA in bulk [23–24]. Being able to degrade into innocuous lactic acid, PLA is used for medical implants in the form of anchors, screws, plates, pins and rods. As a mesh, PLA can be used as a decomposable packaging material, either cast, injection-moulded or spun. It is useful for producing loose-fill packaging, compost bags, food packaging and disposable tableware. In the form of fibres and non-woven fabrics, PLA also has many potential uses, such as upholstery, disposable garments, awnings, feminine hygiene products and diapers. PLA as a matrix for nanocellulose aerogel is yet to be developed.

10.1.3.6 PHA

Polyhydroxyalkanoate (PHA) is a polymer belonging to the class of polyesters that are of interest as bio-derived and biodegradable plastics. It can be produced from various microorganisms, such as bacteria. PHA is composed of 3-hydroxy fatty acid monomers, which form linear, head-to-tail polyester. PHA is typically produced as a polymer of 103–104 monomers, which accumulate as inclusions of 0.2–0.5 μm in diameter [25]. PHA has noticeable ecological advantages since it is non-toxic and 100% bio-based, and is biodegradable even in cold sea water. PHA can be completely biodegraded within a year by a variety of microorganisms. This biodegradation results in carbon dioxide and water, which return to the environment [26–27]. Although, PHA has advantages over conventional plastics, widespread application of PHAs is hampered by high production costs.

10.1.3.7 Starch

Starch is one of the most abundant and low-cost natural polymers on earth and is the main storage supply in botanical resources (cereals, legumes and tubers). Starch has been used in many industrial sectors, such as food, paper, textile and adhesives [28]. Chemically, starch is composed of two polymers with repeating α-d-glucopyranosyl units. These substances are amylase and amylopectin, a linear and a highly-branched polysaccharide respectively [29]. The latter, in general, results in the brittleness of starch-based materials, limiting its further application. Efforts have been made to use starch as a matrix for aerogel development.

10.1.4 Cross-linkers

10.1.4.1 Glutaraldehyde

Glutaraldehyde, a linear, 5-carbon dialdehyde, is a clear, colourless to pale-straw-coloured, pungent oily liquid that is soluble in water and alcohol, as well as in organic solvents. It is mainly available as acidic aqueous solutions (pH 3.0–4.0), ranging in concentration from less than 2% to 70% (w/v). Glutaraldehyde is a commercial and low-cost product in addition to having high reactivity. It reacts rapidly with amine groups at around neutral pH and is more efficient than other aldehydes in generating thermally and chemically stable crosslinks [30]. It has been used a crosslinking agent in a widespread number of fields and has proven to be an effective cross-linker in polymer composites.

10.1.4.2 1,2,3,4-butanetetracarboxylic acid (BTCA)

Polycarboxylic acids have shown high levels of effectiveness for crosslinking cotton, when sodium hypophosphite (NaH_2PO_2) is used as a catalyst. Polycarboxylic acids have also been used as crosslinking agents for wood pulp cellulose to improve paper wet strength. Among the various effective polycarboxylic acids, 1,2,3,4-butanetetracarboxylic acid (BTCA) has proven to be the most effective as a crosslinking agent for cotton fabrics. BTCA can be synthesised by two different methods. The first is to subject the Diels-Alder reaction product of maleic anhydride and 1,3 butadiene to hydrolysis followed by oxidative cleavage. The second method is the electrolytic hydrodimersation of dialkyl maleate followed by hydrolysis of the hydrodimerisation product [31].

10.1.4.3 Polyamide-epichlorohydrin (PAE)

Cationic polyamide-epichlorohydrin (PAE) resins are prepared by reaction of epichlorohydrin with polyamides derived from adipic acid and diethylenetriamine. PAE resins are widely used as wet-strength additives for paper. PAE is the type of polymer used most extensively for shrink-resisting [32]. PAE resins function in neutral/alkaline papermaking processes. They have a high level of wet strength, help improve machine efficiency, and do not adversely affect paper absorbency.

10.1.5 Other additives

10.1.5.1 Silver (Ag) nanoparticles

Silver nanoparticles have unique optical, electrical and thermal properties, and are being incorporated into products ranging from photovoltaics to biological and chemical sensors. Examples include conductive inks, pastes and fillers, which utilise silver nanoparticles for their high electrical conductivity, stability and low sintering temperatures. Additional applications include molecular diagnostics and photonic devices, which take advantage of the novel optical properties of these nanomaterials. An increasingly common application is the use of silver nanoparticles for antimicrobial coatings, and many textiles, keyboards, wound dressings and biomedical devices now contain silver nanoparticles that continuously release a low level of silver ions to provide protection against bacteria [33].

10.1.5.2 Metal-organic frameworks

Metal-organic frameworks, or MOFs, have emerged as an extensive class of crystalline material, with ultra-high porosity (up to 90% free volume) and enormous internal surface areas, extending beyond 6000 $m^2 g^{-1}$. These properties, together with the extraordinary degree of variability for both organic and inorganic components of their structures, make MOFs of interest for potential applications in clean energy, most significantly as storage media for gases such as hydrogen and methane, and as high-capacity adsorbents to meet various separation needs [34].

10.2 Aerogel manufacture

The production of aerogels involves a 3-step process. Production begins with the preparation of a porous 'sol–gel' (a rigid body containing continuous solid and liquid networks), followed by the ageing of the hydrogel, usually by soaking it for a period of 1-2 weeks in organic solvents to displace the water inside to strengthen the gel. Finally, the liquid is replaced with air via drying techniques, leaving a very high porous material [35–36]. Fig. 10.2 illustrates the formed 'aerogel' in which the pores filled with liquid are replaced with air.

10.2.1 Gel preparation by sol-gel process

Sols are dispersions of colloidal particles in a liquid. Colloids are solid particles with diameters of 1–100 nm. A gel is an interconnected rigid network with pores of submicrometre dimensions and polymer chains whose average length is greater than a micrometre. The term 'sol-gel' embraces a diversity of substance combinations, which can be classified in three categories: (1) well-ordered lamellar structures; (2) covalent polymeric networks formed through physical aggregation, predominately disordered; and (3) particularly disordered structures [37]. Three approaches are used to make sol-gel monoliths, namely (1) the gelation of a solution of colloidal powders; (2) hydrolysis and polycondensation of alkoxide or nitrate precursors followed by hypercritical drying of gels; and (3) hydrolysis and polycondensation of alkoxide precursors followed by ageing and drying under ambient atmospheres [38]. The sol-gel

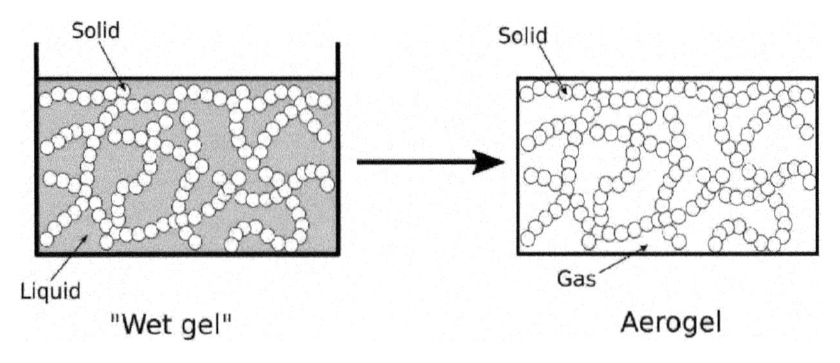

Figure 10.2 Drying process: from a gel (suspension) to a porous solid.

of cellulose aerogel is produced in relation with the drying techniques as well as the desired properties for the intended application.

10.2.2 Ageing of the gel

The 'ageing' process has a great influence on the microstructure, porosity, surface area, pore size and volume shrinkage of the aerogel. The period of ageing has a significant effect on the properties of aerogels, resulting in increased optical transmittance, density and surface area, and hence increased strength and stiffness [39]. The ageing process of cellulose sol-gels usually takes place in acetone or methanol with an average gelation time between 1 and 15 days.

10.2.3 Supercritical drying technologies

In order to remove the solvent from the pores and replace the solvent with air after ageing (Fig. 10.2), the solvent is put into a supercritical state, where the pressure and temperature are raised above their critical point, and therefore there is no liquid/gas interface in the pores during drying. The 'wet gel' is placed into an autoclave and the temperature is slowly raised, thus causing a rise in pressure until the critical point (TC, PC) of the corresponding solvent is reached [40]. Supercritical drying can be divided into two types: (1) high-temperature supercritical drying (HTSCD); and (2) low-temperature supercritical drying (LTSCD). HTSCD may be considered inefficient when using solvents such as water, as their critical temperature and pressure are around 374 °C and 22 MPa respectively [41], which means a high level of energy would be required to reach this state and it could thus be considered a dangerous process, especially when dealing with solvents such as acetone or alcohol that are flammable at high temperatures and pressures. An alternative route is the use of liquid carbon dioxide as it has the advantage of a low critical temperature; however, the time required for exchanging the original pore liquid for liquid CO_2 is determined by the diffusion of carbon dioxide into the solid network and is therefore dependent on the total dimensions of the pores. Another requirement is the miscibility of the pore liquid with carbon dioxide. For example, water and CO_2 are immiscible and therefore an intermediate solvent exchange (e.g. water for acetone) is necessary [42]. CO_2 is harmful, and this method may therefore result in the controversial issue of CO_2 emission when used in large-scale production. However, a modification may be carried out to such a system where the CO_2 is recycled.

10.2.4 Freeze-drying

Freeze-drying processes have been widely used in the food and pharmaceutical industries due to their safety and their minimal effect on both the product and the environment. The fundamental principal of freeze-drying is sublimation: the shift from a solid directly into a gas in order to keep the structure intact. This process may be divided into two steps (Fig. 10.3): first, the solvent is frozen (mostly via liquid nitrogen) to quickly entrap the colloidal particles, and the system is then subjected to high vacuum and progressively increased heat, bringing about sublimation at an optimised rate. Heat transfer and mass transfer occur simultaneously [43–44].

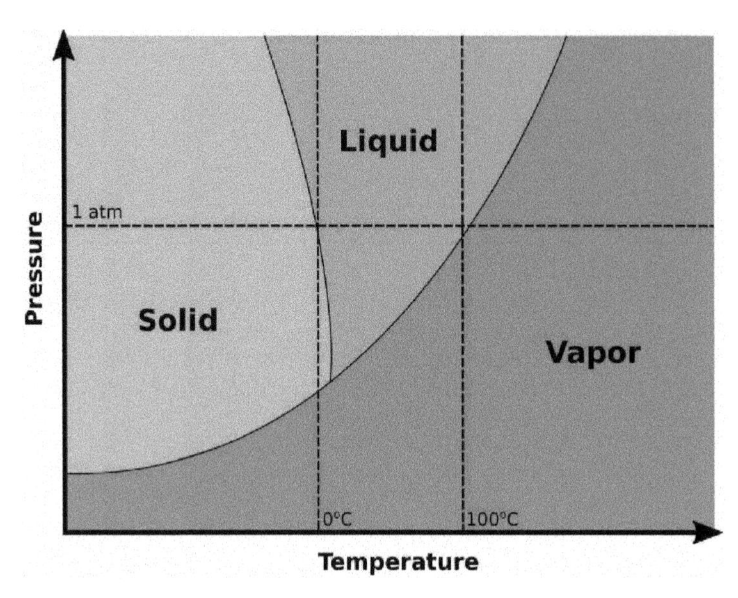

Figure 10.3 Phase change diagram.

The pressure and temperature are the two major factors that determine what phase a substance will take: for instance, water can take liquid form at sea level, where the pressure is equal to 1 atmosphere, and when the temperature is between the freezing (0 °C) and boiling point (100 °C). However, at an atmospheric pressure of less than 0.6 atm. and maintaining the temperature above 0°C, the water is warm enough to thaw but the pressure is not sufficient for the formation of a liquid and thus it becomes a gas. The freeze-drying process controls this procedure.

10.2.5 Ambient pressure drying

Ambient pressure drying (APD) has become attractive as a cheaper alternative process to both HTSCD and LTSCD, which are considered relatively expensive processes. APD involves surface chemical modification, rendering the network hydrophobic, and effectively terminating further interaction between the solid structure and the solvent usually, through the implementation of a silylating agent. Silylation is carried out in the water phase of the gel, which involves the replacement of Si–OH groups by hydrolytically stable Si–R (R-alkyl) groups through oxygen bonding, thereby resulting in a hydrophobic aerogel [45]. The gels are then treated at ambient temperature and drying takes place due to capillary forces. However, when the liquid phase within the gel starts to evaporate, the chemically inert surface silyl groups prevent further infusion of liquid. Realisation of the finished product takes about 4–7 seven days and requires enormous amount of solvent and material. Therefore, it can be costly for large-scale production of aerogels. Further modification and research are required for this method to be implemented efficiently [39].

10.3 Performance and application of nanocellulose aerogel

10.3.1 Aerogel for absorbent

Cellulose aerogel composites have proven to be effective in oil absorption. Microfibrillated cellulose can be used to prepare super-hydrophobic microfibrillated cellulose aerogels through a freeze-drying process, followed by a silanisation reaction that is capable of efficiently absorbing oil and organic pollutants. The aerogel is characterised with a water contact angle of up to 151.8°, with oil absorption up to 159 g g^{-1} and 260 g g^{-1} for chloroform. The microfibrillated aerogels also exhibit a superior re-usability, of at least 30 times, revealing their great potential application in oil/water separation materials [46]. Hydrophobic, flexible and ultra-lightweight nanocellulose aerogels prepared from an aqueous suspension of nanofibrillated cellulose (NFC) containing methyltrimethoxysila (MTMS) as a silylating agent have been fabricated through a freeze-drying process. The developed aerogels are capable of high mass absorption capacities toward a wide variety of organic solvents and oils, up to as much as 100 times their own weight (up to 102 g g^{-1}) and with a water contact angle of up to 130°. They can be used to absorb spilled oil from the surface of water without sinking. The aerogel has a density of $\leq$ 17.3 mg cm^{-3} and can maintain a 96% recovery of the original dimensions after a 50% compression strain [47].

Another approach is to use NFC to prepare aerogels, rendering hydrophobic of a cellulose through a vapour-phase deposition of octyltrichlorosilanes, which gives an aerogel with a contact angle for water of 150° and ability to absorb 45 times its own weight in non-polar liquids [48]. Through a freeze-drying process an aerogel composite with NFC and TiO_2 or TiO_2/SiO_2 has also been developed. When this aerogel is subject to calcination to bring about thermal decomposition, it converts to a ceramic aerogel that is capable of a high absorption efficiency of organic molecules with photocatalytic activity. The ceramic materials showed a high adsorption capacity toward methylene blue and rhodamine B. In particular, specimens containing SiO_2 are effective in removing such molecules from their aqueous solutions [49]. Carbon has also been used along with NFC: the carbon–NFC aerogel achieved approximately 90 g g^{-1} of the normalised oil absorption capacity despite a weight reduction percentage of 82% [50].

10.3.2 Aerogel for insulation

Aerogels of high porosity and with a large internal surface area exhibit outstanding performance as insulators. Using a supercritical drying technique, aerogels with a three-dimensionally ordered nanofibre skeleton of liquid-crystalline nanocellulose (LC-NCell) have been prepared and characterised with a transparent nature, good mechanical toughness and low insulation properties, reaching a thermal conductivity of 0.018 W m^{-1} K^{-1} at a density of 0.017 g cm^{-3} [51]. Bio-based aerogels, prepared through a freeze-process by combining bleached cellulose fibres (BCF) and cellulose nanofibres, have also demonstrated very useful thermal conductivity and mechanical properties. The interest in bio-aerogel insulator encourages the use of bleached cellulose fibres (BCF), which can achieve a thermal conductivity of as low as 0.028 m^{-1} K^{-1} and which can be further reduced to 0.023 m^{-1} K^{-1} by combining both the bleached cellulose fibres and the cellulose nanofibres, as cellulose nanoparticles have a low crystalline index and high surface charge [52].

Nanofibrillated cellulose aerogels can be prepared either by a freeze-drying or a spray freeze-drying process. Their structural, mechanical and thermal insulation properties are comparable to those of NFC aerogels prepared by conventional freeze-drying (CFD). However, by altering the drying method and using spray freeze-drying (SFD), the microstructure can be altered from a two-dimensional sheet-like morphology with macropores to a three-dimensional fibril-nanostructured skeleton morphology with pore size ranging from a few tens of nanometres to a few microns. Reduction of the aerogel pore size and skeleton architecture have been observed (by SEM), which consequently resulted in a lower thermal conductivity of 0.018 m^{-1} K^{-1} [53].

10.3.3 Supercapacitors

Supercapacitors are electrical energy storage devices that provide high capacitance, high power and good cyclability [54]. All-solid-state flexible supercapacitors are fabricated using CNFs/MWCNTs (multiwalled carbon nanotubes) hybrid aerogel film as the electrode material and charge collector. One-dimensional CNFs can effectively prevent the aggregation of MWCNTs, significantly enhance re-wettability, and improve the utilisation efficiency of the mesopores. All-solid-state flexible supercapacitors with CNFs/MWCNTs hybrid aerogel films exhibit excellent electrochemical properties, with the specific capacitance around 178 F g^{-1}. The flexible supercapacitors also exhibit excellent cyclic stability [55]. Crosslinked cellulose nanocrystals (CNCs) and multiwalled carbon nanotubes (MWCNTs) aerogels have been polymerised with polypyrrole. Mechanical robustness, flexibility and low impedance of the current collectors are achieved by the chemical crosslinking of CNC aerogels and efficient dispersion of MWCNTs. The advanced electrode design results in a low contact resistance. A single-electrode areal capacitance of 2.1 F cm^{-2} is obtained at an active mass loading of 17.8 mg cm^{-2} and an active material to current collector mass ratio of 0.57. Large area electrochemical super-capacitor (ES) electrodes and devices show flexibility, excellent compression stability at 80% compression, and electrochemical cyclic stability over 5000 cycles [56]. Cellulose nanocrystal aerogels are also used with in situ incorporation of polypyrrole nanofibres, polypyrrole-coated carbon nanotubes and manganese dioxide nanoparticles to prepare aerogels, and flexible 3D supercapacitor aerogels with excellent capacitance retention, low internal resistance and fast charge-discharge rates [57].

10.4 Outlook

Nanocellulose Cellulose aerogels are highly versatile and adaptive materials with the ability to meet the demands of many applications. Three different types of nanocellulose are currently used, namely cellulose nanocrystals (CNCs), cellulose nanofibrils (CNFs) and bacterial nanocellulose (BC), with each having different physical and mechanical properties. There is a wide range of matrices that can be used to prepare nanocellulose aerogel composites, ranging from organic – such as aliginate, clay and graphene oxide – to inorganic – such as polyvinyl alcohols (PVA), soy protein, lignin and polylactic acid, along with a range of effective cross-linkers and other additives. Nanocellulose aerogel composites are mainly produced through high-temperature supercritical drying (HTSCD) and low-temperature supercritical drying (LTSCD), through freeze-dying (FD) techniques, or, finally, ambient pressure drying (APD). Leading applications range from effective oil and pollutant absorbents to superior thermal insulation and supercapacitors. Many significant breakthroughs have been made regarding the raw materials, processing technologies, new composites and applications of nanocellulose. However, their potential is being further explored and their commercial viability remains yet to be fully realised.

References

[1] Dufresne A. 2013. *Nanocellulose: From Nature to High Performance Tailored Materials.* Berlin: Walter de Gruyter.

[2] Klemm D, Kramer F, Moritz S, Lindstrom T, Ankerfors M, Gray D & Dorris A. 2011. Nanocelluloses: A new family of nature-based materials. *Angewandte Chemie* (International Edition), 50 (24), pp. 5438–5466.

[3] Mahfoudhi N & Boufi S. 2017. Nanocellulose as a novel nanostructured adsorbent for environmental remediation: A review. *Cellulose*, pp. 1–27.

[4] Xu X, Liu F, Jiamg L, Zhu J, Haagenson D & Wiesenborn D P. 2013. Cellulose nanocrystals vs. cellulose nanofibrils: A comparative study on their microstructures and effects as polymer reinforcing agents. *ACS Applied Materials & Interfaces*, 5 (8), pp. 2999–3009.

[5] Bordes R, Van de Ven & Theo G M. 2017. Nanocellulose: What used to be cellulose micelles.

[6] Jozala A F, Delencastre-Novaes L C, Lopes A M, De Carvalho Santos-Ebinuma, V, Mazzpola, P G, Pessoa Jr A, Grotto D, Gereutti M & Chaud M V. 2016. Bacterial nanocellulose production and applications: A 10–year overview. *Applied Microbiology and Biotechnology*, 100 (5), pp. 2063–2072.

[7] Borjesson M & Westman G. 2015. Crystalline nanocellulose: Preparation, modification, and properties. In M. Poletto (ed.), *Cellulose: Fundamental Aspects and Current Trends* (pp. ??–??). London: IntechOpen.

[8] Charreau H L, Foresti M & Vazquez A. 2013. Nanocellulose patents trends: A comprehensive review on patents on cellulose nanocrystals, microfibrillated and bacterial cellulose. *Recent Patents on Nanotechnology*, 7 (1), pp. 56–80.

[9] Silva M A D, Bierhalz A C K & Kieckbusch T G. 2009. Alginate and pectin composite films crosslinked with Ca^{2+} ions: Effect of the plasticizer concentration.

[10] Huq T. Salmieri S, Khan A, Khan RA, Le Tien C, Riedl B, Fraschini C, Bouchard J, Uribecalderon J, Kamal M R & Laroix M. 2012. Nanocrystalline cellulose (NCC) reinforced alginate based biodegradable nanocomposite film.

[11] Bujdak J & Rode B M. 1999. The effect of clay structure on peptide bond formation catalysis. *Journal of Molecular Catalysis A: Chemical*, 144 (1), pp. 129–136.

[12] Pethrick R A. 2002. *Polymer–Clay Nanocomposites*, edited by T J Pinnavaia and G W Beall Wiley 2000 *Polymer International*, 51(5), pp. 464–464.

[13] Marcano D C, Kosynkin D V, Berlin J M, Sinitskii A, Sun Z, Slesarev A, Alemany L B, Lu W & Tour J M. 2010. Improved synthesis of graphene oxide.

[14] Zhu Y, Murali S, Cai W, Li X, Suk J W, Potts J R & Ruoff R S. 2010. Graphene and graphene oxide: Synthesis, properties, and applications. *Advanced Materials*, 22 (35), pp. 3906–3924.

[15] Demerlis C & Schoneker D. 2003. Review of the oral toxicity of polyvinyl alcohol (PVA). *Food and Chemical Toxicology*, 41 (3), pp. 319–326.

[16] Podsiadlo P, Kaushik A K, Arruda E M, Waas A M, Shim B S, Xu J, Nandivada H, Pumplin B G, Lahann J, Ramamoorthy A & Kotov N A. 2007. Ultrastrong and stiff layered polymer nanocomposites. *Science*, 318 (5847), pp. 80–83.

[17] Bandi S A. 2006. High performance blends and composites. Part I: Clay aerogel/polymer composites. Part II: Mechanistic investigation of colour generation in pet/med6 barrier blends. Case Western Reserve University.

[18] Arboleda J C, Hughes M, Lucia L A, Laine J, Ekman K & Rojas O J. 2013. Soy proteinnanocellulose composite aerogels. *Cellulose*, 20 (5), pp. 2417–2426.

[19] Endres J G. 2001. *Soy Protein Products: Characteristics, Nutritional Aspects, and Utilization.* The American Oil Chemists Society. Chicago, IL: AOCS Press.

[20] Ghaffar S H & Fan M. 2013. Structural analysis for lignin characteristics in biomass straw. *Biomass and Bioenergy*, 57, pp. 264–279.

[21] Madyan O A, Fan M, Feo L, Hui D. 2016. Enhancing mechanical properties of clay aerogel composites: An overview. *Composites Part B: Engineering*, 98, pp. 314–329.

[22] Ummartyotin S & Manuspiya H. 2015. A critical review on cellulose: From fundamental to an approach on sensor technology. *Renewable and Sustainable Energy Reviews*, 41, pp. 402–412.

[23] Pang X, Zhuang X, Tang Z & Chen X. 2010. Polylactic acid (PLA): Research, development and industrialization. *Biotechnology Journal*, 5 (11), pp. 1125–1136.

[24] Sosnowski S, Gadzinowski M & Slomkowski S. 1996. Poly (l, l-lactide) microspheres by ring-opening polymerization. *Macromolecules*, 29 (13), pp. 4556–4564.

[25] Reddy C, Ghai R & Kalia V. 2003. Polyhydroxyalkanoates: An overview. *Bioresource Technology*, 87 (2), pp. 137–146.

[26] Madkour M H, Heinrich D, Alghamdi M A, Shabbaj II & Steinbuchel A. 2013. PHA recovery from biomass. *Biomacromolecules*, 14 (9), pp. 2963–2972.

[27] Yu J & Chen L X. 2008. The greenhouse gas emissions and fossil energy requirement of bioplastics from cradle to gate of a biomass refinery. *Environmental Science & Technology*, 42 (18), pp. 6961–6966.

[28] Averous L. 2004. Biodegradable multiphase systems based on plasticized starch: A review. *Journal of Macromolecular Science, Part C: Polymer Reviews*, 44 (3), pp. 231–274.

[29] Imberty A, Buleon A, Tran V & Peerez S. 1991. Recent advances in knowledge of starch structure. *Starch-Stärke*, 43 (10), pp. 375–384.

[30] Migneault I, Dartiguenave C, Bertraand M J & Waldron K C. 2004. Glutaraldehyde: Behavior in aqueous solution, reaction with proteins, and application to enzyme crosslinking. *BioTechniques*, 37 (5), pp. 790–806.

[31] Yang C Q, Lu Y & Lickfield G C. 2002. Chemical analysis of 1, 2, 3, 4-butanetetracarboxylic acid. *Textile Research Journal*, 72 (9), pp. 817–824.

[32] Guise G & Smith G. 1985. The chemistry of a polyamide-epichlorohydrin resin (hercosett 125) used to shrink-resist wool. *Journal of Applied Polymer Science*, 30 (10), pp. 4099–4111.

[33] Oldenburg S J. 2014. Silver nanoparticles: Properties and applications. Sigma-Aldrich Co., nd.

[34] Zhou H, Long J R & Yaghi O M. 2012. Introduction to metal-organic frameworks,.

[35] Bryning M B, Milkie D E, Islam M F, Hough L A, Kikkawa J M & Yodh A G. 2007. Carbon nanotube aerogels. *Advanced Materials*, 19(5), pp. 661–664.

[36] Husing N & Schubert U. 1998. Aerogels – airy materials: Chemistry, structure, and properties. *Angewandte Chemie* International Edition, 37 (1–2), pp. 22–45.

[37] Flory P J. 1953. *Principles of Polymer Chemistry*. Ithaca, NY: Cornell University Press.

[38] Hench L L & West J K. 1990. The sol-gel process. *Chemical Reviews*, 90 (1), pp. 33–72.

[39] Sachithanadam M & Joshi S C. 2016. Silica Aerogel Composites.

[40] Gurav J L, Jung I, Park H, Kang E S & Nadargi D Y. 2010. Silica aerogel: Synthesis and applications. *Journal of Nanomaterials*, 2010, pp. 23.

[41] Yunus C A & Cengel A M. 2002. Boles A. Micahel Termodinámica.

[42] Peterlik H, Rennhofer H, Torma V, Bauer U, Puchberger M, Husing N, Bernstorff S & Schubert U. 2007. Structural investigation of alumina silica mixed oxide gels prepared from organically modified precursors. *Journal of Non-Crystalline Solids*, 353 (16), pp. 1635–1644.

[43] Ratti C. 2001. Hot air and freeze-drying of high-value foods: A review. *Journal of Food Engineering*, 49 (4), pp. 311–319.

[44] Wang Y, Gawryla M D & Schiraldi D A. 2013. Effects of freezing conditions on the morphology and mechanical properties of clay and polymer/clay aerogels. *Journal of Applied Polymer Science*, 129 (3), pp. 1637–1641.

[45] Rao K S, El-Hami K, Kodaki T, Matsushige K & Makino K. 2005. A novel method for synthesis of silica nanoparticles. *Journal of Colloid and Interface Science*, 289 (1), pp. 125–131.

[46] Zhou S, Liu P, Wang M, Zhao H, Yang J & Xu F. 2016. Sustainable, reusable, and superhydrophobic aerogels from microfibrillated cellulose for highly effective oil/water separation. *ACS Sustainable Chemistry & Engineering*, 4 (12), pp. 6409–6416.

[47] Zhang Z, Sebe G, Rentsch D, Zimmermann T & Tingaut P. 2014. Ultralightweight and flexible silylated nanocellulose sponges for the selective removal of oil from water. *Chemistry of Materials*, 26 (8), pp. 2659–2668.

[48] Cervin N T, Aulin C, Larsson P T & Wagberg L. 2012. Ultra-porous nanocellulose aerogels as separation medium for mixtures of oil/water liquids. *Cellulose*, 19 (2), pp. 401–410.

[49] Melone L, Altomare L, Alfieri, I, Lorenzi A, De Nardo L & Punta C. 2013. Ceramic aerogels from TEMPO-oxidized cellulose nanofibre templates: Synthesis, characterization, and photocatalytic properties. *Journal of Photochemistry and Photobiology A: Chemistry*, 261, pp. 53–60.

[50] Meng Y, Wang X, Wu Z, Wang S, Young TM. 2015. Optimization of cellulose nanofibrils carbon aerogel fabrication using response surface methodology. *European Polymer Journal*, 73, pp. 137–148.

[51] Kobayashi Y, Saito T, Isogai A. 2014. Aerogels with 3D ordered nanofiber skeletons of liquid-crystalline nanocellulose derivatives as tough and transparent insulators. *Angewandte Chemie*, 126 (39), pp. 10562–10565.

[52] Seantier B, Bendahou D, Bendahou A, Grohens Y & Kaddami H. 2016. Multi-scale cellulose based new bio-aerogel composites with thermal super-insulating and tunable mechanical properties. *Carbohydrate Polymers*, 138, pp. 335–348.

[53] Jimenez-Saelices C, Seantier B, Cathala B, Grohens Y. 2017. Spray freeze-dried nanofibrillated cellulose aerogels with thermal superinsulating properties. *Carbohydrate Polymers*, 157, pp. 105–113.

[54] Zhang L L & Zhao X. 2009. Carbon-based materials as supercapacitor electrodes. *Chemical Society Reviews*, 38 (9), pp. 2520–2531.

[55] Gao K, Shao Z, Wang X, Zhang Y, Wang W & Wang F. 2013. Cellulose nanofibers/multiwalled carbon nanotube nanohybrid aerogel for all-solid-state flexible supercapacitors. *RSC Advances*, 3 (35), pp. 15058–15064.

[56] Shi K, Yang X, Cranston E D & Zhitomirsky I. 2016. Efficient lightweight supercapacitor with compression stability. *Advanced Functional Materials*, 26 (35), pp. 6437–6445.

[57] Yang X, Shi K, Zhitomirsky I & Cranston E D. 2015. Cellulose nanocrystal aerogels as universal 3D lightweight substrates for supercapacitor materials. *Advanced Materials*, 27 (40), pp. 6104–6109.

Chapter 11
Nanocellulose-based composites

Hassan Ahmad and Professor Mizi Fan

The chapter begins with a discussion of the availability of nanocellulose (NC), whether through top-down or bottom-up approaches, and its properties. A number of processing techniques that have been adopted over the years in the making of different types of NC composites are then introduced. Finally, the chapter presents applications and performance of cellulose nanocomposites in construction, including a number of notable scenarios of nanocellulose incorporation.

11.1 Introduction

The extraction of nano-sized cellulose from biomass and its incorporation into different nanocomposites has the potential to offer enhanced renewable alternatives to many presently used non-renewable materials. The favourable material properties of nanocellulose include high aspect ratios, light weight, biodegradability, sustainability, low cost, and a surface rich in hydroxyl groups that facilitates functionalisation with different reagents to achieve desired surface properties as well as low densities of 1.6 g cm^{-3}. These material properties of nanocellulose have piqued the interests of researchers over the last decade. Currently, the main application of nanocellulose is to be found in its incorporation in polymer matrices to create nanocomposites with enhanced material properties.

11.2 Types and properties of nanocellulose

11.2.1 Nanocellulose via the top-down approach

The type of cellulose extracted varies depending on the material source and the approach to isolation. A top-down approach encompasses the disintegration of lignocellulosic biomass (plant materials), the most abundantly obtainable raw material, and dates back to 1946 when Wuhrmann *et al.* are said to have used sonication methods to break down and separate natural cellulose fibres into finer fibrils, while maintaining the fibrous texture [1]. Contemporary top-down methods toward producing nanocellulose may include the use of grinders to grind wood pulp using high shear fibrillation [2–4]. Nanocellulose produced through such an approach is termed nanofibrillated cellulose (NFC), where it is broken down from the parent microfibrillated cellulose (MFC). However, there are some misconceptions in the literature with regard to distinguishing MFC from NFC fibres and this has somewhat broadened the aspect ratio range, producing overlap between the two. The high aspect ratio of NFC is considered to be at 500–2000 nm in length with a width of 4–20 nm, whereas the aspect ratio of MFC is about 500–10000 nm in length and a width of 50–400 nm. MFC fibres can

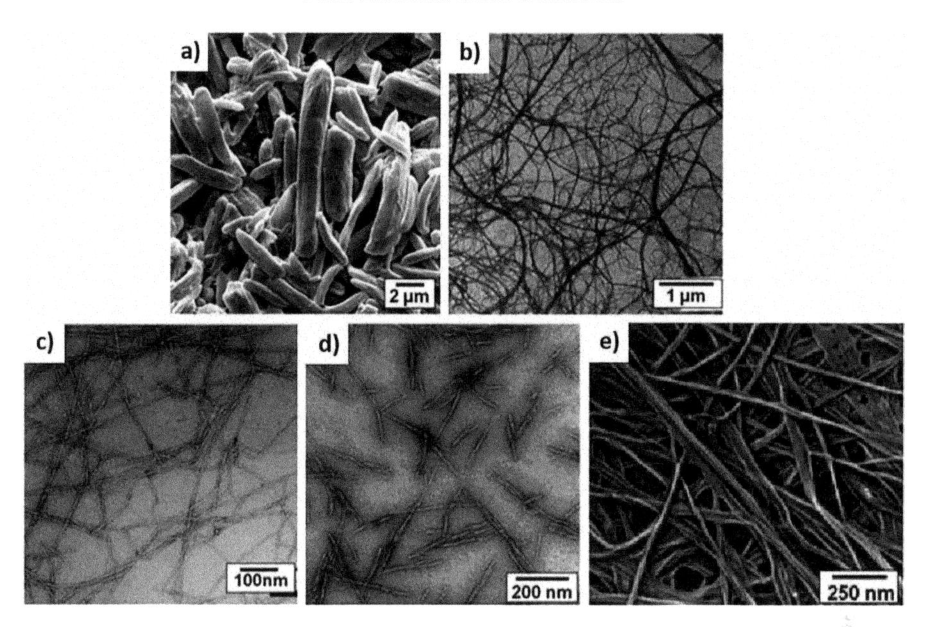

Figure 11.1 Visual comparison via SEM/TEM images of (a) MCC; (b) MFC; (c) NFC; (d) NCC; (e) BC.

be broken down from microcrystalline cellulose (MCC), where its aspect ratio is in the range 10–50 µm in length with a width of 10–50 µm; therefore you can expect the length of MFC to reach a length of up to 50 µm. Cellulose nanocrystals (NCC) with lengths of 50–500 nm and widths of 3–5 nm can also be produced through breaking down NFC fibres, which can be done by further mechanical or chemical procedures (Fig. 11.1) [5–8]. Typically, NCC is achieved through acid hydrolysis of cellulosic fibres. NCC has high crystallinity with hydroxyl accessibility on its surface, giving rise to further functionalisation through surface modification.

11.2.2 Nanocellulose via the bottom-up approach

The bottom-up approach towards producing nanocellulose involves biosynthesis through cellulose-producing bacteria such as from the *Gluconacetobacter* genus. The nanocellulose produced in this way is called bacterial cellulose (BC) and features pure cellulose fibrils without any lignin or pectin content and possesses lateral dimensions between 25 nm and 85 nm, with lengths up to several micrometres (Fig. 11.1). While the dimensions of BC and NFC are similar, differences in the fibrils mainly comprise the purity and crystallinity of the material. Whereas BC is pure cellulose, NFC is often a composite composed of cellulose and hemicellulose and is based on plant cell wall microfibrils. The two types of cellulose also possess different proportions of I_a and I_b crystal structures. BC consists of a high proportion of I_a (one chain triclinic unit cell crystal structure), while NFC is mainly composed of I_b (monoclinic unit cell). The two crystal structures coexist and form the intrinsic structure of natural cellulose.

Bacterial cellulose may be produced by several species of bacteria including (and principally) that of *Gluconacetobacter xylinus* (formerly known as *Acetobacter xylinum*), which is a highly pervasive species that is found in fermentation processes involving sugars and plant carbohydrates. *Gluconacetobacter xylinus* has the ability to use several sugars to synthesise cellulose, of which glucose is commonly used in the laboratory whereby a four-enzymatic step process has been characterised in the conversion. One of the key issues in the commercial production of BC is that a significant portion of the glucose supplied to *Gluconacetobacter xylinus* is converted into gluconic acid, an undesired by-product, while around 50% of glucose is converted into cellulose. *Gluconacetobacter xylinus* can produce nanocellulose with diameters between 10 nm and 30 nm. The bacterium can be guided to produce a network of fibrils of nearly any architecture.

11.2.3 Properties of nanocellulose

In view of the different approaches to producing nanocellulose, there are numerous processing techniques that can be adopted to obtain the desired nanocellulose for the purpose desired (Fig. 11.2). These processing techniques may also complement each other to feature different structural properties, to increase functionality, or even to further reduce the energy consumption in manufacturing nanocellulose. For example, enzymatic treatment could be used before mechanical processing (such as high-pressure homogenisation), so as to reduce energy consumption by up to 15 times. One may therefore expect that the properties of nanocellulose will vary with the approach.

The mechanical properties of nanocellulose have been determined using AFM and spectroscopic techniques. However, the properties measured may be influenced by a number of factors, which must be considered before comparing the properties of different samples. These include crystallinity, defects, the measurement techniques used, crystal structure and anisotropy. A summary of the mechanical properties of some cellulose types is given in Table 11.1. The difficulty associated with testing such small samples can be seen to have had an impact on the results, with a high probability of errors.

The elastic properties of nanocellulose have been the focus of research exploring the mechanical properties of nanocellulose. Differences in the properties are observed for NCC due to its anisotropic nature, causing different properties to be measured at different directions. The elastic properties of cellulose I crystals have been examined both through theoretical estimates and experimentally. The latter has commonly involved measurements using x-ray diffraction (XRD) techniques coupled with in situ tensile tests. These tests involve loading bulk specimens of parallel microfibrils axially and measuring the strain produced using XRD before calculating the elastic modulus in the axial direction (E_A). This allows the measurement of the crystalline properties because XRD only measures the displacement in the crystalline regions, although the method assumes a seamless load transfer and orientation of cellulose crystals within fibrils along the loading axis. These assumptions are unlikely to be true and can thus lead to a lower result for the modulus than is actually the case.

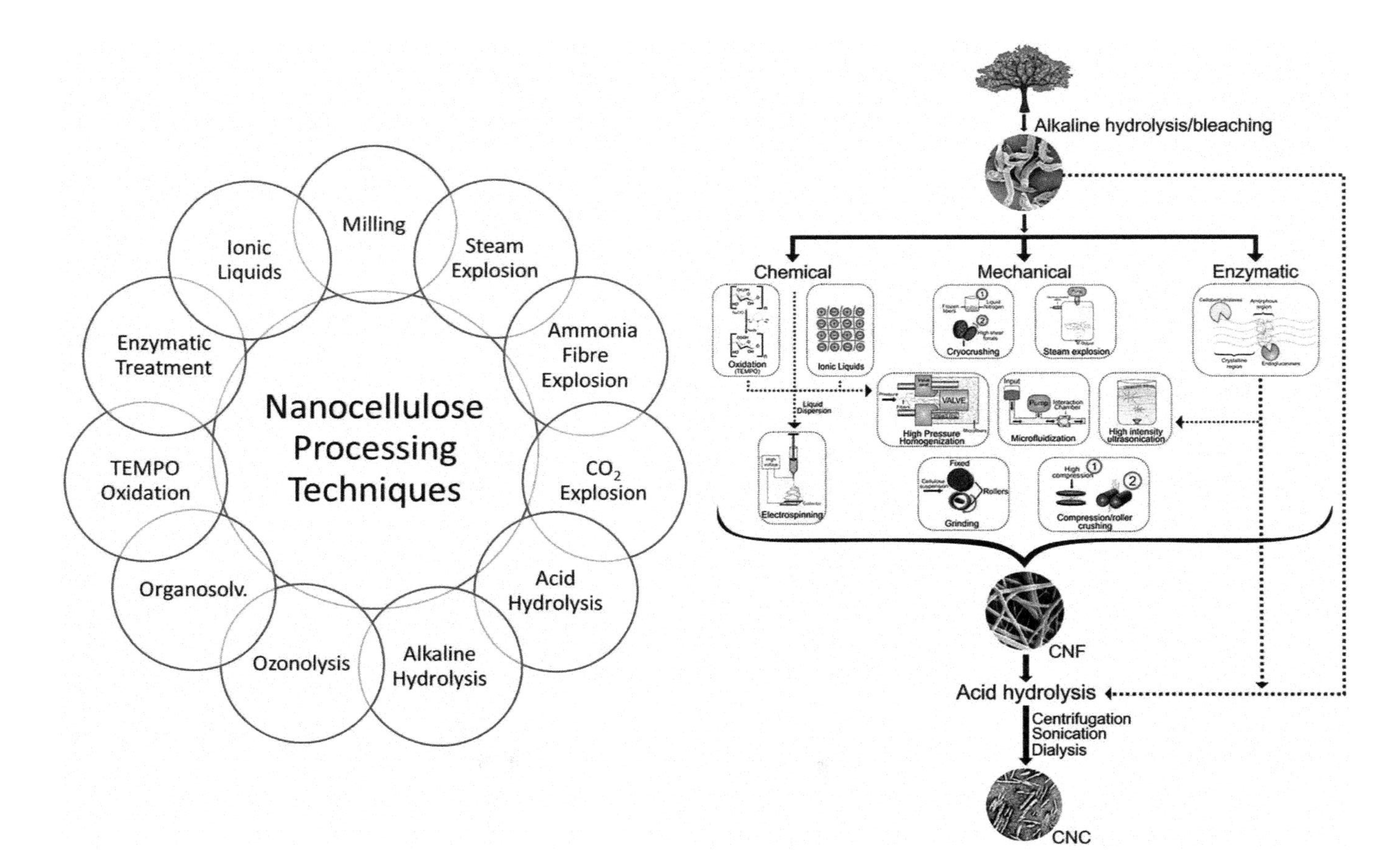

Figure 11.2 Various processing routes to achieving nanocellulose particles [9].

Table 11.1 Mechanical properties of cellulose particles [8].

Material	E_A/E_T (GPa)	Instrument used	Ref
BC	78 ± 17; 114 (E_A)	AFM 3-pt bend; Raman	[10, 11]
MCC	25 ± 4 (E_A)	Raman	[12]
NCC – plant	57, 105 (E_A)	Raman	[13]
t-NCC	143 (E_A)	Raman	[14]
t-NCC – acid	151 ± 29 (E_A)	AFM 3-pt bend	[15]
t-NCC – TEMPO	145 ± 31 (E_A) / 9 ± 3 (E_T)	AFM 3-pt bend; Indentation	[15, 16]
NCC – wood	18 – 50 (E_T)	AFM indentation	[17]

Note: E_A = modulus of elasticity in axial direction; E_T = modulus of elasticity in transverse direction

A different method used to measure axial elasticity, which overcomes these concerns surrounding XRD, involves the use of inelastic X-ray scattering to measure the sound velocity as a function of phonon dispersion within the crystalline portions. This results in much larger elasticity than that given by XRD, with E_A = 220 ± 50 GPa for the X-ray scattering method as compared to E_A = 120 – 138 GPa for the XRD method. It also allows measurement of the elastic modulus in the transverse direction (E_T), which has been found to be E_T = 15 ± 1 GPa.

NCC, which is the crystalline particles that have been isolated from the microfibrils, may possess different properties due to the influence of the extraction process on the particles. NCC derived from plant material is very small in size and individual NCC particles cannot currently be tested for elastic modulus. However, among the natural sources of cellulose, there exists a single known marine animal species that can produce cellulose in its outer tissues. This cellulose is structured in the form of microfibrils similar to plant cellulose, with almost pure monoclinic structured beta-cellulose (I_β) where the particles have longer lengths, greater cross-sections, uniform morphology and higher crystallinity. This allows tunicate NCC, unlike plant-based NCC, to be tested mechanically for the elastic modulus of the individual crystals, which has been found to be E_A = 151 ± 29 GPa for particles extracted using acid hydrolysis and E_A = 145 ± 31 GPa for those extracted using a method called (2,2,6,6-Tetramethylpiperidin-1-yl)oxyl/oxidanyl (TEMPO) mediated oxidation. Atomic force microscopy (AFM) was used to perform a 3-point bending test to determine these values. These measurements are subject to errors associated with the difficulty in calculating the cross-sections of the samples, which is used to determine the modulus. Raman spectroscopy and in situ tensile tests have also been used to determine the axial elastic modulus of tunicate NCC, which was found to be E_A = 143 GPa.

The transverse elastic modulus of individual wood NCC and tunicate NCC has been determined using high-resolution AFM indentation and modelling, which was found to be E_T = 18 – 50 GPa and E_T = 9 ± 3 GPa, respectively. However, these results are subject to large uncertainties due to modelling assumptions and AFM sensitivity limits. The elastic modulus for BC was determined using the same methods as those used for tunicate NCC, which is possible because of the similar properties, including large length, high crystallinity and large cross-section. An axial elastic modulus of

$E_A = 78 \pm 17$ GPa was found when using the AFM 3-point bending test, and $E_A = 114$ GPa using Raman spectroscopy, with the divergence likely being a result of the use of assumptions and difficulties in measuring the properties due to the small nature of the particles [8]. The coefficient of thermal expansion of NCC in the axial direction has been estimated at ~0.1 ppm K^{-1}, which is similar to other anisotropic fibres with a high modulus, like carbon fibre. The thermal chemical degradation of nanocellulose usually occurs at between 200 °C and 300 °C, depending on the type of cellulose, e.g. ~260 °C for freeze-dried samples of NCCs. This degradation temperature also depends on other factors, including surface modification and heating rate [8].

Nanocellulose displays liquid-crystalline behaviour because of the asymmetric rod structure. NCC may exhibit nematic behaviour due to the stiff rod-like structure, where the rods align in certain environments. The interaction between individual crystals is strong, although they easily disperse in solution and exhibit lyotropic phase behaviour as a result. The lyotropic phase is a phase transition from an isotropic liquid into a liquid crystal, which occurs by altering the concentration. The crystals possess a helical twist that results in a chiral nematic structure with a cholesteric phase across multiple planes, with each plane rotated at a phase angle along a perpendicular axis. This causes optical band gaps and creates a fingerprint texture in NCC suspensions. In addition, when nanocellulose is exposed to a magnetic field, its long axis becomes perpendicular to the magnetic field direction and the distance between the rods in a nematic plane is longer than that between nematic planes across the cholesteric axis. This characteristic was explored further, whereby nanocellulose-based composites were found to possess much stronger mechanical properties in the direction perpendicular to an applied magnetic field relative to its parallel counterpart.

The aspect ratio of nanocellulose is a major factor in the liquid crystallinity exhibited as higher aspect ratios, which cause increased anisotropy, resulting in increased liquid crystallinity at lower concentrations. Although, at copious lengths, the stiffness of the particles may be overcome, causing the rod-like structures to behave more like strings thereby eliminating any ordering of the suspension.

Ionic strength is also a major factor, whereby an increase in ionic strength can lead to less ordering, a higher critical lyotropic concentration, a reduction in separation and thus the agglomeration of the particles. The inclusion of an electrolyte in the suspension reduces the repulsion between the particles and decreases the effective diameter. Furthermore, nanocellulose optical properties differ from those of other cellulose materials because of a number of features, including the nanoscale size, the anisotropic morphology, and the ability to display lyotropic liquid-crystalline behaviour depending on the structure and conditions.

NCCs in suspension may develop a chiral nematic phase at a certain concentration whereby the suspension may be evaporated to attain a semi-translucent film that retains the crystal structure formed in the suspension. New applications for NCC films may be adopted, as they display particular optical properties, including iridescence (which is defined by the refractive index) as well as the chiral nematic pitch. NCC films can be red-shifted using ultrasound treatment, whereby the chiral nematic pitch is increased in the suspension with increasing energy. The iridescence colours may be decided from the spectrum, ranging from blue-violet to red, by tuning the sonication treatment and by electrolyte addition, with the effects being permanent. NCCs are also birefringent, with a refractive index of 1.618 in the axial direction and 1.544 in the transverse.

11.3 Production of cellulose nanocomposites

The performance and properties of composites rely on the constituents, i.e. the matrix and reinforcement used, and how their strengths or interfacial mechanisms interact. The property of the composite is also affected by the processing technique used to create the composite structure. The dilemma with the preparation of NC as a reinforcement is the desire to avoid agglomeration, whilst maximising the reinforcing ability of NC composites via the formation of a network structure. Primitive processing procedures need to be adopted to achieve closer proximities of a homogeneous NC dispersion, thereby enhancing composite performance significantly as well as achieving more reproducible results. This also reduces the effective NC particle sizes and thereby widens the interfacial area between the NC and the matrix. These procedures could be adopted through mechanical processing techniques or through chemical means, e.g. enhanced biocompatibility and dispersion could be adopted using organic matrices, by functionalising the surface of NC. However, as this interferes with the reinforcing ability of NC, modification may be unfavourable and thus a water-soluble polymer matrix may be considered, due to the intrinsic hydrophilic nature of NC, to maximise the reinforcement mechanical property.

A number of processing techniques have been adopted over the years in the making of various types of NC composite. Processing techniques such as solution-casting, melt-compounding, partial dissolution and electrospinning have been used to create polymer matrix composites reinforced with NC [18–29]. Here, some of these techniques will be discussed.

The 'solution-casting' processing method is usually used in the making of composite films or aerogels and is dependent on the mixing process. Generally, this technique involves dispersing a known quantity of NC in a specified solvent, e.g. water. NC loadings normally range between 0 and 10 wt% in solution, though this can be varied according to specific applications. Both the NC and polymer solutions are then mixed together under vacuum or ultra-sonication to minimise the presence of air bubbles. The suspension is ordinarily stirred at room temperature, though it can be increased to treat the polymer used and occasionally to enhance the NC/polymer interfacial bonding. The composition of both constituents is subject to property prerequisites and thus the wt% ratio of NC/polymer composites may differ by as high 0/100 and vice versa [30, 31]. These mixtures then undergo different processing techniques to produce the desired composite. For example, if thin films are the desired composite to be produced, the generic process that the mixture undergoes is as follows. The mixture is initially poured into a suitable mould, which is then either evaporated, lyophilised and compression-moulded, or it is first lyophilised, then extruded and finally compression-moulded. Suitable moulds that may be selected include ones made from glass, Teflon or PP so that minimal damage is caused to the film when detaching it from the mould. To strengthen the NC and polymer interactions, cross-linkers are commonly also added as reinforcement agents [32–37].

Alternatively, if aerogels or solid foams are the desired composite to be produced, the mixture can undergo a variety of procedures. One of these procedures might be to rapid freeze the suspension once it is poured into the desired mould. Lyophilisation is used to freeze-dry the frozen mould, extracting out the water in ice form, to achieve the final lightweight aerogel composite. Rapid freezing could include the use of liquid nitrogen for increased control over the direction of freezing, thus establishing a more uniform-layered aerogel composite. Other typical processing techniques can be seen in Fig. 11.3.

Another way of incorporating NC within composites is through the melt-compounding method, whereby the reinforcement, in this case NC, is thermal-mechanically mixed/compounded into thermoplastic polymers followed by extruding the melt suspension. For some applications, further processing may be implemented, whereby compression moulding is used directly after extrusion [19, 22, 23, 28]. In view of this processing technique, the reinforcing ability of NC would be affected by the high temperatures and shear pressures that may be required. Therefore, these parameters may well need to be adjusted/controlled to suit the NC reinforcement, which is otherwise prone to degradation. This processing method also requires some investigation of the dispersion quality of NC in the thermoplastic used. By way of example, the use of NC within polylactic acid has been examined [38, 39].

11.4 Applications of nanocellulose and its polymer nanocomposites

Nanocellulose is a class of material that represents the quest for developing green technology through nanotechnology. Its desirable material properties, along with its minimal environmental impact, makes it ideal in the pursuit to replace environmentally damaging materials with those that cause less harm. It has received much attention as a cost effective and green nanoscale filler alternative to materials, including glass fibre, silica and carbon black, in the reinforcement of polymer composites.

Nanocellulose was initially used to reinforce such polymers as polypropylene and polyethylene before their reinforcing potential became apparent. They have been reported to enhance the properties of a range of different polymer matrices including thermoplastic and thermoset polymers. These polymers include poly(butylene adipate), poly(vinyl chloride), polylactic acid, chitosan and starch among others. There are four primary methods of creating a nanocellulose- reinforced polymer composite including solution-casting, melt-compounding, electrospinning and partial dissolution. The properties of the nanocomposite may depend on several factors, such as the aspect ratio, nanocellulose dispersion homogeneity, the processing method used, interfacial bonding, the type of matrix and the percolation threshold. Presently, NCCs have attracted more attention as a reinforcement than NFCs owing to the desirable properties they confer, such as their excellent mechanical characteristics, optical properties, surface chemistry and crystallinity. The addition of NCC as a reinforcement has been proven to enhance the reinforcing capacity and mechanical properties of various composites for different industrial applications. Some notable examples of nanocellulose incorporation in polymer matrices (i.e. thermoplastics/thermosets/bio-matrices) as well as the property enhancements of the developed composites are given in Table 11.2.

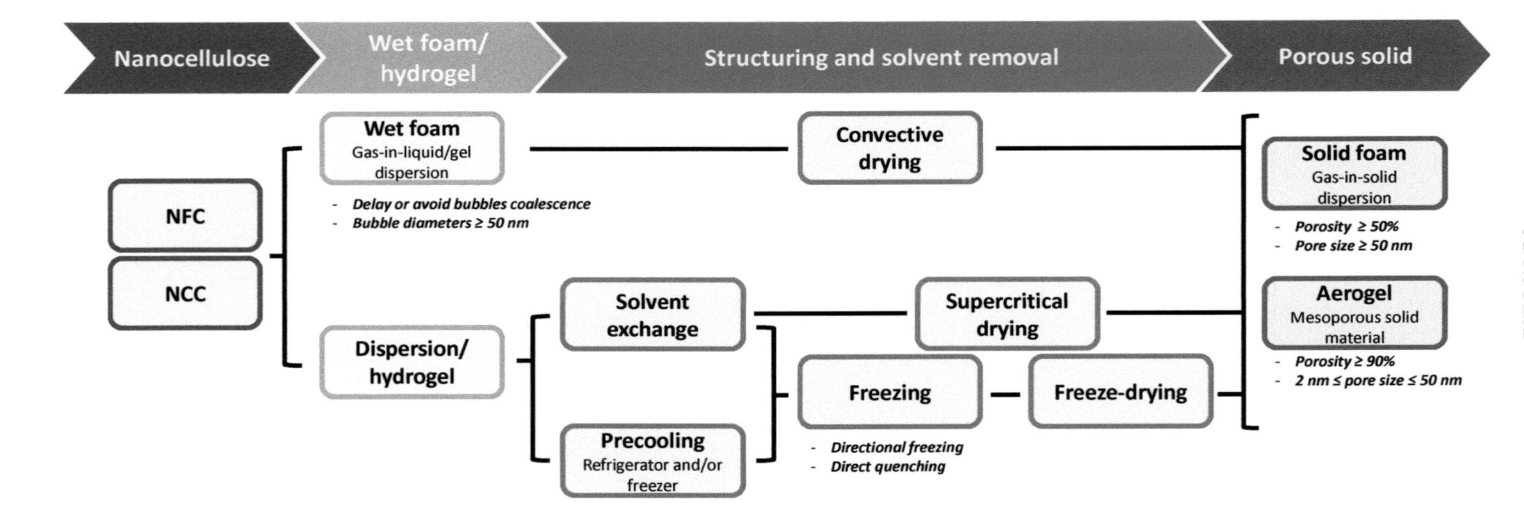

Figure 11.3 Processing routes to achieving aerogels and foams from nanocellulose [37].

Table 11.2 Property enhancements of several nanocellulose-based polymer nanocomposites [40].

NC/polymer composite	Property enhancement	Ref
NCCs/WBPU	Incorporation of NCCs improved the thermomechanical stability and hydrophilicity of developed nanocomposites through the solvent casting technique. It also increases moduli and stress at yield by 3 wt% NCCs loading	[41]
NFCs/Biobased epoxy composites (LCM process)	Fabricated nanocomposites showed improved mechanical and tribological properties with respect to pure epoxy composites	[42]
NC-sugarcane bagasse/semi-IPN of poly(vinyl alcohol)/polyacrylamide	Remarkable improvement in tensile strength, % elongation at break, modulus and toughness of composite films by adding 5 wt% of nanocellulose	[43]
NC-oat husk/WPI	WPI-based nanocomposites containing 5% oat-NC has the best tensile strength, Young's modulus, solubility along with the lowest elongation at break and moisture content. But increasing the content (7.5 wt%) of oat-NC leads to a decrease in the mechanical properties due to growing filler agglomeration	[44]
3-Aminopropyl triethoxysilane silanised NFCs and dodecanoyl chloride esterified NCCs- blue agave bagasse/PLA	Treated NFCs and NCCs addition at 0.5 wt% show better mechanical properties, by more than 20%, than PLA composites	[45]
TEMPO-oxidised NFCs/ PVA and Ppy	Oxidised NFCs incorporation improved the antibacterial activity of composites films, found in food	[46]
NC-wood pulp/graphene/ PVA	Developed hybrid nanocomposite (PVA/NFC/rGO) films show improved mechanical as well as conductivity properties with effective humidity sensors	[47]

(Continued)

Table 11.2 Property enhancements of several nanocellulose-based polymer nanocomposites [40]. (Continued)

NC/polymer composite	Property enhancement	Ref
NC-raw jute fibres/NR	Improved and better morphology, XRD and tensile strength indicate a strong interaction between filler and NR. The rate of biodegradation by vermicomposting is comparatively higher in non-crosslinked composites than in their crosslinked counterparts	[48]
NC-sugarcane bagasse/ PVA (linear and crosslinked state)	TGA indicates higher thermal stability of nanocomposites NC/crosslinked PVA with respect to NC/linear PVA. Tensile strength increases at 5 wt% and 7.5 wt% NC addition of crosslinked PVA and linear PVA, respectively	[49]
NCCs/PLA	TGA shows that decomposition temperature increases by NCCs incorporation at (5 wt%) of PLA/NCCs nanocomposites compared to pure PLA composites	[50]
NC-jute fibres/NR latex	Considerable improvements in Young's modulus and tensile strength of the nanocomposite were observed by adding NC in the NR latex as a matrix	[51]
NFCs/PCL/epoxy	Both healing efficiency (by 26%) and mechanical properties (tensile strength, elongation at break, and impact strength were improved by around 27%, 38%, and 38%, respectively) improved by the homogenous dispersion and bridging effect of added 0.2 wt % of NFCs to the polymer matrix. The glass transition temperature (T_g) of epoxy also increased by 12.8 °C	[52]
NFCs-dry cellulose waste of softwood (*Pinus* sp.) and hardwood (*Eucalyptus* sp.)/UPR	Dynamic mechanical properties and thermal stability improved by the incorporation of NFCs particles	[53]

(*Continued*)

Table 11.2 Property enhancements of several nanocellulose-based polymer nanocomposites [40]. (Continued)

NC/polymer composite	Property enhancement	Ref
NFCs-bleached kraft eucalyptus fibres/ biobased epoxy	Low fracture toughness at interfaces observed for the developed NFCs/bio-based resin composites. Porosity also increased with increasing nanofibre content	[54]
NCC-wood/waterborne epoxy	Wood-NCC/waterborne epoxy nanocomposites show improved storage modulus, loss modulus, tensile strength, Young's modulus and T_g with increasing NCC content	[55]
NCCs-RS, WS, and BS/ CMC	Tensile strength increased by 45.7%, 25.2%, and 42.6%, and the water vapour permeability decreased by 26.3%, 19.1%, and 20.4% by adding 5 wt% of NCCs obtained from RS, WS, and BS, respectively	[56]
NFCs-wood/UPR	T_g increases substantially with NFC content. Modulus and strength of UPR increase about three times at 45 vol% NFCs, whereas ductility and apparent fracture toughness are doubled	[57]
m-CNWs-castor oil-based polyol/ BPU	Incorporation of stiffer and rigid m-CNWs increases the tensile strength and modulus of BPU composites compared with the BPU composites. DMA results showed increased in storage modulus and loss tangent peak shifted toward higher temperatures by incorporation of m-CNWs	[58]

WBPU = waterborne polyurethanes; NFC = nanofibrillated cellulose; LCM = liquid composite moulding; semi-PIN = semi-interpenetrating polymer network; WPI = whey protein isolate; PLA = polylactic acid; PVA = polyvinyl alcohol; PPy = polypyrrole; NR = natural rubber; NCC = nanocrystalline cellulose; NC = nanocellulose; MCC = microcrystalline cellulose; PCL = poly(ε-caprolactone); NFC = nanofibrillated cellulose; UPR = unsaturated polyester resin; RS = rice straw; WS = wheat straw; BS = barley straw; CMC = carboxymethyl cellulose; BPU = bio-based polyurethane; m-CNWs = modified cellulose nanowhiskers.

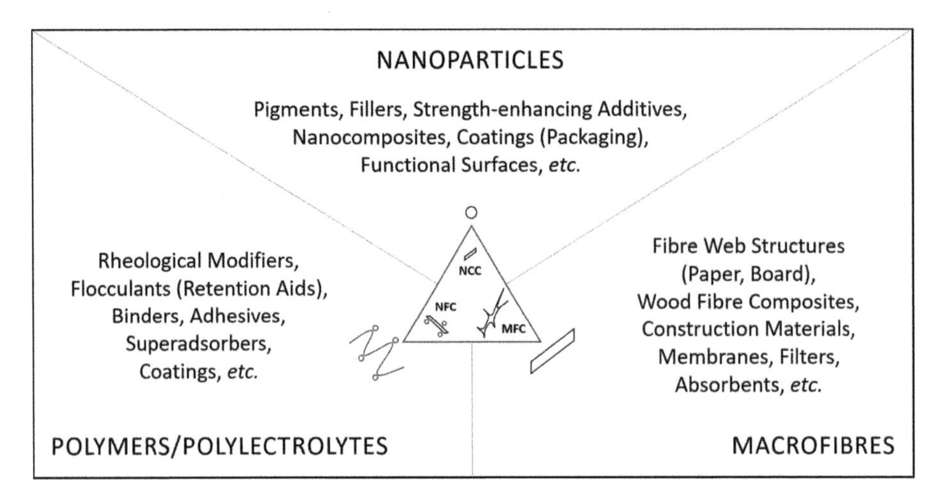

Figure 11.4 Potential application areas of different types of nanocelluloses based on their specific structure–properties relationships [77].

Nanocellulose refers to a class of nanomaterials obtained from natural sources with potential for use in a wide set of applications (Fig. 11.4). Commercialisation of these applications has yet to reach any considerable scale, with the applications normally consisting of high-volume products such as automotive components and construction materials, or low-value products and novel applications. The latter applications may involve photonics, where a chiral mesoporous material can be formed by compositing NCCs with a material of interest [59, 60]. These materials have the potential to be applied as optical filters, chiral plasmonics, soft actuators, antireflective coatings and flexible electronics [61–66]. The liquid crystalline behaviour of NCCs may also be used to create iridescent and optically transparent films and NFCs may be used as films for optoelectronics and as coatings [67–69]. Nanocellulose may also be used as a film or a foam where NCC films have been suggested for a wide variety of applications from electronics to packaging. NCC foams and aerogels could potentially be used as insulation in construction, as supercapacitors, in water filtration and other interesting applications; however, such foams have yet to be investigated as thoroughly as NCC films [70]. Moreover, nanocellulose has been the subject of intense research regarding its use in medical applications due to its excellent biocompatibility and inherent low toxicity [71–74]. BC is generally considered non-toxic, although NCC and NFC may exhibit toxicity depending on the particle size, purity and surface chemistry [71]. Nanocellulose composites have been used to create tissue-engineered scaffolds and wound dressings using BC due to its suitability for cell adhesion and minimal rejection during cellular contact [75, 76].

Nanocellulose is regarded as a renewable alternative to expensive, non-renewable reinforcement fillers and has been used to reinforce polymers, paper and other materials to improve the properties and biodegradability of composites. Nanocellulose may be interwoven to create mechanically strong bulk materials, including paper and films [78, 79]. The global market for nanocellulose has been projected to reach $808.3 million by 2022 due to its potential as an alternative material to synthetic polymers and chemicals that require energy-intensive processes as petroleum costs are projected to increase.

11.4.1 Incorporation and performance of nanocellulose in construction

There is a wide variety of cementitious materials for different industries, such as roads, buildings and waste management, with a large quantity of cement produced annually at around 4 billion metric tons. The desired qualities in these cements for the different industries are usually lower costs, higher durability, increasing sustainability, higher strength and faster setting times [80]. Additives are usually the primary means of achieving these qualities, including fibre reinforcements, chemical admixtures, and supplementary cementitious materials where the use of natural fibres has seen increasing attention in the push towards sustainability and renewable materials [81, 82]. Examples of this may include the use of fibres in non-structural building material such as cladding and fibre cement boards, and functionalised fibres in shrinkage cracking control and internal curing [83, 84]. The use of natural fibres as an alternative to metals for reinforcement is disadvantaged by the reduced mechanical abilities in reinforcement, although they prove favourable in other respects, including the lower costs and sustainability.

Compared to cellulose of other size scales, nanocellulose possesses properties that are more favourable for use in cementitious products, such as the higher surface area to volume ratio and higher stiffness compared to microscale cellulose [8]. Using nanocellulose as an additive in cement has been found to modify the setting times, mechanical properties and rheology of cement, where low volumes of nanocellulose additive were found to increase the strength and stiffness of the composite due to changes in the hydration reactions [85, 86]. This offers the possibility of improving the mechanical performance of cements by perfecting the hydration kinetics. However, thus far, no commercial cementitious product containing nanocellulose has been successfully developed and introduced to the market.

Considering the patent literature regarding nanocellulose applications in cements, most of the literature revolves around the use of nanocellulose as an additive in concrete and cementitious composites (Fig. 11.5a). The incorporation of nanocellulose in fibre-cement boards has been a focus among patents where nanocellulose is added for the purpose of internal curing and using NFC within the composite where an even dispersal of the NC is accomplished [87]. Such composites were found to possess a greater toughness and higher elastic modulus, lower shrinkage during cement setting, and reduced water adsorption [87–89]. Nanocellulose may also be used as an additive to improve concrete and cement properties, which has also been a focus of many patents. The addition of nanocellulose may improve the strength, provide stability to self-compacting concrete (SCC), and modify the viscosity and water retention of different cementitious products [90–93]. It has been found that the use of NCC additives to cement pastes may improve the flexural strength, while NFC has been found to be able to raise the solidification delay time while sustaining good liquidity, which reduces the porosity of the paste by enabling the release of entrapped gas [90, 91]. This led to the enhancement of the mechanical properties of the slurry. Moreover, nanocellulose may be used as a substitute for cellulose ethers in cementitious materials [92].

Nonetheless, nanocellulose must overcome a number of challenges if it is to penetrate the cementitious material industry in any significant way. Currently, the long-term durability of cementitious materials incorporating cellulose products has not been studied and requires further understanding, as cellulose fibres are known to degrade in high alkali conditions. Furthermore, the production of nanocellulose products needs to be increased significantly if it is to become a viable additive and satisfy the potential demand in the cement industry due to the enormous scale of the industry. At present, global annual production output of cement is 4 billion metric tons, while the upper estimates of nanocellulose potential for use in cement are thought to be around 4.1 million metric tons [94, 95]. To put this in perspective, the global production of NCC is presently 650 tons per year and 1700 tons per year for NFC [96]. Standards also need to be developed to quantify the material enhancements that would result as a consequence of using nanocellulose as an additive in cementitious composites, along with supporting metrology, which is difficult to achieve considering the nanoscale size of the cellulose. These developments will accelerate the adoption of nanocellulose in the cement industry.

The addition of nanocellulose in cementitious products is intended to attain a preferred particle packing and achieve desired properties due to the high specific surface area of the nanocellulose, which can promote chemical reactions and act as nuclei within cement to increase the degree of hydration [97]. NFCs show potential as a reinforcement in cementitious products because of their high aspect ratios, tensile strength and moduli of elasticity. However, dispersion procedures need to be established for nanocellulose, and for other nanomaterials to be incorporated, as there is a tendency for nanoparticles to agglomerate, which can cause stress concentrations when loading is applied.

Cement is the primary binder in concrete and its production is responsible for 2–3% of energy consumption and 5% of CO_2 emissions globally [98]. Nanocellulose may improve the sustainability of cementitious materials as an additive whereby the resulting composite shows improved mechanical properties and durability. Nanocellulose may potentially be produced by the paper industry where a high volume of product can be produced at a low-cost using renewable sources. NFC from softwood pulp added to general-use limestone cement at 0.05%, 0.1%, 0.2%, and 0.4% loadings of cement weight was found to increase the setting time and degree of hydration in all loadings. These results were attributed to internal curing of NFC causing an increase in the degree of hydration and an increase in setting time because of alkaline hydrolysis of the cellulose producing organic acids and other products [99]. A 0.1% NFC loading demonstrated an optimum increase in flexural strength with microscopic data showing a lower fibre agglomeration than for higher loadings. NFC produced by high-pressure homogenisation from eucalyptus pulp was incorporated within Portland cement and studied at 0.01%, 0.05%, 0.1%, 0.2%, 0.3% and 0.5% loadings of cement weight [100]. The resulting composite at 0.3% loading showed a desirable strength increase of 43% in comparison with a control sample lacking NFC.

An investigation into the autogenous and drying shrinkage of cement paste for high-performance fibre-reinforced cementitious composites found that NFC does not have the internal curing capacity compared to other larger cellulose fibres because of its nano-sized nature [101], although it was theorised that the porous and hydrophilic nature of NFC may encourage self-healing behaviour within the cementitious composites. NFC produced in laboratory conditions with high numbers of hydroxyl groups (1.85 mmolg^{-1}) and a high-water retention capacity (3700%) was incorporated within Type I cement paste of 0.50 w/c at 0.15 wt% loading of NFC [90]. The resulting compressive strength increased by 20% and flexural strength increased by 15% with an increase in the degree of hydration and a delay in the hydration reaction. The enhancement in the composite properties was attributed to the small size of the NFC facilitating crack-bridging and to the hydroxyl groups increasing the interaction between the NFC and the cement.

NCCs were added to Type V cement paste with 0.04%, 0.1%, 0.2%, 0.5%, 1.0% and 1.5% volume loadings and the resulting composites showed increased DOH with increased loadings [86]. The composite also displayed a 20–30% increase in flexural strength compared to cement paste lacking NCC, with those at 0.2% volume loading showing optimum results. The strength was also found to decrease with increasing loading, which was likely due to increased agglomeration at increased loading.

The dispersion of NCC within a cement composite is an important factor in trying to achieve ideal property enhancements. The critical concentration of NCC in a simulated cement pore solution to achieve the lowest possible yield stress was determined to be 0.18% of volume loading and the use of ultrasonication to improve the dispersion and reduce agglomeration was found to increase the critical concentration to 1.38 vol% [102]. The ultrasonication treatment increased the flexural strength of the composite by up to 50% and energy-dispersive X-ray spectroscopy attributed these results to the increased concentration of NCC at interfacial regions between the cement paste and the particles with increased NCC dispersion in the paste. In addition, nanoindentation tests found higher elastic moduli in cement paste regions with higher NCC concentrations.

11.4.2 Other notable applications of nanocellulose incorporation

11.4.2.1 Insulation

Nanocellulose may also be incorporated into aerogel structures for many different potential applications, including insulation in the construction industry (Fig. 11.5b). Aerogels are materials with very low densities (0.01–0.4 g cm^{-3}), high surface areas (30–600 m^2 g^{-1}), high porosities and low thermal transport [40, 107]. The abundance of surface hydroxyl groups found in nanocellulose and nano-structure parameters/dimensions serves as an ideal building block for functionalising and constructing novel hybrid nanomaterials. Aerogel composites composed of graphene oxide (GO) and TEMPO oxidised NFC have been shown to have superior compressive strength

Figure 11.5 (a) Use of cellulosic fibres as rheological modifiers and stabilisers in cement; (b) Various cellulose-based panels; (c) Cellulosic fibre based composites used in Ford vehicles; (d) Potential use of nanocellulose for packaging [103–106].

than that of individual GO (more than 15-fold) and NFC (more than 5-fold) aerogels. With the addition of polyvinyl alcohol (PVA) and crosslinks, crosslinked PVA/GO/NFC aerogels have been reported to have more than quadruple the strength of GO/NFC aerogels. Nanocellulose degradation at elevated temperatures, and a reduction in mechanical properties at such temperatures, limits its applicability. Nanocellulose usually degrades thermally at temperatures between 200 °C and 300 ±1 °C but can be chemically modified to alter the onset of thermal degradation.

11.4.2.2 Automotive

Nanomaterials including nanocellulose-based materials are being adopted for use in the automotive industry to improve performance and meet regulatory requirements [108]. Thermoplastic matrices used in the industry are being modified with nanocellulose materials whereby plastic composites reinforced by nanocellulose are being applied as dashboards, bumper beams and underbody shields in automotives (Fig. 11.5c). The use of these composites reinforced by nanocellulose in automotives has increased considerably over the last few years due to their low costs and weight, corrosion resistance and design flexibility, with the most common use of these composites being in exterior body panels [109, 110].

11.4.2.3 Packaging

Nanocellulose has also been used in composites for the packaging of different food types, including beverages, frozen food and fresh food (Fig. 11.5d) [111]. This is in addition to their use in pharmaceutical packaging [112, 113]. Ideal packaging materials are those that offer durability, recyclability and non-contaminants. Nanocellulose has been found to improve the gas barrier properties and heat stability when incorporated into polymer composites, which may be used in packaging [8, 114–118].

11.4.2.4 Medical

Nanocellulose shows potential for use as reinforcement in biocomposites because of the performance enhancement that is offered [119]. NFCs incorporated in biodegradable polymers have been found to be truly versatile and have been applied to a variety of medical applications, including tissue-engineered scaffolds, articular cartilage repair, artificial skin, blood bags, urethral catheters, adhesion barriers, among many other (Fig. 11.6) [120]. A report found that reinforcing polyvinyl alcohol with BC

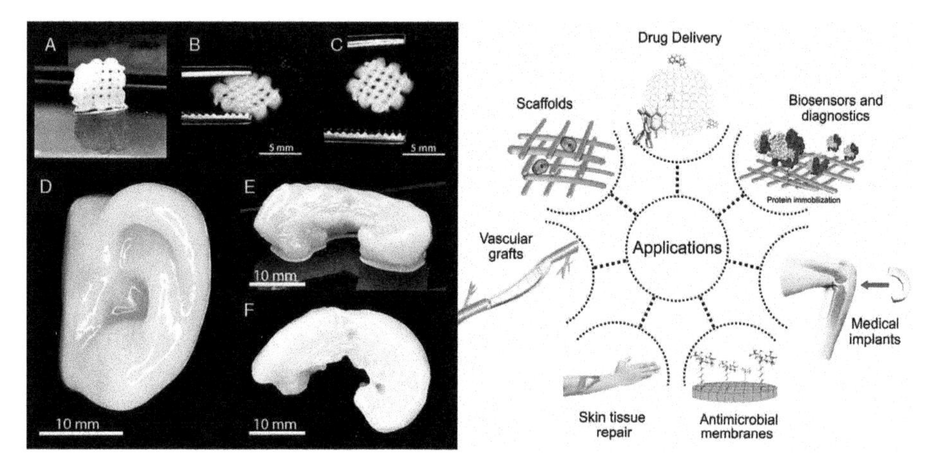

Figure 11.6 3D bioprinted cartilage structures using nanocellulose (left), and other applications where nanocellulose can be incorporated (right) [9, 132].

forms biocompatible nanocomposites similar to that of cardiovascular tissue [121]. Another study found that polyurethane with 5 wt% of nanocellulose increases the strength and stiffness of the composite, while nanocellulose polyurethane vascular grafts have also reportedly been developed [122, 123]. Bionanomaterials, such as those incorporating nanocellulose, have been applied in the development of applications across multiple medical sectors, including the veterinary, dental and pharmaceutical industries [124–126]. They may be applied in drug delivery, wound dressings, medical implants and tissue engineering [127–131].

11.5 Outlook

Nano-sized celluloses from biomass and their incorporation into various nanocomposites hold potential to offer enhanced renewable alternatives to many presently used non-renewable materials. The favourable material properties of nanocellulose include high aspect ratios, light weight, biodegradability, sustainability, low cost and a surface rich in hydroxyl groups that facilitates functionalisation with different reagents to achieve desired surface properties as well as low densities. Numerous processing techniques could be adopted to utilise the desired nanocellulose in the making of various types of NC composites. Processing techniques such as solution-casting, melt-compounding, partial dissolution and electrospinning have been used to create polymer matrix composites reinforced with NC. Nanocellulose and cellulose nanocomposites refer to a class of nanomaterials obtained from natural sources with the potential for use in a wide set of applications. The commercialisation of these applications has yet to reach any considerable scale, with applications normally consisting of high-volume products such as automotive components and construction materials, or low-value products and novel applications, such as photonics, optics, electronics, supercapacitors, and the medical and surface chemistry industries.

References

[1] Wuhrmann, K., Heuberger, A. & Mühlethaler, K. Elektronenmikroskopische Untersuchungen an Zellulosefasern nach Behandlung mit Ultraschall. *Experientia* **2**, 105–107 (1946).

[2] Turbak, A. F., Snyder, F. W. & Sandberg, K. R. Microfibrillated cellulose, a new cellulose product: Properties, uses, and commercial potential. *J. Appl. Polym. Sci. Appl. Polym. Symp.* **37,** 459–94 (1983).

[3] Herrick, F. W., Casebier, R. L., Hamilton, J. K. & Sandberg, K. R. Microfibrillated cellulose: Morphology and accessibility. *J. Appl. Polym. Sci. Appl. Polym. Symp.* **37,** 797–813 (1983).

[4] Iwamoto, S., Nakagaito, A. N., Yano, H. & Nogi, M. Optically transparent composites reinforced with plant fiber-based nanofibers. *Appl. Phys. A* **81,** 1109–1112 (2005).

[5] Dufresne, A., Cavaille, J.-Y. & Vignon, M. R. Mechanical behavior of sheets prepared from sugar beet cellulose microfibrils. *J. Appl. Polym. Sci.* **64,** 1185–1194 (1997).

[6] Saito, T., Kimura, S., Nishiyama, Y. & Isogai, A. Cellulose nanofibers prepared by TEMPO-mediated oxidation of native cellulose. *Biomacromolecules* **8,** 2485–2491 (2007).

[7] Shinsuke Ifuku *et al.* Surface modification of bacterial cellulose nanofibers for property enhancement of optically transparent composites: Dependence on acetyl-group DS. *Biomacromolecules* **8,** 1973–1978 (2007).

[8] Moon, R. J., Martini, A., Nairn, J., Simonsen, J. & Youngblood, J. Cellulose nanomaterials review: Structure, properties and nanocomposites. *Chem. Soc. Rev.* **40,** 3941 (2011).

[9] Rojas, J., Bedoya, M. & Ciro, Y. Current trends in the production of cellulose nanoparticles and nanocomposites for biomedical applications. In *Cellulose - Fundamental Aspects and Current Trends* 194–228 (2015). doi:10.5772/61334

[10] Hsieh, Y.-C., Yano, H., Nogi, M. & Eichhorn, S. J. An estimation of the Young's modulus of bacterial cellulose filaments. *Cellulose* **15,** 507–513 (2008).

[11] Guhados, G., Wan, W. & Jeffrey L., H. Measurement of the elastic modulus of single bacterial cellulose fibers using atomic force microscopy. *Langmuir* **21,** 6642–6646 (2005).

[12] Eichhorn, S. J. & Young, R. J. The Young's modulus of a microcrystalline cellulose. *Cellulose* **8,** 197–207 (2001).

[13] Rusli, R. & Eichhorn, S. J. Determination of the stiffness of cellulose nanowhiskers and the fiber-matrix interface in a nanocomposite using Raman spectroscopy. *Appl. Phys. Lett.* **93,** 033111 (2008).

[14] Šturcová, A., Davies, G. R. & Eichhorn, S. J. Elastic modulus and stress-transfer properties of tunicate cellulose whiskers. *Biomacromolecules* **6,** 1055–1061 (2005).

[15] Iwamoto, S., Kai, W., Isogai, A. & Iwata, T. Elastic modulus of single cellulose microfibrils from tunicate measured by atomic force microscopy. *Biomacromolecules* **10,** 2571–2576 (2009).

[16] Postek, M. T. *et al.* Development of the metrology and imaging of cellulose nanocrystals. *Meas. Sci. Technol.* **22,** 024005 (2011).

[17] Lahiji, R. R. *et al.* Atomic force microscopy characterization of cellulose nanocrystals. *Langmuir* **26,** 4480–4488 (2010).

[18] My Ahmed Said Azizi Samir, Fannie Alloin & Alain Dufresne. Review of recent research into cellulosic whiskers, their properties and their application in nanocomposite field. *Biomacromolecules* **6,** 612–626 (2005).

[19] Hubbe, M. A., Rojas, O. J., Lucia, L. A. & Sain, M. Cellulosic nanocomposites: A review. *BioResources* **3,** 929–980 (2008).

[20] Peresin, M. S., Habibi, Y., Zoppe, J. O., Pawlak, J. J. & Rojas, O. J. Nanofiber composites of polyvinyl alcohol and cellulose nanocrystals: Manufacture and characterization. *Biomacromolecules* **11,** 674–681 (2010).

[21] Olsson, R. T. *et al.* Extraction of microfibrils from bacterial cellulose networks for electrospinning of anisotropic biohybrid fiber yarns. *Macromolecules* **43,** 4201–4209 (2010).
[22] Habibi, Y., Lucia, L. A. & Rojas, O. J. Cellulose nanocrystals: Chemistry, self-assembly, and applications. *Chem. Rev.* **110,** 3479–3500 (2010).
[23] Eichhorn, S. J. *et al.* Review: Current international research into cellulose nanofibres and nanocomposites. *J. Mater. Sci.* **45,** 1–33 (2010).
[24] Rojas, O. J., Montero, G. A. & Habibi, Y. Electrospun nanocomposites from polystyrene loaded with cellulose nanowhiskers. *J. Appl. Polym. Sci.* **113,** 927–935 (2009).
[25] Zoppe, J. O., Peresin, M. S., Habibi, Y., Venditti, R. A. & Rojas, O. J. Reinforcing poly (ε-caprolactone) nanofibers with cellulose nanocrystals. *ACS Appl. Mater. Interfaces* **1,** 1996–2004 (2009).
[26] Park, W.-I., Kang, M., Kim, H.-S. & Jin, H.-J. Electrospinning of poly(ethylene oxide) with bacterial cellulose whiskers. *Macromol. Symp.* **249–250,** 289–294 (2007).
[27] Magalhães, W. L. E., Cao, X. & Lucia, L. A. Cellulose nanocrystals/nellulose core-in-shell nanocomposite assemblies. *Langmuir* **25,** 13250–13257 (2009).
[28] Kamel, S. Nanotechnology and its applications in lignocellulosic composites: A mini review. *Express Polym. Lett.* **1,** 546–575 (2007).
[29] Lu, P. & Hsieh, Y.-L. Cellulose nanocrystal-filled poly(acrylic acid) nanocomposite fibrous membranes. *Nanotechnology* **20,** 415604 (2009).
[30] Anna J. Svagan, My A. S. Azizi Samir, A. & Berglund, L. A. Biomimetic polysaccharide nanocomposites of high cellulose content and high toughness. *Biomacromolecules* **8,** 2556–2563 (2007).
[31] Anna J., S., Mikael S., H. & Lars, B. Reduced water vapour sorption in cellulose nanocomposites with starch matrix. *Compos. Sci. Technol.* **69,** 500–506 (2009).
[32] Samir, M. A. S. A., Alloin, F., Sanchez, J.-Y., Kissi, N. El & Dufresne, A. Preparation of cellulose whiskers reinforced nanocomposites from an organic medium suspension. *Macromolecules* **37,** 1386–1393 (2004).
[33] Shweta A., P., John, S. & John, L. Poly(vinyl alcohol)/cellulose nanocrystal barrier membranes. *J. Memb. Sci.* **320,** 248–258 (2008).
[34] Marcovich, N. E., Auad, M. L., Bellesi, N. E., Nutt, S. R. & Aranguren, M. I. Cellulose micro/nanocrystals reinforced polyurethane. *J. Mater. Res.* **21,** 870–881 (2006).
[35] Lee, G., Aji, M., Kristiina, O., Paul, G. & Arthur J., R. A novel nanocomposite film prepared from crosslinked cellulosic whiskers. *Carbohydr. Polym.* **75,** 85–89 (2009).
[36] My Ahmed Saïd Azizi Samir, Fannie Alloin, Jean-Yves Sanchez & Alain Dufresne. Cross-linked nanocomposite polymer electrolytes reinforced with cellulose whiskers. *Macromolecules* **37,** 4839–4844 (2004).
[37] Klemm, D. *et al.* Nanocelluloses: A new family of nature-based materials. *Angew. Chemie* (International Edition) **50,** 5438–5466 (2011).
[38] K., O., A.P., M., D., B. & I., K. Manufacturing process of cellulose whiskers/polylactic acid nanocomposites. *Compos. Sci. Technol.* **66,** 2776–2784 (2006).
[39] Bondeson, D. & Oksman, K. Dispersion and characteristics of surfactant modified cellulose whiskers nanocomposites. *Compos. Interfaces* **14,** 617–630 (2007).
[40] Jawaid, M., Boufi, S. & Khalil, A. *Cellulose-reinforced Nanofibre Composites: Production, Properties and Applications.* Cambridge, UK: Woodhead Publishing (Elsevier) (2017).
[41] Arantzazu, S.-E. *et al.* Cellulose nanocrystals reinforced environmentally-friendly waterborne polyurethane nanocomposites. *Carbohydr. Polym.* **151,** 1203–1209 (2016).
[42] Barari, B. *et al.* Mechanical, physical and tribological characterization of nano-cellulose fibers reinforced bio-epoxy composites: An attempt to fabricate and scale the 'Green' composite. *Carbohydr. Polym.* **147,** 282–293 (2016).

[43] Arup, M. & Debabrata, C. Characterization of nanocellulose reinforced semi-interpenetrating polymer network of poly(vinyl alcohol) & polyacrylamide composite films. *Carbohydr. Polym.* **134,** 240–250 (2015).

[44] Zeinab, Q. & Mahdi, K. Properties of whey protein isolate nanocomposite films reinforced with nanocellulose isolated from oat husk. *Int. J. Biol. Macromol.* **91,** 1134–1140 (2016).

[45] Eduardo, R., Iñaki, U., Jalel, L. & Luis, S. Surface-modified nano-cellulose as reinforcement in poly(lactic acid) to conform new composites. *Ind. Crops Prod.* **71,** 44–53 (2015).

[46] Benoit, B., Julien, B., Seema, S., Claude, D. & Eric, L. Mechanical and antibacterial properties of a nanocellulose-polypyrrole multilayer composite. *Mater. Sci. Eng. C* **69,** 977–984 (2016).

[47] Shuman, X., Wenjin, Y., Xuelin, Y., Qin, Z. & Qiang, F. Nanocellulose-assisted dispersion of graphene to fabricate poly(vinyl alcohol)/graphene nanocomposite for humidity sensing. *Compos. Sci. Technol.* **131,** 67–76 (2016).

[48] Eldho, A. *et al.* X-ray diffraction and biodegradation analysis of green composites of natural rubber/nanocellulose. *Polym. Degrad. Stab.* **97,** 2378–2387 (2012).

[49] Arup, M. & Debabrata, C. Studies on the mechanical, thermal, morphological and barrier properties of nanocomposites based on poly(vinyl alcohol) and nanocellulose from sugarcane bagasse. *J. Ind. Eng. Chem.* **20,** 462–473 (2014).

[50] Khoo, R. Z., Ismail, H. & Chow, W. S. Thermal and morphological properties of poly (lactic acid)/nanocellulose nanocomposites. *Procedia Chem.* **19,** 788–794 (2016).

[51] Martin George, T. *et al.* Nanocelluloses from jute fibers and their nanocomposites with natural rubber: Preparation and characterization. *Int. J. Biol. Macromol.* **81,** 768–777 (2015).

[52] Zhang, Y. *et al.* Effects of probiotic type, dose and treatment duration on irritable bowel syndrome diagnosed by Rome III criteria: A meta-analysis. *BMC Gastroenterol.* **16,** 62 (2016).

[53] Alessandra, L., Lisete Cristine, S. & Ademir José, Z. Dynamic-mechanical and thermomechanical properties of cellulose nanofiber/polyester resin composites. *Carbohydr. Polym.* **136,** 955–963 (2016).

[54] R., M., R.F., E.-H., K.M., P. & R., S. Mechanical characterization of cellulose nanofiber and bio-based epoxy composite. *Mater. Des.* **36,** 570–576 (2012).

[55] Shanhong, X., Natalie, G., Gregory, S., Meisha L., S. & J. Carson, M. Mechanical and thermal properties of waterborne epoxy composites containing cellulose nanocrystals. *Polymer* **54,** 6589–6598 (2013).

[56] Ahmed A., O. & Jong-Whan, R. Isolation of cellulose nanocrystals from grain straws and their use for the preparation of carboxymethyl cellulose-based nanocomposite films. *Carbohydr. Polym.* **150,** 187–200 (2016).

[57] Farhan, A., Mikael, S. & Lars, B. Nanostructured biocomposites based on unsaturated polyester resin and a cellulose nanofiber network. *Compos. Sci. Technol.* **117,** 298–306 (2015).

[58] Sang Ho, P., Kyung Wha, O. & Seong Hun, K. Reinforcement effect of cellulose nanowhisker on bio-based polyurethane. *Compos. Sci. Technol.* **86,** 82–88 (2013).

[59] Hamad, W. Y. Photonic and semiconductor materials based on cellulose nanocrystals. *Advances in Polymer Science Cellulose Chemistry and Properties: Fibers, Nanocelluloses and Advanced Materials* 287–328 (2015). doi:10.1007/12_2015_323

[60] Kelly, J. A., Giese, M., Shopsowitz, K. E., Hamad, W. Y. & MacLachlan, M. J. The development of chiral nematic mesoporous materials. *Acc. Chem. Res.* **47,** 1088–1096 (2014).

[61] Schlesinger, M., Giese, M., Blusch, L. K., Hamad, W. Y. & MacLachlan, M. J. Chiral nematic cellulose–gold nanoparticle composites from mesoporous photonic cellulose. *Chem. Commun.* **51,** 530–533 (2015).

[62] Qi, H., Shopsowitz, K. E., Hamad, W. Y. & MacLachlan, M. J. Chiral nematic assemblies of silver nanoparticles in mesoporous silica thin films. *J. Am. Chem. Soc.* **133,** 3728–3731 (2011).

[63] Querejeta-Fernández, A., Chauve, G., Methot, M., Bouchard, J. & Kumacheva, E. Chiral plasmonic films formed by gold nanorods and cellulose nanocrystals. *J. Am. Chem. Soc.* **136,** 4788–4793 (2014).

[64] Chu, G. *et al.* Free-standing optically switchable chiral plasmonic photonic crystal based on self-assembled cellulose nanorods and gold nanoparticles. *ACS Appl. Mater. Interfaces* **7,** 21797–21806 (2015).

[65] Querejeta-Fernández, A. *et al.* Circular dichroism of chiral nematic films of cellulose nanocrystals loaded with plasmonic nanoparticles. *ACS Nano* **9,** 10377–10385 (2015).

[66] Lukach, A. *et al.* Coassembly of gold nanoparticles and cellulose nanocrystals in composite films. *Langmuir* **31,** 5033–5041 (2015).

[67] Mu, X. & Gray, D. G. Droplets of cellulose nanocrystal suspensions on drying give iridescent 3-D 'coffee-stain' rings. *Cellulose* **22,** 1103–1107 (2015).

[68] Mu, X. & Gray, D. G. Formation of chiral nematic films from cellulose nanocrystal suspensions is a two-stage process. *Langmuir* **30,** 9256–9260 (2014).

[69] Dumanli, A. G. *et al.* Digital color in cellulose nanocrystal films. *ACS Appl. Mater. Interfaces* **6,** 12302–12306 (2014).

[70] Yang, X., Shi, K., Zhitomirsky, I. & Cranston, E. D. Cellulose nanocrystal aerogels as universal 3D lightweight substrates for supercapacitor materials. *Adv. Mater.* **27,** 6104–6109 (2015).

[71] Roman, M. Toxicity of cellulose nanocrystals: A review. *Ind. Biotechnol.* **11,** 25–33 (2015).

[72] Martínez Ávila, H. *et al.* Biocompatibility evaluation of densified bacterial nanocellulose hydrogel as an implant material for auricular cartilage regeneration. *Appl. Microbiol. Biotechnol.* **98,** 7423–7435 (2014).

[73] Jia, B. *et al.* Effect of microcrystal cellulose and cellulose whisker on biocompatibility of cellulose-based electrospun scaffolds. *Cellulose* **20,** 1911–1923 (2013).

[74] Helenius, G. *et al.* In vivo biocompatibility of bacterial cellulose. *J. Biomed. Mater. Res. Part A* **76A,** 431–438 (2006).

[75] Domingues, R. M. A., Gomes, M. E. & Reis, R. L. The potential of cellulose nanocrystals in tissue engineering strategies. *Biomacromolecules* **15,** 2327–2346 (2014).

[76] Li, Y. *et al.* Evaluation of the effect of the structure of bacterial cellulose on full thickness skin wound repair on a microfluidic chip. *Biomacromolecules* **16,** 780–789 (2015).

[77] Harlin, A. & Vikman, M. *Developments in Advanced Biocomposites.* Espoo, Finland: VTT Technical Research Centre of Finland (2010).

[78] Cai, J. *et al.* Cellulose-silica nanocomposite aerogels by in situ formation of silica in cellulose gel. *Angew. Chemie* **124,** 2118–2121 (2012).

[79] Sehaqui, H., Liu, A., Zhou, Q. & Berglund, L. A. Fast preparation procedure for large, flat cellulose and cellulose/inorganic nanopaper structures. *Biomacromolecules* **11,** 2195–2198 (2010).

[80] Kurtis, K. E. Innovations in cement-based materials: Addressing sustainability in structural and infrastructure applications. *MRS Bull.* **40,** 1102–1109 (2015).

[81] Fernando, P.-T. & Said, J. Cementitious building materials reinforced with vegetable fibres: A review. *Constr. Build. Mater.* **25,** 575–581 (2011).

[82] Ridi, F., Fratini, E. & Baglioni, P. Cement: A two thousand year old nano-colloid. *J. Colloid Interface Sci.* **357,** 255–264 (2011).
[83] L.R., B., C., O. & S.P., S. Fiber-matrix interaction in microfiber-reinforced mortar. *Adv. Cem. Based Mater.* **2,** 53–61 (1995).
[84] Mezencevova, A., Garas, V., Nanko, H. & Kurtis, K. E. Influence of thermomechanical pulp fiber compositions on internal curing of cementitious materials. *J. Mater. Civ. Eng.* **24,** 970–975 (2012).
[85] Ardanuy, M. *et al.* Nanofibrillated cellulose (NFC) as a potential reinforcement for high performance cement mortar composites. *BioResources.* **7,** 3883–3894 (2012).
[86] Yizheng, C., Pablo, Z., Jeff, Y., Robert, M. & Jason, W. The influence of cellulose nanocrystal additions on the performance of cement paste. *Cem. Concr. Compos.* **56,** 73–83 (2015).
[87] Sherry L., T., David J., O., John A., W. & Bing, S. Method of making a fiber cement board with improved properties and the product. (2012).
[88] S. Ananda, W. & Harshadkumar M., S. Fiber for fiber cement and resulting product. (2013).
[89] S. Ananda, W. & David J., O. Internally curing cement based materials. (2012).
[90] Dai, H., Jiao, L., Zhu, Y. & Pi, C. Nanometer cellulose fiber reinforced cement-based material. (2015).
[91] Jeffrey Paul, Y., Pablo Daniel, Z., Robert John, M., William Jason, W. & Yizheng, C. Cellulose nanocrystal additives and improved cementious systems. (2014).
[92] Shaul, L., Oded, S., Tord, G. & Lea, C.-G. Nano crystalline cellulose in construction applications. (2015).
[93] Antti, L. *et al.* Material to be used as a concrete additive. (2010).
[94] Cowie, J., Bilek, E. M. T., Wegner, T. H. & Shatkin, J. A. Market projections of cellulose nanomaterial-enabled products. Part 2: Volume estimates. *Tappi Journal* **13 (6),** 57–69 (2014).
[95] Miller, J. *Nanocellulose: Technology, Applications and Markets.* (RISI, 2014).
[96] Miller, J. *Nanocellulose State of the Industry.* (2015).
[97] Florence, S. & Konstantin, S. Nanotechnology in concrete – A review. *Constr. Build. Mater.* **24,** 2060–2071 (2010).
[98] J.S., D., J., L., D., H., D., S. & E.M., G. Sustainable development and climate change initiatives. *Cem. Concr. Res.* **38,** 115–127 (2008).
[99] Obinna, O., Daman K., P. & Mohini, S. Properties of nanofibre reinforced cement composites. *Constr. Build. Mater.* **63,** 119–124 (2014).
[100] Mejdoub, R. *et al.* Nanofibrillated cellulose as nanoreinforcement in Portland cement: Thermal, mechanical and microstructural properties. *J. Compos. Mater.* **51,** 2491–2503 (2017).
[101] Ferrara, L. *et al.* Effect of cellulose nanopulp on autogenous and drying shrinkage of cement based composites. In K. Sobolev & S. Shah (eds), *Nanotechnology in Construction* (pp. 325–330). Basel, Switzerland: Springer International (2015). doi:10.1007/978-3-319-17088-6_42
[102] Cao, Y., Zavattieri, P., Youngblood, J., Moon, R. & Weiss, J. The relationship between cellulose nanocrystal dispersion and strength. *Constr. Build. Mater.* **119,** 71–79 (2016).
[103] Karppinen, A. What makes microfibrillated cellulose stable over wide ph range. http://blog.exilva.com/microfibrillated-cellulose-stable-over-wide-ph-range (2016). doi:10.1021/bm061215p.
[104] Turunen, H. Nanocellulose samples. Available at: https://www.buildingcentre.co.uk/page/nanocellulose-samples (accessed 6 November 2017).
[105] Ford Media Center. What's Super Strong, Fast Growing, and Potentially Part of Your Next Car? Bamboo! | Great Britain | Ford Media Center. (2017). Available at: https://media.ford.

com/content/fordmedia/feu/gb/en/news/2017/04/25/what-s-super-strong--fast-growing--and-potentially-part-of-your-.html# (accessed 6 November 2017).
[106] Aandacht voor lekdetectie kan derving voorkomen - VMT. (2015). Available at: http://www.vmt.nl/Nieuws/Aandacht_voor_lekdetectie_kan_derving_voorkomen-150529151132 (accessed 6 November 2017).
[107] Heath, L. & Thielemans, W. Cellulose nanowhisker aerogels. *Green Chem.* **12,** 1448 (2010).
[108] Kiziltas, A., Nazari, B., Gardner, D. J. & Bousfield, D. W. Polyamide 6-cellulose composites: Effect of cellulose composition on melt rheology and crystallization behavior. *Polym. Eng. Sci.* **54,** 739–746 (2014).
[109] S.D., B. & U.K., V. Performance of long fiber reinforced thermoplastics subjected to transverse intermediate velocity blunt object impact. *Compos. Struct.* **67,** 263–277 (2005).
[110] Ayrilmis, N., Jarusombuti, S., Fueangvivat, V., Bauchongkol, P. & White, R. H. Coir fiber reinforced polypropylene composite panel for automotive interior applications. *Fibers Polym.* **12,** 919–926 (2011).
[111] Ahmed M., Y., Magda Ali, E.-S. & Mona H. A. R. Preparation of conductive paper composites based on natural cellulosic fibers for packaging applications. *Carbohydr. Polym.* **89,** 1027–1032 (2012).
[112] Abdorreza Mohammadi, N., Abd Karim, A., Shahrom, M. & Marju, R. Antimicrobial, rheological, and physicochemical properties of sago starch films filled with nanorod-rich zinc oxide. *J. Food Eng.* **113,** 511–519 (2012).
[113] Zadbuke, N., Shahi, S., Gulecha, B., Padalkar, A. & Thube, M. Recent trends and future of pharmaceutical packaging technology. *J. Pharm. Bioallied Sci.* **5,** 98–110 (2013).
[114] H. P. S., A. K. *et al.* Production and modification of nanofibrillated cellulose using various mechanical processes: A review. *Carbohydr. Polym.* **99,** 649–665 (2014).
[115] Kalia, S., Boufi, S., Celli, A. & Kango, S. Nanofibrillated cellulose: Surface modification and potential applications. *Colloid Polym. Sci.* **292,** 5–31
[116] Lewis, H. C., Wichman, O. & Duizer, E. Transmission routes and risk factors for autochthonous hepatitis E virus infection in Europe: A systematic review. *Epidemiol. Infect.* **138,** 145 (2010).
[117] Freire, M. G. *et al.* Electrospun nanosized cellulose fibers using ionic liquids at room temperature. *Green Chem.* **13,** 3173 (2011).
[118] Zhang, X., Tu, M. & Paice, M. G. Routes to potential bioproducts from lignocellulosic biomass lignin and hemicelluloses. *BioEnergy Res.* **4,** 246–257 (2011).
[119] Chakraborty, A., Sain, M. & Kortschot, M. Cellulose microfibers as reinforcing agents for structural materials. In K. Oksman & M. Sain (eds), *Cellulose Nanocomposites: Processing, Characterization and Properties* (pp.169–186). Washington, DC: ACS Publications (2006). doi:10.1021/bk-2006-0938.ch012.
[120] Maya Jacob, J. & Sabu, T. Biofibres and biocomposites. *Carbohydr. Polym.* **71,** 343–364 (2008).
[121] Millon, L. E. & Wan, W. K. The polyvinyl alcohol–bacterial cellulose system as a new nanocomposite for biomedical applications. *J. Biomed. Mater. Res. Part B Appl. Biomater.* **79B,** 245–253 (2006).
[122] Bibin Mathew, C. *et al.* Isolation of nanocellulose from pineapple leaf fibres by steam explosion. *Carbohydr. Polym.* **81,** 720–725 (2010).
[123] Costa, L. M. M. *et al.* Bionanocomposites from electrospun PVA/pineapple nanofibers/Stryphnodendron adstringens bark extract for medical applications. *Ind. Crops Prod.* **41,** 198–202 (2013).

[124] Nuruddin, M. *et al.* Extraction and characterization of cellulose microfibrils from agricultural wastes in an integrated biorefinery initiative. *Cell. Chem. Technol.* **45 (5–6),** 347–354 (2011).

[125] M. Douglas, B. *et al.* An injectable drug delivery platform for sustained combination therapy. *J. Control. Release* **138,** 205–213 (2009).

[126] Nadagouda, M. N. & Varma, R. S. Green synthesis of silver and palladium nanoparticles at room temperature using coffee and tea extract. *Green Chem.* **10,** 859 (2008).

[127] B.L, S., T.C, O. & A, P. Polymeric biomaterials for tissue and organ regeneration. *Mater. Sci. Eng. R Reports* **34,** 147–230 (2001).

[128] Capes, J. S., Ando, H. Y. & Cameron, R. E. Fabrication of polymeric scaffolds with a controlled distribution of pores. *J. Mater. Sci. Mater. Med.* **16,** 1069–1075 (2005).

[129] Dieter, K., Dieter, S., Ulrike, U. & Silvia, M. Bacterial synthesized cellulose – artificial blood vessels for microsurgery. *Prog. Polym. Sci.* **26,** 1561–1603 (2001).

[130] Ikada, Y. Challenges in tissue engineering. *J. R. Soc. Interface* **3,** 589–601 (2006).

[131] Rajendran, K., Selvaraj Mohana, R., Arunachalam, P., Venkatesan Gopiesh, K. & Subhendu, C. Agricultural waste Annona squamosa peel extract: Biosynthesis of silver nanoparticles. *Spectrochim. Acta Part A Mol. Biomol. Spectrosc.* **90,** 173–176 (2012).

[132] Subrata; Mondal. Preparation, properties and applications of nanocellulosic materials. *Carbohydr. Polym.* **163,** 301–316 (2017).

Chapter 12
Bioresin–natural fibre composites

Dr Nairong Chen

This chapter begins with an overview of the production and properties of the natural bioresins of vegetable protein, lignin, vegetable oil and starch-based resources. There follows a discussion of the formulation and performance of natural resin–plant fibre (all natural) composites of both plant-fibre reinforced composites – including vegetable protein, lignin, vegetable oil and starch-based natural resin matrix natural fibre composites – and bioresin-bonded composites, including bioresin-bonded wood fibre, bamboo fibre and non-food crop fibre composites.

12.1 Classification of natural resin–plant fibre composites

Natural resin–plant fibre (all natural) composites are mainly composed of all-natural materials, including natural resin and natural plant fibre. Natural resins, such as rosin, pectin and rubber, can be directly obtained from nature, and can also be modified from natural materials such as vegetable protein, lignin, vegetable oil and carbohydrates. Pretreatment or modification of natural resin is generally able to equip the material with better physical and chemical properties (e.g. compatibility and reactivity). According to the natural resin content, all-natural composites can be classified into natural resin matrix plant fibre composites (plant fibre reinforced composites) or plant fibre-based natural resin composites (bioresin bonded composites).

12.2 Preparation and properties of natural resins

12.2.1 Vegetable protein-based natural resin

Vegetable protein-based resins are mainly derived from low-cost biomass materials with high vegetable protein content, e.g. soybeans. Soybeans consist of approximately 25% carbohydrate, 48% protein and 0.4–0.9% starch. The protein in soybeans is mainly globular protein with a diameter of 5–20 μm [1, 2]. Protein is in general vulnerable to high temperature, namely denaturation of protein easily occurs under high temperature, leading to a decrease on the nitrogen solubility index (NSI) or the protein dispersibility index (PDI). Therefore, the treatment temperature of raw materials for deriving vegetable protein-based resins should not be higher than 70°C [3]. The residues after defatting soybean could be used as raw material for developing vegetable protein-based natural resins, which include defatted soybean flour, concentrated soy protein and isolated soy protein.

Defatted soybean flour is the product obtained from grinding the residue of soybean oil extraction and has the advantage of low cost. Pressing and solvent extraction are the major methods used in the extraction process of soybean oil, and the corresponding product is named defatted soybean meal. Soybean protein with less denaturation and higher solubility can be obtained via the cold pressing method, but the resulting high oil content gives rise to rancidity, and so this has not been extensively used. By contrast, the hot pressing method leads to less soybean oil and higher protein denaturation, and the resulting product is a suitable raw material for food [4]. Extraction of soybean oil can be carried out at high or low temperature by using hydrophobic solvent (e.g. hexane) [5]. Soybean meal obtained at low temperature has little soybean oil and protein denaturation. Generally, the NSI of such meal is higher than 80% , and it contains 44–52% protein, 30% carbohydrate, 5–6% ash and less than 10% water [3, 6]. After low-temperature processing, the defatted soy meal can be further pulverised into fine defatted soy flour (≥325 mesh), suitable for the preparation of soy-based resins [7].

Soybean protein concentrate is usually obtained by employing a specific method (such as hot water rinsing, filtration with ethanol, and acid precipitation) to remove some components (such as soluble sugars, inorganic salts) of defatted soybean flour so as to increase the protein content. The protein content of soybean protein concentrate is between 65% and 70% depending on the extraction method and is higher than that of defatted soy flour. However, soybean protein concentrate extracted from defatted soybean meal has a larger particle size (≥100 mesh) with higher cost. Therefore, preparation of soy-based resins using soybean protein concentrate is rarely reported.

Soybean protein isolate also uses defatted soy flour as a raw material, and its production process on the one hand eliminates soluble sugars and salts and on the other hand removes insoluble polysaccharides and other substances in order to further improve the protein content (≥90%) [8]. However, the reduced yield (only 35–40%) increases the cost of soybean protein isolate. The methods for producing soybean protein isolate mainly include alkali extraction followed by acid precipitation and ion exchange accompanied with membrane separation. The method of alkali extraction and acid precipitation has been widely used in industry, and the production process is shown in Fig. 12.1. Soybean protein isolate has been used as a substitute for pure soy protein in the preparation of soy-based resin in the laboratory due to its high soy protein content, and the resulting soy-based resins showed outstanding performance. However, its industrial applications are hampered by high cost.

Soy protein-based natural resins could be produced by:

1. physical methods, such as heating, refrigeration, magnetic, mechanical and electromagnetic waves, and radiation, through breaking the hydrogen bonding of protein molecules and stretching it while exposing internal amino groups – hence the denatured soy protein has low cost and simple operation characteristics;
2. chemical treatments, such as acylation, crosslinking, copolymerisation and blending, which are the most widely used methods in the structural modification of proteins; all able to improve the performance of cured soy-based resins to some extent with different efficiency; and

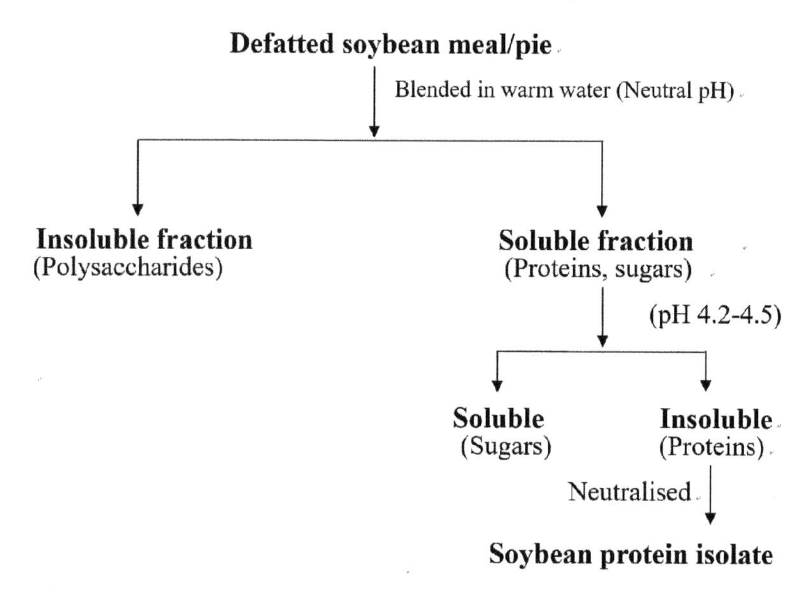

Figure 12.1 Schematic of the preparation process of soybean protein isolate.

3. enzyme (e.g. protease) treatment, which is another approach to degrade soy protein to increase or graft functional groups within the molecules. Therefore, this treatment may alter the functional properties of protein.

In addition to the preparation methods, there are other parameters that influence the physical and chemical properties of soy protein-based natural resins.

12.2.1.1 Appearance

The appearance of soy protein-based natural resin is closely related to the raw material and pH during processing. The colour of the soy protein-based natural resin will be beige or light brown at pH 5–8. The initial colour of all the resins is close to beige, decreasing pH leads to the resin turning black-brown, while increasing pH results in orange or even brown resins depending on the extent to which the pH is increased. It should be pointed out that at the same pH (>7), the colour of the resin with higher soy protein content will be slightly darker compared to the counterpart.

12.2.1.2 pH value

Bearing functional amino and carboxyl groups, soy protein-based resins can be adapted to a wide pH range, and have been comprehensively investigated at pH <7 or >7 [9–11]. However, the residue of acid or base in acidic or basic resin products impairs their shelf life. In addition, acidic resin can cause precipitation of soy protein (the isoelectric point of the protein is pH 4.2–4.5) and result in delamination in the stored resin. Thus, soy protein-based natural resin at a pH of 5–9 is the most commonly used in both research and industry.

12.2.1.3 Solid content

The solid content of soy-based resins is related to the solubility of raw materials in different solvents. For example, the solid content decreases with an increase in the protein content of the raw materials when water is used as the solvent. The solid content of defatted soy flour based resin is usually less than 30% (mostly 18–25%), and the solid content of soy protein concentrate and soy protein isolate based resins is usually less than 15%. Lower solid content brings the benefit of a reduction in the overall cost of the resin, but it may lead to defects in the final composites, since the removal process for the resin solvent may result in a large amount of internal stress, thereby affecting the lifetime of the composites.

12.2.1.4 Storage time

Storage of soybean protein-based resin is affected by the microbial and natural resin component. The main component of resin, namely protein, is a nutrient for microbial growth. The structure of modified soybean protein has a certain degree of durability but it can still be degraded by microorganisms. The storage time of unpreserved resin is usually less than one week (2–4 days in summer) but it can be extended to a month or more by adding preservatives (such as BIT, salicylic acid, dehydroacetic acid, benzoic acid, etc.). Because the resin contains a large number of reactive functional groups (such as amino, carboxyl and hydroxyl), which are inclined to combine with each other, giving rise to aggregation of resin molecules and the formation of further precipitation after long-time storage, the performance of the resin decreases.

12.2.1.5 Viscosity

The viscosity of resin is related to the solid content and the preparation process, e.g. higher solid content means higher resin viscosity and, under the same conditions, the viscosity of the resin prepared at pH 2.0–3.5 and pH 7.5–9.5 is higher than the viscosity of resin prepared at other pH levels. Moreover, the reaction time and the temperature in the preparation process can affect the viscosity of the resin (Fig. 12.2

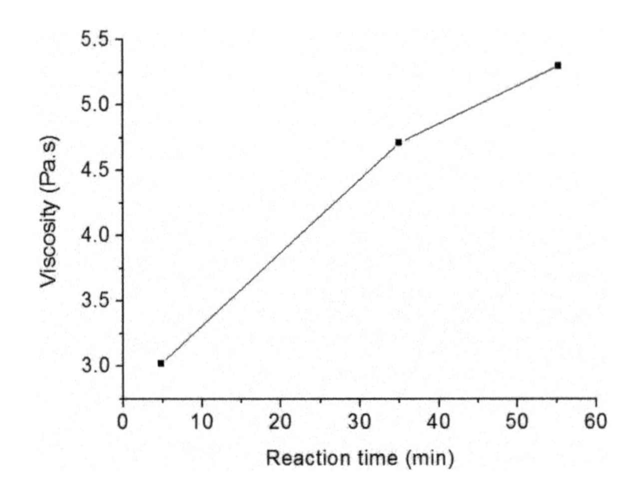

Figure 12.2 Correlation of viscosity and reaction time in soy protein-based resin.

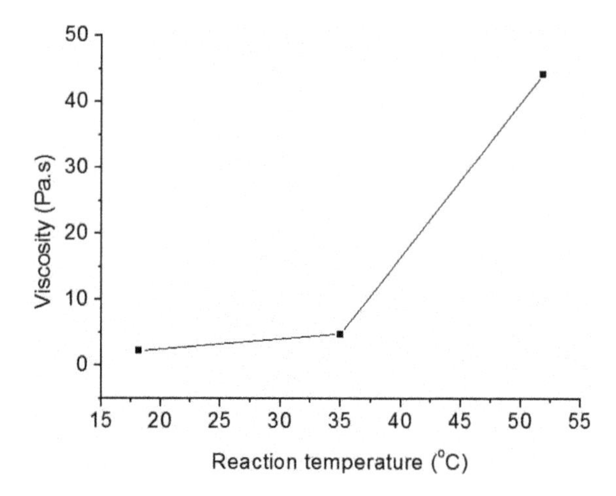

Figure 12.3 Correlation of viscosity and reaction temperature in soy protein-based resin.

and Fig. 12.3). The viscosity of resin increases with an increase in reaction time and temperature.

12.2.1.6 Water resistance

The water resistance of soy protein-based natural resin is an important factor in extending its field of application. Since soybean protein contains a large number of hydrophilic functional groups (e.g. amino, carboxyl and hydroxyl), the resulting soy protein resins would display poor water resistance if a crosslinked structure was not formed after curing. There are numerous studies suggesting that the water resistance of soy protein-based natural resin could be effectively improved through various modification methods such as grafting, blending, etc. [12–14]. In addition, although the Maillard reaction of soy protein and carbohydrates occurs at high temperatures, it improves the water resistance of the natural resins [11, 15]. For example, mixtures containing 53.4%, 60.0%, 70.0%, 80.0% and 90.8% protein consisting of defatted soybean flour and soybean protein isolate were prepared, respectively. Aqueous dispersions with protein concentrations of 9.9% were obtained by blending 20 g, 17.76 g, 15.17 g, 13.22 g and 11.60 g of mixtures into 80 g distilled water, and then adding sucrose or glucose and stirring for 40 minutes at 35°C. The dispersions were placed in a 100°C temperature and 91% humidity chamber for 3 hours to test the water absorption (Fig. 12.4). The results showed that water absorption of soy protein-based natural resin decreased with decreasing carbohydrate content (Fig. 12.4a); the water absorption of cured natural resins increased with an increase in sucrose content (Fig. 12.4b) or decrease in glucose content (Fig. 12.4c), while the total carbohydrates in the resins were maintained at 7.3% consistently.

The results illuminated that the water resistance of soybean protein-based natural resin was affected by its carbohydrate content and component. Therefore, soy protein natural resin with excellent water resistance could be formulated by the rational use of carbohydrates in defatted soybean flour.

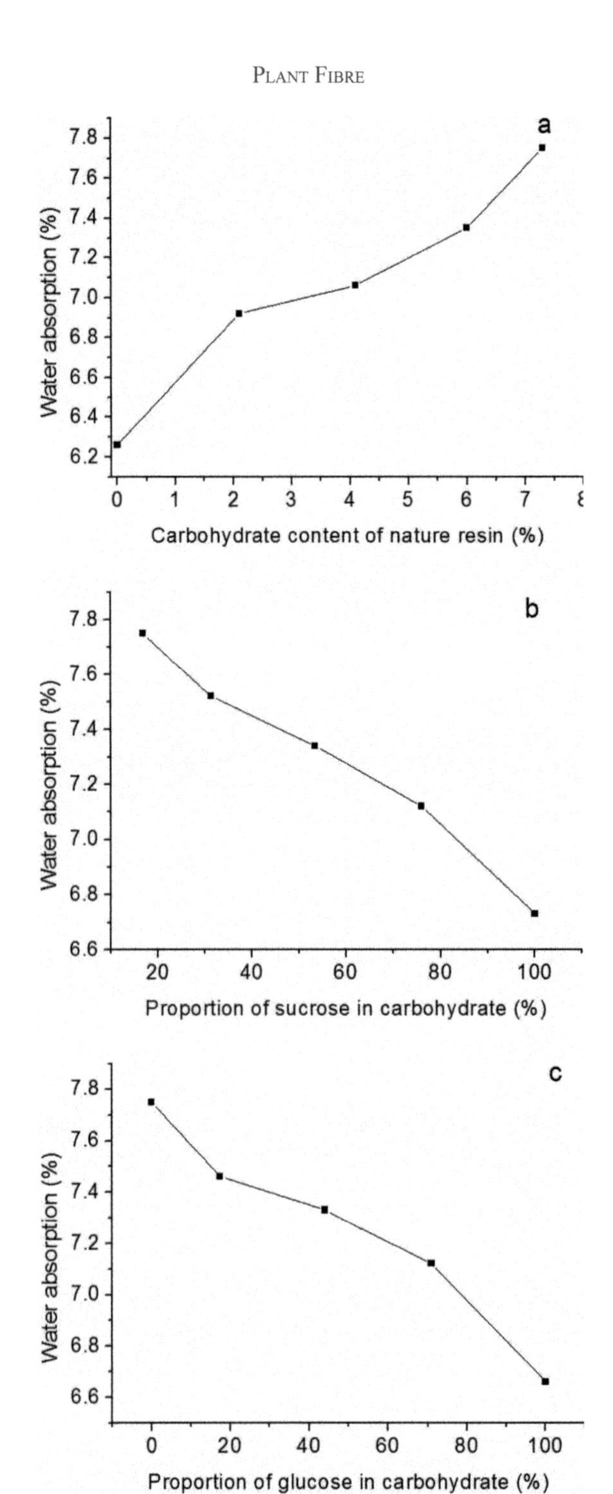

Figure 12.4 Correlation of water resistance and carbohydrate in soy protein-based resin.

12.2.2 Lignin-based natural resin

Phenylpropene is the main unit in lignin structure; specifically, it includes guaiacylpropane, syringylpropane and *p*-hydroxyphenylpropane (Fig. 12.5).

There exist various functional groups (e.g. methoxy, phenolic hydroxyl, alcoholic hydroxyl, and carboxyl) in lignin molecules which provide a basis for carrying out a set of applicable reactions such as reduction, oxidation, condensation, and graft copolymerisation. Among these reactions, polycondensation and copolymerisation are the major reactions occurring in the process of natural resin preparation. Polycondensation between lignin and formaldehyde occurs under acidic or basic conditions, wherein the acid-catalysed polycondensation reaction occurs at the C6 position of the guaiacol ring, while base catalysis occurs at the C5 position. During acid catalysis, α-carbon in the side chain of lignin can also form C-C linkages with phenolic.

Replacing some phenol with lignin in synthetic phenolic resins has been widely used in industry. There are no significant differences in the mechanical properties and thermal stability when an appropriate amount of lignin is introduced into phenolic resin, but the modulus and high temperature insulation can be dramatically improved. However, the bulky and sterical lignin molecules will hinder the reactivity of the aromatic ring to the phenol and formaldehyde, or even impede the condensation of the phenol and formaldehyde. Typically, post-demethylation and hydroxymethyl modification of lignin can considerably increase the reactivity. For example, the phenolic resins prepared by partially replacing phenol with hydroxymethyl lignin are generally able to meet the quality requirements, but the replacement rate of lignin barely reaches 50% [16]. In addition, the active hydroxyl in lignin can also be made to react with an isocyanate for the preparation of isocyanate-based resin. Preparation of iso-cyanate-based natural resins using different types of lignin and its hydroxylated or alkylated derivatives requires the addition of some catalysts, such as dibutyltin dilaurate, triethylamine, and ethyl caproatetin. Ultimately, the final performance of the product is affected by the composition and molecular structure of lignin (i.e. lignin type, content and molecular weight), isocyanate type, NCO/OH molar ratio, etc. Natural resins prepared from lignin and its derivatives, can be used as engineering plastics, adhesives, foams or films, depending on their properties and performance.

The brittle nature of lignin-based polyurethane resin needs to be refined or overcome prior to maximising its application. A small portion of rigid diisocyanate or soft polyethylene glycol (PEG) can be introduced to deal with this issue, leading to excellent mechanical properties, while continuously increasing the amount of PEG results in a decrease in T_g and the compression modulus of the materials [17, 18].

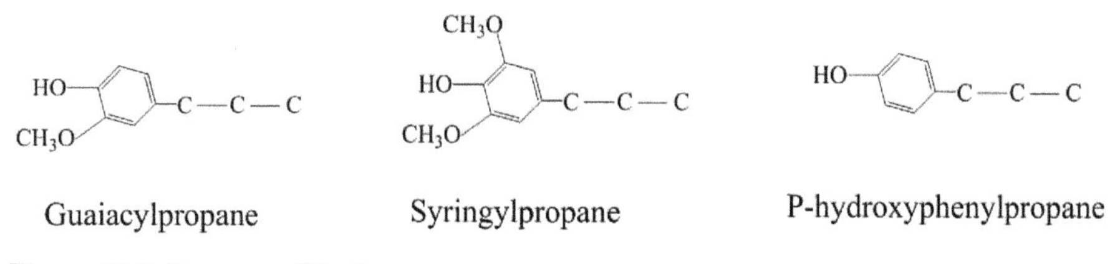

Figure 12.5 Structure of lignin.

In addition, a tough polyurethane material can be obtained by employing more functional alcohols such as polyester triol, due to the formation of a crosslinked structure between the alcohols and lignin under a suitable molar ratio of NCO/OH. The lignin molecule acts as a crosslinker and hard segment in this situation. It has been reported that flexible polyurethane can be achieved when the content of low molecular weight lignin is less than 30% (by mass) [19, 20]. However, the thermal instability of lignin may result in a decreased thermal stability of the material after serving as part of the component in the hard segment. Therefore, it may be reckoned that an appropriate content of moderate molecular weight lignin with a relatively lower molar ratio of functional groups NCO/OH, and synergy with the soft segment of polyethylene glycol and isocyanate, results in fine polyurethane material [21, 22]. The key is to promote the chemical reactions in these materials and increase the number of alcoholic hydroxyl groups by hydroxyalkylate or caprolactone derivatised with formaldehyde, ethylene oxide or epoxypropane [17, 19]. The aim of improving the mechanical properties of the material can be achieved when lignin participates in the curing reaction of polyurethane by compounding lignin and polyurethane. The mechanical properties of the composites depend on the extent of the reaction between polyurethane and lignin, and the tensile modulus is determined by the adhesion balance of the two-phase section.

The Young's modulus and mechanical properties of the material were improved when different types of lignin were incorporated into polyurethane. The molecular weight between crosslinked bonds of hydroxypropyl lignin polyurethane/poly methyl methacrylate composite decreased with an increase in the lignin content, and an inter-penetrating polymeric network structure was formed when the lignin content was more than 25% (by mass). The variations in properties (such as tensile, dynamic mechanical, and thermal) were in line with the characteristics of a bi-continuous phase: therefore the lignin fully became part of the polyurethane [23, 24]. Composite material made of polybutylene succinic anhydride and lignin has similar characteristics [25]. Tensile strength and elongation of the material were significantly improved when a small amount of lignin nitrate was grafted with a polyurethane to form an interpenetrating polymer network structure [26].

It may occur that the material does not show the rubbery plastic yield point of transition after adding lignin, indicating that the composites have superior toughness to pure polyurethane. Moreover, the strength of the composite was higher than that of pure polyurethane when the lignin nitrate (NL) content was less than 5.5% (by mass) [26]. This is due to the grafting reaction of nitrification lignin molecules to the -NCO of polyurethane, and thus the formation of a network connected with the centre of the large star-shaped structure. The entangled and penetrated structure of the polyurethane molecule and its networks plays an important role in improving the strength and elongation of the materials. One of the suitable conditions to form the crosslinking network structure inside the material is 2.8% (by mass) of nitrated lignin, MDI, a crosslinker (trimethylolpropane) and NCO/OH molar ratio of 1.20, which maximises the physical crosslinking of hydrogen bonds in the hard segment of polyurethane. Tensile strength and elongation at break can be increased 3 and 1.5 times, respectively, by introducing nitrification lignin.

In another polyurethane system (less than 9.3% lignin content by mass), the strength and elongation of the material also appeared to increase simultaneously. Its strength, toughness, and elongation were increased by 370%, 470% and 160%, respectively, and it achieved the best thermal-mechanical property when the lignin content was 4.2% (by mass) [27].

Lignin is also used in development of epoxy resin in the following forms [28–30]: lignin derivative blended with general epoxy resin, epoxy modified lignin derivatives, and modified lignin to improve the reactivity before epoxidation. The composite is prepared by compounding the lignin and epoxy with the presence of a curing agent to obtain the interpenetrating polymer network with high compatibility [31]. The bonding strength between the lignin and epoxy is outstanding when compared to the unmodified epoxy resin. However, most lignin epoxy materials still have disadvantages such as poor solubility and workability.

12.2.3 Vegetable oil-based natural resin

Vegetable oils are derived from various types of oil plants, and are specifically named after their vegetable origins, e.g. soybean oil. The main component of vegetable oil is esters formed by glycerol and various fatty acids (Fig. 12.6), which contain various functional groups in the molecules for further chemical reactions. There are numerous studies on the modification of C=C groups in vegetable oil molecules [32].

Vegetable natural resins can be prepared as paints and coatings by free radical and cationic polymerisation reactions. For instance, boron trifluoride-ethyl ether complex can be used as an initiator to develop a variety of vegetable oils by reacting with petroleum polymers through cationic copolymerisation. This reaction was applied to prepare rubber and hard plastics or green composite with excellent performance [33–35]. Larock reported a green resin via the free radical polymerisation of vegetable oil and petroleum-based comonomers by using t-butyl peroxide free radical as initiator [33, 36–38]. Comonomer was cured for at least 12 hours under 160 °C or at 180°C for around 4 hours, and then a thermoset resin was obtained. The T_g of the obtained resin was below 60 °C, and the thermal stability temperature was 350 °C. Various petroleum-based monomers such as styrene, divinyl benzene, cyclopentadiene, 4-vinylphenyl boronic acid, and n-butyl methacrylate can be used as raw materials [39]. Mitsuhiro Shibata reported that the thermoset natural resin can be prepared by tung

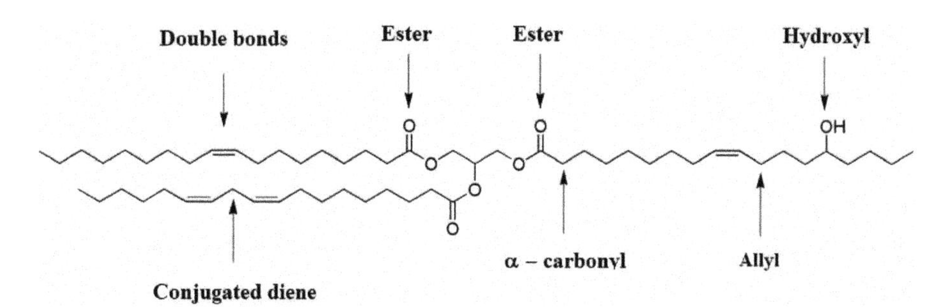

Figure 12.6 Structure of vegetable oil.

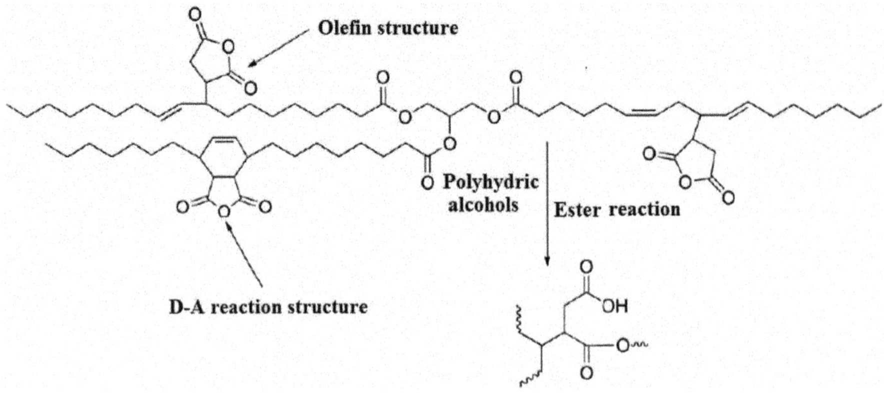

Figure 12.7 Crosslinked polymerisation of maleic anhydride-modified soybean oil and polyester polyols.

oil and methyl-2-imidazolidinone under various reactions (e.g. D-A reaction, the reaction alkenes, hereinafter). Specifically, the resin was prepared at 150 °C for 4 hours to obtain a prepolymer and then cured at 200°C for 2 hours to obtain the crosslinked polymer; its T_g can be up to 150 °C [40]. Richard Vendamme used the strongest base as a catalyst to prepare maleic anhydride-modified soybean oil. It was of adjustable viscosity and could crosslink with polyester polyol by curing reaction (Fig. 12.7) [35].

The steric hindrance of long flexible molecular chains of vegetable oil itself and chain transfer effects of free radicals make the cationic and radical polymerisation of vegetable oils more difficult, and most vegetable oils have relatively low functionality and crosslinked density, which leads to relatively poor thermal performance and mechanical strength of the resins [41]. Thus, natural resins with relatively better performance can be copolymerised by using unmodified vegetable oil and rigid petroleum-based monomers.

High-quality resin can be prepared by using conventional polymerisation methods through the chemical modification (e.g. triglyceride alcoholysis and transesterification, etc.) of functional groups (e.g. ester group) in the vegetable oil. For example, the application of tung oil in thermoset polymer has been studied by Liu *et al.* (Fig. 12.8). Alcoholysis of tung oil was initiated by pentaerythritol, and then reacted with maleic anhydride. The reaction of tung oil monoglyceride maleic anhydride with styrene by free radical copolymerisation was carried out under the following conditions: the curing temperature was increased from room temperature to 120°C at the rate of 5°C per minute, then kept at 120°C for 3 hours, then cured at 150°C for 1 hour. Finally, a rigid thermoset polymer with superb mechanical properties was obtained [42]. DA Echeverri *et al.* reported a similar study of chemically modified soybean oil, i.e. soybean oil reacted with maleic anhydride after alcoholysis by glycerol, and then reacted with polythiol (TMP-3MP, Penta-3MP4) to prepare the elastic polymer with a T_g lower than room temperature [43]. A similar method was also used to synthesise monoglycerides [44]. With the presence of an initiator, a terpolymer product with

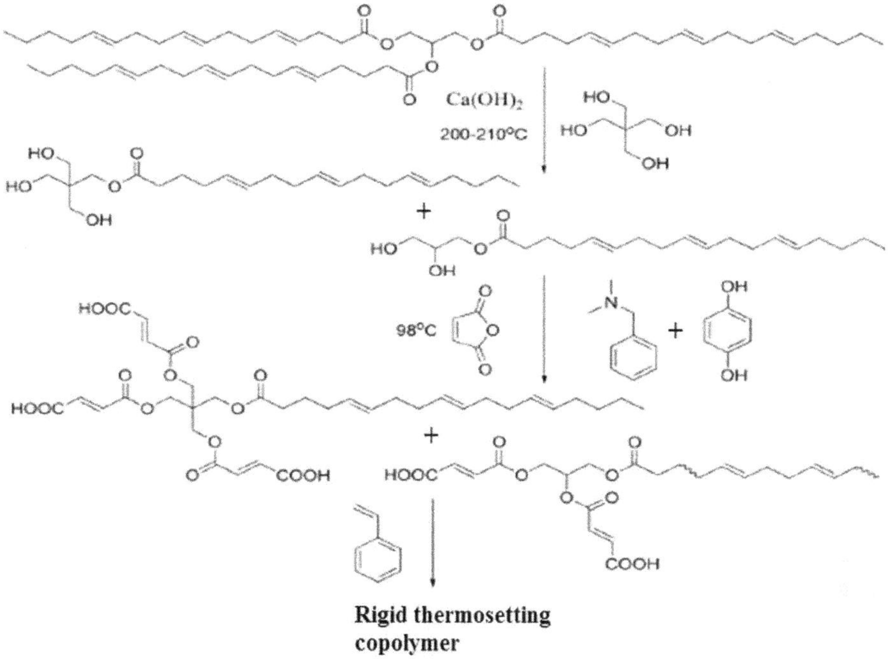

Figure 12.8 Synthetic of tung oil based thermoset natural resin.

outstanding mechanical properties could be derived from maleic anhydride castor monoglycerides, styrene acrylate, epoxidised oil methyl ester through the homopolymerisation of maleic anhydrided castor oil monoglycerides or reaction with styrene and methyl acrylate epoxidised oil.

Epoxidisation is another solution to conducting chemical modification of vegetable oils. The process is shown in Fig. 12.9. As an inexpensive and renewable raw material for industrial application, epoxidised vegetable oil is of great interest and has great advantages [32]. The epoxidation methods of vegetable oil can be either solvent-required or solvent-free, in which metal catalysis, chemical cyclooxygenase catalysis, ion-exchange resin catalysis, peracetic acid oxidation and phase-transfer catalysis oxidation are involved [45].

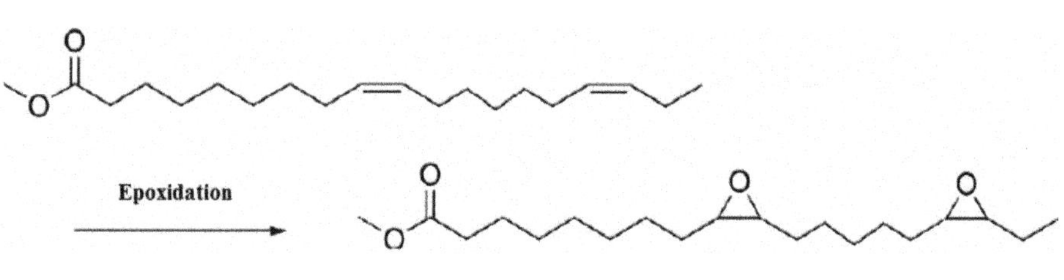

Figure 12.9 Epoxidation of vegetable oil.

Compared to other epoxidised vegetable oils still being investigated – such as epoxidised castor oil, epoxidised rapeseed oil or epoxidised palm oil – epoxidised soybean oil (ESO) is the most industrialised product. There are numerous studies on the conversion of epoxidised vegetable oil to thermoset resin through a crosslinking polymerisation reaction. Similar to the curing process of epoxy resins, the curing of epoxidised vegetable oil also uses a curing agent, such as amines, anhydrides, acids, phenols. Chow *et al.* studied epoxidised soybean oil using methyl hexahydrophthalic anhydride (MHHPA) as the curing agent and 2-ethyl-4-methylimidazole (EMI) as the catalyst. Results showed that with an increase in the catalyst content, the monomer conversion, the degree of polymerisation, the crosslinking density of the product, and the glass transition temperature and storage modulus of the ESO curing system increased, while the mechanical performance decreased [46]. Further study on the reaction kinetics by FTIR suggested that an autocatalytic phenomenon was observed in the curing process of the ESO-MHHPA-EMI system. The EMI content and curing temperature considerably affected the catalysis rate of the system: e.g. as the content of the EMI catalyst was increased, the activation energy of the reaction system was reduced. Zhao *et al.* synthesised the linear copolymer of epoxy soybean oil and capric diamine with the aid of succinic anhydride. It was found that the linear elastomers produced had a low T_g range (from –30 °C to –17 °C), and the crosslinked elastomers had excellent damping properties and low water absorption [47]. Judith *et al.* compared the properties of amines (such as polyether diamine, bis (4-aminocyclohexyl) methane, benzyltriethylammonium chloride, and a mixed amine) and anhydrides (such as methyl tetrahydrophthalic anhydride, hexahydrophthalic anhydride, and methyl hexa-hydrophthalic anhydride) cured epoxy and epoxidised soybean and rapeseed oil. DSC analysis indicated that the exothermic peak of the anhydride curing system was higher than that in the amines system [48].

12.2.4 Starch-based natural resin

Starch is the main energy storage material of plants, widely found in roots, stems, seeds and other plant tissues in nature. Starch is a polysaccharide comprising glucose monomers joined in α-1,4 linkages, and the basic chemical formula is $(C_6H_{10}O_5)_n$. According to the chemical structure, it can be classified into straight and branched chain starch. The former is soluble in water at 70–80°C, whilst the latter is insoluble. Straight and branched chains have different particle structures and crystalline regions, resulting in a starch granule state that is difficult to process and use. Thus, the grain structure and crystalline regions are destructured, and then diffused in a medium to convert the starch to a resinous solution [49]. For example, the most common way is to gelatinise starch by dispersing the starch in water. After being stirred into a suspension, the starch is heated to form a transparent or semi-transparent resin-like starch solution. However, the properties of rosin status starch solution are unstable, and prone to retro gradation, and these shortcomings severely limit its industrial application.

Chemical modification is the main way to expand its application. Peroxides, hypochlorite and other oxidants may be used to oxidise starch molecules to carboxyl and carbonyl groups. The decreased hydroxyl group will diminish the role of hydrogen bonding in starch molecules and weaken the retrogradation of starch resin solution.

On the other hand, part of glycosidic-bond connected anhydrous glucose unit will be broken under the action of an oxidising agent, further degrading the molecules. The decrease in molecular weight helps to improve the solubility of the starch, and hence an oxidised starch resin solution with low viscosity and high solid content can be obtained [50, 51]. Autio *et al.* studied the oxidation process of potato starch. The results showed that with increased oxidation degree, so the polymerisation degree, molecular weight, and gelatinisation temperature decreased, while the viscosity increased [52]. The disadvantage of this method for producing oxidised starch-based natural resin is the low oxidation degree. It is difficult to obtain a high oxidisation degree by a simple and low-cost technology, because the introduced carboxyl group with single regiment molecular structure impairs the water resistance of oxidised starch-based natural resin.

Esterification reaction of hydroxyl and carboxyl groups can also be used to prepare starch-based natural resin, and the properties of the native starch can be significantly changed at a low degree of substitution. There are three hydroxyl groups in each glucose unit of starch molecule, and the formation of monoester, diester and triester compounds is dependent on the number of hydroxyl groups participating in the reaction process. Starch esters can be classified into inorganic and organic acid esters. Phosphates starch, xanthates starch, sulphates starch, etc., are the most used inorganic acid esters, and the main organic acid ester is acetate starch, also known as acetylated starch. In the esterification process, the hydrogen bonding associated with starch molecules can be prevented or reduced by introducing some functional groups into the starch, which results in a decreased gelatinisation temperature, weakened retrogradation, and better gelatinised starch solution transparency and film-forming properties. Chi *et al.* studied starch acetate of different degrees of substitution prepared from corn starch with acetic anhydride. It was found that the starch granules formed a new crystalline region, and that the hydrophilic starch decreased with an increase in the degree of substitution [53]. Compared to the oxidation process, higher costs and lower degree of substitution are the disadvantages of the esterification process for the preparation of starch-based natural resins. Although the oxidised starch possesses better properties, it still cannot meet the higher requirements necessary for extensive application.

Crosslinked starch is another type of starch, in which the hydroxyl molecules are crosslinked with chemical agents via the etherified or esterified bond, and the derivatives with multi-dimensional network structure. After gelatinisation, the starch may also be used to prepare natural resin. There is acylated crosslinking, esterification crosslinking and etherification crosslinking, depending on the different crosslinking reaction processes. The improved conjunction in the starch molecule after the cross-linking reaction strengthened the stability of the starch granules. In addition, grafting followed by crosslinking is also an important method for preparing starch-based natural resin. To obtain products with desired properties, the selected monomers with different functional groups were grafted onto starch molecules, e.g. most of the vinyl monomers can react with starch, and the resulting copolymer has the advantageous properties of both natural and synthetic polymers. The preparation of starch-based natural resin via crosslinking or graft-crosslinking methods has been used in various composite products [54, 55].

12.3 Preparation and properties of natural resin plant fibre composites

It is generally believed that a composite is defined as a material with two or more raw materials combined to produce a system, through physical methods, so that the resulting system will have different properties from the raw materials. Composite not only retains the main features of the original materials (components) but it also gains enhanced performance from the combination of the raw materials. Composites are multiphase materials in microcosmic, with the continuous phase as the matrix and the dispersed phase as the reinforcement or function phase (functional component).

Natural resin plant fibre composite is a new environmentally friendly composite consisting of plant fibres (such as hemp fibre or wood fibre) and a thermoplastic or thermoset natural resin as the matrix. In the composite, plant fibre is the main component-bearing load: on one hand it improves the strength, modulus and heat distortion temperature of the material; on the other hand, it reduces the shrinkage. In addition, compared to synthetic fibre, plant fibre has the advantageous features of low price, low density and being readily biodegradable. The incorporation of plant fibre as a reinforcement in resin composite has been intensively studied over recent decades. Bioresin–natural fibre composites are normally considered as natural fibre reinforced composites (>50 wt% natural resin as matrix) or bioresin bonded plant fibre composites (<20 wt% bioresin as binder), depending on the amount of natural resin in the composites.

12.3.1 Natural resin matrix plant fibre composites

12.3.1.1 Vegetable protein-based natural resin matrix plant fibre composites

Of the various vegetable proteins, soy protein has attracted the greatest attention from researchers and engineers. Soy protein is biodegradable but has poor mechanical properties. The use of plant fibres in soy protein is expected to produce a composite material with better mechanical properties that is completely biodegradable. For instance, Indian grass fibres can be used to enhance soybean protein composite [56]. Alkali treatment removes the hemicellulose and lignin, leading to reduced area and weakened adhesion of the fibrils in the fibres. Treated Indian grass fibres, therefore, have a higher aspect ratio of length to breadth, giving the benefit of uniform distribution in the matrix and thus improvement of mechanical properties of the composites. Moreover, the alkaline treatment also increases the amount of hydroxyl groups on the fibre surface as well as the interaction between the fibre and the matrix. Attempts have been made to produce composites by blending soy flour with pineapple leaf fibres (PALF) [57]. It was found that the mechanical properties, including the tensile, flexural and impact properties of the composite, were improved with the addition of vegetable fibre and compatibiliser. The fibres appeared in the matrix in two different ways, i.e. one of small fibre size, representing good fibre distribution, the other of large-size due to fibre aggregation, representing poor fibre distribution. With an increase in the fibre content, the fibre distribution in the matrix decreased, but it was improved after adding compatibiliser. Since the fibres became more decentralised after the addition of compatibiliser, and the interfacial adhesion between the fibre and the matrix was enhanced, the stress can be smoothly transferred to the fibres from the matrix, thus increasing the mechanical properties of the material. In addition, fibres of ramie [58, 59], hemp [60], flax [61–63], kenaf [64], etc., can also be used to prepare soybean protein composites. Wu *et al.* investigated a blended material of cellulose/soy protein prepared using ionic liquids as the co-solvent of cellulose and soy protein [65]. The blend films showed good compatibility, decreased water vapour permeability, and excellent barrier properties against oxygen and carbon dioxide. Also, the tensile strength, elongation at break, water resistance, and thermal stability increased with an increase in the cellulose content. The surface and fracture sections of the blend films have shown dense and uniform structure with no obvious phase separation, indicating good compatibility between the protein and cellulose.

In addition to being blended with natural polymers, soy protein may also be blended with natural polymer derivatives to prepare composites. For example, the protein may be blended with cellulose derivatives such as methyl cellulose, hydroxyethyl cellulose, and hydroxypropyl cellulose, to prepare a biodegradable material [66]. Su *et al.* prepared soy protein isolate/carboxymethyl cellulose films. Compared to pure protein material, blended film has a higher tensile strength and elongation at break [67]. With an increase in the CMC mixture, the water sensitivity of the material is reduced, which can be attributed to the Maillard reaction between SPI and CMC (Fig. 12.10). Soy protein contains 18 kinds of amino acids, and the peptide bonds are the main structure. Therefore, the Maillard reaction between CMC and protein readily occurs. Dastidar and Netravali also prepared a soy protein based resin plant fibre composite via the Maillard reaction between soy protein and carbohydrates [68]. The study noted that isolated carbohydrates from defatted soy flour and oxidation increase the aldehyde group content in carbohydrate, then a good water resistance of soy protein natural resins can be obtained after mixing the oxidised carbohydrates with soy protein. The elastic modulus and fracture strength of the composite were increased by 476% and 530% respectively when 5% of original micro fibre was added to the resin. Moreover, a 4.3 GPa Young's modulus was achievable when nanocellulose, nanoclay, and flax fibre were added into the soy protein. This was due to the formation of a nanocomposite structure between the soy protein and the reinforcing phase [69–71].

Numerous studies show that soy protein natural resins have a major impact on the performance of these composites. Soy protein was treated with hot alkali and acid to

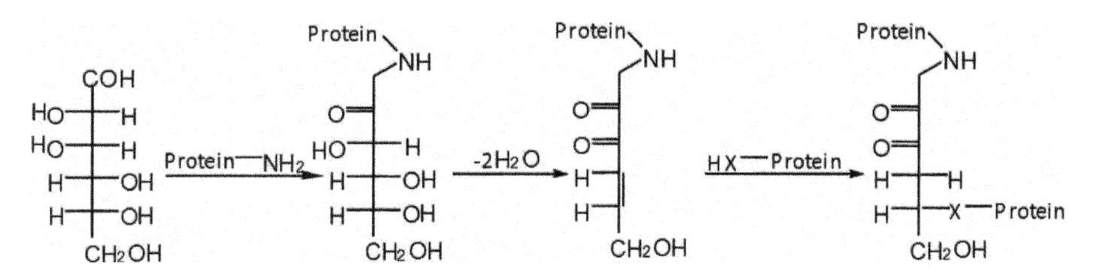

Figure 12.10 Schematic of the Maillard reaction in carbohydrate and protein.

prepare natural resin, and then mixed with wood flour to formulate composites. The treatment improved the tensile strength and water resistance of the composites. The properties could be further improved by the addition of some crosslinking agents [72, 73]. Hot acid and alkaline treatments of soy protein mainly improved the content of active functional groups and gave rise to a better dry cohesion after curing. Crosslinking agent improved the interfacial interaction between wood flour and natural resin through the formation of chemical bonds, resulting in improved overall performance of the composites.

12.3.1.2 Lignin-based natural resin matrix plant fibre composites

The interface between matrix resin and plant fibre affects the properties of the composites. Lignin as the matrix resin will decrease the flowability and wettability of plant fibre, thereby affecting the final properties of the composites [74]. Thus, lignin is mainly used as a compatibiliser between the base resin and the plant fibres to improve performance. Adding lignin also affects the viscosity of the matrix resin (Fig. 12.11). As can be seen in Fig. 12.11, when 1% lignin was added, the viscosity of the matrix resin (polypropylene) was increased by 35%; when 15% of lignin was added, the viscosity was further increased by 503%. Lignin as a compatibiliser of coir and polypropylene can significantly improve the mechanical properties of the composite, but the results are affected by the combination of lignin and coir. Rozman *et al.* reported that 5% (by mass) of lignin added in plant fibre considerably increased the modulus of elasticity and tensile strength of the composite, while decreasing the hydrophilicity and thickness swelling of the composite [75].

The effect of lignin as a compatibiliser on different plant fibres and matrix resins varies. The impact resistance of epoxy/hemp fibre composite can be enhanced

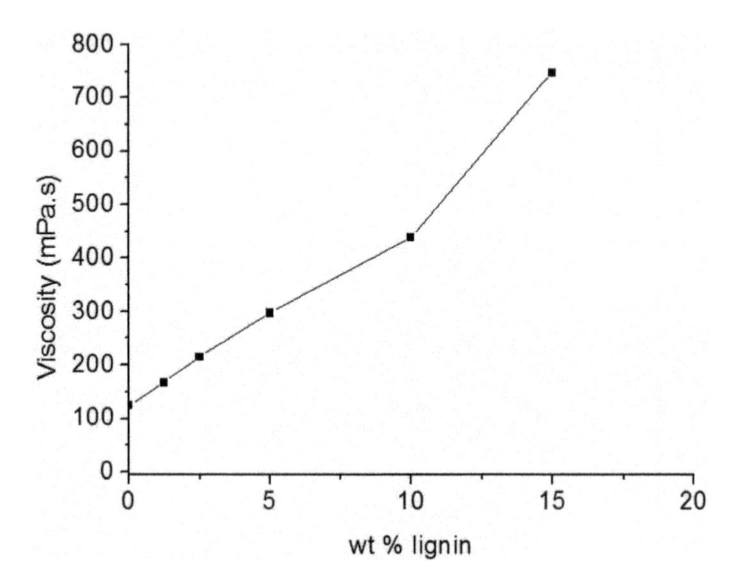

Figure 12.11 Effect of lignin content in styrene on the viscosity.

significantly by adding lignin into the system [76]. In addition, the performance of such composite materials is also related to the distribution and size of the plant fibres: usually, shorter fibre and uniform distribution promote the wetting of resin in the surface of plant fibre. Studies of sulphate lignin blended with unsaturated thermoset resins (epoxy acrylate styrene and soybean oil system), and then moulded into a flax fibre composite, have shown that the interface between the thermoset resin and the plant fibre has a significant influence on the performance of the composite, and the combination effect of plant fibre can be increased by the lignin, e.g. the addition of 5% (by mass) lignin into the composite will increase the anti-bending strength of the composite by 40% [77]. There are minor effects of lignin on the mechanical performance of composites, but the water resistance and heat resistance of the material can be enhanced when lignin is used as a compatibiliser for phenolic resin-based plant fibre composites [78, 79].

12.3.1.3 Vegetable oil based natural resin matrix plant fibre composites

Generally, vegetable oil based thermoset is the matrix resin for preparing vegetable oil based natural resin plant fibre composites. Different raw materials result in different composite properties, e.g. vegetable oil or oil (such as tung oil and castor oil) are with much distinct contents of reactive functional groups, the crosslinking density of the composite fabricated with chemically active matrix resin is relatively higher. Plant fibre composites have the features of high brittleness, strength and hardness. Vegetable oil (e.g. soybean oil) with less functional groups has good flexibility. Thus, the curing of modified vegetable oil requires a curing agent due to the formation of new functional groups in the modification process.

Acrylate copolymer epoxidised soybean oil is a common matrix resin for the development of vegetable oil fibre composite, and has been widely used in the preparation of composites with sisal, flax, hemp, and pulp fibres [80]. The curing process of the double bond within epoxidised soybean oil was affected by the curing agent styrene. The storage modulus and glass transition temperature (T_g) of the matrix resin were enhanced by an increase in styrene content. The storage and loss moduli of the composites increased with the content of plant fibre when the dosage of styrene is one-third of the resin matrix. The storage modulus of the composites can be increased from 1.1 GPa to 5.0 GPa, and the loss modulus can be increased from nearly 190 MPa to 430 MPa with the addition of 50% of plant fibres (such as waste paper) [81].

Physical or chemical treatments are important methods for improving the interfacial compatibility and thus performance of the composites. The tightness between the matrix and reinforcement phases can be seen by the variation in T_g of the composites. Substances (such as maleic anhydride, aluminum acetate, titanate, and isocyanate) containing multifunctional groups can react with both the matrix resin and the plant fibre and thus enhance the compatibility of the composite materials. Pfister and Larock studied flaxseed oil switchgrass reinforced composites using maleic anhydride as the compatibiliser. Results showed that maleic anhydride reacted with the matrix resin through C-C double bond, and acid anhydride formed ester bond with the hydroxyl of plant fibres, thereby improving the bonding strength between the resin and plant fibres [82]. In addition, the heat resistance of the composites was governed by the plant fibre, because the three-dimensional structure of the cured adhesive in general suggested better performance than the fibre.

12.3.1.4 Starch-based natural resin matrix plant fibre composites

The strength of starch-based natural resin is inferior to that of vegetable fibres; therefore, the mechanical properties of the composites mainly depend on the plant fibres. The tensile strength and elastic modulus of the composites are proportional to the content and length of the plant fibres when they are completely covered by starch-based resins [83, 84]. Since starch is a polymer made by plants to store energy, it contains a considerable number of hydroxyl groups, which results in shortcomings such as poor water resistance and durability when it is used as resin material. These problems cannot be avoided when starch is employed in preparing plant fibre composites. Methods such as closuring, reducing or eliminating the hydroxyls of starch and plant fibres are used to improve the water resistance of the composites, and hydrophobic material or chemical substances capable of reacting with hydroxyl groups can be used as the modifiers.

Girones *et al.* blended glycerin, natural rubber latex and corn starch first, and then moulded it with sisal or hemp to prepare composite [85]. The results showed that the water absorption of the composites significantly decreased after 200 hours of treatment. The tensile and bending strength of the composites increased with an increase in plant fibre content. Carboxylic acid was able to react with the hydroxyl in the starch or plant fibres to form esters and also improve the water resistance of the natural resin. Fig. 12.12 shows the water-resistance of the composites in which corn starch was crosslinked by acid 4 n-butane. It can be seen from Fig. 12.12 that the water-resistance of the starch increased with an increase in the amount of 4-carboxylic acid n-butane.

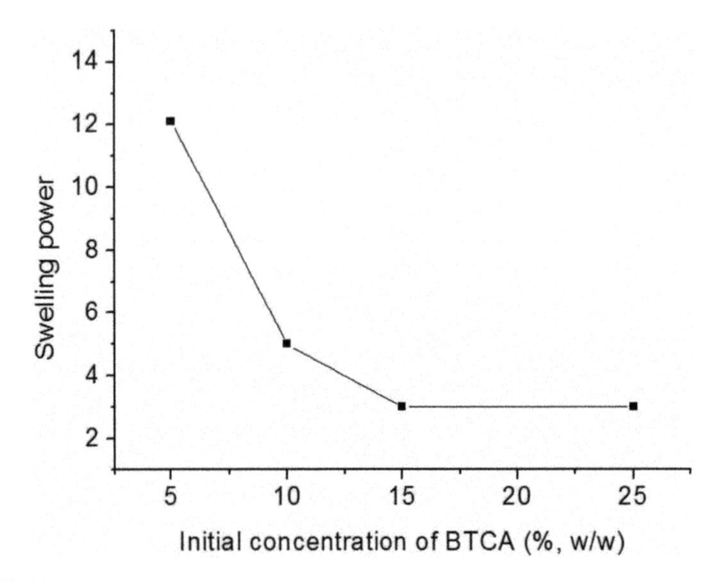

Figure 12.12 Relationship between the amount of n-4-carboxylate and the water resistance of starch corn.

Mechanical properties such as elastic modulus and tensile strength could be considerably increased if the original micro plant fibre were used to prepare composites. These phenomena should not be ascribed to the reinforcement of plant fibre on the resin alone: n-4-carboxylate also works with plant fibres to form ester bonds, and improves the interfacial compatibility of the composites [55].

12.3.2 Plant fibre based natural resin composite

12.3.2.1 Wood fibre based natural resin composite

Wood fibre, as a form of wood, represents the largest production of renewable resources on Earth. There are a number of reasons for using wood fibre with a variety of resins (such as aldehyde resin) to develop composites. Wood-based panels (such as plywood, particle board, and fibreboard) are typical examples of this type of composite, and the performance is mainly related to the resin (e.g. adhesive), wood materials, and production process. Water resistance is a common parameter used to evaluate the application of wood-based panels, and it can be classified into low, medium, and high water resistance. The matrix resin is a decisive factor affecting the performance of wood-based panels.

Aldehyde resin is one of the most commonly used resins in wood-based panels, but the production or application process of aldehyde resin releases formaldehyde and other toxic substances, which are harmful to the health of humans. Moreover, the raw material of these resins is mainly derived from unrenewable fossil resources. Therefore, the development of environmentally friendly natural resin deriving from renewable raw material for the production of wood-based panels has become a trend, wherein the resin derived from soybean protein and starch are the most extensively studied alternatives.

The natural resins obtained through milder methods, such as acid-, alkali- and surfactants-treated soy protein and starch, often have poor water resistance. High humidity or cold-water immersion drying cycle is commonly used to detect the resin's resistance to cold water, and to assess its suitability for the production of low water resistant wood-based panels [86, 87]. For example, medium-density fibreboard prepared from SDS-modified soy protein natural resin can be used in interior dry environments [88]. Natural resin prepared from urea, choline chloride, and glycerin modified corn starch was mixed with the corresponding wood material to fabricate particleboard, fibreboard and other composite materials [89]. Such wood-based panels with poor water resistance and high dry strength are suitable for interior application. By increasing the content of the functional group of the raw material, the cured natural resin will have warm water resistance [90]. Usually, the wood-based panel bonded with this adhesive will pass the warm-water (e.g. 63°C water) immersion test, and it is mainly used in humid indoor environments (e.g. bathrooms) [91]. Gui *et al.* blended soy flour with modifiers (polyamide resin and maleic anhydride) to prepare poplar plywood, and the highest bonding strength of the plywood reached 0.82 MPa [92]. But the bonding strength can be improved to 0.95 MPa when the modifier is changed to itaconic acid polyamide epichlorohydrin resin [93]. Lei *et al.* studied the water resistance of epoxy resin/MF resin mixture and the resin mixture blended with defatted soy flour. The results suggested that the mixture of epoxy resin and MF was the best modifier for soy protein natural resin; it can be used for preparing plywood type II (warm-water resistance) [94].

Medium water-resistant wood-based panels have a wider range of applications and they meet the requirements of home safety that end users expect. Therefore, this type of panel has an expanding market and research is deepening. Increasing the functional groups of the raw material, and then blending with other resin, results in excellent water resistance. The prepared natural resin has a three-dimensional structure after curing, and wood-based panels bonded with this resin are resistant to boiling water. To determine the water resistance of wood-based panel, the sample is usually soaked in boiling water for 4 hours, dried in a chamber at 63°C for 20 hours, and then soaked in boiling water for another 4 hours [14, 95, 96]. The wood-based panels that pass the test are suitable for high-humidity and exterior environments.

Yang *et al.* studied the natural resin prepared from defatted soy flour treated with sodium bisulphite and alkaline, and then blended with phenolic resin at a weight ratio of 1:1. The resin was then used to prepare southern pine plywood and tested under the American Standard PS 1-95 for exterior plywood requirements. The results showed that the bonding strength of the plywood is 172 psi (1.19 MPa) and 140 psi (0.97 MPa), but that that of the control PF is only 106 psi (0.73 MPa) [97]. Liu and Li reported a soy protein isolated resin modified with 10% maleic anhydride, and then mixed it with 20% polyethyleneimine. The resin was applied in preparing poplar plywood, which showed a shear strength of about 1.75 MPa, but the control PF plywood has a shear strength of 4.0 MPa [98]. The maximum shear strength of the plywood was just 70 psi (0.48 MPa) when soy protein isolate was replaced with defatted soy flour [99].

Zhang *et al.* prepared a natural resin by first degrading defatted soy flour using alkali, and then crosslinking by formaldehyde, and finally reacting with phenol. The poplar plywood was bonded by this plywood and was able to reach the requirements of plywood for type I (≥0.7 MPa) [100]. Gao and Gu degraded the soy protein isolate with 9% sodium hydroxide solution at 90–92°C, and then blended the slurry with MF. The birch plywood bonded with this resin was soaked in boiling water for 4 hours, and the bonding strength was 0.70 MPa, while the control group was completely delaminated [101]. Highly water-resistant wood-based panels have better performance and their application is also wider than other kinds of wood-based panel.

The effect of the wood on the properties of wood-based composite is mainly reflected in aspects that include density, moisture content, texture and fibre, extracts, and natural defects [102]. The strength of the wood is positively correlated with its own density: therefore, high-density wood-based panels usually have a glue layer or glued interface destruction, whereas the destruction of low-density plywood is the opposite. Moisture content will directly affect the final properties of the composites [88]. Generally, the fast absorption of resin by dry timber will easily lead to a lack of resin at the interface and decrease the bonding strength of wood-based panels. However, a high moisture content of wood can easily dilute the resin, resulting in decreased viscosity, a slow curing process, and some defects in the panel.

Wood is an anisotropic material, and the mechanical properties in its three sections (radial section, string section, cross-section) display great differences. Thus, the texture of the glued surface and the fibre in different directions will produce different internal stress, and further affect the durability of wood-based panels. There are complex compositions in timber extraction, the impact of which is mainly hampered moistening and curing of natural resin, ultimately weakening the performance of wood-based panels. Natural defects are mainly generated from the growth process of the trees. Some flaws (such as knots, stress wood, resin ducts) will weaken the bonding quality of natural resin, resulting in decreased strength and mechanical properties of the final product.

The production process of wood-based panels mainly involves a compression moulding process and a curing process of the natural resin. The curing status of natural resins is closely related to the properties of wood-based panels. Natural resin easily and completely cured under high temperature results in the preferred mechanical properties of wood-based panels. There are correlations between the processing time and the temperature. A high temperature guarantees the properties of the wood-based panels and decreases the production time. But high temperature and long processing time can easily lead to excessively compressed wood-based panels and decrease the production yield. Proper pressure can compact the wood slabs and promote natural resin to flow into the pores of wooden materials. The resin generates gel nail after curing, and further improve the mechanical properties of wood-based panels.

12.3.2.2 Bamboo fibre based natural resin composites

Bamboo fibre, powder or stick are the major reinforcements of bamboo-based natural resin composites. The mixture of bamboo fibre or powder and natural resin can produce composites by the moulding or extruding process, and the mixture of bamboo stick and natural resin is generally used to produce compression moulding composite. The composite compression moulded from bamboo fibre and soy protein showed improved fracture stress and Young's modulus with no significant effect on fracture strain. This indicated significantly improved toughness of the composite [70]. In order to further improve the properties of the composite, not only modification of the natural resin was used to enhance the cohesive strength after curing: adding a coupling agent in the preparation of the composite also enabled enhancement of the interfacial properties. For example, lysine diisocyanate-treated bamboo fibres can lead to the isocyanate group reacting with hydroxyl in bamboo fibre and producing a urethane bond, and the lysine may form a bond structure with the matrix resin containing carboxyl amide, ultimately improving the binding properties of the interface between the bamboo fibre material and the matrix resin [103].

There are also studies suggesting that coupling agents did not significantly increase the fracture strength and Young's modulus of the composite material, but did improve the fracture toughness of the material [70]. Typically, the density and longitudinal toughness of bamboo materials are stronger than wood, but the internal stress is also higher than for wood. Therefore, the matrix of bamboo-based composite is mainly petroleum-based resin (e.g. PF, polyester, and epoxy resin), and bamboo-reinforced natural resin composites are not as common as wood-reinforced natural resin composites [104, 105].

12.3.2.3 Non-food crop fibre-based natural resin composites

There are more than 100 species of non-food crops. Non-food crop based fibre mainly refers to annual or perennial herbaceous dicotyledon fibres (e.g. sesame, flax.) and monocot leaf fibres (e.g. sisal, hemp, pineapple), with high specific strength and modulus (Table 12.1) [106, 107]. Although non-food crop fibre reinforced synthetic

Table 12.1 Physico-mechanical properties of fibres.

Fibre	Density (g cm^{-3})	Tensile strength (MPa)	Young's modulus (GPa)	Elongation at break (%)
Ramie	1.5	400–938	61.4–128	1.2–3.8
Jute	1.3–1.45	393–773	13–26.5	7–8
Flax	1.5–3	45–1100	27.6	2.7–3.2
Hemp	1.48	690	30.1	1.6
Sisal	1.45	486–640	9.4–22	3–7

resin composite materials have been widely used in packaging, cushioning and automotive interiors [108, 109], a number of fibre-reinforced natural thermoplastic and thermoset composites are emerging. For instance, the composites can be produced by moulding a mixture of hemp fibre and starch-based resin [84]. The studies suggested that the moulding temperature should not exceed 160 °C and that the tensile and flexural strength of the composites increased with an increase in the hemp fibre content when hemp fibre amount was less than 70%. Tensile failure is mainly caused by breakage of the hemp fibre, and bending damage is initiated by separation of the fibre and the matrix resin, resulting in a much lower bending strength than tensile strength.

A study of composites of jute fibre and starch resin shows that the interface between the matrix resin and the reinforcement was vitally important for the composites. The properties of glutaraldehyde crosslinked composite prepared from jute fibres and starch can be improved by more than 50% with good interface [110]. Guessasma *et al.* used both experimental and theoretical techniques to study the interfacial behaviour of starch natural resin and hemp fibre in the preparation process of composites [111]. Mechanical testing revealed a loss of stiffness of the composites due to limited load transfer with the increase of fibre content. The two-phase theories predict a reinforcement trend due to the contrary phase properties. Le *et al.* prepared a cementitious composite from marijuana and starch resin. Studies have shown that the porous material can be used for building filler, the density of which is 158.9–170.8 kg m^{-3}, being lower than that of lime-hemp based porous material, but the composites have a low compressive strength and modulus of elasticity [112]. Currently, there are many reports on non-food crop fibre reinforced natural resin composites, but their commercialisation has rarely been achieved [113, 114].

12.4 Conclusion and future trends

Natural plant fibre reinforced bio-resin composites and bioresin-bonded natural plant fibre composites are two different types of bioresin natural fibre composites, depending on the proportion of bioresin content. The preparation and properties of bioresin derived from vegetable protein, lignin, vegetable oil, and starch have been studied. Bioresin with reactive functional groups can readily react with modification agents and has been widely studied due to it being renewable and environmentally benign. Natural plant fibre reinforced bioresin matrix composites are mainly prepared by extrusion moulding. Bioresin-bonded natural plant fibre composites can be prepared via extrusion and compression moulding. Whilst the literature on bioresin-natural fibre composites is extensive, the industrialisation or commercialisation of composites remains in its infancy. Thus, recent advances necessitate the speedy development of bioresin-natural fibre composites with enhanced properties and reduced cost over petroleum-based resin composites.

References

[1] Lee A L & Wand A J. (2001). Microscopic origins of entropy, heat capacity and the glass transition in proteins. *Nature*, 411, 501–504.

[2] Réat V, Dunn R, Ferrand M, Finney JL, Daniel R M & Smith J C. (2000). Solvent dependence of dynamic transitions in protein solutions. *Proceedings of the National Academy of Sciences*, 97, 9961–9966.

[3] Lambuth A L. (2003). Protein adhesives for wood, In: (ed.). *Handbook of Adhesive Technology, Revised and Expanded Edition*, Taylor & Francis Group, LLC,

[4] Moure A, Sineiro J, Domínguez H & Parajó J C. (2006). Functionality of oilseed protein products: A review. *Food Research International*, 39, 945–963.

[5] Morrisey P, Olbrantz K & Greaser M. (1982). A simple, sensitive enzymatic method for quantitation of soya proteins in soya-meat blends. *Meat Science*, 7, 109–116.

[6] Keshun L. (1997). *Soybeans: Chemistry, Technology, and Utilization.* London: Chapman & Hall.

[7] Kumar R, Choudhary V, Mishra S, Varma I & Mattiason B. (2002). Adhesives and plastics based on soy protein products. *Industrial Crops and Products*, 16, 155–172.

[8] Frihart C R & Birkeland M J. (2014).. Soy properties and soy wood adhesives, In: (ed.). *Soy-based Chemicals and Materials*, ACS Publications.

[9] Chen N, Lin Q & Bian L. (2012). Curing mechanism of modified soy-based adhesive and optimized plywood hot-pressing technology. *T. Chinese Soc. Agri. Eng.*, 28, 248–253.

[10] Lin Q, Chen N, Bian L & Fan M. (2012). Development and mechanism characterization of high performance soy-based bio-adhesives. *International Journal of Adhesion and Adhesives*, 34, 11–16.

[11] Chen N, Lin Q, Rao J & Zeng Q. (2013). Water resistances and bonding strengths of soy-based adhesives containing different carbohydrates. *Industrial Crops and Products*, 50, 44–49.

[12] Frihart C R & Wescott J M. (2004). Improved water resistance of bio-based adhesives for wood bonding. In: (Ed.). *Proceedings, 1st International Conference on Environmentally Compatible Forest Products, September*,

[13] Chen N, Lin Q, Zeng Q, Rao J. (2013). Optimization of preparation conditions of soy flour adhesive for plywood by response surface methodology. *Ind. Crop. Prod.*, 51, 267–273.

[14] Li K, Peshkova S & Geng X. (2004). Investigation of soy protein–Kymene® adhesive systems for wood composites. *Journal of the American Oil Chemists' Society*, 81, 487–491.

[15] Chen N, Zeng Q, Lin Q & Rao J. (2015). Development of defatted soy flour based bio-adhesives using Viscozyme L. *Ind. Crop. Prod.*, 76, 198–203.

[16] Pacheco-Torgal F, Ivanov V, Karak N & Jonkers H. (2016). *Biopolymers and Biotech Admixtures for Eco-efficient Construction Materials.* Cambridge: Woodhead Publishing (Elsevier Science).

[17] Stewart D. (2008). Lignin as a base material for materials applications: Chemistry, application and economics. *Industrial Crops and Products*, 27, 202–207.

[18] Saraf V P & Glasser W G. (1984). Engineering plastics from lignin. III. Structure property relationships in solution cast polyurethane films. *Journal of Applied Polymer Science*, 29, 1831–1841.

[19] Chung H & Washburn N R. (2013). Chemistry of lignin-based materials. *Green Materials*, 1, 137–160.

[20] Murch R M & Wood L L. (1982). Flexible polyurethane foams from polymethylene polyphenyl isocyanate containing prepolymers.

[21] Liu W, Jiang H & Yu H. (2015). Thermochemical conversion of lignin to functional materials: A review and future directions. *Green Chemistry*, 17, 4888–4907.

[22] Cinelli P, Anguillesi I & Lazzeri A. (2013). Green synthesis of flexible polyurethane foams from liquefied lignin. *European Polymer Journal*, 49, 1174–1184.

[23] Kelley S S, Glasser W G & Ward T C. (1989). Multiphase materials with lignin: 9. Effect of lignin content on interpenetrating polymer network properties. *Polymer*, 30, 2265–2268.

[24] Kelley S S, Ward T C & Glasser W G. (1990). Multiphase materials with lignin. VIII. Interpenetrating polymer networks from polyurethanes and polymethyl methacrylate. *Journal of Applied Polymer Science*, 41, 2813–2828.

[25] Sahoo S, Misra M & Mohanty A K. (2011). Enhanced properties of lignin-based biodegradable polymer composites using injection moulding process. *Composites Part A: Applied Science and Manufacturing*, 42, 1710–1718.

[26] Zhang L & Huang J. (2001). Effects of nitrolignin on mechanical properties of polyurethane-nitrolignin films. *Journal of Applied Polymer Science*, 80, 1213–1219.

[27] Ciobanu C, Ungureanu M, Ignat L, Ungureanu D & Popa V. (2004). Properties of lignin-polyurethane films prepared by casting method. *Industrial Crops and Products*, 20, 231–241.

[28] Sasaki C, Wanaka M, Takagi H, Tamura S, Asada C & Nakamura Y. (2013). Evaluation of epoxy resins synthesized from steam-exploded bamboo lignin. *Industrial Crops and Products*, 43, 757–761.

[29] Liu W, Zhou R, Goh HLS, Huang S & Lu X. (2014). From waste to functional additive: Toughening epoxy resin with lignin. *ACS Applied Materials & Interfaces*, 6, 5810–5817.

[30] Asada C, Basnet S, Otsuka M, Sasaki C & Nakamura Y. (2015). Epoxy resin synthesis using low molecular weight lignin separated from various lignocellulosic materials. *International Journal of Biological Macromolecules*, 74, 413–419.

[31] Nonaka Y, Tomita B & Hatano Y. (1997). Synthesis of lignin/epoxy resins in aqueous systems and their properties. *Holzforschung – International Journal of the Biology, Chemistry, Physics and Technology of Wood*, 51, 183–187.

[32] Fox N & Stachowiak G. (2007). Vegetable oil-based lubricants – a review of oxidation. *Tribology International*, 40, 1035–1046.

[33] Xia Y & Larock R C. (2010). Vegetable oil-based polymeric materials: Synthesis, properties, and applications. *Green Chemistry*, 12, 1893–1909.

[34] Ronda J C, Lligadas G, Galià M & Cádiz V. (2011). Vegetable oils as platform chemicals for polymer synthesis. *European Journal of Lipid Science and Technology*, 113, 46–58.

[35] Sharma V, Banait J & Kundu P. (2008). Spectroscopic characterization of linseed oil based polymers. *Industrial & Engineering Chemistry Research*, 47, 8566–8571.

[36] Lu Y & Larock R C. (2009). Novel biobased plastics, rubbers, composites, coatings and adhesives from agricultural oils and byproducts. *Polymer Preprints*, 50, 56.

[37] Quirino R L & Larock R C. (2011). Bioplastics, biocomposites, and biocoatings from natural oils.

[38] Pfister D P & Larock R C. (2012). Cationically cured natural oil-based green composites: Effect of the natural oil and the agricultural fiber. *Journal of Applied Polymer Science*, 123, 1392–1400.

[39] Pfister D P & Larock R C. (2010). Green composites from a conjugated linseed oil-based resin and wheat straw. *Composites Part A: Applied Science and Manufacturing*, 41, 1279–1288.
[40] Shibata M, Teramoto N & Nakamura Y. (2011). High performance bio-based thermosetting resins composed of tung oil and bismaleimide. *Journal of Applied Polymer Science*, 119, 896–901.
[41] Sharma V & Kundu P. (2006). Addition polymers from natural oils – A review. *Progress in Polymer Science*, 31, 983–1008.
[42] Çakmakli B, Hazer B, Tekin İ Ö & Cömert F B. (2005). Synthesis and characterization of polymeric soybean oil-g-methyl methacrylate (and n-butyl methacrylate) graft copolymers: Biocompatibility and bacterial adhesion. *Biomacromolecules*, 6, 1750–1758.
[43] Echeverri D, Cádiz V, Ronda J & Rios L. (2012). Synthesis of elastomeric networks from maleated soybean-oil glycerides by thiol-ene coupling. *European Polymer Journal*, 48, 2040–2049.
[44] Campanella A, La Scala J J & Wool R P. (2009). The use of acrylated fatty acid methyl esters as styrene replacements in triglyceride-based thermosetting polymers. *Polymer Engineering & Science*, 49, 2384–2392.
[45] Tan S & Chow W. (2010). Biobased epoxidized vegetable oils and its greener epoxy blends: A review. *Polymer-Plastics Technology and Engineering*, 49, 1581–1590.
[46] Tan S & Chow W. (2011). Curing characteristics and thermal properties of epoxidized soybean oil based thermosetting resin. *Journal of the American Oil Chemists' Society*, 88, 915–923.
[47] Wang Z, Zhang X, Wang R, Kang H, Qiao B, Ma J, Zhang L & Wang H. (2012). Synthesis and characterization of novel soybean-oil-based elastomers with favorable processability and tunable properties. *Macromolecules*, 45, 9010–9019.
[48] Espinoza-Perez J D, Nerenz B A, Haagenson D M, Chen Z, Ulven C A & Wiesenborn D P. (2011). Comparison of curing agents for epoxidized vegetable oils applied to composites. *Polymer Composites*, 32, 1806–1816.
[49] Whistler R L, BeMiller J N & Paschall E F. (2012). *Starch: Chemistry and Technology*, Elsevier Science,
[50] Kuakpetoon D & Wang Y J. (2001). Characterization of different starches oxidized by hypochlorite. *Starch-Stärke*, 53, 211–218.
[51] Ali S & Kempf W. (1986). On the degradation of potato starch during acid modification and hypochlorite oxidation. *Starch-Stärke*, 38, 83–86.
[52] Jiranuntakul W, Puttanlek C, Rungsardthong V, Puncha-Arnon S & Uttapap D. (2011). Microstructural and physicochemical properties of heat-moisture treated waxy and normal starches. *Journal of Food Engineering*, 104, 246–258.
[53] Chi H, Xu K, Wu X, Chen Q, Xue D, Song C, Zhang W & Wang P. (2008). Effect of acetylation on the properties of corn starch. *Food Chemistry*, 106, 923–928.
[54] Das K, Ray D, Bandyopadhyay N, Gupta A, Sengupta S, Sahoo S, Mohanty A & Misra M. (2010). Preparation and characterization of cross-linked starch/poly (vinyl alcohol) green films with low moisture absorption. *Ind. Eng. Chem. Res.*, 49, 2176–2185.
[55] Ghosh Dastidar T & Netravali A. (2013). Cross-linked waxy maize starch-based 'green' composites. *ACS Sustain. Chem. Eng.*, 1, 1537–1544.
[56] Liu W, Mohanty A K, Askeland P, Drzal L T & Misra M. (2004). Influence of fiber surface treatment on properties of Indian grass fiber reinforced soy protein based biocomposites. *Polymer*, 45, 7589–7596.
[57] Liu W, Misra M, Askeland P, Drzal L T & Mohanty A K. (2005). 'Green' composites from soy based plastic and pineapple leaf fiber: Fabrication and properties evaluation. *Polymer*, 46, 2710–2721.

[58] Kumar R & Zhang L. (2009). Aligned ramie fiber reinforced arylated soy protein composites with improved properties. *Composites Science and Technology*, 69, 555–560.

[59] Lodha P & Netravali A. (2002). Characterization of interfacial and mechanical properties of 'green' composites with soy protein isolate and ramie fiber. *Journal of Materials Science*, 37, 3657–3665.

[60] Mohanty A K, Tummala P, Liu W, Misra M, Mulukutla P V & Drzal L T. (2005). Injection molded biocomposites from soy protein based bioplastic and short industrial hemp fiber. *J. Polym. Environ.*, 13, 279–285.

[61] Chabba S & Netravali A N. (2005). 'Green'composites part 1: Characterization of flax fabric and glutaraldehyde modified soy protein concentrate composites. *Journal of Materials Science*, 40, 6263–6273.

[62] Chabba S, Matthews G & Netravali A. (2005). 'Green'composites using cross-linked soy flour and flax yarns. *Green Chemistry*, 7, 576–581.

[63] Huang X & Netravali A. (2007). Characterization of flax fiber reinforced soy protein resin based green composites modified with nano-clay particles. *Composites Science and Technology*, 67, 2005–2014.

[64] Liu W, Drzal L T, Mohanty A K & Misra M. (2007). Influence of processing methods and fiber length on physical properties of kenaf fiber reinforced soy based biocomposites. *Composites Part B: Engineering*, 38, 352–359.

[65] Wu R, Wang X, Wang Y, Bian X & Li F. (2009). Cellulose/soy protein isolate blend films prepared via room-temperature ionic liquid. *Industrial & Engineering Chemistry Research*, 48, 7132–7136.

[66] Zhou Z, Zheng H, Wei M, Huang J & Chen Y. (2008). Structure and mechanical properties of cellulose derivatives/soy protein isolate blends. *Journal of Applied Polymer Science*, 107, 3267–3274.

[67] Su J, Huang Z, Yuan X, Wang X & Li M. (2010). Structure and properties of carboxymethyl cellulose/soy protein isolate blend edible films crosslinked by Maillard reactions. *Carbohydrate Polymers*, 79, 145–153.

[68] Dastidar T G & Netravali A N. (2013). A soy flour based thermoset resin without the use of any external crosslinker. *Green Chemistry*, 15, 3243–3251.

[69] Huang X & Netravali A N. (2006). Characterization of nano-clay reinforced phytagel-modified soy protein concentrate resin. *Biomacromolecules*, 7, 2783–2789.

[70] Huang X & Netravali A. (2009). Biodegradable green composites made using bamboo micro/nano-fibrils and chemically modified soy protein resin. *Composites Science and Technology*, 69, 1009–1015.

[71] Huang X, Jiang P, Kim C, Ke Q & Wang G. (2008). Preparation, microstructure and properties of polyethylene aluminum nanocomposite dielectrics. *Composites Science and Technology*, 68, 2134–2140.

[72] Zhang Y, Zhu W, Gao Z & Gu J. (2015). Effects of crosslinking on the mechanical properties and biodegradability of soybean protein-based composites. *Journal of Applied Polymer Science*, 132, n/a–n/a.

[73] Gao Z, Zhang Y, Fang B, Zhang L &, Shi J. (2015). The effects of thermal-acid treatment and crosslinking on the water resistance of soybean protein. *Industrial Crops and Products*, 74, 122–131.

[74] Sen S, Patil S & Argyropoulos D S. (2015). Thermal properties of lignin in copolymers, blends, and composites: A review. *Green Chemistry*, 17, 4862–4887.

[75] Rozman H, Tan K, Kumar R, Abubakar A, Ishak Z M & Ismail H. (2000). The effect of lignin as a compatibilizer on the physical properties of coconut fiber-polypropylene composites. *European Polymer Journal*, 36, 1483–1494.

[76] Wood B M, Coles S R, Maggs S, Meredith J & Kirwan K. (2011). Use of lignin as a compatibiliser in hemp/epoxy composites. *Composites Science and Technology*, 71, 1804–1810.

[77] Thielemans W & Wool R P. (2004). Butyrated kraft lignin as compatibilizing agent for natural fiber reinforced thermoset composites. *Composites Part A: Applied Science and Manufacturing*, 35, 327–338.

[78] Megiatto J D, Silva C G, Rosa D S & Frollini E. (2008). Sisal chemically modified with lignins: correlation between fibers and phenolic composites properties. *Polymer Degradation and Stability*, 93, 1109–1121.

[79] Morandim-Giannetti A A, Agnelli J A M, Lanças B Z, Magnabosco R, Casarin S A & Bettini S H. (2012). Lignin as additive in polypropylene/coir composites: Thermal, mechanical and morphological properties. *Carbohydrate Polymers*, 87, 2563–2568.

[80] Zini E & Scandola M. (2011). Green composites: An overview. *Polymer Composites*, 32, 1905–1915.

[81] O'Donnell A, Dweib M & Wool R. (2004). Natural fiber composites with plant oil-based resin. *Composites Science and Technology*, 64, 1135–1145.

[82] Pfister D P & Larock R C. (2013). Green composites using switchgrass as a reinforcement for a conjugated linseed oil-based resin. *Journal of Applied Polymer Science*, 127, 1921–1928.

[83] Takagi H, Ichihara Y. (2004). Effect of fiber length on mechanical properties of ' green' composites using a starch-based resin and short bamboo fibers. *JSME International Journal Series A*, 47, 551–555.

[84] Ochi S. (2006). Development of high strength biodegradable composites using Manila hemp fiber and starch-based biodegradable resin. *Composites Part A: Applied Science and Manufacturing*, 37, 1879–1883.

[85] Girones J, Lopez J, Mutje P, Carvalho A, Curvelo A & Vilaseca F. (2012). Natural fiber-reinforced thermoplastic starch composites obtained by melt processing. *Composites Science and Technology*, 72, 858–863.

[86] Mo X, Sun X & Wang D. (2004). Thermal properties and adhesion strength of modified soybean storage proteins. *Journal of the American Oil Chemists' Society*, 81, 395–400.

[87] Wang Y, Wang D & Sun X. (2005). Thermal properties and adhesiveness of soy protein modified with cationic detergent. *Journal of the American Oil Chemists' Society*, 82, 357–363.

[88] Li X, Li Y, Zhong Z, Wang D, Ratto J A, Sheng K & Sun X S. (2009). Mechanical and water soaking properties of medium density fiberboard with wood fiber and soybean protein adhesive. *Bioresource Technology*, 100, 3556–3562.

[89] Abbott A P, Conde J P, Davis S J & Wise W R. (2012). Starch as a replacement for urea-formaldehyde in medium density fibreboard. *Green Chemistry*, 14, 3067–3070.

[90] Yingying Q, Yue Z, Cheng L, Shifeng Z & Jianzhang L. (2012). Preparation and application of modified soy-based protein adhesive. *China Adhesives*, 21, 27–30.

[91] Tong L, Lin Q, Weng X & Chen N. (2008). Environment-friendly wood adhesive preparation from soy. *Chinese Journal of Eco-Agriculture*, 16, 957–962.

[92] Gui C, Liu X, Wu D, Zhou T, Wang G & Zhu J. (2013). Preparation of a new type of polyamidoamine and its application for soy flour–based adhesives. *Journal of the American Oil Chemists' Society*, 90, 265–272.

[93] Gui C, Wang G, Wu D, Zhu J & Liu X. (2013). Synthesis of a bio-based polyamidoamine-epichlorohydrin resin and its application for soy-based adhesives. *Int. J. Adhes. Adhes.*, 44, 237–242.

[94] Lei H, Du G, Wu Z, Xi X & Dong Z. (2014). Cross-linked soy-based wood adhesives for plywood. *Int. J. Adhes. Adhes.*, 50, 199–203.

[95] Wang W & Xu G. (2006). Effect of soy adhesive/UF mixture on the shear strength of plywood. *China Forest Products Industry*, 33, 30–32.

[96] Wang W & Xu G. (2007). Effect of soy adhesive/UF mixture on the shear strength of plywood. *Journal of Northeast Forestry University*, 32, 57–58.

[97] Yang I, Kuo M & Myers D J. (2006). Bond quality of soy-based phenolic adhesives in southern pine plywood. *Journal of the American Oil Chemists' Society*, 83, 231–237.

[98] Liu Y & Li K. (2007). Development and characterization of adhesives from soy protein for bonding wood. *International Journal of Adhesion and Adhesives*, 27, 59–67.

[99] Huang J & Li K. (2008). A new soy flour-based adhesive for making interior type II plywood. *Journal of the American Oil Chemists' Society*, 85, 63–70.

[100] Zhang Y, Yu W & Zhu R. (2008). Study on preparation and bond properties of soy-based protein adhesive modified by phenol. *Chemistry and Adhesion*, 30, 13–16.

[101] Gao Z, Gu H. (2010). Preparation and characterization of wood adhesives with strong alkali-degraded soybean proteins. *Polymer Materials Science & Engineering*, 26, 126–129.

[102] Youngquist J A. (1999). Wood–based composites and panel products.

[103] Lee S H & Wang S. (2006). Biodegradable polymers/bamboo fiber biocomposite with bio-based coupling agent. *Composites Part A: Applied Science and Manufacturing*, 37, 80–91.

[104] Khalil H A, Bhat I, Jawaid M, Zaidon A, Hermawan D & Hadi Y. (2012). Bamboo fibre reinforced biocomposites: A review. *Materials & Design*, 42, 353–368.

[105] Liu D, Song J, Anderson D P, Chang P R & Hua Y. (2012). Bamboo fiber and its reinforced composites: Structure and properties. *Cellulose*, 19, 1449–1480.

[106] Thomas S & Pothan L A. (2009). *Natural Fibre Reinforced Polymer Composites: From Macro to Nanoscale.* Philapelphia, PA: Édition des Archives Contemporaines, Old City Publishing.

[107] Kalia S, Kaith B & Kaur I. (2009). Pretreatments of natural fibers and their application as reinforcing material in polymer composites – A review. *Polymer Engineering & Science*, 49, 1253–1272.

[108] Ramzy A, Beermann D, Steuernagel L, Meiners D & Ziegmann G. (2014). Developing a new generation of sisal composite fibres for use in industrial applications. *Composites Part B: Engineering*, 66, 287–298.

[109] Koronis G, Silva A & Fontul M. (2013). Green composites: A review of adequate materials for automotive applications. *Composites Part B: Engineering*, 44, 120–127.

[110] Iman M & Maji T K. (2012). Effect of crosslinker and nanoclay on starch and jute fabric based green nanocomposites. *Carbohydrate Polymers*, 89, 290–297.

[111] Guessasma S, Bassir D & Hedjazi L. (2015). Influence of interphase properties on the effective behaviour of a starch-hemp composite. *Materials & Design*, 65, 1053–1063.

[112] Le A, Gacoin A, Li A, Mai T, Rebay M & Delmas Y. (2014). Experimental investigation on the mechanical performance of starch-hemp composite materials. *Construction and Building Materials*, 61, 106–113.

[113] Shahzad A. (2012). Hemp fiber and its composites – A review. *J. Compos. Mater.*, 46, 973–986.

[114] Akil H, Omar M, Mazuki A, Safiee S, Ishak Z M & Bakar A A. (2011). Kenaf fiber reinforced composites: A review. *Materials & Design*, 32, 4107–4121.

Chapter 13
Wood–plastic composites

Dr Yonghui Zhou and Dr Wendi Liu

This chapter discusses advanced wood plastic composites (WPC). It begins with an outline of both the physical and chemical modification of natural fibres for the production of WPC. There follows an investigation of the bonding of natural and plastic matrices, and the interface structure of WPC. Finally the chapter discusses the bulk static and dynamic mechanical performance and the in situ *mechanical properties of WPC. Some suggestions for the future development of WPC materials are also made.*

13.1 Introduction

Wood plastic composites (WPC) are considered one of the most advanced materials and their use has grown consistently in the last decade in many industrial sectors, such as decking, automotives, siding, fencing and outdoor furniture. This is mainly because of the advantages that wood material possesses: namely, ubiquitous availability at low cost and in a variety of forms, bio-renewability and biodegradability, low density, non-toxicity, flexibility during processing, and acceptable specific strength properties [1–7]. However, the inherently highly polar and hydrophilic nature of wood flour or natural fibre makes it incompatible with hydrophobic and non-polar matrices, especially hydrocarbon matrices (e.g. polyethylene (PE), polypropylene (PP) and polyester) [8, 9]. This may cause problems in the composite processing and material performance, such as uneven distribution of the filler in the matrix during the compounding process and insufficient wetting of wood by the matrix, which results in weak interfacial adhesion and strength [1, 2, 10–13].

The interface of WPC is a heterogeneous transition zone extending from nanometres to microns with different morphological features, chemical composition and mechanical properties [14, 15]. Interfacial adhesion plays a fundamental role in the global performance of a composite. To formulate a reasonable WPC with optimum interface bonding, various modifications, including both physical (e.g. corona and plasma) and chemical approaches (e.g. alkaline, acetylation, silane and maleated coupling agent treatments), have been attempted to decrease the hydrophilicity of wood flour, enhance the wettability of wood by matrix polymer and eventually promote the interfacial adhesion of the constituents within the composite. With respect to the commercial production of WPC, incorporating coupling agents is probably the best available and feasible strategy for its interface optimisation [16].

The major factors affecting the mechanical performance of WPC include: (1) the strength and modulus of fibrous reinforcement; (2) the strength and chemical stability of the polymer matrix; and (3) the effectiveness of the load transfer across the interface [17].

Investigation of the interface is of special significance in understanding the macro-behaviour of composites [18]. Recently, direct determination of the property and size of the interface of WPC has been achieved with the advent of the depth-sensing indentation technique, which is generally referred to as nanoindentation. This technique allows the penetration of an indenter into the material surface with controlled force and synchronous recording of the force applied as a function of the indentation depth, thus providing considerable information on local mechanical properties [19].

13.2 Pretreatment of natural fibre

Extensive work has been done on the modification of natural fibres to make them more hydrophobic, thus rendering them more compatible with hydrophobic polymer resins. The modifications include both physical and chemical modifications.

13.2.1 Physical modification of natural fibres

13.2.1.1 Plasma treatment of natural fibres

Plasma is a low-density gas in which some of the individual atoms or molecules are ionised while the total number of positive charges is equal to the total number of negative charges. Plasma is typically generated by passing an electric current through a gas. With the electrical current inside a reactor (vessel) containing the gas, the electrons move very quickly and collide with the gas to dissociate the gas molecules into atoms. Further collisions between the electrons and the atoms lead to ionisation, forming ions and free radicals. Therefore, plasma contains active particles such as electrons, ions, radicals, and photons, that may initiate chemical and physical changes on the surfaces of materials such as natural fibres when they are exposed to the plasma.

There are three possible effects when natural fibres are treated with plasma [20, 21]:

1. cleaning and etching;
2. surface reactions; and
3. polymerisation.

The physical sputtering of the plasma can easily remove volatile compounds on the fibre surfaces, which is called cleaning. The plasma can also break down the molecules on the fibre surfaces, thus removing a greater amount of material on the fibre surfaces than cleaning itself, which is called etching. Functional groups are generated from the reactions between the active particles in the plasma and the molecules on the fibre surfaces. The crosslinking of molecules on the fibre surfaces may also occur through the coupling of the newly generated free-radicals by the active particles in the plasma. A thin film may also form on the fibre surfaces in the presence of an organic monomer.

Plasma has been extensively used for the surface modification of polymers. The plasma treatment of natural fibres affects the surfaces only within a few tens of nano-metres and thus does not affect the bulk properties of the fibres [22].

It has been reported that plasma treatment can induce dramatic changes in the surface morphology of plant fibres [23]. More specifically, some tiny grains, cracks and longitudinal grooves appeared on the surfaces of plasma-treated flax fibres, indicating that plasma treatment causes degradation and increases the surface roughness of the flax fibres. The surface roughness of the fibres increased with an increase in the incident power of the plasma treatment. The increase in the surface roughness enhances mechanical interlocking between the plasma-treated fibres and the polymer resins, thus increasing the interfacial adhesion. It was found that the plasma from ambient air was more effective than that from argon for the treatments of flax fibres in terms of the improvement of interfacial adhesion between the flax fibres and the polymer matrix [23]. The plasma treatment may generate reactive free radicals, double bonds and peroxides on the fibre surfaces when ambient air is used for the generation of the plasma [24]. These reactive functional groups may participate in the free-radical polymerisation of polymer resins and form covalent linkages between the fibres and the polymer resins, thus improving the interfacial adhesion between the fibres and matrices [24].

13.2.1.2 Heat treatment of natural fibres

The heat treatment of natural fibres is a commonly used pretreatment method for the reduction of their moisture content. Microstructures of the fibres have been known to change significantly along with the removal of the moisture during the heat treatment. A complete removal of moisture from natural fibres is essential if the superior properties of WPC are to be realised, because the residual water may evaporate to create bubbles and voids within the composites and weaken the interfacial adhesion during the preparation of the composites. In addition, the heat treatment may also affect the morphology and composition of the fibres to some extent. Depending on the temperature of the heat treatment, lignin may melt and flow, thus rearranging the molecular structures and making the fibre surfaces more hydrophobic and more compatible with polymer resins. Some oxidation reactions may also occur.

The treatment of hemp fibres at 100 °C or 150 °C, was found to significantly improve the tensile strength and tensile modulus of hemp fibre unsaturated polyester (UPE) composites [25, 26]. However, significant improvements in the tensile strength and modulus were not obtained when the fibres were treated at 200 °C. It was speculated that hemp fibres started to degrade at 200 °C, which resulted in a reduction of the fibre strength, thus offsetting the benefits of the improved interfacial adhesions brought by the heat treatment. In a separate study, the heat treatment of hemp fibres was also found to improve the tensile strength and tensile modulus of the hemp fibre-UPE composites [27]. However, treatment temperatures of 100 °C, 150 °C and 200 °C did not make much difference in the tensile strength of the composites, whereas the tensile modulus increased with an increase in temperature [27].

13.2.1.3 Other physical treatments of natural fibres

It has been demonstrated that the irradiation of jute and glass fibres with ultraviolet light at the optimum intensity significantly increases the mechanical (tensile, bending and impact) properties of the mixed jute/glass fibre–UPE composite [28]. It was found that the irradiation of henequen fibres with electron beams improved the interfacial adhesion between the fibres and UPE resins as well as the dynamic mechanical properties of henequen fibre-UPE composites [29]. Steam explosion was found to have the capacity to significantly improve the density, thermal stability, water resistance, and thermomechanical properties of sugarcane bagasse reinforced polymer composites [30].

13.2.2 Chemical treatment

Natural fibres are lignocellulosic materials that are inherently hydrophilic and incompatible with hydrophobic polymer matrix. One strategy for improving the strength and water resistance of WPC is to reduce the number of hydroxyl groups on the fibre surfaces through chemical modification of the fibres, thus making the fibres more hydrophobic and more compatible with the matrix. Another strategy is to impart the fibres with active functional groups that can form covalent linkages with polymer resins during the preparation of WPC whilst reducing the number of hydroxyl groups. Most of the chemical treatments fall into these two strategies.

13.2.2.1 Silane coupling agents

Silane coupling agents have been used extensively in glass fibre reinforced polymer composites, all of which contain a $–Si(OR)_3$ group that can react with the Si-OH groups in silica to form Si-O-Si bonds through the release of ROH. Natural fibres contain abundant C-OH groups that may be able to form C-O-Si bonds with the $–Si(OR)_3$ in the silane coupling agent. However, the required reaction conditions are so harsh that a direct formation of C-O-Si bonds between C-OH groups and the $–Si(OR)_3$ group is often accompanied by a substantial degradation of the natural fibres [31]. The $–Si(OR)_3$ group of a silane coupling agent has to be hydrolysed to form an $–Si(OH)_3$ group before the coupling agent can be used in WPC [31]. The $–Si(OH)_3$ groups can condense with each other to form oligomeric structures. The $–Si(OH)_3$ groups and their condensed structures are physically adsorbed onto fibre surfaces by forming hydrogen bonds with the hydroxyl groups of natural fibres. The hydrogen bonds can be converted into covalent -Si-O-C- bonds through the release of water at an elevated temperature. The hydrolysis of $–Si(OR)_3$ depends on the structure of the organic part of the silane and the hydrolytic conditions, such as pH, temperature and solvents, and will eventually affect the interaction between the silanols and natural fibres [31, 32]. The formation of Si-O-C bonds between natural fibres and the hydrolysed silane coupling agents is typically less efficient than that of Si-O-Si bonds between glass fibres and silane coupling agents, because less harsh reaction conditions are required for the treatment of natural fibres with silane coupling agents. The Si-O-C bonds in natural fibres treated with a silane coupling agent are more susceptible to hydrolysis with water than the Si-O-Si bonds in glass fibres. Therefore, natural fibres treated with silane coupling agent are typically not as water resistant as glass fibres treated with a silane coupling agent. An effective silane coupling agent requires a second functional group that can produce strong interactions, preferably forming covalent linkages, with the polymer matrix in fibre-reinforced polymer composites. For instance, a silane coupling agent containing an amino group is often used in the epoxy-based matrix because the amino group can effectively react with the epoxy group [33].

13.2.2.2 Anhydrides

The esterification of natural fibres with an anhydride is an effective method for substituting the hydroxyl groups of the fibres with new functional groups that are more hydrophobic and thus more compatible with the polymer matrix. The reaction of natural fibres with acetic anhydride introduces the acetyl functional group onto the fibres, thus reducing the number of hydroxyl groups and improving the hydrophobicity of the fibres, which makes the fibres more compatible with the hydrophobic matrix. Acetylation of empty fruit bunch (EFB) and coconut fibres significantly enhanced the interfacial shear strength between the fibres and resins [34]. Modifications of empty fruit bunch fibres with propionic and succinic anhydrides also significantly increased the density, tensile, flexural and impact properties, as well as the water resistance of the empty fruit bunch fibre–UPE composites at different fibre loadings [35].

Maleic anhydride (MA) can react with the hydroxyl groups of natural fibres to form ester linkages and also has a C=C bond that can form covalent linkages with polymer matrix, and thus it is considered as an effective coupling agent for WPC. MA was indeed found to increase the interfacial shear strength between hemp fibres and polymer matrix [36]. Treatment of short sisal fibres and wood flour with MA increased the static and dynamic mechanical properties as well as the water resistance of the resulting composites [37]. Treatment of wood flours from different wood species was found to improve the dispersion of these wood flours in the polymer matrix and improved the flexural and compression properties of the resulting WPC [38]. Alkenyl succinic anhydride (ASA) contains an anhydride group that can react with the hydroxyl groups of natural fibres, a long hydrocarbon chain that can significantly increase the hydrophobicity of natural fibres, and a C=C bond that can potentially form covalent linkages with polymer matrix. Treatment of wood fibres with ASA improved the dispersion of wood fibres in polyester matrix and increased the stiffness and resistance to creep of the wood fibre–polyester composites [39].

13.2.2.3 Alkaline treatment

Alkaline treatment of natural fibres is one of the commonly used chemical methods in the preparation of WPC. Alkaline treatment can remove the low-molecular weight materials that typically form weak layers between the natural fibres and the polymer matrix, and thus increase the porosity of the fibres and the roughness of fibre surfaces which can facilitate the penetration of the resins and the formation of mechanical interlocking between the fibres and the matrix. In the literatures, alkaline treatment has proven to be an effective method for increasing the mechanical properties, including tensile modulus, and the tensile and impact strength of WPC [40].

13.2.2.4 Benzoylation

Treatment of natural fibres with benzoyl chloride has been investigated for reducing the hydrophilicity of natural fibres [41–44]. It was found that the esterification of hemp fibres with benzoyl chloride greatly improved the mechanical and thermal properties of the resulting hemp fibre composites [41]. The benzoylation of palm leaf stalk fibres resulted in increased mechanical (tensile, flexural and impact) properties, water resistance and thermal properties of palm leaf stalk fibre–UPE composites [42]. The treatment of isora, sisal, and bamboo fibres with benzoyl chloride was also found to improve the interfacial adhesion and bonding between the natural fibres and polymer resins in WPC [43, 44].

13.2.2.5 Acrylic acid, acrylonitrile and styrene treatment

Agave and alfa fibres were treated with acrylic acid in a mixed water/benzene solution in the presence of benzoyl peroxide as an initiator [45–47]. During the treatment, acrylic acid was polymerised and entangled with the fibres, and the treated fibres were found to have lower overall water uptake than the untreated fibres [45–47]. These results are surprising because poly(acrylic acid) is known to be very hydrophilic and tends to absorb a large amount of water. The treated fibre/polyester composites had much lower tensile strength and tensile modulus than their untreated counterparts [47]. In other words, the treatment had a detrimental effect on the tensile properties of the resulting composites. By way of contrast, the treatment of mercerised wood fibres with acrylic acid in the presence of benzoyl peroxide as a free radical initiator was shown to increase the interfacial adhesion between the wood fibres and the polymer matrix [48].

Acrylonitrile has also been used to modify the surface polarity of natural fibres and thus to improve the wettability of the fibres by polymer resins [49, 50]. The cyanoethylation of natural fibres can be presented as follows: Fibre-OH + $H_2C = CHCN \rightarrow$ Fibre-O-CH_2CH_2CN. During the cyanoethylation, fibres are usually first swollen with a NaOH solution and then impregnated with an acrylonitrile solution. Another reaction between natural fibres and acrylonitrile is involved in the generation of free radicals on natural fibres typically with ammonium cerium (IV) nitrate followed by the free radical polymerisation of acrylonitrile, thus resulting in poly(acrylonitrile)-grafted fibres, which certainly reduces the hydrophilicity of natural fibres. The graft-polymerisation can be shown as follows: Fibre-OH + $H_2C = CHCN \rightarrow$ Fibre-O-$(CH_2CHCN)_n$.

13.3 Interface structure of WPC

The interface region formed between the wood flour and the matrix is in fact a zone of compositional, structural and property gradients, generally varying in width from a single atom layer to micrometres. Fig. 13.1 shows the effect of silane (bis(triethoxysilylpropyl) tetrasulfide (Si69) and vinyltrimethoxysilane (VTMS)) and maleic coupling agent (maleic anhydride grated polyethylene (MAPE)) treatments on the interfacial structure of wood flour/PE composites. It can be seen that a number of clear cracks or boundaries between the wood particles and the matrix occurred in the untreated WPC (Fig. 13.1a), which indicates an inferior compatibility between the untreated raw materials. It was also observed that there were a few cell lumens partially filled by the polymer resin. However, the majorly unfilled cell lumens along with the existence of micro-cracks between the wood and the PE indicated the poor interfacial adhesion of the untreated WPC.

By contrast, the wood flour in the treated samples was completely coated by the matrix and firmly bonded to the polymer resin (Fig. 13.1b–1d), demonstrating superior interfacial adhesion with resin impregnation throughout the interface. More importantly, a large number of cell lumens of these samples were discerned to be partially or utterly filled by the polymer, which again confirmed the enhanced interfacial adhesion and also the compatibility and wettability.

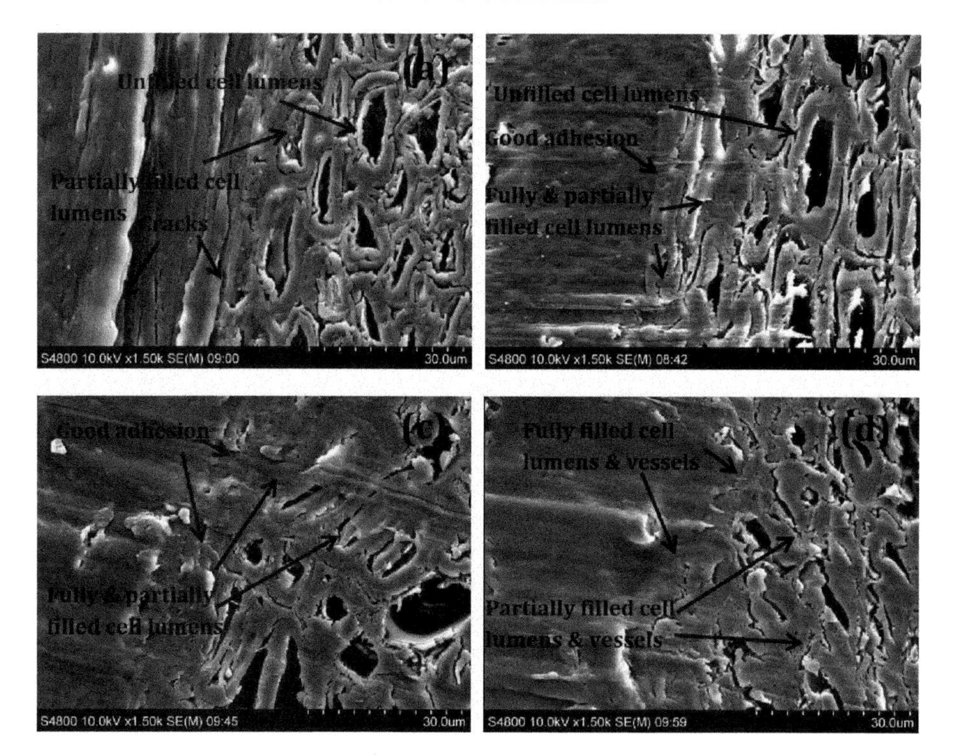

Figure 13.1 SEM photographs of cross-sections of wood flour/PE composites: (a) untreated; (b) MAPE treated; (c) Si69 treated; (d) VTMS treated.

It is interesting that, apart from the cell lumens, the vessels of the wood particles in the treated samples, especially the VTMS-treated sample, were also completely or partially impregnated with PE polymer. This phenomenon may be related to the substantially deformed and damaged cell lumens and vessels (Fig. 13.1d), which were inclined to facilitate the flow of the polymer in the composite. Hydrodynamic flow of molten PE resin in the composites may be initiated by an external compression force through the vessels, which then proceeds into the interconnected network of the cell lumens and the pits in the interface region, with the flow moving primarily along the paths of least resistance [51, 52]. The flow paths along any direction were, in general, a combination of open-cut lumens and vessels, as well as of large pits.

13.4 Bulk mechanical property of WPC

13.4.1 Static mechanical property

The static mechanical properties of WPC were systematically investigated in our previous studies. Fig. 13.2 presents the tensile and flexural properties of bamboo fibre–UPE composites treated with 1,6-diisocyanatohexane (DIH) and 2-hydroxyethyl acrylate (HEA) for improving the strength and hydrophobicity of the composites. It was found that the tensile strength of the orginal long bamboo fibre (LC)–UPE composites was

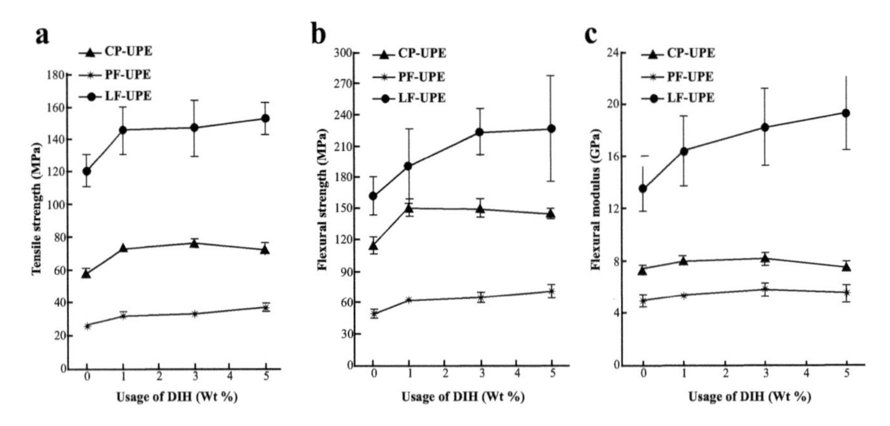

Figure 13.2 Mechanical properties of bamboo fibre–UPE composites: (a) tensile strength; (b) flexural strength; and (c) flexural modulus. (The data reported are the mean values of five replicates and the error bars represent standard deviations of the means.)

much higher than that of chemical pulp bamboo fibre (CP)–UPE and bamboo particle (PF)–UPE composites, and that CP–UPE composites had significantly greater tensile strength than PF–UPE composites. All the resulting bamboo fibre–UPE composites displayed a significant increase in tensile strength (Fig. 13.2a) after treatment of the fibres. Taking CP–UPE composites as an example, the resulting DIH-1 (1 wt% DIH usage in the composite), DIH-3 (3 wt% DIH usage), and DIH-5 (5 wt% DIH usage) composites had higher tensile strength (25.3%, 30.5% and 23.9% respectively) compared with the untreated CP-UPE composites. However, the tensile strengths of composite DIH-1, DIH-3, and DIH-5 are comparable: namely, the tensile strength of the composites did not increase significantly with an increase in DIH–HEA usage.

The untreated bamboo fibre–UPE composites had the greatest flexural strength compared with other untreated fibre–UPE composites (Fig. 13.2b). The resulting DIH-1, DIH-3, and DIH-5 bamboo fibre-UPE composites had greater flexural strengths of 17.4, 37.4 and 39.3%, respectively, compared with the untreated bamboo fibre–UPE composites. The CP–UPE and PF–UPE composites had similar trends for flexural strength after fibre treatment with DIH–HEA. The resulting DIH-5 PF–UPE composites had a significantly greater flexural strength than the resulting DIH-1 and DIH-3 PF–UPE composites. However, the flexural strength of CP–UPE and LF–UPE composites did not increase significantly with increased usage of DIH–HEA.

The LF–UPE composites had a much greater flexural modulus than the CP-UPE and PF-UPE composites (Fig. 13.2c). The flexural modulus of the resulting DIH-1, DIH-3, and DIH-5 LF–UPE composites increased by 21.2, 35.0, and 42.8%, respectively, compared with that of the untreated LF–UPE composites. All the treated fibre–PE composites had a greater flexural modulus than their respective untreated fibre–UPE composites. However, the flexural modulus of the composites did not increase significantly with increased usage of DIH–HEA.

The bamboo fibres were also treated with isocyanatoethyl methacrylate (IEM) in the presence of dibutyltin dilaurate as a catalyst to improve the interfacial adhesion

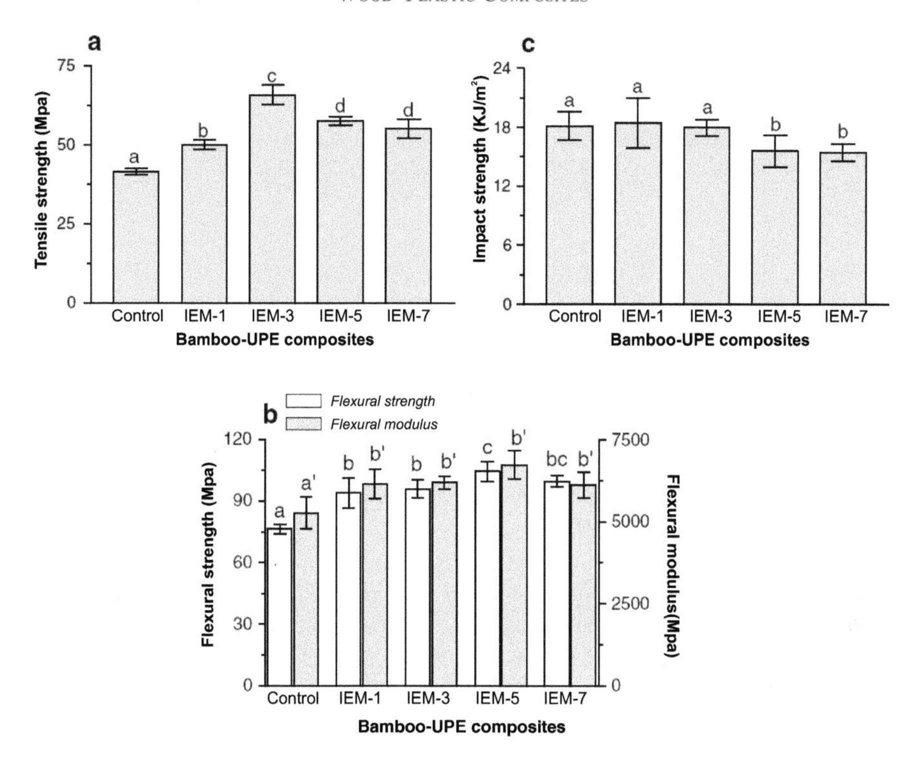

Figure 13.3 Effects of treatments of bamboo fibres on (a) tensile strength; (b) flexural properties; and (c) impact strength of bamboo–UPE composites. (The number after IEM is the weight percentage based on the dry weight of the fibres. The error bar at the top of each column represents the standard deviation of the mean value. The groups do not differ significantly from one another when they have the same letter(s) at the top of the columns, and vice versa.)

between the fibres and the UPE resins. As shown in Fig. 13.3, the tensile strength, flexural strength, flexural modulus, and impact strength of the bamboo–UPE composites were 42–66 MPa, 76–105 MPa, 5,300–6,800 MPa and 15–19 kJ m^{-2}, respectively. Similar tensile strength (40–50 MPa) but lower flexural strength (65–100 MPa) and flexural modulus (3500–5000 MPa) of bamboo–UPE composites were reported by Tran *et al.* [53]. Bamboo–UPE composites with comparable mechanical properties were also investigated by Kushwaha and Kumar, Kumar *et al.* [54, 55] and Kim *et al.* [56]. In addition, the mechanical properties of the composites were comparable with those of most natural fibres–UPE composites [57, 58]. For example, Du *et al.* [59, 60] reported similar tensile strength (40–55 MPa) and flexural strength (75–85 MPa), yet a greater flexural modulus (8–9 GPa) of kenaf–UPE composites than those of the bamboo–UPE composites in our work (Fig. 13.3).

Fig. 13.3a indicates that composites IEM-1, IEM-3, IEM-5 and IEM-7 had significantly greater tensile strengths than the composites of untreated fibres (control), indicating that the IEM treatments effectively improved the tensile strength of the composites. The tensile strength of the resulting IEM-1, IEM-3, IEM-5 and IEM-7 composites was greater than that of the control by 20.6%, 58.6%, 38.7% and 32.6 %, respectively.

Composite IEM-1 had a significantly greater flexural strength than the control (Fig. 13.3b). The flexural strength of the composites also significantly increased when IEM usage increased from 1 wt% to 3 wt% and from 3 wt% to 5 wt%. However, a slight decrease in flexural strength occurred in the composite IEM-7. Thus, composite IEM-5 had the greatest flexural strength, i.e. a 36.7% increment compared with the control. The flexural moduli of the composites with IEM-treated fibres were essentially higher than that of the control (Fig. 13.3b). However, when IEM usage was raised from 1 wt% to 7 wt%, the flexural moduli of the composites did not change significantly.

The impact strengths of composites IEM-1 and IEM-3 were comparable and did not significantly differ from that of the control (Fig. 13.3c). However, further increasing IEM usage to 5 wt% or to 7 wt% resulted in a dramatic decline in the impact strength of the composites. The impact strengths of composites IEM-5 and composites IEM-7 were significantly lower than that of the control.

The tensile strength, flexural strength, and flexural modulus of the bamboo–UPE composites were significantly increased by the chemical treatment of the bamboo fibres with IEM. This contributes to improved interfacial adhesion between the bamboo fibres and the UPE resins. However, the tensile strength of the composites decreased when the addition of IEM was more than 3 wt%, probably because the excess IEM molecules that were not covalently bonded onto the bamboo fibres might be loosely trapped in the interface of the composites, forming weak interfacial layers between the fibres and the resins, and affecting the stress transfer from the matrices to the fibres. Similar observations have appeared in previous studies of hemp–UPE composites using DIH-HEA and N-methylol acrylamide as the coupling agents [57, 61].

On the other hand, the impact strength of the composites is a result of the total energy absorption during the processing of the composites subjected to impact loading [62]. This energy is the sum of the absorption of impact energy up to maximum force (initiation energy) and the energy absorption after the maximum loading (propagation energy). The effect of the fibre/matrix interface bonding of composites on the total energy absorption is complicated because the effect on the initiation energy and the propagation energy is opposite. This explains why several studies have reported contradictory results regarding the influence of fibre modification on the impact strength of the resulting WPC [58, 63]. In our study, the reduction in the impact strength of composite IEM-5 and composite IEM-7 might be due to the formation of weak interfacial layers in the composites, which cause a reduced total energy absorption.

13.4.2 Dynamic mechanical properties

Polymers composed of long molecular chains have unique viscoelastic properties that combine the characteristics of elastic solids and Newtonian fluids. Dynamic mechanical analysis (DMA) is the most commonly used technique for the study and characterisation of the viscoelastic behaviour of polymers.

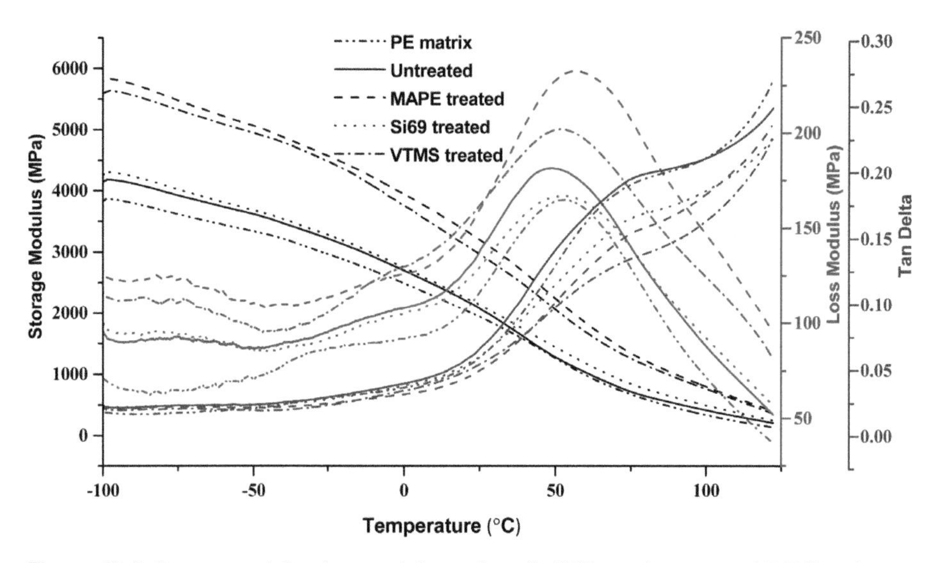

Figure 13.4 Storage modulus, loss modulus and tan δ of PE matrix, untreated WPC, and treated WPC, as a function of temperature.

The storage modulus is closely related to the load-bearing capacity of a material [64, 65]. The variation in the storage modulus of untreated and coupling-agent (MAPE, Si69 and VTMS) treated wood flour–PE composites as a function of temperature is shown in Fig. 13.4. It was observed that the coupling agents MAPE and VTMS treated WPC had much higher storage moduli than the untreated one, primarily attributed to improved interfacial adhesion between the wood flour and the PE matrix. The storage moduli in all composites decreased with an increase in temperature, and the differentiation of the modulus between the treated and untreated WPC gradually diminished from the transition region (from –100 °C to 75 °C) to the plateau region (75–120 °C).

The loss modulus vs temperature curves in Fig. 13.4 were used to investigate the transition behaviour of the composites. There was one major transition (45–60 °C) in these curves, corresponding to the α transition of the PE matrix, which was associated with the chain segment mobility in the crystalline phase due to the reorientation of the defect area in the crystals [5, 64, 65]. In comparison to the relaxation peak of untreated WPC (47.93 °C), the counterparts of MAPE- and VTMS-treated WPC shifted towards the higher temperature regions (i.e. 57.35 °C and 51.79 °C for MAPE and VTMS, respectively), and the corresponding moduli of the treated WPC (i.e. 232.40 MPa and 201.93 MPa for MAPE and VTMS, respectively) were accordingly higher than that of the untreated one (181.43 MPa). These performances can be attributed to the generated constraints on the segmental mobility of polymeric molecules at the relaxation temperatures during the coupling agent treatment [66, 67].

It is worth noticing that, compared to the untreated WPC, albeit the α transition appeared at higher temperature (52.94 °C), the Si69-treated composite showed a reduced loss modulus (166.72 MPa), indicating that Si69 treatment did not lead to comparable segmental immobilisation of the matrix chains on the wood surface as the MAPE and VTMS treatments had. This might be related to the comparatively limited crosslinking between the polysulphides of Si69 and the PE molecules compared to that between the grafted PE on MAPE or the unsaturated C=C groups of VTMS and the PE matrix molecules. However, it was reported that the transition peak broadens and the peak position shifts if there is an interaction between the filler or reinforcement and the matrix polymer [68]. Therefore, the broadening of the transition regions from around 10 °C to 85 °C that was discerned in all the treated composites (especially MAPE- and VTMS-treated composites) might be an indication of the existence of the interaction between the coupling agents and the constituents of the composites.

The ratio of the loss modulus to storage modulus, defined as the damping factor tan δ, was also determined (Fig. 13.4) for the benefit of better understanding the damping behaviour and interface properties of the composites. As can be seen in Fig. 13.4, the tan δ amplitude of the composites decreased with the addition of the coupling agents. This was expected since the enhanced interfacial adhesion provided the treated WPC with an interface of greater stiffness, which in turn restrained the segmental mobility of the polymer molecules leading to a decrease in the magnitude of tan δ [66–68]. In addition, the composite with poorer interfacial bonding between the wood flour and the matrix was inclined to dissipate more energy, showing greater tan δ amplitude than the composite with a firmly bonded interface [67, 69, 70].

The adhesion factor, A, was determined from the mechanical damping factor (tanδ), to further investigate the effect of the coupling agent treatments on the interfacial adhesion between the filler and the matrix of the composites, in accordance with Eq. 1 reported by Correa *et al.* [71] and Kubát *et al.* [72]:

$$A = \frac{\tan \delta_c}{(1 - V_w)\tan \delta_m} - 1 \qquad \text{(Eq. 1)}$$

where, the subscripts c, m and w refer to composite, matrix, and wood flour respectively, and V_w is the volume fraction of wood flour in the composite. As Fig. 13.5 shows, the untreated composite had a higher adhesion factor than the coupling-agent treated composites, which was consistent with the tan δ results. According to Correa *et al.* [71] the molecular mobility of the polymer surrounding the filler was reduced at high levels of interfacial adhesion, and consequently low values of the adhesion factor suggested improved interactions at the filler-matrix interface. The determined lower levels of the adhesion factor in coupling-agent treated WPC evidently confirmed the enhanced interfacial interactions after the treatments, which resulted in the decreased molecular mobility and mechanical damping (tan δ) [73, 74]. In addition, the adhesion factor of Si69-treated composite was higher than that for MAPE- and VTMS-treated composites. This result, along with its relatively greater tan δ amplitude, again substantiated the inferior interfacial adhesion improvement induced by Si69 treatment, resulting in the comparatively lower storage modulus and loss modulus of the composite.

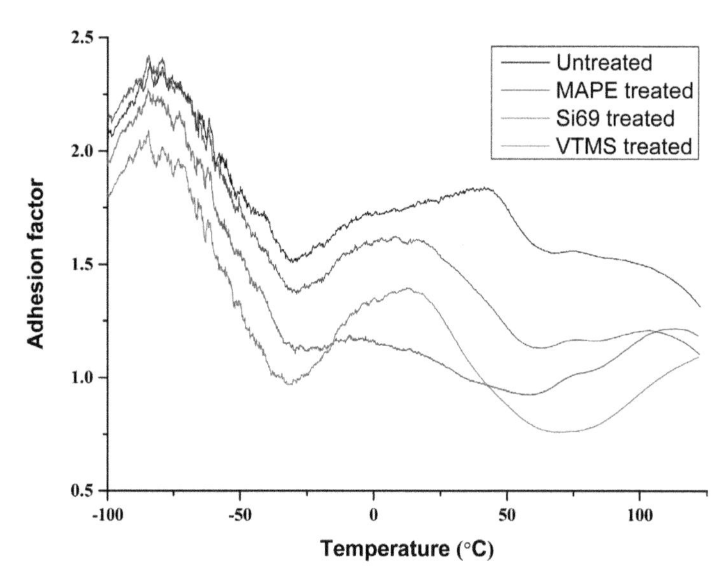

Figure 13.5 Adhesion factor of untreated and treated composites as a function of temperature.

13.5 *In situ* mechanical properties of WPC

The performance and behaviour of WPC not only relies on the reinforcing wood and the polymer matrix, but also critically depends on the effectiveness of the load/stress transfer across the interface [17, 75]. The interface of WPC formed when the wood flour is embedded in the polymer matrix during fabrication of the composite is a heterogeneous transition zone with chemical composition, morphological features and mechanical properties distinct from those of the reinforcing phase and the bulk polymer [14, 15, 75]. An appropriately engineered interface could considerably improve the strength and toughness of the composite as well as the environmental stability. Therefore, the determination of the interfacial properties and characteristics is of utmost importance in evaluating the overall property of the composite and enabling its optimal design [76]. Numerous techniques, such as the single fibre fragmentation test, single fibre pull-out test, and microbond test, have been developed for characterising the fibre/matrix interfacial strength. However, very few studies have been carried out to determine the size and relative mechanical properties of the interface due to the lack of unequivocally established experimental techniques.

The nanoindentation technique has proven to be an effective method in determining material surface properties at the nanoscale. This is achieved by monitoring a probe inserted in the specimen surface and synchronously recording the penetration load and depth [18, 77]. It has recently been found that its application is feasible for wood, natural fibres and plastics [14, 18, 78–85]; however, it has rarely been employed in measuring the interface properties and performance of WPC materials.

A study on the effect of water absorption on the nanomechanical properties of woven fabric flax fibre-reinforced bioresin-based epoxy biocomposites observed approximately 35% and 12% reductions in nanohardness and reduced modulus respectively, after exposure to water, indicating that the strength and elastic modulus were fibre-sensitive properties in the composites and that the interface suffered from the water absorption [86]. Nanoindentation was also conducted to measure the hardness and elastic modulus in the interface region of cellulose fibre-reinforced PP composite. There was a gradient of hardness and modulus across the interface region, and the distinct properties were revealed by 1–4 indents depending on the nanoindentation depth and spacing [14].

Fig. 13.6 demonstrates the nanomechanical properties of the wood cell walls of untreated and coupling-agent (MAPE, Si69 and VTMS) treated wood flour–PE composites determined by nanoindentation. It is most notable that the untreated composite

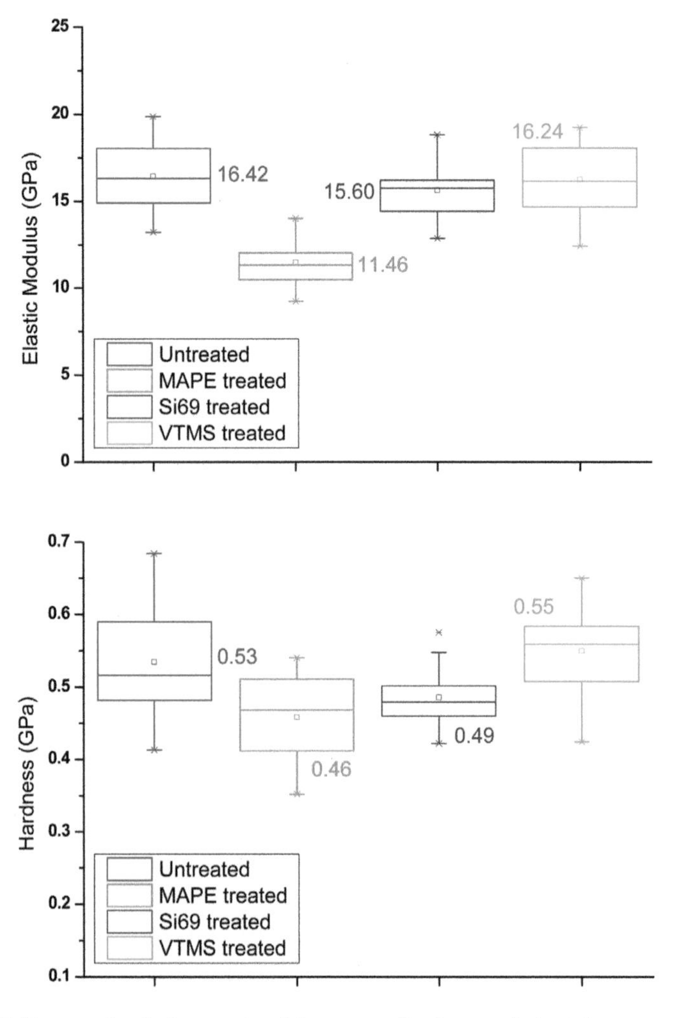

Figure 13.6 Nanomechanical property of the composites by nanoindentation.

showed an elastic modulus of 16.42 GPa and a hardness of 0.53 GPa, while the counterparts of the MAPE- and Si69-treated composites interestingly decreased to 11.46 GPa and 0.46 GPa (MAPE treated), and 15.60 GPa and 0.49 GPa (Si69 treated) respectively. This is in disagreement with that of the bulk properties tested (Fig. 5). The reductions in the modulus and hardness were assumed to be partially a result of the fibre weakening or softening impact of the treatments, namely the chain scission (cleavage of β-1,4-glycosidic bonds between two anhydroglucose units) and weakening of the interfibrillar interaction in cellulose occurring with the presence of maleic and silane coupling agents under high pressure and temperature during processing [87, 88]. On the other hand, it might be associated with the degree of crystalline orientation in the wood flour. The crystalline carbohydrates in the wood flour might undergo transformation into an amorphous form during the treatments, which facilitates the nanoindentation loading along the fibre direction, and thus lower modulus and hardness are measured.

Although the VTMS treatment had an analogous effect on the composites, there was no noticeable difference between the untreated and VTMS-treated composites in terms of their elastic modulus and hardness. This suggested that there were other critical factors governing the *in situ* mechanical properties of the composites. Due to the relatively lower molecular weight (M_W) of Si69 (M_W = 539) and VTMS (M_W = 148) compared to MAPE (500 cps viscosity at 140 °C), silanes might be more capable of diffusing into cell walls and reacting with the structural components by forming hydrogen and covalent bonding, giving rise to more outstanding nanomechanical properties. In addition, the maleic coupling agent was found to be able to create a thin, soft and ductile interfacial layer in the composite [89], which might impart an adverse impact on the nanomechanical properties of the composite. More importantly, the better nanaomechanical properties of VTMS-treated composite over MAPE- and Si69-treated composites should be ascribed to the severe deformation and damage to its cell walls due probably to the compression under high pressure and temperature. Interfacial structure analysis (Fig. 13.1) has shown that the VTMS-treated composite possesses many more deformed and damaged cell lumens and vessels than the untreated and MAPE- and Si69-treated composites, which promotes the wetting of wood flour by the polymer and the polymer penetration into wood particles. The higher level of penetration is assumed to compensate the loss of the *in situ* mechanical properties due to the aforementioned fibre weakening or softening impact and the crystalline structure transformation. Therefore, the nanomechanical properties of the treated composites falls into the sequence: VTMS > Si69 > MAPE.

It should be pointed out that the coupling-agent treatments employed enhanced the interfacial adhesion and bonding of the composites by strengthening the chemical, physical, and mechanical bonding. The optimised interface contributed to the increase in the bulk mechanical properties of the composites. Although both the MAPE- and VTMS-treated composites demonstrated a significant increase in bulk mechanical properties (i.e. storage and loss modulus (Fig. 13.4)) after treatment, the corresponding *in situ* mechanical properties were different: MAPE-treated composite possessed a softer interface by showing lower *in situ* elastic modulus and hardness than the VTMS-treated composite. In comparison to that of VTMS-treated composite, the comparatively softer and tougher interface of MAPE treated composite might provide it with greater resistance to fracture. In addition, owing to the inherently tough and ductile nature of the PE matrix, the softer and tougher interface might benefit the continuity in load transfer from the matrix to the wood flour, promoting the composite to function as a mechanical entity. It was evident that the extent to which the bulk mechanical properties of the composite could be improved was predominantly reliant upon the level of the enhancement of compatibility, the interfacial adhesion and the bonding after different coupling-agent treatments rather than individual local properties within the interface. The distinct nature and characteristics of the interfaces might, however, play some other fundamental role in determining the global properties of the composites.

13.6 Conclusion and outlook

The major drawback of using wood flour or natural fibres as a reinforcement in plastics – namely the incompatibility and poor adhesion between the constituents, which hampers the commercial exploitation of the composites – could be overcome by implementing various physical and chemical modification or pretreatment methods. Thus, the static and dynamic mechanical properties of the composites are improved by a refined/optimised interface and bonding quality. Investigation of the *in situ* mechanical properties of WPC has benefited our understanding of the relationship between the structure and performance of the composites as well as the further tailoring of functionalised WPC.

It is evident that the use of cellulosic fillers is of great significance both with regard to reducing reliance on petroleum-based and non-renewable resources (i.e. polymer resins) and exploiting the additional advantageous features provided by WPC. However, the speedy development of bioplastics such as poly(lactic acid) and thermoplastic starch has led to the emergence of fully biodegradable cellulosic polymer composite, and the replacement of traditional polymers with these renewable substitutes. Thus, in order to realise a significantly improved environmental impact, biodegradable polymers must be considered more in the future development of WPC. Significant effort is required in the selection of the most suitable biodegradable matrix, optimisation of processing technologies, application in different industrial sectors and minimisation of economic impact. Although the application of nanoscience in lignocellulosic polymer composites is currently at the infant stage, nano-enhanced polymer composites will be inevitable in the near future. Nanofillers, such as nanorubber, nanocellulose and nanographene, should be taken into account in the further development of WPC, which might equip the resulting material with more attractive properties.

References

[1] Xie Y, Hill C A S, Xiao Z, Militz H & Mai C. Silane coupling agents used for natural fiber/polymer composites: A review. *Composites Part A: Applied Science and Manufacturing* (2010), 41 (7): 806–819.

[2] Belgacem M N & Gandini A. The surface modification of cellulose fibres for use as reinforcing elements in composite materials. *Composite Interfaces* (2005), 12 (1–2): 41–75.

[3] Bengtsson M, Stark N M & Oksman K. Durability and mechanical properties of silane cross-linked wood thermoplastic composites. *Composites Sci. Technol.* (2007), 67 (13): 2728–2738.

[4] Bengtsson M & Oksman K. The use of silane technology in crosslinking polyethylene/wood flour composites. *Composites Part A: Applied Science and Manufacturing* (2006), 37 (5): 752–765.

[5] Bengtsson M, Gatenholm P & Oksman K. The effect of crosslinking on the properties of polyethylene/wood flour composites. *Composites Sci. Technol.* (2005), 65 (10): 1468–1479.

[6] Kuo P, Wang S, Chen J, Hsueh H & Tsai M. Effects of material compositions on the mechanical properties of wood–plastic composites manufactured by injection molding. *Mater. Des.* (2009), 30 (9): 3489–3496.

[7] Zhang H. Effect of a novel coupling agent, alkyl ketene dimer, on the mechanical properties of wood–plastic composites. *Mater. Des.* (2014), 59: 130–134.

[8] Bledzki A K, Gassan J & Theis S. Wood-filled thermoplastic composites. *Mechanics of Composite Materials* (1998), 34 (6): 563–568.

[9] Cantero G, Arbelaiz A, Llano-Ponte R & Mondragon I. Effects of fibre treatment on wettability and mechanical behaviour of flax/polypropylene composites. *Composites Sci. Technol.* (2003), 63 (9): 1247–1254.

[10] Azwa Z N, Yousif B F, Manalo A C & Karunasena W. A review on the degradability of polymeric composites based on natural fibres. *Mater. Des.* (2013), 47: 424–442.

[11] Araújo J R, Waldman W R & De Paoli M A. Thermal properties of high-density polyethylene composites with natural fibres: Coupling agent effect. *Polym. Degrad. Stab.* (2008), 93 (10): 1770–1775.

[12] Dittenber D B & Gangarao H V S. Critical review of recent publications on use of natural composites in infrastructure. *Composites Part A: Applied Science and Manufacturing* (2012), 43 (8): 1419–1429.

[13] Dhakal H N, Zhang Z Y, Richardson M O W & Errajhi O A Z. The low velocity impact response of non-woven hemp fibre reinforced unsaturated polyester composites. *Composite Structures* (2007), 81 (4): 559–567.

[14] Lee S, Wang S, Pharr G M & Xu H. Evaluation of interphase properties in a cellulose fiber-reinforced polypropylene composite by nanoindentation and finite element analysis. *Composites Part A: Applied Science and Manufacturing* (2007), 38 (6): 1517–1524.

[15] Kim J & Mai YW. *Engineered Interfaces in Fiber Reinforced Composites.* Elsevier Sciences, 1998.

[16] Pickering K. *Properties and Performance of Natural-Fibre Composites.* 1st edn. Woodhead Publishing (Elsevier), 2008.

[17] Kumar R, Cross W M, Kjerengtroen L & Kellar J J. Fiber bias in nanoindentation of polymer matrix composites. *Composite Interfaces* (2004), 11 (5–6): 431–440.

[18] Zhang T, Bai S L, Zhang Y F & Thibaut B. Viscoelastic properties of wood materials characterized by nanoindentation experiments. *Wood Sci. Technol.* (2012), 46 (5): 1003–1016.

[19] Yedla S B, Kalukanimuttam M, Winter R M & Khanna S K. Effect of shape of the tip in determining interphase properties in fiber reinforced plastic composites using nanoindentation. *Journal of Engineering Materials and Technology* (2008),130 (4): 041010–041010.

[20] Chan C, Ko T & Hiraoka H. Polymer surface modification by plasmas and photons. *Surface Science Reports* (1996), 24 (1): 1–54.

[21] Pethrick R A. Review of *Plasma Surface Modification of Polymers: Relevance to Adhesion.* Edited by M. Strobel, C. S. Lyons and K. L. Mittal. VSP, Zeist, the Netherlands, 1994. *Polym. Int.* (1996), 39 (1): 79.

[22] Acda M N, Devera E E, Cabangon R J & Ramos H J. Effects of plasma modification on adhesion properties of wood. *Int. J. Adhes. Adhes.* (2012), 32: 70–75.

[23] Bozaci E, Sever K, Sarikanat M, Seki Y, Demir A, Ozdogan E & Tavman I. Effects of the atmospheric plasma treatments on surface and mechanical properties of flax fiber and adhesion between fiber–matrix for composite materials. *Composites Part B: Engineering* (2013), 45 (1): 565–572.

[24] Marais S, Gouanvé F, Bonnesoeur A, Grenet J, Poncin-Epaillard F, Morvan C & Métayer M. Unsaturated polyester composites reinforced with flax fibers: Effect of cold plasma and autoclave treatments on mechanical and permeation properties. *Composites Part A: Applied Science and Manufacturing* (2005), 36 (7): 975–986.

[25] Shahzad A. Effects of fibre surface treatments on mechanical properties of hemp fibre composites. *Composite Interfaces* (2011), 18 (9): 737–754.

[26] Ishikawa H, Takagi H, Nakagaito A N, Yasuzawa M, Genta H & Saito H. Effect of surface treatments on the mechanical properties of natural fiber textile composites made by VaRTM method. *Composite Interfaces* (2014), 21 (4): 329–336.

[27] Tong Y J & Xu L H. Hemp fiber reinforced unsaturated polyester composites. *Advanced Materials Research* (2006), 11–12: 521–524.

[28] Abdullah-Al-Kafi M Z, Abedin M D H, Beg K L & Khan M A. Study on the mechanical properties of jute/glass fiber-reinforced unsaturated polyester hybrid composites: Effect of surface modification by ultraviolet radiation. *J. Reinf. Plast. Compos.* (2006), 25 (6): 575–588.

[29] Pang Y, Cho D, Han S O & Park W H. Interfacial shear strength and thermal properties of electron beam-treated henequen fibers reinforced unsaturated polyester composites. *Macromolecular Research* (2005), 13 (5): 453–459.

[30] Brugnago R J, Satyanarayana K G, Wypych F & Ramos L P. The effect of steam explosion on the production of sugarcane bagasse/polyester composites. *Composites Part A: Applied Science and Manufacturing* (2011), 42 (4): 364–370.

[31] Xie Y, Hill C A S, Xiao Z, Militz H & Mai C. Silane coupling agents used for natural fiber/polymer composites: A review. *Composites Part A: Applied Science and Manufacturing* (2010), 41 (7): 806–819.

[32] Castellano M, Gandini A, Fabbri P & Belgacem M N. Modification of cellulose fibres with organosilanes: Under what conditions does coupling occur? *J. Colloid Interface Sci.* (2004), 273 (2): 505–511.

[33] Salon M C B & Belgacem M N. Hydrolysis-condensation kinetics of different silane coupling agents. *Phosphorus, Sulfur, and Silicon and the Related Elements* (2011), 186 (2): 240–254.

[34] Khalil H P S A, Ismail H, Rozman H D & Ahmad M N. The effect of acetylation on interfacial shear strength between plant fibres and various matrices. *European Polymer Journal* (2001), 37 (5): 1037–1045.

[35] Abdul H, Khalil H, Ismail M, Ahmad A & Hassan K. The effect of various anhydride modifications on mechanical properties and water absorption of oil palm empty fruit bunches reinforced polyester composites. *Polym. Int.* (2001), 50 (4): 395–402.

[36] Sawpan M A, Pickering K L & Fernyhough A. Effect of fibre treatments on interfacial shear strength of hemp fibre reinforced polylactide and unsaturated polyester composites. *Composites Part A: Applied Science and Manufacturing* (2011), 42 (9): 1189–1196.

[37] Marcovich N E, Ostrovsky A N, Aranguren M I & Reboredo M M. Resin–sisal and wood flour composites made from unsaturated polyester thermosets. *Composite Interfaces* (2009), 16 (7–9): 639–657.

[38] Aranguren M I, Marcovich N E & Reboredo M M. Composites made from lignocellulosics and thermoset polymers. *Molecular Crystals and Liquid Crystals Science and Technology Section A: Molecular Crystals and Liquid Crystals* (2000), 353 (1): 95–108.

[39] Marcovich N E, Reboredo M M & Aranguren M I. Lignocellulosic materials and unsaturated polyester matrix composites: Interfacial modifications. *Composite Interfaces* (2005), 12 (1–2): 3–24.
[40] Li K, Qiu R & Liu W. Improvement of interfacial adhesion in natural plant fiber-reinforced unsaturated polyester composites: A critical review. *Reviews of Adhesion and Adhesives* (2015), 3 (1): 98–120.
[41] Singha A S & Rana A K. Improvement of interfacial adhesion in cannabis indica/unsaturated polyester biocomposites through esterification reaction. *International Journal of Polymer Analysis and Characterization* (2012), 17 (8): 590–599.
[42] Thiruchitrambalam M & Shanmugam D. Influence of pre-treatments on the mechanical properties of palmyra palm leaf stalk fiber–polyester composites. *J. Reinf. Plast. Compos.* (2012), 31 (20): 1400–1414.
[43] Kushwaha P K & Kumar R. Influence of chemical treatments on the mechanical and water absorption properties of bamboo fiber composites. *J. Reinf. Plast. Compos.* (2011), 30 (1): 73–85.
[44] Joseph S P A & Thomas U G. Surface-modified sisal fiber-reinforced eco-friendly composites: Mechanical, thermal, and diffusion studies. *Polymer Composites* (2011), 32 (1): 131–138.
[45] Bessadok A, Marais S, Roudesli S, Lixon C & Métayer M. Influence of chemical modifications on water-sorption and mechanical properties of agave fibres. *Composites Part A: Applied Science and Manufacturing* (2008), 39 (1): 29–45.
[46] Bessadok A, Marais S, Gouanvé F, Colasse L, Zimmerlin I, Roudesli S & Métayer M. Effect of chemical treatments of alfa (*Stipa tenacissima*) fibres on water-sorption properties. *Composites Sci. Technol.* (2007), 67 (3–4): 685–697.
[47] Bessadok A, Roudesli S, Marais S, Follain N & Lebrun L. Alfa fibres for unsaturated polyester composites reinforcement: Effects of chemical treatments on mechanical and permeation properties. *Composites Part A: Applied Science and Manufacturing* (2009), 40 (2): 184–195.
[48] Zhang H, Cui Y & Zhang Z. Chemical treatment of wood fiber and its reinforced unsaturated polyester composites. *Journal of Vinyl and Additive Technology* (2013), 19 (1): 18–24.
[49] Saha A K, Das S, Bhatta D & Mitra B C. Study of jute fiber reinforced polyester composites by dynamic mechanical analysis. *J. Appl. Polym. Sci.* (1999), 71 (9): 1505–1513.
[50] Singha A S & Rana A K. Preparation and characterization of graft copolymerized *Cannabis indica L.* fiber-reinforced unsaturated polyester matrix-based biocomposites. *J. Reinf. Plast. Compos.* (2012), 31 (22): 1538–1553.
[51] Grmusa I G, Dunky M, Miljkovic J & Momcilovic M D. Influence of the degree of condensation of urea-formaldehyde adhesives on the tangential penetration into beech and fir and on the shear strength of the adhesive joints. *European Journal of Wood and Wood Products* (2012), 70 (5): 655–665.
[52] Grmusa I G, Dunky M, Miljkovic J & Momcilovic M D. Influence of the viscosity of UF resins on the radial and tangential penetration into poplar wood and on the shear strength of adhesive joints. *Holzforschung* (2012), 66 (7): 849–856.
[53] Tran D T, Nguyen D M, Ha C N & Dang T T. Effect of coupling agents on the properties of bamboo fiber-reinforced unsaturated polyester resin composites. *Composite Interfaces* (2013), 20 (5): 343–353.
[54] Kushwaha P K & Kumar R. Studies on water absorption of bamboo-epoxy composites: Effect of silane treatment of mercerized bamboo. *J. Appl. Polym. Sci.* (2010), 115 (3): 1846–1852.
[55] Kumar S, Choudhary V & Kumar R. Study on the compatibility of unbleached and bleached bamboo-fiber with LLDPE matrix. *Journal of Thermal Analysis and Calorimetry* (2010), 102 (2): 751–761.

[56] Kim H, Okubo K, Fujii T & Takemura K. Influence of fiber extraction and surface modification on mechanical properties of green composites with bamboo fiber. *J. Adhes. Sci. Technol.* (2013), 27 (12): 1348–1358.

[57] Qiu R, Ren X & Li K. Effect of fiber modification with a novel compatibilizer on the mechanical properties and water absorption of hemp-fiber-reinforced unsaturated polyester composites. *Polymer Engineering & Science* (2012), 52 (6): 1342–1347.

[58] Sreekumar P A, Thomas S P, Saiter J M, Oseph K J, Unnikrishnan G & Thomas S. Effect of fiber surface modification on the mechanical and water absorption characteristics of sisal/polyester composites fabricated by resin transfer molding. *Composites Part A: Applied Science and Manufacturing* (2009), 40 (11): 1777–1784.

[59] Du Y, Zhang J, Wang C, Lacy T E, Xue Y, Toghiani H, Horstemeyer M F & Pittman C U. Kenaf bast fiber bundle-reinforced unsaturated polyester composites. II: Water resistance and composite mechanical properties improvement. *For. Prod. J.* (2010), 60 (4): 366–372.

[60] Du Y, Zhang J, Yu J, Lacy T E, Xue Y, Toghiani H, Horstemeyer M F & Pittman C U. Kenaf bast fiber bundle-reinforced unsaturated polyester composites. IV: Effects of fiber loadings and aspect ratios on composite tensile properties. *For. Prod. J.* (2010), 60 (7-8): 582–591.

[61] Qiu R, Ren X, Fifield L S, Simmons K L & Li K. Hemp-fiber-reinforced unsaturated polyester composites: Optimization of processing and improvement of interfacial adhesion. *J. Appl. Polym. Sci.* (2011), 121 (2): 862–868.

[62] Park R & Jang J. A study of the impact properties of composites consisting of surface-modified glass fibers in vinyl ester resin. *Composites Sci. Technol.* (1998), 58 (6): 979–985.

[63] Ray D, Sarkar B K & Bose N R. Impact fatigue behaviour of vinylester resin matrix composites reinforced with alkali treated jute fibres. *Composites Part A: Applied Science and Manufacturing* (2002), 33 (2): 233–241.

[64] Mohanty S, Verma S K & Nayak S K. Dynamic mechanical and thermal properties of MAPE treated jute/HDPE composites. *Composites Sci. Technol.* (2006), 66 (3–4): 538–547.

[65] Mohanty S & Nayak S K. Interfacial, dynamic mechanical, and thermal fiber reinforced behavior of MAPE treated sisal fiber reinforced HDPE composites. *J. Appl. Polym. Sci.* (2006), 102 (4): 3306–3315.

[66] López-Manchado M A, Biagitti J & Kenny J M. Comparative study of the effects of different fibers on the processing and properties of ternary composites based on PP-EPDM blends. *Polymer Composites* (2002), 23 (5): 779–789.

[67] Ou R, Xie Y, Wolcott M P, Sui S & Wang Q. Morphology, mechanical properties, and dimensional stability of wood particle/high density polyethylene composites: Effect of removal of wood cell wall composition. *Mater. Des.* (2014), 58: 339–345.

[68] Lai S, Yeh F, Wang Y, Chan H & Shen H. Comparative study of maleated polyolefins as compatibilizers for polyethylene/wood flour composites. *J. Appl. Polym. Sci.* (2003), 87 (3): 487–496.

[69] Felix J M & Gatenholm P. The nature of adhesion in composites of modified cellulose fibers and polypropylene. *J. Appl. Polym. Sci.* (1991), 42 (3): 609–620.

[70] Ashida M, Noguchi T & Mashimo S. Dynamic moduli for short fiber-CR composites. *J. Appl. Polym. Sci.* (1984), 29 (2): 661–670.

[71] Correa C A, Razzino C A & Hage E. Role of maleated coupling agents on the interface adhesion of polypropylene: Wood composites. *Journal of Thermoplastic Composite Materials* (2007), 20 (3): 323–339.

[72] Kubát J, Rigdahl M & Welander M. Characterization of interfacial interactions in high density polyethylene filled with glass spheres using dynamic-mechanical analysis. *J. Appl. Polym. Sci.* (1990), 39 (7): 1527–1539.

[73] Poletto M, Zeni M & Zattera A J. Dynamic mechanical analysis of recycled polystyrene composites reinforced with wood flour. *J. Appl. Polym. Sci.* (2012), 125 (2): 935–942.
[74] Ornaghi H L, Bolner A S, Fiorio R, Zattera A J & Amico S C. Mechanical and dynamic mechanical analysis of hybrid composites molded by resin transfer molding. *J. Appl. Polym. Sci.* (2010), 118 (2): 887–896.
[75] Downing T D, Kumar R, Cross W M, Kjerengtroen L & Kellar J J. Determining the interphase thickness and properties in polymer matrix composites using phase imaging atomic force microscopy and nanoindentation. *J. Adhes. Sci. Technol.* (2000), 14 (14): 1801–1812.
[76] Graham J F, McCague C, Warren O L & Norton P R. Spatially resolved nanomechanical properties of Kevlar® fibers. *Polymer* (2000), 41 (12): 4761–4764.
[77] Oliveira G L, Costa C A, Teixeira S C S & Costa M F. The use of nano- and micro-instrumented indentation tests to evaluate viscoelastic behavior of poly(vinylidene fluoride) (PVDF). *Polym. Test.* (2014), 34: 10–16.
[78] Gindl W, Konnerth J & Schöberl T. Nanoindentation of regenerated cellulose fibres. *Cellulose* (2006), 13 (1): 1–7.
[79] Ghomsheh M Z, Spieckermann F, Polt G, Wilhelm H & Zehetbauer M. Analysis of strain bursts during nanoindentation creep of high-density polyethylene. *Polym. Int.* (2015), 64 (11): 1537–1543.
[80] Lee J & Deng Y. Nanoindentation study of individual cellulose nanowhisker-reinforced PVA electrospun fiber. *Polymer Bulletin* (2013), 70 (4): 1205–1219.
[81] Oliver W C & Pharr G M. An improved technique for determining hardness and elastic modulus using load and displacement sensing indentation experiments. *J. Mater. Res.* (1992), 7 (6): 1564–1583.
[82] Tze W T Y, Wang S, Rials T G, Pharr G M & Kelley S S. Nanoindentation of wood cell walls: Continuous stiffness and hardness measurements. *Composites Part A: Applied Science and Manufacturing* (2007), 38 (3): 945–953.
[83] Hobbs J K, Winkel A K, McMaster T J, Humphris A D L, Baker A A, Blakely S, Aissaoui M & Miles M J. Nanoindentation of polymers: An overview. *Macromolecular Symposia* (2001), 167 (1): 15–43.
[84] Wu Y, Wang S, Zhou D, Xing C & Zhang Y. Evaluation of elastic modulus and hardness of crop stalks cell walls by nano-indentation. *Bioresour. Technol.* (2010), 101 (8): 2867–2871.
[85] Xing C, Wang S & Pharr G M. Nanoindentation of juvenile and mature loblolly pine (*Pinus taeda L.*) wood fibers as affected by thermomechanical refining pressure. *Wood Sci. Technol.* (2009), 43 (7–8): 615–625.
[86] Dhakal H, Zhang Z, Bennett N, Lopez-Arraiza A & Vallejo F. Effects of water immersion ageing on the mechanical properties of flax and jute fibre biocomposites evaluated by nanoindentation and flexural testing. *J. Composite Mater.* (2014), 48 (11): 1399–1406.
[87] Sawpan M A, Pickering K L & Fernyhough A. Effect of various chemical treatments on the fibre structure and tensile properties of industrial hemp fibres. *Composites Part A: Applied Science and Manufacturing* (2011), 42 (8): 888–895.
[88] Ganser C, Hirn U, Rohm S, Schennach R & Teichert C. AFM nanoindentation of pulp fibers and thin cellulose films at varying relative humidity. *Holzforschung* (2013), 68 (1): 53–60.
[89] Hristov V & Vasileva S. Dynamic mechanical and thermal properties of modified poly(propylene) wood fiber composites. *Macromolecular Materials and Engine*ering (2003), 288 (10): 798–806.

Chapter 14
Long-fibre strong composites

Dr Lundeng Hou and Professor Mizi Fan

This chapter investigates long natural fibres as reinforcements for the production of strong natural fibre composites (NFC). The chapter focuses on bast long fibres. It begins with a review of the structure, composition and mechanical properties of natural fibres (NF). There follows a detailed discussion of bast fibre reinforcements of NF mats, yarns and fabrics, and NFC matrix systems of thermosets, thermal plastics and bio-based resins. The chapter then discusses long fibre NFCs, their affecting parameters and performance in use. A review of NFC models and modelling, including the prediction modelling of NF yarn strength and NFC properties concludes the chapter. The application of NFCs is also explored.

14.1 Introduction

As a result of a growing awareness of global environmental factors, the principles of sustainability, industrial ecology, eco-efficiency, green chemistry and engineering are being integrated into the development of the next generation of materials, products and processes. There is a renewed interest in bio-based materials, e.g. biocomposites, for numerous industry applications due to the merits of thermal insulation, thermal dimensional stability, specific modulus of elasticity, biodegradability, renewability, abundance, comparable low-cost and natural appearance.

There are six types of natural fibre (NF) based on classification by botanical type, including bast, leaf, seed, core, grass, and others like wood and root fibres [1]. All NF characteristics can be utilised and enhanced when used as reinforcements for biocomposites [2]. Taking into account the nature of fibres, and the requirements of processing and products, bast fibres are considered the most suitable materials for the production of long-fibre strong composites, in particular flax and hemp fibres.

NF is a highly anisotropic and complex material. The properties of plant materials are affected by multiple factors, such as plant species, breeding region and conditions, fibre position in the plant, and fibre extraction techniques. Numerous investigations have been conducted, aimed at optimising NF material for consistency in properties, by improving, for example, growth conditions, harvesting and the fibre extraction process [3]. From the material point of view, the physical properties of NFs are mainly determined by the chemical and physical composition, such as the structure of the fibres, cellulose content, angle of fibrils, cross-section, and by the degree of polymerisation. Each of these reasons give rise to a great challenge, which is to modify the fibres in order to realise their full potential [4].

Composites are currently mainly processed out of fossil resin matrices, which are widely available on the commercial market. A combination of bio-based reinforcement and bio-based resin, in the form of an NF composite, giving high strength, durability and multiple areas of application, is the ultimate goal of researchers working in the field. For the efficient utilisation of NF properties, the processing routes of composites need to be developed and optimised. Composite material is composed of two or more different materials, each of which displays inferior properties compared with the composite. Fusion of the faces between the matrix and reinforcement plays a crucial role in load transfer and the overall durability of the material. The interface between hydrophilic plant fibres and a hydrophobic polymeric resin is of special concern in terms of wettability and adhesion, which need to be sufficiently high to utilise the strength of fibres. Research has shown that adhesion in natural fibre composites can be improved by chemical treatment of the fibres, resin modification, or the method of processing used for modification. However, a repeatable and reliable approach has yet to be developed [1–2].

In contrast to manmade fibres, NF has a complex structure and its overall properties are a function of it composition. A better understanding of fibres, their processing routes and the novel architectural design of their composites, is required if NFs are to be used as raw materials for the development of structural products. Predictions about composite properties based on manmade fibres is an important and well researched topic, but solutions to the modelling of long natural fibre composites have yet to be established in order to create strong, durable and affordable bio-based composite material for load-bearing applications.

14.2 Long fibres–bast fibres

14.2.1 Structure and composition

Bast fibres are made up of the same elements as wood fibres, namely cellulose, hemicellulose and lignin. Most natural bast, leaf and seed fibres contain around 65% cellulose, which is the reinforcing element of the plant (Table 14.1) [4–5].

Table 14.1 Composition of natural fibres (%).

	Hemp	Flax	Jute	Sisal	Ramie	Cotton
Cellulose	67.0	62.1–64.1	64.4	65.8	68.8	82.7–92.7
Hemicellulose	16.1	16.7	12.0	12.0	13.1	5.7
Pectin	0.8	1.8	0.2	0.8	1.9	0.0
Lignin	3.3	2.0	11.8	9.9	0.6	0.0
Water soluble	2.1	3.9	1.1	1.2	5.5	1.0
Wax	0.7	1.5	0.5	0.3	0.3	0.6
Water	10	10	10	10	10	10
Microfibril angle	6.2	10.0	8.0	–	7.5	–

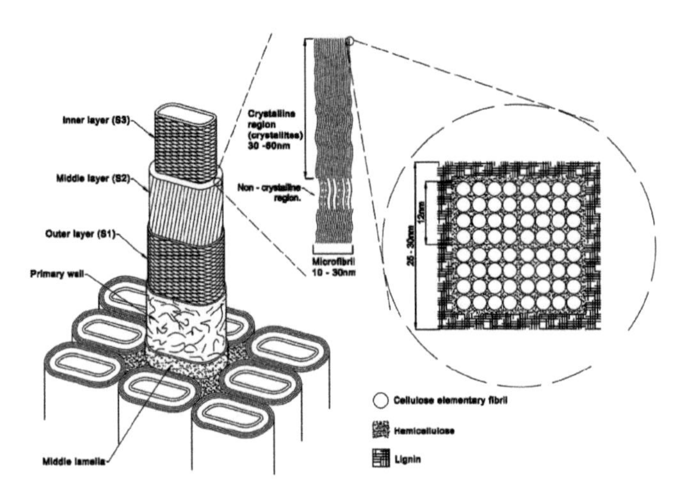

Figure 14.1 Bast fibre microstructure [7].

Cells of bast fibres are usually uneven, e.g. 16–7 μm in breath and 1–35 mm in length for hemp fibres. The cross-section of bast fibre plants reveals a wooden core with a fibrous stem. Fibre bundles are located close to the plant surface. Single technical fibres have a hierarchical structure composed of several distinguishable layers. Each layer has its own substructure. The middle lamella, composed mainly of pectin with macrofibrils, is located in the outer layer and binds the fibres together. This is followed by a thin cellulose network making up the primary wall [6].

Regarding the mechanical properties, the middle layer of the bast fibre is responsible for the plant reinforcement due to its cellulose content and microfibril arrangement. It is the biggest and makes up to 98% of the fibre wall. It consists of three layers, each with a different microfibril arrangement.

The arrangement of cellulose fibrils in the middle (S2) layer is almost longitudinal, and thus is the layer most responsible for plant stiffness and fibre strength. The S2 layer consists of microfibrils that are composed of elementary cellulose fibrils bound together with hemicelluloses and amorphous lignin (Figure 14.1). The inner fibre lumen consists of proteins and pectin [7].

14.2.2 Mechanical properties

For mechanical properties, including tensile, flexural and compression properties, fibres are often measured macroscopically, using a variety of test methods. Usually bundles of technical fibres are measured, since single fibres are difficult to separate physically. However, this may give rise to major variations, which may not represent the real strength of NFs [8].

The properties of NFs are inherently related to the plant properties. Chand and Hashmi [9] investigated the correlation between the age of a plant and the tensile properties of its fibres and found a correlation between the mechanical properties and the time of harvest in relation to plant life cycle. Barkakaty [8] found a correlation between the mechanical properties of the fibres and the plant growth conditions.

The properties of NFs vary according to the different parts of the plant. Fibres in the stem will usually have higher mechanical properties in comparison to those in the leaf. Additionally, the position within the plant affects mechanical properties. In the case of the hemp, the strongest fibres are found in the middle of the stem and the weakest at the top. This led to the development of harvesting machines that automatically alter the point at which stem is cut. Moreover, the type of fibre separation from the stem has a great influence on fibre performance [7].

When investigating the influence of microstructure on the mechanical properties of NFs, it was shown that the strength is dependent on the cellulose content, crystallinity index and microfibril angle [4, 10]. In general, higher cellulose content and low microfibril angle give rise to higher mechanical properties of fibres. Table 14.2 summarises the tensile properties of hemp and flax single fibres. Those of other natural and conventional reinforcement fibres are also included for comparison.

The strength and stiffness distributions for hemp and flax fibres vary considerably; this might be related to the procedure with which the fibres were tested [11], the treatment of the fibres, and other aforementioned factors. Short fibres can be converted into mats and used as randomly distributed reinforcement in composite production, such as compression moulding. The density of hemp and flax fibres is within the same range as carbon and aramid fibres, but half the density of glass reinforcement [12–13]. Hemp and flax fibres have similar density, ranging between 1.35 g cm^{-3} and 1.52 g cm^{-3}. Hemp has a wider fibre length, ranging from 5 mm to 110 mm, in comparison with the flax fibre, which ranges from 10 mm to 70 mm in length.

Flax fibres have been reported to have the highest Young's modulus among natural fibres, which is 100 GPa, and is only exceeded by that of carbon fibres. The highest Young's modulus for hemp fibres is 70 GPa, which is the same as that of E-glass. The difference is that the stiffness of E-glass is consistent whilst the reported stiffness of natural fibres reported starts from 5.5 GPa. Both hemp and flax fibres have the highest reported tensile strength values, above 1 GPa, which is comparable with that of soft wood kraft fibres and up to three times lower than that of E-glass or aramid fibres. This information may be used when predicting the mechanical properties of the composite

Table 14.2 Tensile properties of cellulosic fibres.

Fibre	Density ρ (g cm^{-3})	Fibre length (mm)	Ø (μm)	Elong. (%)	Young's modulus E_L (GPa)	Tensile strength σ_L (MPa)	Ref.
Hemp (*Cannabis sativa*)	1.35–1.50	5.6–110	10–51	1.6–4.2	5.5–70	690–1040	[1, 17, 18]
Flax (*Linum usitatissium*)	1.38–1.52	10–70	5–38	1.5–3.2	12–100	345–1100	[1, 4, 16]
Wool	1.20–1.32	38–150	12–45		3.9–5.2	40–200	[14]
Jute (*Corchorus capsularis*)	1.23–1.45	0.8–6.0	5–25	1.5–1.8	13.0–55.0	393–773	[1, 4, 15]
Soft Wood Kraft	1.50				40.0	1000	[4]
E-glass	2.50			2.5	70.0–72.0	2000–3500	[4, 15]
Aramid	1.40			3.3–3.7	63.0–67.0	3000–3150	[4]
Carbon	1.40			1.4–1.8	230–240	4000	[4]

laminate element but, because of the wide range of the properties, it is necessary to test the raw materials that are used for composite production, in order to achieve reliable mechanical performance data.

When considering natural fibres, wood may be used as an example. The bulk Young's modulus of wood is around 10 GPa. After the pulping process, single pulp fibre stiffness is around 40 GPa. When pulp fibres are hydrolysed and mechanically disintegrated, microfibrils are exposed, which have a stiffness of about 70 GPa. There is as yet no technology in existence that can break down and test microfibrils into crystallites, which have a stiffness value approaching the 250 GPa level [4].

14.2.3 Long NF reinforcements

Reinforcement fibre can be categorised by its form (Figure 14.2). Filaments can be continous and used directly, or processed into various forms of fabric. Discontinous, shorter fibres can be used in the form of mats or woven into continus yarn.

The structure and type of NF reinforcement have an influence on the mechanical properties of natural fibre composites (NFC). NFs are inherently short and, in order to use them as continuous reinforcement, various forms of mat and yarn need to be prepared.

14.2.3.1 NF mats

The term NF mat refers to a non-woven material, which is made from NFs by chemical, thermal, mechanical or other processes, but does not involve knitting or weaving the fibres (Fig. 14.3). Examples of non-woven materials include felts, insulation mats, packaging materials, and reinforcement fibre mats. NF reinforcement mats can be produced to specified forms and dimensions. The main factors used to describe fibre mats are fibre length and the level of processing. Mats are prepared with a wide range

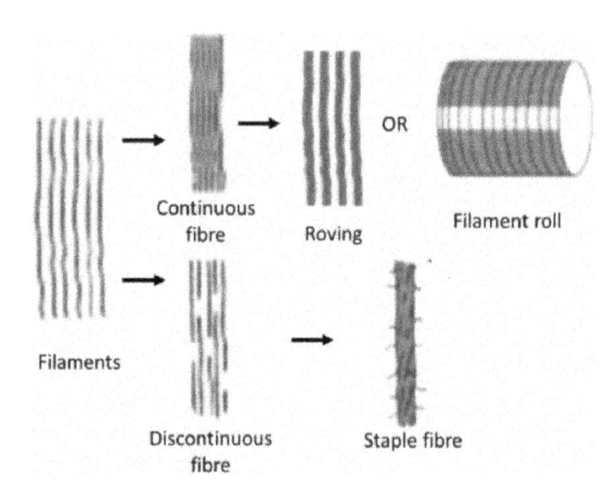

Figure 14.2 Composites: fibre forms.

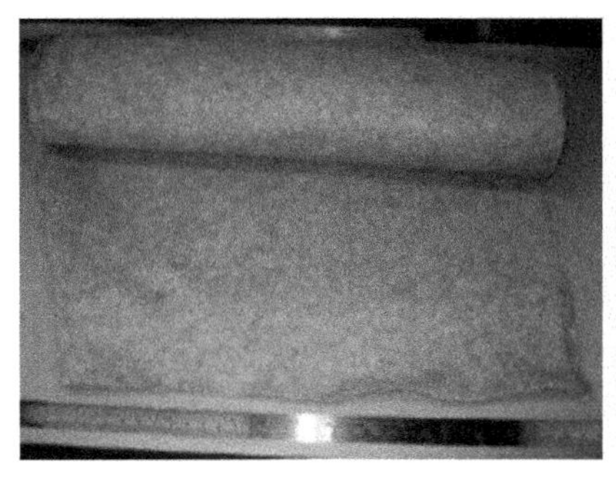
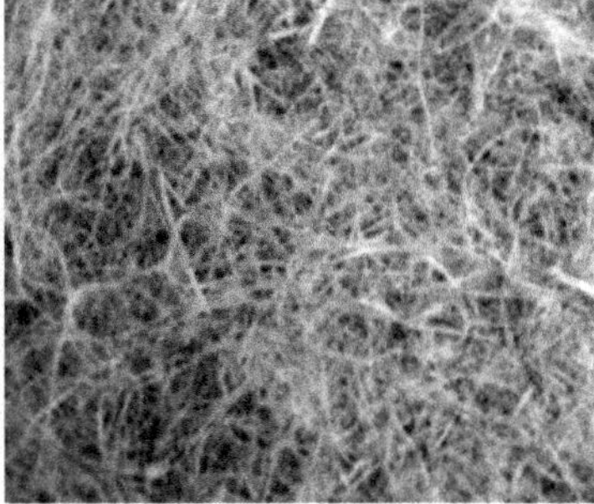

Figure 14.3 Hemp mat roll (left); and hemp mat surface macro-image revealing a wavy shape (right).

of fibre lengths. Short fibres are not usually incorporated into composite reinforcement mats but can be used to produce mats for other applications with the use of binders or techniques like hydro-entanglement. Processes involving combing or water jets can create the preferred direction of the majority of fibre lengths [19].

There are multiple solutions available to process NF reinforcement mats. The most basic process includes chopping the fibres and separating them into specific lengths. Then, fibres with selected lengths are used to create mats with various properties. This is done by pressing, or in a continuous process which uses forming machines. Most of the mats have randomly orientated fibres, which are within the x–y plane of the mat. Equal layers of fibres are spread over the belt and then transported in order to be formed into a mat by pressing on the continuous belt forming press. Processes involve the use of mechanical force, chemical pretreatments, and heat or binders to create fibre–fibre bonds of a mechanical or chemical nature. In order to increase mechanical interlocking between fibres, the hydroentanglement process can be used. It is a continuous process, which uses high-pressure water jets to induce interlocking of the fibres. A web of fibres is entered into the jet area and the water passes through it. The outcome is a thinner layer of non-woven mat with increased mechanical properties. Synthetic and natural fibres can be used in this process. As with pressed mats, the hydroentangled mats can be used for composite laminate impregnation, since they do not include binders [20–21].

14.2.3.2 NF yarns

A yarn is composed of short or long fibres, which are held together by means of a mechanical interlocking. The main uses of yarns include fabric processing, weaving and rope-processing. NF yarns can be used as the composite reinforcement. Using the yarn, instead of the mat, to reinforce the composites, allows for higher control over fibre orientation, increased fibre loading, and a continuous production process [22]. With regard to the structure of yarn, there are two types: twisted fibres and non-twisted fibres (Figure 14.4).

With twisted yarns, NFs are held together by shear forces created by fibre twisting, as in conventional textile yarn or rope. In this type of yarn, the fibres are aligned at an

Figure 14.4 Images comparing ring-spun yarns, with the twisted fibres (left), and wrapped yarns, with no twist and the fibres aligned in the main yarn direction (right).

angle to the main direction of the yarn. This type of yarn is produced by a spinning process and is mainly used in fabric processing. Technical fibres are stored in bales, which is the starting point for the yarn-spinning process. At first the bales are opened by a mechanical or manual process and transferred to a picker, which loosens and cleans the fibres. In the next step, a carding machine aligns the fibres, which are passed through a funnel and the outcome parallel strand of fibres in form of a strand and is called sliver. Then, a set of rollers elongates and slightly twists the sliver, which is then transferred to a container [23].

The spinning of hemp fibres differs from the spinning processes of other fibres, due to the mechanical properties of hemp. In order to spin long hemp fibres, dedicated processing equipment is needed. Spinning machines which are used to process cotton yarns can be used for the processing of short fibre hemp yarn, whence the hemp will go through the 'cottonising' process [24]. Slivers produced in aforementioned stage are spun into yarns, which can be composed out of one or multiple twisted slivers. Yarns can be produced continuously in S or Z arrangements, which correspond to two opposite twist directions.

Non-twisted yarns were designed especially for composite reinforcement. The fibres have no twist and are aligned in the main direction of the yarn. They are held together by frictional forces created with the polymeric wrapping of yarn. Wrapping yarn is made out of a continuous synthetic polymer fibre. The first stage in processing this type of yarn is the same as that of twisted yarn. After creation of the sliver, the fibres are not twisted. The sliver is divided into multiple strands of slivers with the required linear density. In the next stage, the slivers are wrapped around with polymeric filament, which applies pressure to the fibres to create frictional forces, which hold the fibres together. This type of yarn has no application in the textile industry, due to its dry tenacity, which causes its load-bearing capacities to be very low. After impregnation, the fibres in non-twisted yarn should have a higher degree of alignment when compared with twisted yarns. It is possible to process yarns with various linear densities from 200 tex to over 2000 tex. Lower values of the linear densities are not usually practical, since the ratio between artificial yarn and natural fibres is too high. It is possible to process higher tex values, but compaction of the fibres in the thick yarn may obstruct the resin from penetrating the yarn [25].

14.2.3.3 NF fabrics

After spinning or wrapping, yarns are woven together to form fabrics. There are multiple types of fabric, which vary in the type of weave, as well as the reinforcement direction (Figure 14.5). Fabric can have yarns aligned in two, three and even four directions [26], in the case of a three-dimensional preform. Two dimensional include weaves, braids and knitted fabrics. Three-dimensional fabrics include weaved, braided, stitched and knitted fabrics [27]. The fabrics most commonly used for composite reinforcement are the ones with two-dimensional biaxial patterns. Fabrics with plane and twill arrangements are one of the typical examples.

In general, fabrics allow for close packing of the composite and can be easily arranged when processing laminates. Processing techniques, such as lay-up, depend mainly on the use of woven reinforcement fabrics. Various laminate shapes can be processed, such as curvatures. The drawback is the waviness, which changes the axial filament arrangement, thus influencing the mechanical properties of the composite. There is a difference between the mechanical properties of the laminates reinforced with various forms of knitted and waved fabrics, and this subject has been investigated thoroughly for synthetic reinforcements. An increase in the waviness of the individual yarns within the fabric creates higher stress concentrations. In-plane mechanical properties are reduced, but transverse and through thickness laminate properties can be increased depending on the type of fabrics used [27–29].

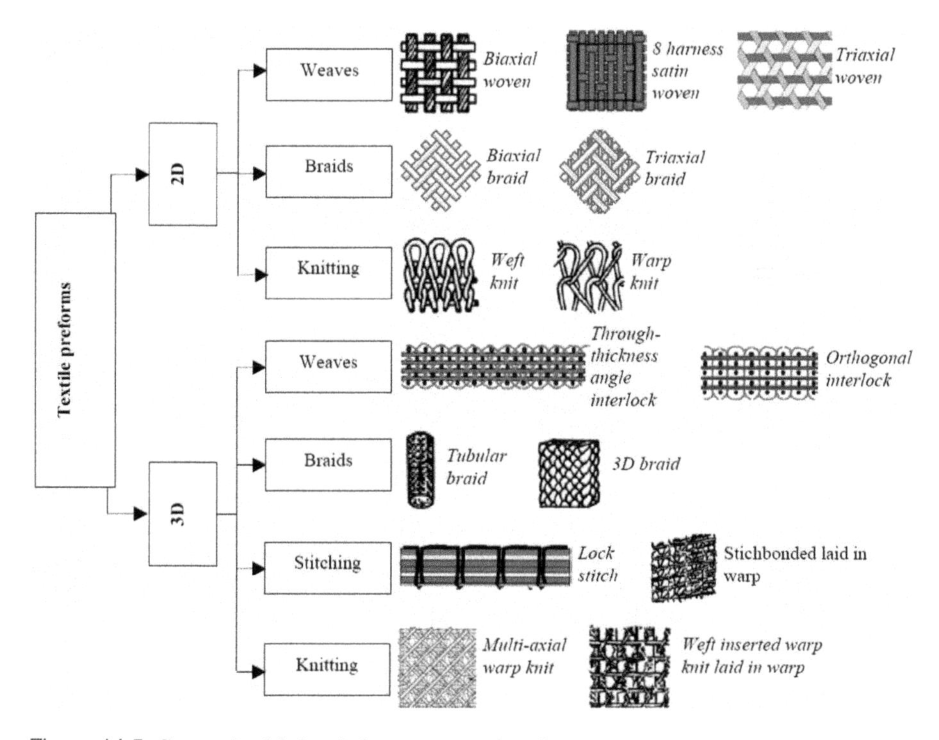

Figure 14.5 Composite fabric reinforcements and preforms.

14.3 NFC resin systems

NFCs based on bast fibres mainly use polymeric matrices, which are classified by the curing mechanism, namely thermoset and thermoplastic. The curing mechanism influences the processing techniques used, which are used for the processing of NFCs and improving their mechanical performance.

14.3.1 Thermosets

Thermoset resins include phenolics, polyesters, melamines, silicones, epoxies and polyurethanes. While curing, thermosets undergo a crosslinking reaction until almost all the molecules are crosslinked, making three-dimensional networks. After setting, thermosets cannot be melted but the shape cannot be changed. They are usually supplied partially polymerised or monomer–polymer mixtures. The crosslinking reaction can be realised by the application of heat, oxidisers or ultraviolet radiation. The most frequently used thermosetting resins in processing are polyester, phenolic and epoxy. Due to the low cost and low processing temperatures, unsaturated polyester is the most popular resin.

Figure 14.6 shows the steps involved in thermoset matrix laminate processing. At first, an appropriate mould, the exact volume of the matrix, and the reinforcement are prepared. The other procedures depend on the impregnation technique selected. In lay-up processing, resin mix is first prepared and then, after that, reinforcement layers are placed one after another and impregnated with resin by using brushes or rollers. Resin mix is prepared by mixing resin with the catalyst, which starts an exothermic curing reaction. Additionally, other substances are added, such as accelerator, fillers, pigments and solvents, which are added prior to the catalyst. The required pressure and heat can then be applied to create the desired shape and accelerate the curing reaction. Pressure can be applied by mechanical means (hot press), atmospheric pressure (vacuum bagging) or elevated pressure (autoclave). The heat can be applied to the mould by conduction or by the use of microwaves.

The degree of wetting during the production process is important for good adhesion between the fibre and the matrix. An uncured thermoset resin can have very low

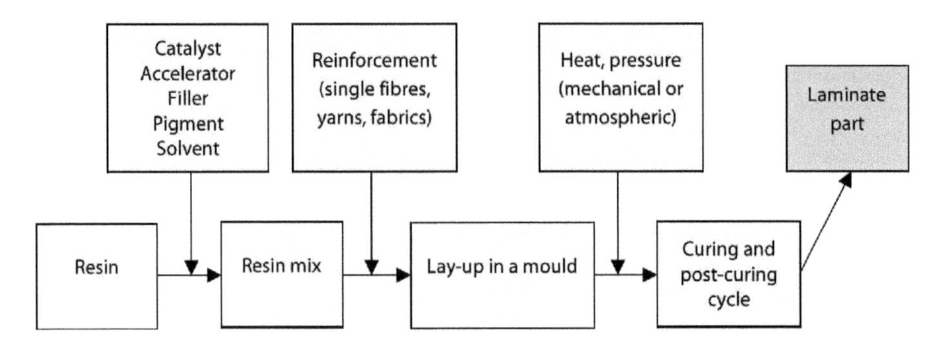

Figure 14.6 Composite laminate preparation with thermoset resin systems like polyester or epoxy.

viscosity, which eases the wetting and the penetration of the resin into the reinforcement during the impregnation process.

Unsaturated polyester polymers have several carbon–carbon double bond sites, which are called unsaturation sites. The reaction is started with a catalyst in form of peroxide with the presence of the styrene. When the peroxide initiates the reaction, it combines with the carbon–carbon double bond, which forms a free radical at the same time. The free radical is very reactive and can create a bridge between two polyester chains by reacting with the styrene. When the bridge is formed, a free radical is created on the other polyester chain. Reaction occurs as long as there are free radicals, styrene and unsaturation sites present. The chain reaction stops when the free radical reacts with something else than styrene, such as oxygen, contaminants, or peroxide molecules. Therefore, the reaction can stop prematurely and an insufficient number of crosslinks will be created [30].

The curing process in thermoset resins is affected by multiple additives, which can slow or accelerate the crosslinking process. Additives which slow down the reaction include inhibitors absorbing free radicals, styrene, fillers, oxygen, flame retardants, reinforcement; also, an increase in a mould heat capacity can slow down the reaction. Additives which accelerate the process include initiators, heat, ultraviolet radiation, accelerators, waxes and films; also, an increase in the part thickness can accelerate the process. Other factors, like resin grade and type, water, pigments and contaminants, can affect the crosslinking process speed either way [31–32].

The advantages of using a thermoset matrix system are that it is easy to process and laminate, there is no need to apply pressure or heat, it is inexpensive, has superior mechanical properties in comparison with thermoplastic matrices and has a higher resistance to elevated temperatures. On the other hand, the use of thermoset matrix systems involves some disadvantages, including emissions of VOCs during processing, a short workable and pot life and expiry date. The cured composites cannot be recycled or reclaimed easily and, depending on the technique, can have a low-quality surface finish [33].

Table 14.3 presents a comparison of the physical and mechanical properties of the three most popular thermoset resins used in the lamination of composites, namely polyester, epoxy and vinylester. Epoxy resin has the highest stiffness and strength, up to 6 GPa and 100 MPa, respectively. Polyester resin can have a compressive strength

Table 14.3 Comparative properties of the three main synthetic thermoset resins [34–35].

Properties	Polyester resin	Vinylester resin	Epoxy resin
Density (g cm^{-3})	1.1–1.5	1.1–1.4	1.2–1.4
Young's modulus (GPa)	2–4.5	3–6	3.1–3.8
Tensile strength (MPa)	40–90	35–100	69–83
Compressive strength (MPa)	90–250	100–200	–
Barcol hardness	20–60	35–40	
Tensile elongation to break (%)	2–6	1–6	4–7
Cure shrinkage (%)	4–8	1–2	–
Water absorption 24h at 20°C	0.1–0.3	0.1–0.4	–
Fracture energy (kPa)	–	–	2.5

of up to 250 MPa. Most thermoset resins are inherently brittle but modifications are possible to elevate elongation at break. Moreover, polyester resin has the highest cure shrinkage: up to 8% [34–35].

14.3.2 Thermoplastics

Thermoplastic resin includes polyphenylene, polypropylene, sulfone, polyamide, polyetherketone. This group of polymers exhibits little crosslinking. This feature enables them to be repeatedly softened with temperature and re-shaped. Fabrication techniques for thermoplastic matrices include moulding and various extrusion techniques. It is relatively difficult to achieve an effective wetting of the fibres due to a high viscosity, which results in a lower performance. Coupling agents, such as maleic-anhydride-polypropylene copolymer (MAPP or MAH-PP), are often required to increase the interface properties [33, 36].

Due to the higher processing viscosity of thermoplastic polymers, it is difficult to achieve a proper wetting of the fibres. High temperatures can also cause unwanted changes in the fibre surface or even destroy the fibres altogether. Nevertheless, its low price, reasonable processing temperatures and recyclability are the reason for a growing interest in polypropylene (PP). Unmodified PP, however, will not undergo proper adhesion with the fibres through the application of consolidation forces alone. The mechanical properties are hardly improved, as the fibres simply act as a filler. Natural fibres will only act as a reinforcement if a compatibiliser is used. An interface between the fibre and the matrix should correct the natural rejection of both materials. A compatibiliser that is often used is MAPP, which is a modification of a PP chain with maleic anhydride. A small amount of MAPP is added to the PP, which leads to significantly higher strength properties of the material.

Another promising development of thermoplastic prepregs is the use of latex emulsions or dispersions to enhance the wetting of the reinforcement. Advantages of thermoplastic systems include high impact strength, an attractive surface finish, recyclability, reusability of the processing scrap, and virtually zero emissions. They can be sintered with other thermoplastics and moulded or shaped by reheating. Disadvantages of thermoplastics include the softening with heat and the fact that the composite elements are more difficult to prototype [33]. Polypropylene (PP) has a density lower than water, which makes it suitable for marine applications, but at the same time PP presents a relatively low Young's modulus of 1200 MPa in comparison to 4000 MPa of polyphenylene sulphide (PPS) and polyether-ether-ketone (PEEK) matrices (Table 14.4). The highest working temperature for PP is within the range of 70 °C and 140 °C.

14.3.3 Bio-based resin systems

The bio-based polymers, or bioplastics, are a group of materials derived from bio resources, as opposed to fossil-fuel based polymers. Source materials for bioplastic production can be virgin or come from food industrial waste [37]. Bioplastics can be biodegradable under weathering conditions and degraded by microorganisms [38]. Some of the bioplastics are synthesised by microorganisms and have an increased biocompatibility [39–40]. Some of the commonly used bio-based polymer systems

Table 14.4 Properties of selected resin systems.

Properties	PP	PPS	PPE
Density (kg m^{-3})	900	1300	1300
Young's modulus (MPa)	1200	4000	4000
Tensile strength (MPa)	30	65	90
Tensile elongation (%)	20–400	100	50
Working temperature (°C)	70–140	130–250	140–250
Coeff. Of thermal expansion, α (20 °C)	9×10^{-5}	5×10^{-5}	5×10^{-5}

include polylactic acid (PLA), polyhydroxyaldehyde (PHA), polyhydroxybutyrate (PHB), polyester TP, furan resin and wood-based epoxy resin [41]. Existing systems have relatively poor mechanical properties.

The main applications for the bio-based polymers are in the packaging industry and in insulation, which are mainly thermoplastic polymers. With the development of new thermoset biopolymers with enhanced mechanical properties, such as phenolics, epoxy, polyester and polyurethane resins, other applications are possible [42]. Thermoset biopolymers can be reinforced with natural fibres to create a fully bio-resourced composite [43]. Plant-based polymers have also been developed, such as proteins [38], oils, carbohydrates, starch [44] and cellulose. The cost of biopolymers is relatively high and in some areas the production of biopolymers for industrial applications can compete with food production for raw materials. Shortages in the area of cultivation, together with increasing global population, have also led to the increased research into sea-plant and microorganism sources for biopolymers. Biodegradability is an increasingly important factor in the research of bioplastics. Fast-food utensils, packaging containers and trash bags produced from biodegradable plastics, can reduce environmental problems by saving energy and diminishing land contamination. It should be noted that bio-based polymer may not be biode gradable and that biodegradable polymers may be developed from petroleum raw materials (Table 14.5).

Table 14.5 Bio-based and biodegradable polymers.

	Bio-based		
Polymers	Synthesised	Natural	Petroleum-based
Biodegradable	PLA, PHA/PHB Corn-based polyesters Starch-based polymers Soy-based polymers	Gluten resin Starch-based emulsion	PBS, PCL, PVA
Non-biodegradable	Furan, Oil PUR Epoxy Ethanol-based PE	Cashew nut shell	PP, PE, Nylon Epoxy

14.4 Long natural fibres composites

14.4.1 Long natural fibre composites

Long natural fibre composites (NFC) are by no means new, but it is only recently that developments have defined better performance and new fields of application. A composite is a material composed of two or more constituents that are distinguishable on a macroscopic scale. The properties of the composites are superior to those of the constituent materials acting separately. Usually one or more phases act as reinforcement and the other binds them together.

Composites are usually developed to meet a specific design performance. It is possible to improve and control properties – such as strength, stiffness, weight, toughness, fatigue, temperature-dependent behaviour, thermal and electrical conductivity, and acoustical insulation – by manipulation of the composite constituents and by processing methodology [45]. Principal fibres used in the reinforcement of composite include glass, aramid (Kevlar), carbon, boron, silicon carbide and cellulose natural fibres. In general, high mechanical properties render composite material suitable for structural applications. Unidirectional composites have their highest mechanical properties in the fibre direction and usually the weakest in a direction perpendicular to the fibre. Anisotropic composite layers can be stacked in the form of a laminate at various orientations in order to fulfil design requirements in other directions. The most common types of laminate layer are presented in the Figure 14.7, namely randomly orientated mat composite, unidirectional layer in three orientations, and biaxial fabric [11].

NFCs can be defined as materials based on organic or inorganic fibres from a natural source, bonded together with various types of polymeric or ceramic matrices.

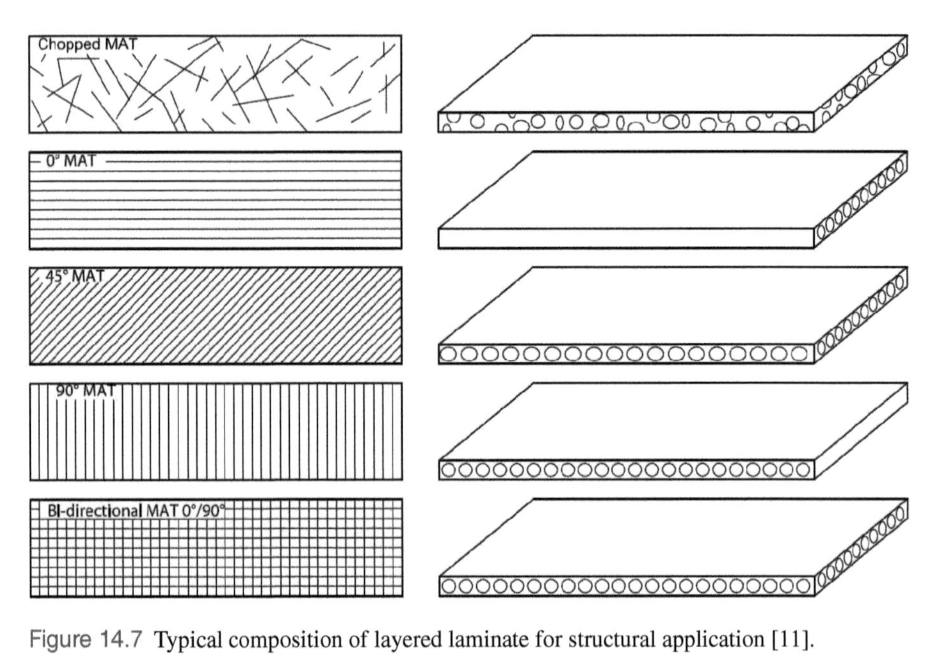

Figure 14.7 Typical composition of layered laminate for structural application [11].

NFs are divided into three categories based on the source of the fibre, namely animal, mineral and plant. Composites based on wood fibres fall within the category of NFCs but are normally termed wood-based composites, such as particleboard, or wood plastic composites (WPC). Fibres from flax, hemp, jute, sisal, kenaf and others act as a structural reinforcement for the plant. Extracted and processed natural fibres can be embedded into the matrix in order to create various forms of laminate. The processing techniques used are the same as those for synthetic composites. Due to the low density and high mechanical properties of cellulose, the specific properties of NFC are comparable to glass-fibre reinforced plastics (GFRP). The mechanical properties, combined with cost, production capacity and environmental superiority, render NFC an attractive material.

Developments in synthetic composites and natural fibre composites are often linked. Key developments in processing methodologies and matrix materials have been made for both groups of composite. For obvious reasons, the use of the natural fibre composites predates the development of synthetic composites, which were developed after breakthroughs in organic chemistry which led to the creation of synthetic bonding materials.

The first natural composites used by humanity from time immemorial were wood, bamboo and bone, where cellulose or mineral apatite played the role of the reinforcements, which were not manmade. Natural materials were used because of their abundance, renewability and the lack of alternatives. For two centuries, wood and bast fibres were used for the construction of houses, bridges, ships and weaponry. Bricks reinforced with straw fibre in ancient Egypt are considered to be the first natural fibre composite created by people. It may, at the same time, be considered the first application of NFCs in civil engineering. At the beginning of the twentieth century, asbestos fibres bonded with phenolic resin were used [33].

Looking at the recent history, NFCs based on non-wood, bast fibre reinforcement have generated much interest in research for more than a century. In fact, flax fibre reinforced phenolic resin composite, developed and described by Norman de Bruyne in 1937, is considered a milestone leading to the development of glass fibre composites. Moreover, in the 1930s Henry Ford experimented with multiple natural fibres as a reinforcement for car bodies [46]. These developments date back to the need for light and strong materials for the automobile and aviation industries during the World War II [47]. A scarcity of light metal alloys, created by excessive military use and economic warfare between countries, led to work on NFCs and further developments. This need was less urgent at the end of the war by the development of low energy alloy production, emergence of synthetic composites and growing performance requirements for the automobile and aviation materials. Renewed interest in natural fibre composite materials was noticeable two decades ago. This was related to a better understanding of NFC properties, as well as growing awareness of environmental benefits. NFCs are now seen as potential materials for the internal components of cars, for civil engineering and other applications, because of their strength-to-weight properties, possible low cost, and their reduced environmental footprint among other benefits [2]. The properties of NFCs are directly related to many factors, e.g. fibre and matrix performance, processing parameters, and the fibre–matrix interphase.

14.4.2 Maximum and minimum volume fraction

The volume fraction of the reinforcement is one of the most important factors determining the mechanical properties of composites. Theoretically, there are two arrangements, which can give the highest packing of fibres with circular cross-sections, namely hexagonal packing and square packing [48]. It can be shown that the maximum volume fraction for hexagonal packing is:

$$V_{fmh} = \frac{\pi}{2\sqrt{3}}\left(\frac{r}{R_{min}}\right)^2 \quad (1)$$

where, R is the fibre spacing and r is the fibre radius. For the case of closest fibre packing, this becomes:

$$V_{fmh} = \frac{\pi}{2\sqrt{3}} \quad (2)$$

The maximum volume fraction for the square packing arrangement is:

$$V_{fmh} = \frac{\pi}{4}\left(\frac{r}{R_{min}}\right)^2 \quad (3)$$

For the case of closest fibre packing, this becomes

$$V_{fmh} = \frac{\pi}{4} \quad (4)$$

There are minimum (V_{min}) and maximum (V_{max}) critical fibre volume fractions. Below V_{min} and above V_{max}, the properties of the composite do not fully follow the rule of mixtures and reinforce the composite efficiently. In the case of long fibres, reinforcing the composite in one direction, Kelly and Macmillan [48]derived an expression for the minimum critical volume fraction, given by Eq. (5):

$$V_{min} = \frac{\sigma_{bm} - \sigma_{fm}}{\sigma_{bf} + \sigma_{bm} - \sigma_{fm}} \quad (5)$$

Where, σ_{bf} is the fibre tensile strength, σ_{bm} is the matrix tensile strength and σ_{fm} is the stress on the matrix at the fibre maximum tensile stress.

The maximum critical volume fraction, V_{max}, is more important because it is a value that determines the maximum mechanical properties for a particular matrix and fibre combination. There are a couple of factors which practically influence this value, such as fibre geometry, resin mechanical properties, interphase properties, reinforcement arrangement and the processing technique used.

V_{max} can be theoretically derived based on the shear lag theory developed by Cox [49] and later Rosen [50]. Pan [51] derived an expression for fibres above the critical length, linking the fibre spacing ratio and the fibre–matrix properties, as displayed in Eq. (6):

$$\frac{R_{min}}{r} = e^{\left(\frac{\sigma_{bf}}{\tau_s}\right)^2 \frac{G_m}{2E_f}} \quad (6)$$

where, τ_s is the shear strength of the matrix, G_m is the shear modulus of the matrix and E_f is the fibre modulus.

Pan [51] presented relationships using typical ratios between properties in composites for both short and long fibres. It was shown that with an increase in the matrix shear properties, and taking into consideration the fibre strength, it is possible to pack a higher number of fibres into the material more efficiently. The maximum volume fraction for the fibres with 50% higher tensile strength than the matrix shear strength is around 75%. Looking at the modulus relationship, the highest volume fraction – calculated for 0.02 of the matrix shear modulus to fibre modulus ratio – is 70%.

14.4.3 Compression strength of NFCs

Compression strength in composites has been investigated for various types of reinforced composite system. Both fibre and matrix have significant effects on the compressive strength of NFCs. The factors relating to fibres include fibre type, properties, volume fraction, diameter, length and orientation, while the factors related to the matrix include mechanical properties, additives, void content, fibre–matrix interface and bonding properties.

The composite compressive strength for different carbon and glass fibres as a function of volume fraction shows a linear relationship between fibre volume fraction and compressive strength [52]. Figure 14.8 shows the effect that varying the matrix has on the compressive strength of various composites [53–54]. Boron fibre reinforcing

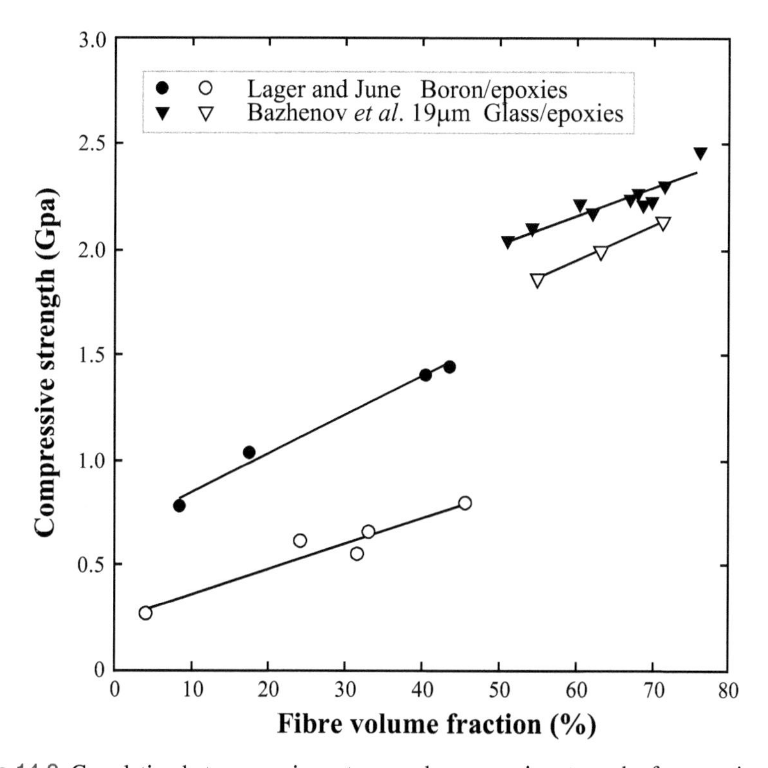

Figure 14.8 Correlation between resin systems and compressive strength of composites.

epoxy systems AF-126 and BP-907 (left two lines) are presented in the lower graph and the upper graph respectively. Glass fibre reinforcing epoxy systems EDT-10 and EDT-15e (right two lines) are presented in the lower and the upper graph respectively. There is a clear correlation between the resin system used and the compressive strength.

Another factor that significantly influences compressive strength is the confining pressure. It is tested by applying hydrostatic pressure to the sample during the test. A carbon fibre reinforced epoxy system with 36% Vf was tested by Weaver and Williams [55]. Glass fibre reinforced epoxy system with 60% Vf was studied by Wronski and Parry [56]. Glass fibre reinforced polyester was studied by Sigley *et al.* [57]. They all concluded that increasing the confining pressure has a linear relationship with compressive strength of the composites.

14.4.4 Tensile properties of NFC

The tensile and flexural properties of composites are influenced by multiple factors, which are related to fibre properties, matrix properties and reinforcement arrangement [58]. Fibre properties include the mechanical properties of the fibres, fibre morphology and fibre type. Matrix properties include the mechanical properties, additives content, void and impurities content, and the fibre–matrix interface properties. The influence of the reinforcement arrangement can be tested by changing the angle of the reinforcement in relation to the test direction and by changing the type of reinforcing fabric, which includes the waviness of the yarns within the fabric. It is well known that the greatest potential for orthotropic fibrous composites lies in their tensile properties parallel to the direction of the fibres. Any alteration of the reinforcement type and hence the fibre arrangement or morphology of the fibre will result in a decrease in tensile strength. Cellulosic bast fibres used as reinforcement have a finite length and varying properties, what is illustrated with fibre tensile properties (Table 14.2). Fibre properties are inherently influenced by multiple factors, namely plant species, region of growth, the position of the fibre in the plant, and the extraction and processing route [59–60]. Hemp and flax are easily grown in temperate climate regions and are seen as having high potential as fibres for NFC reinforcement.

A review of the reported tensile properties of hemp and flax short fibre randomly reinforced unsaturated polyester, epoxy and other matrix systems is presented in Table 14.6. The reported tensile properties vary from 0.6 GPa to 10.0 GPa for tensile stiffness and from 19 MPa to 90 MPa for ultimate tensile strength, respectively. This is due to the various processing routes used, altering volume fractions and matrices. Additionally, the tensile properties of flax biaxial fabric reinforced epoxies and polyester composites are included for a comparison. It was reported that the yarn twist and type of fabric reinforcement have an influence on the tensile and flexural properties [18]. The results obtained by Hepworth *et al.* [3] for hemp and epoxy composties are higher compared to others for short-fibre reinforced composites. Although the research used short fibres, the composites are made with the changed orientation by combing in order to align fibres prior to impregnation.

Most reported investigations were done on compression moulded samples. This technique is especially convenient for randomly orientated fibres and allows for the use of high pressures, thus increasing the volume fraction of the fibres inside composites.

Table 14.6 Tensile properties of composites reinforced with hemp and flax mat or fabric.

Materials	Vf (%)	UTS (MPa)	Modulus (GPa)	Elongation (%)	Processing	Ref.
Hemp/UP	10–26	20–60	0.6–1.2	6–11%		[61]
Hemp/UP	20	33	1.4			[62]
Hemp/UP	35	60	1.7			[62]
Hemp/UP	16–44	20–53	3.7–6.2	0.52–1.39	CM	[63]
Hemp/UP	11–21	19–44	–		RTM	[64]
Hemp/UP	15–45	35–50	5.0–10.0		CM	[65]
Hemp/UP	30	38–65	6.0–7.0		CM	[66]
Hemp raw/Epoxy	26	23	–			[67]
Biaxial Hemp combed/ Epoxy	20	50–90	5.5–8.0		Lay-up	[3]
Biaxial Hemp/Euphorbia resin	19–21	23–35	2.3– 3.1		CM	[9]
Flax/Epoxy	15–35	52–58	2.9–9.8		Auto.	[68]
Flax/PP	20–40	35–60	4.5–8.5		CM	[69]
Flax biaxial fabric/Epoxy	28	160	15		Lay-up	[70]
Flax biaxial fabric/UP	28–31	140–143	14		Lay-up RTM	[70]
Kraft Wood Fibre/PP	40	25–41	3.9–4.5		CM	[71–72]
GFRP (0,±45,90)/EP	40–55	138–241	15–28	0.85–0.95		[73]

UP – Unsaturated polyester resin; EP – Epoxy resin; CM – Compression moulding; RTM – Resin transfer moulding; Auto. – Autoclave processing

Alternatively, vacuum bagging related processes, using atmospheric pressure, yield lower volume fractions of the reinforcement. George and Verpoest [68] used an autoclave process in order to increase pressure and compaction.

Taking advantage of the longitudinal properties of natural fibres requires alignment, which can be facilitated by the use of combed mats of either the twisted or untwisted yarns or a hybrid of both. Aligning fibres in yarns allows for the preparation of composites with a higher volume fraction. The consistency of properties can be enhanced by blending fibres from different harvests, of specific lengths and from selected stem region, since it has been found that mechanical properties vary in accordance with the position of the cellulosic fibre in the plant [59–60].

The tensile properties of hemp and flax NFCs with aligned reinforcement in the form of yarns are summarised in Table 14.7. Most research focuses on epoxy thermoset resin systems or using polypropylene thermoplastic as a matrix. The volume fractions reported range between 30% and 60%, which is considered to be a barrier above which the achievement of consistent fibre wetting becomes difficult [74]. Winding of the yarns improves consistency in fibre distribution, and consecutive pressing allows further consolidation and compaction, hence achieving higher volume fraction. The reported tensile properties for hemp yarn reinforced composites varied between 39 MPa and 277 MPa for the ultimate strength, and 5.9 GPa and 27 GPa

Table 14.7 Tensile properties of composites with hemp or flax reinforcement aligned in the axial direction.

Materials	Vf (%)	UTS (MPa)	Modulus (GPa)	Elongation (%)	Processing	Ref.
Hemp/Epoxy	35	43			Winding	[67]
Hemp/Epoxy	35	39			Lay-up	[67]
Hemp/Epoxy	35	230	21.3		HP	[76]
Hemp/PET	33	205	17.6		Winding/HP	[74]
Hemp/PE	32	161	15.8		Winding/HP	[74]
Hemp/PP	32	173	15.4		Winding/HP	[74]
Hemp/PP	30–35	148–228	5.9–12.4		Winding/HP	[77]
Hemp/PP		122–125			HP	[77]
Hemp/PET	47	277	27.0		Winding/HP	[74]
Flax/UP	55–60	165–304	27.8–29.9	0.8–1.7	Vac/HP	[78]
Flax/UP	40	199	19.5	1.3		[60]
Flax/Epoxy		132–280	15.0–39.0	0.8–1.2	RTM	[75]
Flax/Epoxy	40	65–150	16.0–30.0		Winding/Auto.	[59]
Flax/Epoxy	40	210–328	22.5–22.9	1.2–1.6		[60]
Flax/Epoxy	32	132	15.0	1.2		[60]
Flax/Epoxy	50	325	24.0			[60]
Flax/PP	51	287	28.7			[60]
Glass UD/EP	40–60	300–1100	3545	2–3		[73]

UP – Unsaturated polyester resin; Winding – filament winding; HP – Hot press; RTM – Resin transfer moulding; Auto. – Autoclave processing.

for the modulus of elasticity. Similarly, for flax yarn reinforced composites, it was reported that ultimate tensile strength ranged between 65 MPa and 328 MPa and the modulus of elasticity between 15 GPa and 39 GPa, which are higher than those for hemp. One of the results for the stiffness, obtained for aligned unidirectional arctic flax composite processed by resin transfer moulding, described by Oksman [75], was 39 GPa for the tensile stiffness and 279 MPa for ultimate tensile strength, which is higher than most of the results.

It is apparent that the mechanisms of the tensile behaviour of NFCs are complex and there could be a range of influencing factors. It is important to understand why there is wide variation in the reported results and if it can be optimised. For this reason, an investigation into the relationship between hemp and flax reinforcement structure and natural fibre composite properties has been conducted by Weclawski [79]. This study systematically assessed the influence of the type and structure of the reinforcements on the tensile properties of NFCs with a bio-derived unsaturated polyester resin as a matrix. The influence of the fibre volume fraction on the tensile properties for each type of reinforcement was analysed. The study then went one step further to explain the effect of unidirectional yarn wrapping on the build-up of internal stress concentrations, fibre alignment and fracture mode.

14.4.5 Flexural properties of NFC

Flexural properties together with tensile properties are often evaluated for NFCs. In tensile and flexural loading cases, fibrous reinforcement can be utilised in the most efficient way along the length, due to its anisotropic properties [33]. However, it was reported that designing elements made out of NFCs or other fibrous composite material, working mainly under a compression load, yields the lowest performance [80]. Since the flexural test is relatively simple to conduct, flexural coupons can be used to determine the influence of factors such as the environmental degradation of the mechanical properties of NFC material. For this reason, a number of researchers have used flexural testing as a benchmark to compare different materials.

Table 14.8 summarises the flexural strength and modulus of hemp and flax NFCs, together with the volume fraction and processing methods. The volume fraction of reinforcement ranges from 10% to 70%, the flexural modulus is between 0.8 GPa and 25 GPa, and the flexural strength is in the range from 31 MPa to 219 MPa. The large variation reported in the results from the literature is due to NFC properties being dependent on multiple factors, namely fibre aspect ratio, fibre type, surface morphology, fibre treatment, structure, arrangement, resin type and processing route [33]. Recently, knitted fabric reinforced flax composites were revisited by Sawpan *et al.* [81], where the flexural properties of epoxy matrix composites processed by the lay-up technique were evaluated. The specific flexural properties reported were low in comparison

Table 14.8 Flexural properties of composites unidirectionally reinforced with hemp or flax.

Materials	Vf (%)	MOR (MPa)	MOE (GPa)	Processing	Ref.
Rand Flax hackled–PP/MAPP	20	70	4.0	HP 200° 40 bar	[69]
Rand Flax hackled–PP/MAPP	40	90	7.0	HP 200°40 bar	[69]
Rand Flax scutched–PP/MAPP	40	80	6.0	HP 200°40 bar	[69]
UD Hemp Epoxy	35	148–219	5.9–12.4	FW	[77]
UD Flax PP	35	77–149	–	FW	[77]
UD Flax non-hackled UP	25	168	19.4	PU (lab.)	[70]
UD Flax hackled UP	25	182	19.5	PU (lab.)	[70]
0/90 Fabric Flax UP	31	198	17.0	RTM	[70]
0/90 Fabric Flax EP	28	190	16.0	Hand Lay-up	[70]
Fabric Flax EP	18–34	*31–106	*0.8–2.9	Hand Lay-up	[82]
Treated Hemp UP	56	101	10.0	HP 6 MPa	[83]
Hemp mat rand UP	10–40	40–110	4.0–7.0	RTM	[84]
Manila Hemp STBR	30–70	100–200	10.0–25.0	HP 10 MPa	[85]
Alcali UD Hemp long PLA	10–40	90–68	6.4–5.7	HP 5 MPa	[81]
Alcali & Silane UD Hemp UP	10–50	97–88	4.9–6.8	HP 5 MPa	[81]
Glass UP	25	450±24	20.8±1.4	PU (lab.)	[70]

HP – Hot press; Rand – Random mat reinforcement; MAPP – , PP – polypropylene; FW – Filament winding; UD – unidirectional; PU – pultrusion; EP – epoxy; UP – unsaturated polyester; RTM – Resin transfer moulding; * – values in MPa/g.cm^3; STBR –Starch biodegradable resin to the findings of Goutianos *et al.* [70] for the same hand lay-up processing technique. Mechanical processing, such as hot pressing, filament winding and pultrusion, allows for higher fibre orientation precision and increased volume fractions. The selection of the processing route directly influences the mechanical properties and the performance of NFCs [69, 77].

14.4.6 Environmental degradation of NFs, matrices and NFCs

Natural fibres can be used to reinforce composites due to their excellent mechanical properties: for example, cellulose provides the strength in plants. It is estimated that cellulose is synthesised globally at a rate of 4×10^{10} tonnes per year [84], which makes it the most abundant carbohydrate of plant biomass available. Many organisms and microorganisms use cellulose as a source of energy by breaking it down into glucose via enzymatic processes [85]. Cellulosic bast fibres are prone to various types of environmental degradation, such as exposure to moisture, ultraviolet radiation, bacteria, fungus and termite attack, which will all degrade cellulosic fibres [41]. The natural degradation of cellulose allows for biomass circulation within the system, but at the same time the properties of cellulose in manmade materials also degrade. In order to prevent this deterioration, various treatments, additives and inhibitors are used. The material can also be sealed from the surrounding environment, which is more difficult to achieve for construction materials; in fact it is only possible to achieve this for indoor applications.

When considering the structure of natural fibres, among the natural fibre components lignin is the most prone to photochemical degradation by ultraviolet light. The colour change during its exposure to light is mostly due to lignin degradation. Treatments that reduce the lignin content of fibres are important for extending the life of the NFC when it is exposed to ultraviolet radiation [18]. The polymeric matrix material usually deteriorates due to the same factors, but at a slower rate. Moreover, matrix deterioration is not uniform. Gu *et al.* [86] studied the degradation of polyester films and coatings in alkaline solutions by examining the surfaces with an atomic force microscope before and after exposure. It was concluded that the surface does not degrade uniformly and that certain distinguished regions were more susceptible to hydrolysis, which explained the creation of pits in exposed surface.

NFCs with polyester and other matrices exposed to weathering conditions have been reported [87–88]. Composites of polyester reinforced with banana fibre with randomly orientated mats of fibre lengths 20–40 mm gave a decrease of 6% for thermal ageing in tensile strength, 9% for 6-month natural weathering and 32% for the accelerated water ageing test. That is, a tensile strength 903 MPa for non-aged composite, 845 MPa for thermally-aged composite, 614 MPa for water-aged composite, and 821 MPa for the 6-month weathered weathering test. Hybrid composites of polyester reinforced with cotton and kapok under accelerated degradation exposure had no deteriorative effect on flexural properties. The results presented show a 20% reduction in flexural strength on average, and minimal change in the flexural modulus.

Weathering studies on jute alkali treated fibres in unsaturated polyester resin showed a decrease in flexural stiffness, ultimate tensile strength, Young's modulus, and strain to failure of the matrix material [89]. The high sensitivity to weathering conditions of the polyester matrix natural composites restricts their applications to dry and low-humidity environments. The environmental degradation of polypropylene composites randomly reinforced with short sisal fibres was also examined [90]. The research used water immersion and ultraviolet radiation treatments, and urethane derivative treatments of fibres PPG, PMPPIC and MAPP were found to reduce the water uptake of the fibres. Ultraviolet irradiation caused tensile the mechanical properties to deteriorate by photo oxidation regardless of the treatment.

The reaction with water of composites of polypropylene reinforced with various natural fibre composites was also investigated [91]. It was found that the diffusion of water follows the Fickian diffusion process. Three factors were found to influence the process, namely volume fraction of fibres, temperature, and matrix type. The influence of cyclic water sorption on the degradation of epoxy and vinyl ester composites reinforced with sisal fibres indicated that immersion cycles of 9 days resulted in a decrease in fracture toughness properties at levels of 50% for epoxy and 60% for vinyl ester composites. The same deterioration levels were observed after continuing the immersion for 400 days [36]. This was also the case for composites of polypropylene matrix reinforced with sisal short fibres under weathering conditions. The mechanical properties continuously reduced as the immersion time increased. This can be explained by the plasticisation effect of the interphase [92].

In order to understand the behaviour of NFCs during immersion, the value of the Fickian diffusivity constant, moisture equilibrium and correction factor were investigated for various materials [93]. A significant difference was found in the water absorption values for the samples with higher volume fractions of fibres and, therefore, numbers of fibres exposed at the surface (Table 14.9). This was concluded as the main reason for the water absorption rate difference. The surface of the NFC needs to be coated or given a gel-coat layer in order to minimise the fibre water uptake on the surface.

Table 14.9 NFC property before and after weathering.

Fibre	Matrix	Test [Property symbol]	Initial Property Value	Exposure Type/Duration [remarks]	Post-exposure Property Value	Ref.
Woven sisal	Epoxy based	Fracture toughness K_{IC} [MPa m$^{1/2}$]	2.20	Water immersion/ 45 days	1.11	[36]
Woven sisal	Vinyl–ester based	Fracture toughness K_{IC} [MPa m$^{1/2}$]	1.41	Water immersion/ 45 days	0.59	[36]
Cotton/ Kapok	Polyester based	Flexural strength [MPa]	55.3	Accelerated weathering	34.4	[88]
Cotton/ Kapok	Polyester based	Flexural modulus [GPa]	0.699	Accelerated weathering	0.676	[88]

Unsaturated polyester resin reinforced with hemp fibre composites were also examined for environmental degradation with fluid immersions [18]. The mechanical properties of randomly reinforced composites gradually decreased with immersion time before reaching equilibrium, 16% saturation point. The tensile strength and stiffness were reduced by 35% and 60%, respectively, after immersion for 3700 hours. By comparison, GFRP composites with a UV-stabilised polyester matrix were tested for environmental weathering for three years. After exposure, the composite exhibited significantly higher water sorption, leading to increased swelling and internal stresses, therefore reducing mechanical properties [94]. The level of deterioration was still lower than that exhibited by the aforementioned NFC.

14.5 Composites modelling

14.5.1 Modelling basics

Composite strength can be modelled with a high confidence level. There are multiple ways of dealing with the subject, starting with generalised methods such as the basic rule of mixtures, and ending with more comprehensive solutions such as finite element analysis. The mechanical properties of NFCs can be modelled with some degree of confidence by adopting the solutions used for laminates reinforced by synthetic fibres. NF reinforcement differs from synthetic reinforcement, the properties of which are more consistent and repeatable, and are therefore easier to use in prediction modelling. A structure of natural fibre yarns, used as the reinforcement is composed out of short fibres, which is different from a synthetic continuous reinforcement. This needs to be taken into account when modelling and predicting the strength of NFCs. Kelly and Tyson [95] proposed four situations for the modelling of composite materials, namely elastic deformation for both the fibre and the matrix; plastic deformation of the matrix while the fibres deform elastically; plastic deformation of both the fibre and the matrix; and fracturing of the fibres with consecutive fractures in the composite.

The rule of mixtures (ROM) can be used to predict the axial properties of the composite, and the basic form is given by Eq. (7):

$$\sigma_c = \sigma_f V_f + \sigma_m (1 - V_f) \tag{7}$$

where, σ_f and σ_m are the tensile stresses of the fibre and the matrix respectively, and V_f is the volume fraction of the fibre in the composite. As mentioned previously, NFs vary in strength significantly. This version of the ROM has multiple assumptions, e.g. a linear change of the composite stiffness or the ultimate strength properties with a change in reinforcement quantity, and does not give reliable results when used with a wide range of variations in NF properties. An improvement to this method for the calculation of specific properties needs to be made.

14.5.2 Prediction modelling of NF yarn strength

Yarns are composed of multiple short fibres. As mentioned before, there are two types of yarn used as composite reinforcement: twisted fibre yarns and non-twisted fibre yarns. Twisted fibre yarns are the same classic yarns that have been used in the textile industry since time immemorial. The twist induces a frictional force between the fibres, keeping them together to form a continuous yarn. Increasing the fibre twist increases the strength of the yarn, because a higher percentage of individual fibres will break, rather than slip, when the yarn is loaded. However, after exceeding the optimum level of the twist, fibres start to deform and the yarn strength is reduced.

Non-twisted fibre yarns are a new invention and were specifically developed for use as composite reinforcement. Individual fibres are aligned along the main axis of the yarn and a polymeric yarn is wrapped around them to hold the fibres in place before processing. An advantage of this type of structure is the more precise alignment of individual fibres in the designated direction, increasing the tensile modulus. Therefore, when modelling composites reinforced with non-twisted fibre yarns, the reinforcement should be treated as having aligned short fibres.

The modelling of twisted yarn reinforced composites differs for two reasons: (1) the fibres are misaligned from the main yarn direction and (2) the presence of higher frictional forces between the fibres. Yarn strength modelling based on surface twist angle was analysed by Gegauff [96] and later by Platt [97]. More recently, Ghosh *et al.* [98] compiled a review of predictive models for yarn properties, dividing them into four groups: mechanistic, regression analysis, simulation, and neural network models. Yarn modelling is used to optimise the properties of the production process, in order to achieve a product designed for the desired application. The selection of the appropriate model depends on the resources and the application. Some models require deep insight into fibre mechanics and require multiple experiments to find the required variables in order to minimise the error of prediction. The mechanistic model described by Hearle *et al.* [99] assumed yarn to be a circular entity with single fibres lining in the same distance from the centre of the yarn. In reality the fibre position changes from the surface to the centre of the yarn along the length. A comparison of the mechanistic, statistical and neural network approaches for the modelling of the tenacity of yarns showed that the statistical and neural network approaches were superior to the mechanistic approach, with errors of 2.2%, 1.1% and 8% respectively for ring-spun polyester yarns [100].

There are two types of twisted yarn, one with long continuous filaments and one with short fibres, such as natural fibre staple yarns. There are several factors that need to be taken into account when analysing staple yarns. The fibres are not evenly distributed across yarn diameter and the core of the yarn is more densely packed. Also, fibres migrate across the diameter. Furthermore, individual fibres have variable mechanical and physical properties; therefore they do not break at the same time when the yarn is loaded.

In the textile industry the term tenacity relates to the breaking tenacity, the tensile stress at the rupture of the specimen, which can be the fibre, yarn, filament or cord. It is expressed in Newtons per tex, gram force per tex or gram force per denier [45]. For composite applications the value of the tenacity for the yarn is important, since yarn needs to withstand tensile loads during processing, using single yarn as a form of reinforcement, such as in filament winding, pultrusion or similar techniques. In order to predict tensile strength of the composite, the mechanical properties of the fibre need to be identified. Conventional reinforcing fibres are usually straight, continuous and uniform, which makes it easy to test and use these values for calculations. This is not the case when the composite is reinforced with natural fibre yarns. Individual fibres in the yarn are short, have different lengths and cross-sections, and therefore present varying mechanical properties. Moreover, yarn tenacity value is a combination of fibre fracture and fibre slippage along 250 mm length. This property can be used directly in designing a composite manufacturing process, such as knowing the filament winding to predict the allowed tensioning force.

If the interface, based on chemical bonding or mechanical interlocking between the fibres and the matrix, is strong enough, the aspect ratio of the fibre is the main factor that dictates whether the fibre will break or the interface will fail in the shear mode. Yarns for application in composite materials should be tested using small gauge lengths in order to minimise the influence of weak links in the test. In an ideal test situation, where no weak links are present along the gauge length, only the stress of the bundle of continuous fibres is tested. The tenacity depends on multiple variables and the precise modelling of this value requires a knowledge of the physical properties of the constituents of the yarn, as well as knowledge of the statistical data variability.

Madsen *et al.* [74] investigated the properties of natural fibre spun yarns for the reinforcement of thermoplastic composites. They analysed single fibre orientation within the yarn structure. Within the yarn structure, single fibres are oriented with decreasing twist angle measured from the outer surface to the centre, with respect to the longitudinal direction of the yarn. The mean twist angle as a function of the angle of the outer measurable fibre can be written Eq. (8):

$$\theta_{mean} = \theta_r + \frac{\theta_r}{\tan^2 \theta_r} - \frac{1}{\tan \theta_r} \tag{8}$$

where, θ_{mean} is the mean angle of fibre twist within the yarn and θ_r is the outer measured twist angle. The function assumes that during spinning, the fibres simultaneously spin around the yarn axis, that the yarn cross-section is circular, that the twist of the fibres is uniform and that the fibre stays the same distance from the centre of the yarn throughout the entire length of the fibre.

14.5.3 Modelling of NFC properties

Neagu, *et al.* [101] conducted a mechanical analysis of wood-fibre reinforced thermoplastic composites, where they calculated the Young's modulus of the fibres from tensile tests using micromechanical models and classical lamination theory. The approach was justified by a high statistical variation in the single-fibre tests due to fibril angle, kinks and wall thickness. They presented a mixed analytical–experimental methodology using the global stiffness of the composite to calculate fibre stiffness. The basis of their methodology was used to obtain the statistical distribution in the form of a normalised Fourier series expansion of the probability density function. Each designated orientation distribution range was treated as a separate laminate ply and stacked together to form a symmetric laminate. Based on the assumption of ideal fibre–matrix interface in terms of load transfer, the method was divided into three parts: (1) measuring the Young's modulus, fibre volume fraction and fibre orientation distribution in the composite; (2) modelling the Young's modulus of the composite based on laminate plate theory as a function of the fibre modulus; and (3) calculation of the fibre modulus with the tested composite values.

Marklund and Varna [102] undertook an analysis of elementary natural fibres and composites by using the concentric cylinder assembly (CCA) model, which is based on classical lamination theory (CLT). The CCA model was described by Sutcu [103], who was studying a theoretical composite with aligned coated fibres. The main difference between the models is the coordinate system, where classical lamination theory is based on Cartesian coordinates (x,y,z), whilst in CCA it is based on the concentric cylinder assembly.

14.6 Applications of NFC

Due to its mechanical performance and price, NFC can be used to process elements, which can utilise its benefits. One of the first reported NFC applications, in the later 1940s, was a plane fuselage made out of NFC [47]. This experimental product was under development in response to alloy shortages. The project did not go through to the production stage and was dropped.

Another interesting application for NFC is a bike made out of bamboo framing and hybrid joints. NFC is used as a bonding material in the bike joints, which are reinforced with alloy or CFRP elements inside. Fibres immersed in epoxy resin are wrapped around the joints and are subsequently cured [104].

In the automotive industry there are non-structural elements processed with NFC. The elements processed include interior door panels, seats, and dashboard and trunk elements. The most important benefits of NFC are low density and lack of harmful shrapnel or sharp edges during collision, in comparison with glass-fibre composites. The Lotus Elise was promoted as having elements of the car body made out of NFC, and the benefit was a weight loss at the level of 32 kg, which is significant for an already lightweight car (Fig. 14.9) [105]. The experimental door elements made out of sandwich panels are prepared with faces made with hybrid short natural fibre mats or fabrics and glass-fibre reinforcement. The core is made out of various types of foam. NFC panels are prepared by compression moulding or pultrusion [106].

Figure 14.9 Lotus (left) and Mercedes (right) car panels produced from natural fibres and biocomposites.

Figure 14.10 Natural fibre door frame testing set up in dual chambers, 60 °C and –20 °C.

NFCs have been extensively researched for doors and window frames [107]. The frames can be prepared with fabric-reinforced NFC with a filling of foam or other natural materials. The experiment carried out by author at the Building Research Establishment, UK (BRE), showed that doors framed with natural hemp fibres were more thermally stable compared to those with glass-fibre reinforcement (GRP) when subjected to dual chambers exposure regimes (Fig. 14.10).

Various types of panel have been manufactured from NFC (Fig. 14.11). One of the foreseeable applications is roof panels. Hybrid composition panels with sisal and glass fibres can be processed by compression moulding with epoxy resin. The reported tensile properties of the panels were 57 MPa and 2.6 GPa for strength and modulus, respectively [108].

With regard to civil engineering applications, there was an idea to build an NFC sandwich roof. It was intended to be a light and low-cost solution for roofing in

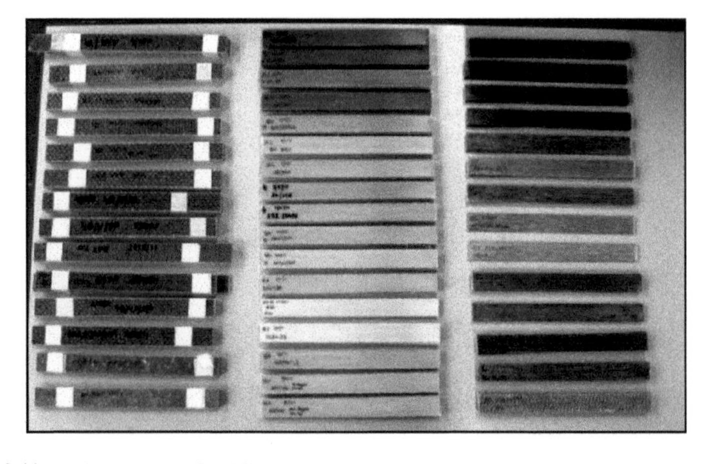

Figure 14.11 Various types of NFC panel.

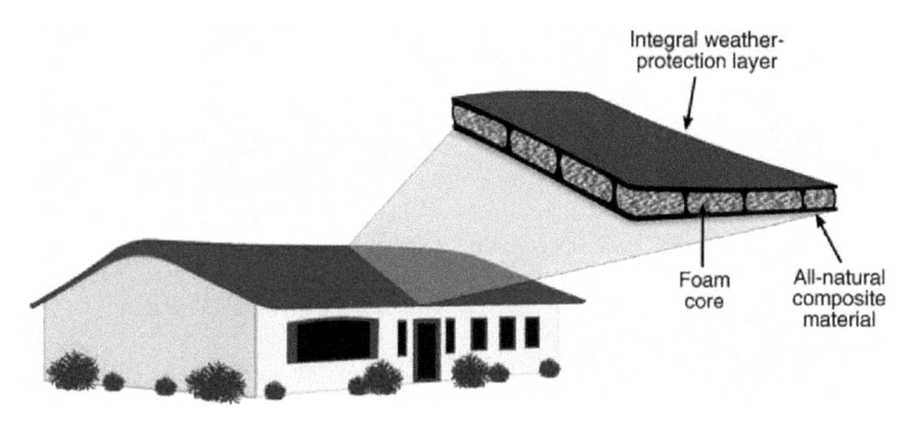

Figure 14.12 Monolithic house roof made of NFC [109].

hurricane affected areas. According to researchers, whole roofs can be processed with the resin-transfer moulding technique. As reinforcement for laboratory scale panels, researchers used recycled paper sheets, layered with chicken feathers, corrugated paper or glass fibre. The reported flexural rigidity and strength tests for these composites gave results of 12–20 GPa and 24–26 MPa respectively, which is within the range of stiffness for wood beams and double the wood strength (Fig. 14.12) [109]

NFCs for furniture are another major application. Natural fibres can either act as a filler or a reinforcement or both. Chairs, tables and other elements can be processed with the same techniques as glass-fibre reinforced furniture. Figure 14.13 shows chairs made with soy-oil based resin, reinforced with flax mats. Chairs can be made with the resin-transfer moulding process or by compression moulding.

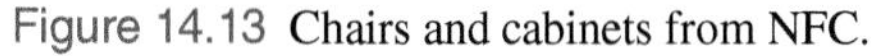

Figure 14.13 Chairs and cabinets from NFC.

14.7 Outlook

Long natural fibres hold the potential to various strong composites for enhanced renewable alternatives to many presently used non-renewable materials. A number of favourable material properties of natural fibres, e.g. high aspect ratios and a surface rich in hydroxyl groups, facilitate functionalisation with different reagents to achieve desired performance for strong composite products. The processing techniques of long natural fibre composites have been studied but have yet to be further developed for the efficient manufacturing of various types of NFC. The commercialisation of both effective long natural fibres and strong composites has yet to reach the scale normally associated with applications for high-volume products such as automotive components and construction materials.

References

[1] Pickering K. (2008). *Properties and Performance of Natural Fibre Composites*. Cambridge, UK: Woodhead Publishing.

[2] Mohanty A K, Manjusri M & Drzal L T. (2005). *Natural Fibres, Biopolymers and Biocomposites*. Boca Raton, FL: CRC Press.

[3] Hepworth D G. (2000). The use of unretted hemp fibre in composite manufacture. *Composites Part A: Applied Science and Manufacturing*, 31, pp. 1279–1283.

[4] Blendzki A K & Gassan J. (1999) Composites reinforced with cellulose based fibres. *Progress in Polymer Science*, 24, pp. 221–274.

[5] Krassig H. (2002). Cellulose, in *Ulmann's Encyclopedia of Industrial Chemistry*. Wiley-VCH Verlag.

[6] Schafer T & Honermeier B. (2006). Effect of sowing date and plant density on the cell morphology of hemp (*Cannabis sativa L.*). *Industrial Crops and Products*, 23, pp. 88–98.

[7] Fan M & FU F. (2017). Introduction: A perspective. Natural fibre composites in construction. In M Fan and F Fu (eds), *Advanced High Strength Natural Fibre Composites in Construction* (pp. 1–20). Cambridge, UK: Woodhead Publishing (Elsevier).

[8] Barkakaty B C. (19760. Some structural aspects of sisal fibres. *Journal Applied Polymer Science*, 20 (11), pp. 2921–2940.

[9] Chand N & Hashmi S A R. (1993). Mechanical properties of sisal fibre at elevated temperatures. *Journal of Materials Science*, 28 (24), pp. 6724–6728.

[10] Mwaikambo L Y, Tucker N & Clark A J. (2007). Mechanical properties of hemp-fibre-reinforced Euphorbia composites. *Macromolecular Materials and Engineering*, 292, pp. 993–1000.

[11] Fan M. (2009). Sustainable fibre-reinforced polymer composites in construction. In V. Goodship (ed.) *Management, Recycling and Reuse of Waste Composites* (pp. 518–569). Cambridge: Woodhead Publishing.

[12] Peng X & Fan M. (2011). Properties of natural fibre composites made by pultrusion process. *Journal of Composite Materials*, 44 (2), pp. 237–246.

[13] Mohanty A K, Misra M & Drzal L T. (2002). Sustainable bio-composites from renewable resources: Opportunities and challenges in the green materials world. *Journal of Polymers and the Environment*, 10, pp. 19–26.

[14] Morton W E & Hearle J W S. (1993). *Physical Properties of Textile Fibres*. Manchester, UK: Textile Institute.

[15] Saheb D N & Jog J P. (1999). Natural fiber polymer composites: A review. *Advances in Polymer Technology*, 18, pp. 351–363.

[16] Hagstrand P O & Oksman K. (2001). Mechanical properties and morphology of flax fiber reinforced melamine-formaldehyde composites. *Polymer Composites*, 22, pp. 568–578.
[17] Mueller D H & Krobjilowski A. (2003). New discovery in the properties of composites reinforced with natural fibers. *Journal of Industrial Textiles*, 33, pp. 111–130.
[18] Shahzad A. (2011). Hemp fibre and its composites – A review. *Journal of Composite Materials*, 46 (8), pp. 973–986.
[19] Umer R, Bickerton S & Fernyhough A. (2007). Characterising wood fibre mats as reinforcements for liquid composite moulding processes. *Composites Part A: Applied Science and Manufacturing*, 38 (2) pp. 434–448.
[20] Acar M & Harper J F. (2000). Textile composites from hydro-entangled non-woven fabrics. *Computers & Structures*, 76 (1–3), pp. 105–114.
[21] Ghassemieh E, Acar M & Versteeg H K. (2001). Improvement of the efficiency of energy transfer in the hydro-entanglement process. *Composites Science and Technology*, 61 (12), pp. 1681–1694.
[22] Madsen B, Hoffmeyer P & Lilholt H. (2007). Hemp yarn reinforced composites – II. Tensile properties. *Composites Part A: Applied Science and Manufacturing*, 38, pp. 2204–2215.
[23] Tortora P G. (1987). Making fibers into yarns – Understanding textiles. London: Macmillan.
[24] Tang H B *et al.* (2011). Mathematical modeling and numerical simulation of yarn behavior in a modified ring spinning system. *Applied Mathematical Modelling*, 35, (1), pp. 139–151.
[25] Needles H L. (1986). *Textile Fibres, Dyes, Finishes and Processes: A Concise Guide.* Norwich, NY: William Andrews (Elsevier). Originally published by Noyes Publications.
[26] Cox B & Flanagan G. (1997). *Handbook of Analytical Methods for Textile Composites*, NASA contractor report 4750.
[27] Ko F K. (2004). From textile to geotextiles. In Seminar in honour of professor Robert Koerner.
[28] Luo Y & Verpoest I. (2002). Biaxial tension and ultimate deformation of knitted fabric reinforcements. *Composites Part A: Applied Science and Manufacturing*, 33 (2), pp. 197–203.
[29] Hivet G & Boisse P. (2008) Consistent mesoscopic mechanical behaviour model for woven composite reinforcements in biaxial tension. *Composites Part B: Engineering,* 39 (2), pp. 345–361.
[30] Yang Y S & Lee L J. (1988). Microstructure formation in the cure of unsaturated polyester resin. *Polymer*, 29 (10) pp. 1793–1800.
[31] Ton-That M T *et al.* (2000). Polyester cure monitoring by means of different techniques. *Polymer Composites*, 21 (4).
[32] Gay D & Hoa S V. (2007). *Composite Materials, Design and Application*. Boca Raton, FL: CRC Press.
[33] Biron M. (2007). Detailed accounts of thermoset resins In: *Thermoplastics and Thermoplastic Composites* (pp. ??–??). Oxford: Elsevier Ltd.
[34] Sreekumar P A. (2007). A comparative study on mechanical properties of sisal-leaf fibre-reinforced polyester composites prepared by resin transfer and compression moulding techniques. *Composite Science and Technology*, 67, pp. 453–461.
[35] Kim H J & Seo D W. (2006). Effect of water absorption fatigue on mechanical properties of sisal textile-reinforced composites. *International Journal of Fatigue*, 28, pp. 1307–1314.
[36] Yu P H *et. al.* (1998). Conversion of food industrial wastes into bioplastics. *Applied Biochemistry and Biotechnology*, pp. 603–614.
[37] Domenek S *et al.* (2004). Biodegradability of wheat gluten-based bioplastics. *Chemosphere*, 54 (4), pp. 551–559.

[38] Witholt B & Kessler B. (1999). Perspectives of medium chain length poly (hydroxyalkanoates), a versatile set of bacterial bioplastics. *Current Opinion in Biotechnology*, 10 (3), pp. 279–285.

[39] Luengo J M *et al.* (2003). Bioplastics from microorganisms. *Current Opinion in Microbiology*, 6 (3), pp. 251-260.

[40] Wool R P & Sun XS. (2005). *Bio-based Polymers and Composites.* London: Elsevier.

[41] Raquez J M. (2010). Thermosetting (bio)materials derived from renewable resources: A critical review. *Progress in Polymer Science*, 35 (4), pp. 487–509.

[42] Jeroen van Soest J G *et al.* (1996). Crystallinity in starch bioplastics. *Industrial Crops and Products*, 5 (1), pp. 11–22.

[43] Mehta G *et al.* (2004). Biobased resin as a toughening agent for biocomposites. *Green Chemistry*, 6, pp. 254–258.

[44] Harris B. (2003). Fatigue in composites: Science and technology of the fatigue response of fibre-reinforced plastics. Cambridge, UK: Woodhead Publishing.

[45] Rosato D V & Rosato M G. (2000). *Concise Encyclopedia of Plastics Vol. 1.* Norwell, MA: Kluwer Academic Publishers.

[46] Norman B. (1939). Some further developments in the manufacture and use of synthetic materials for aircraft construction. *Flight: The Aircraft Engineer*, 12 January, pp. a-c.

[47] Kelly, A & Macmillan N H. (1986). *Strong Solids.* 3rd edn. Oxford: Oxford University Press.

[48] Cox H L. (1952). The elasticity and strength of paper and other fibrous materials. *British Journal of Applied Physics*, 3 (3) p. 72.

[49] Rosen B W. (1965). Mechanics of composite strengthening: Fibre Composite Materials, 37±75. Am. Soc. Metals Seminar, Metals Park, Ohio.

[50] Pan N. (1993). Theoretical determination of the optimal fiber volume fraction and fiber-matrix property compatibility of short fiber composites. *Polymer Composites*, 14 (2), pp. 85–93.

[51] Waas A M & Schultheisz C R. (1996). Compressive failure of composites, part II: Experimental studies. *Progress in Aerospace Sciences*, 32, pp. 43–78.

[52] Bazhenov S L, Kuperman A M, Zelenskii E S & Berlin A A. (1992). Compression failure of unidirectional glass-fibre-reinforced plastics. *Composite Science and Technology*, 45 (3), pp. 201–208.

[53] Lager J R & June R R. (1969). Compressive strength of boron–epoxy composites. *J. of Composite Materials*, 3 (1), pp. 48–56.

[54] Weaver C W & Williams J G. (1975). Deformation of a carbon-epoxy composite under hydrostatic pressure. *J. of Materials Science*, 10 (8), pp. 1323–1333.

[55] Wronski A S & Parry T V. (1982). Compressive failure and kinking in uniaxially aligned glass-resin composite under superposed hydrostatic pressure. *J. of Materials Science*, 17 (12), pp. 3656–3662.

[56] Sigley R H, Wronski A S & Parry T V. (1992). Axial compressive failure of glass fobre polyester composites under superposed hydrostatic pressure: Influence of fibre bundle size. *Composite Science and Technology*, 43 (2), pp. 171–183.

[57] Weclawski B, Fan M & Hui D. (2014). Compressive behaviour of natural fibre composite. *Composite Part B: Engineering*, 67, pp. 183–191.

[58] Van de Weyenberg I *et al.* (2003). Influence of processing and chemical treatment of flax fibres on their composites. *Composites Science and Technology*, 63, pp. 1241–1246.

[59] Charlet K *et al.* (2007). Characteristics of Hermès flax fibres as a function of their location in the stem and properties of the derived unidirectional composites. *Composites Part A: Applied Science and Manufacturing*, 38, pp. 1912–1921.

[60] Dhakal H N, Zhang Z Y & Richardson M O W. (2007). Effect of water absorption on the mechanical properties of hemp fibre reinforced unsaturated polyester composites. *Composites Science and Technology*, 67, pp. 1674–1683.

[61] Rouison D, Sain M & Couturier M. (2006). Resin transfer molding of hemp fiber composites: Optimization of the process and mechanical properties of the materials. *Composites Science and Technology*, 66, pp. 895–906.

[62] Yuanjian T & Isaac D H. (2007). Impact and fatigue behaviour of hemp fibre composites. *Composites Science and Technology*, 67, pp. 3300–3307.

[63] Rouison D, Sain M & Couturier M. (2004). Resin transfer molding of natural fiber reinforced composites: Cure simulation. *Composites Science and Technology*, 64, pp. 629–644.

[64] Hughes M, Hill C A S & Hague J R B. (2002). The fracture toughness of bast fibre reinforced polyester composites Part 1: Evaluation and analysis. *Journal of Materials Science*, 37, pp. 4669–4676.

[65] Mehta G *et al.* (2006). Effect of fiber surface treatment on the properties of biocomposites from nonwoven industrial hemp fiber mats and unsaturated polyester resin. *Journal of Applied Polymer Science*, 99, pp. 1055–1068.

[66] Thygesen A *et al.* (2007). Comparison of composites made from fungal defibrated hemp with composites of traditional hemp yarn. *Industrial Crops and Products*, 25, pp. 147–159.

[67] George J & Verpoest J I I. (1999). Mechanical properties of flax fibre reinforced epoxy composites. *Die Angewandte Makromolekulare Chemie*, 272 (1), pp. 41–45.

[68] Van den Oever M J A, Bos H L & Van Kemenade M J J M. (2000). Influence of the physical structure of flax fibres on the mechanical properties of flax fibre reinforced polypropylene composites. *Applied Composite Materials*, 7, pp. 387–402.

[69] Goutianos S *et al.* (2006). Development of flax fibre based textile reinforcements for composite applications. *Applied Composite Materials*, 13, pp. 199–215.

[70] Beg M D H & Pickering K L. (2008). Accelerated weathering of unbleached and bleached Kraft wood fibre reinforced polypropylene composites. *Polymer Degradation and Stability*, 93, pp. 1939–1946.

[71] Beg M D H & Pickering K L. (2008). Mechanical performance of Kraft fibre reinforced polypropylene composites: Influence of fibre length, fibre beating and hygrothermal ageing. *Composites Part A: Applied Science and Manufacturing*, 39, pp. 1748–1755.

[72] Talreja R & Manson J. (2000). *Comprehensive Composite Materials, Volume 2: Polymer Matrix Composites.* Amsterdam: Elsevier.

[73] Madsen B *et al.* (2007a). Hemp yarn reinforced composites: I. Yarn characteristics. *Composites Part A: Applied Science and Manufacturing*, 38, pp. 2194–2203.

[74] Oksman K. (2001). High quality flax fibre composites manufactured by the resin transfer moulding process. *Journal of Reinforced Plastics and Composites*, 20, pp. 621–627.

[75] Thygeson A, Thomson, A B, Daniel, G. & Lilholt H. (2006). Comparison of hemp fibres with glass fibres for wind power turbines: Effect of composite density. In H Lilholt, B Madsen, T L Andersen, L P Mikkelsen and A Thygesen (eds), *Proceedings of the 27th Riso International Symposium on Materials Science, 2006.* Roskilde, Denmark: Riso National Laboratory.

[76] Bledzki A K, Fink H P & Specht K. (2004). Unidirectional hemp and flax EP- and PP-composites: Influence of defined fiber treatments. *Journal of Applied Polymer Science*, 93, pp. 2150–2156.

[77] Hughes M, Carpenter J & Hill C. (2007). Deformation and fracture behaviour of flax fibre reinforced thermosetting polymer matrix composites. *Journal of Materials Science*, 42, pp. 2499–2511.

[78] Weclawski B T. (2015). The potential of bast natural fibres as reinforcement for polymeric composite materials in building applications. PhD thesis, Brunel University, London.
[79] Schultheisz C R & Waas A M. (1996). Compressive failure of composites. Part 1: Testing and micromechanical theories. *Progress in Aerospace Sciences*, 32, pp. 1–42.
[80] Sawpan M A, Pickering K L & Fernyhough A. (2012). Flexural properties of hemp fibre reinforced polyactide and unsaturated polyester composites. *Composites: Part A*, 43, 519–526.
[81] Muralidhar B A, Giridev V R & Raghunathan K. (2012). Flexural and impact properties of flax woven, knitted and sequentially stacked knitted/woven preform reinforced epoxy composites. *Journal of Reinforced Plastics & Composites*, 31 (6), pp. 379–388.
[82] Aziz S H & Ansell M P. (2004). The effect of alkalization and fibre alignment on the mechanical and thermal properties of kenaf and hemp bast fibre composites: Part 1 – polyester resin matrix. *Composites Science and Technology*, 64, pp. 1219–1230.
[83] Sebe G *et al.* (2000). RTM Hemp fibre-reinforced polyester composites. *Applied Composite Materials*, 7, pp. 341–349.
[84] Coughlan M P. (1985). The properties of fungal and bacterial celulases with comment on their production and application. *Biotechnology & Genetic Engineering Reviews*, 3, pp. 39–109.
[84] Ochi S. (2006). Development of high strength biodegradable composites using Manila hemp fiber and starch-based biodegradable resin. *Composites Part A*, 37, pp. 1879–1883.
[85] Beguin P. (1990). Molecular biology of cellulose degradation. *Annual Review of Microbiology*, 44, pp. 219–248.
[86] Gu X *et al.* (2001). Characterization of polyester degradation using tapping mode atomic force microscopy: Exposure to alkaline solution at room temperature. *Polymer Degradation and Stability*, 74, pp. 139–149.
[87] Pothan L A, Thomas S & Neelakantan N R. (1997) Short banana fiber reinforced polyester composites: Mechanical failure and aging characteristics. *Journal of Reinforced Plastics and Composites*, 16 (8), pp. 744–765.
[88] Mwaikambo L Y & Bisanda E T N. (1998). The performance of cotton-kapok fabric polyester composites. *Polymer Testing*, 18, pp. 181–198.
[89] Dash B N *et al.* (2000). Novel low-cost jute–polyester composites. III: Weathering and thermal behaviour. *Journal of Applied Polymer Science*, 78 (9), pp. 1671–1679.
[90] Joseph P V *et al.* (2002). Environmental effects on the degradation behaviour of sisal fibre reinforced polypropylene composites. *Composite Science and Technology*, 62, pp. 1357– 1372.
[91] Espert A, Vilaplana F & Karlsson S. (2004). Comparison of water absorption in natural cellulosic fibres from wood and one-year crops in polypropylene composites and its influence on their mechanical properties. *Composites: Part A*, 35, pp. 1267–1276.
[92] Chow C P L, Xing X S & Li R K Y. (2007). Moisture absorption studies of sisal fibre reinforced polypropylene composites. *Composites Science and Technology*, 67, pp. 306–313.
[93] Blaga A. (1981). Water sorption characteristics of GRP composite: Effect of outdoor weathering. *Polymer Composites*, 2, pp. 13 – 17.
[94] Leman Z *et al.* (2007). Moisture absorption behaviour of sugar palm fibre reinforced composites. *Material Design*, doi:10.1016/j.matdes.2007.11.004.
[95] Kelly A & Tyson W R. (1965). Tensile properties of fibre-reinforced metals: Copper/ tungsten and copper/molybdenum. *Journal of Mechanics and Physics of Solids*, 13 (6), pp. 329–338.
[96] Gegauff G. (1907). Force et élasticité des files en cotton. *Bulletin de la Société Industrielle De Mulhouse*, 77, p. 153.

[97] Ghosh A *et al.* (2005) Predictive models for strength of spun yarns: An overview. *Autex Research*, 5 (1).
[98] Platt M M. (1950). Mechanics of elastic performance of textile materials, Part III: Some aspects of stress analysis of textile structures – continuous filament yarns. *Textile Research Journal*, 20 (1).
[99] Hearle J W S, Grosbery P & Backer S. (1969). *Theory of the Extension of Continuous Filament Yarns. Vol. 1.* New York: Wiley.
[100] Guha A, Chattopadhyay R & Jayadeva A. (2001). Predicting yarn tenacity: A comparison of mechanistic, statistical, and neural network models. *Journal of the Textile Institute*, 92 (2), pp. 139–145.
[101] Neagu R C, Gamstedt E K & Berthold F. (2006). Stiffness contribution of various wood fibres to composite materials. *Journal of Composite Materials*, 40 (8), pp. 663–699.
[102] Marklund E & Varna J. (2009). Modeling the hygroexpansion of aligned wood fiber composites. *Composites Science and Technology*, 69, pp. 1108–1114.
[103] Sutcu M. (1991). A recursive concentric cylinder model for composites containing coated fibres. *International Journal of Solids and Structures*, 29 (2), pp. 197–213.
[104] Automoto (2013). Bamboo bike [Online image] Available from: http://www.automotto.com/entry/10-bamboo-bikes-giving-style-performance-dimension/ [accessed 30/04/2013].
[105] Allworldcars (2013). Lotus with biomaterial components [Online image] Available from: http://allworldcars.com/wordpress/wp-content/uploads/2008/07/lotus_eco_elise.jpg [accessed 30/04/2013].
[106] Hanf Info (2013). Mercedes bio-material componenets [Online image] Available from: http://www.chanvre-info.ch/info/en/local/cache-vignettes/L320xH221/industrie_mercedes 62553.jpg [accessed 30/04/2013].
[107] Fan M. (2002). Exterior door frame made from all natural fibre reinforced polyester composites. *BRE Report*, 2002, Building Research Establishment, UK.
[108] Singh B & Gupta M. (2005). Performance of pultruded jute fibre reinforced phenolic composites as building materials for door frame. *J. Polym Environ.*, 13, pp. 127–137.
[109] Dweib M A, Hu B, O'Donnell A, Shenton, H W & Wool R P (2004). All natural composite sandwich beams for structural applications. *Composite Structures*, 63 (2), pp. 147–157.

Index